Plant Nucleotide Metabolism – Biosynthesis, Degradation, and Alkaloid Formation

Plant Nucleotide Metabolism – Biosynthesis, Degradation, and Alkaloid Formation

Hiroshi Ashihara
Ochanomizu University
Bunkyo-ku, Tokyo, Japan

Iziar A. Ludwig
University Rovira I
Virgili
Reus, Spain

Alan Crozier
University of California
Davis, CA, USA

University of Glasgow
Glasgow, UK

WILEY

This edition first published 2020

Registered Office(s)
John Wiley & Sons Ltd, The Atrium, Southern Gate, Chichester, West Sussex, PO19 8SQ, UK

Editorial Office
9600 Garsington Road, Oxford, OX4 2DQ, UK

For details of our global editorial offices, customer services, and more information about Wiley products visit us at www.wiley.com.

Wiley also publishes its books in a variety of electronic formats and by print-on-demand. Some content that appears in standard print versions of this book may not be available in other formats.

Library of Congress Cataloging-in-Publication Data

Name: Ashihara, Hiroshi, author.
Title: Plant nucleotide metabolism : biosynthesis, degradation, and alkaloid formation / Hiroshi Ashihara, Emeritus Professor, Ochanomizu University, Tokyo, Japan, Iziar A. Ludwig, University Rovira I, Virgili, Spain, Alan Crozier, University of California, Davis and University of Glasgow, UK.
Description: First edition. | Hoboken : Wiley, 2020. | Includes bibliographical references and index.
Identifiers: LCCN 2019047672 (print) | LCCN 2019047673 (ebook) | ISBN 9781119476122 (hardback) | ISBN 9781119476108 (adobe pdf) | ISBN 9781119476078 (epub)
Subjects: LCSH: Plants–Metabolism. | Nucleotides–Metabolism.
Classification: LCC QK881 .A74 2020 (print) | LCC QK881 (ebook) | DDC 572/.42–dc23
LC record available at https://lccn.loc.gov/2019047672
LC ebook record available at https://lccn.loc.gov/2019047673

Cover Design: Wiley
Cover Image: © Mazur Travel/Shutterstock

Set in 10/12pt WarnockPro by SPi Global, Chennai, India

10 9 8 7 6 5 4 3 2 1

Contents

Preface

Almost all organisms produce nucleobases, nucleosides and nucleotides of purines and pyrimidines. There have been a number of books on nucleotide metabolism in microorganisms and humans. However, this is the first to focus on plants which exhibit important differences to other organisms in key areas of nucleotide metabolism and function.

The book covers the metabolism of purine, pyrimidine and pyridine nucleotides and nucleotide alkaloids in higher plants and points out differences from that occurring in other organisms. Likewise, differences in the salvage pathways and diversity of interconversions in plants, fungi and bacteria are highlighted. Various physiological aspects of these processes are covered along with their involvement in the control of plant growth and development. Among the topics covered are the purine alkaloids caffeine, theobromine and their metabolites which, in species including coffee, tea and cocoa, accumulate in quantity. There is also discussion of the function of purine alkaloids and the potential allelopathic role of caffeine. Studies, some making use of genetically-modified plants, have indicated that caffeine can play a role in a variety of plant defence strategies. Other investigations have provided evidence that trigonelline, found principally in coffee and legumes, has a role in resistance to salt stress and can act as a natural pesticide to reduce insect infestations. Finally, the book explores the absorption, metabolism and potential impact on health of dietary caffeine, theobromine and trigonelline

The book provides comprehensive information on nucleotide structures and metabolic pathways and is a unique resource on a diversity of topics and as such is essential reading for students, researchers, and lecturers in plant biochemistry, physiology, chemistry, agricultural sciences, nutrition and the associated applied fields of research.

We owe special thanks Professors Tatsuhito Fujimura, Claudio Stasolla and Takao Yokota for their help with some of the figures and advice on genes encoding key enzymes and chemical structures for the book.

Hiroshi Ashihara
Iziar A. Ludwig
Alan Crozier

Part I

General Aspects of Nucleotide Metabolism

1

Structures of Nucleotide-Related Compounds

1.1 Introduction

The chemistry of purine, pyrimidine, and pyridine nucleobases, nucleosides, and nucleotides constitute one of the oldest topics in biochemistry. In this chapter, the nomenclature and structures of nucleotides are briefly described.

1.2 Nomenclature and Abbreviations of Nucleotide-Related Compounds

The nucleotide nomenclature and abbreviations employed in the text are those used by Henderson and Paterson (1973) in their textbook '*Nucleotide Metabolism – An Introduction*'. The terms 'nucleoside' and 'nucleotide' in the strictest sense refer, respectively, to *N*-glycosides and phosphorylated *N*-glycosides derived from nucleic acids. However, they are now used in a wider context. *N*-Ribosides, such as nicotinamide mononucleotide (NMN), are called nucleotides only by extension and analogy, and nicotinamide adenine dinucleotide (NAD) and nicotinamide adenine dinucleotide phosphate (NADP) are referred to as dinucleotides. Flavin mononucleotide (FMN) is a step further removed, as it contains ribitol, a pentose alcohol formed by the reduction of ribose, instead of ribose, while flavin adenine dinucleotide (FAD) similarly extends the meaning of dinucleotide. *N*-Glycosides such as orotidine 5′-monophosphate (OMP) and adenylosuccinate (SAMP) are called nucleotides through their close relationship to the 'true' nucleotides. The terms ribonucleoside and ribonucleotide are used in preference to riboside and ribotides. The IUPAC-IUB Combined Commission on Biochemical Nomenclature has abbreviations and symbols for nucleotides and related compounds. However, as argued by Henderson and Paterson (1973), while they are appropriate for polynucleotides, the distinction between bases and nucleosides is not always immediately obvious, and this has limited their use. The abbreviations used here are more intuitive and better suited to the portrayal of reaction schemes in which the addition or removal of substituent groups occurs.

In Table 1.1, the abbreviations for the major nucleotides, ribo- and deoxyribonucleotides and nucleobases are presented. For readers convenience, the styles used both in this book (style #1) and those recommended by IUPAC (style #2) are shown.

Plant Nucleotide Metabolism – Biosynthesis, Degradation, and Alkaloid Formation, First Edition.
Hiroshi Ashihara, Iziar A. Ludwig and Alan Crozier.

Table 1.1 Nomenclature and abbreviations of purine and pyridine ribo- and deoxyribonucleotides and related compounds.

Ribonucleotides		Ribonucleosides	#1	#2	Nucleobases	#1	#2
Adenosine-5′-monophosphate	AMP	Adenosine	AR	Ado	Adenine	A	Ade
Guanosine-5′-monophosphate	GMP	Guanosine	GR	Guo	Guanine	G	Gua
Inosine-5′-monophosphate	IMP	Inosine	IR	Ino	Hypoxanthine	H	Hyp
Xanthosine-5′-monophosphate	XMP	Xanthosine	XR	Xao	Xanthine	X	Xan
Uridine-5′-monophosphate	UMP	Uridine	UR	Urd	Uracil	U	Ura
Cytidine-5′-monophosphate	CMP	Cytidine	CR	Cyd	Cytosine	C	Cyt
Orotidine-5′-monophosphate	OMP	Orotidine	OR	Ord	Orotic acid	O	Oro

Deoxyribonucleotides		Deoxyribonucleosides	#1	#2	Nucleobases	#1	#2
Deoxyadenosine-5′-monophosphate	dAMP	Deoxyadenosine	AdR	dAdo	Adenine	A	Ade
Deoxyguanosine-5′-monophosphate	dGMP	Deoxyguanosine	GdR	dGuo	Guanine	G	Gua
Deoxyuridine-5′-monophosphate	dUMP	Deoxyuridine	UdR	dUrd	Uracil	U	Ura
Deoxycytidine-5′-monophosphate	dCMP	Deoxycytidine	CdR	dCyd	Cytosine	C	Cyt
Thymidine-5′-monophosphate	dTMP	Thymidine	TdR	dThd	Thymine	T	Thy

Two types of symbols are used for nucleoside and nucleobases. Style #1: recommended in *Nucleotide Metabolism* (Henderson and Paterson 1973). Style #2: recommended by the IUPAC-IUB Commission on Biochemical Nomenclature (1970). In this book, style #1 is adopted.

Abbreviations for ribonucleosides and 2-deoxyribonucleosides are derived from those used for the bases plus those for the ribosyl or 2′-deoxyribosyl groups. Thus AR stands for adenosine and AdR is the abbreviation for deoxyadenosine. In the case of inosine, (hypoxanthine + ribose) HR may be possible, but IR is often used. The latter is used in this text.

For nucleotides, the traditional abbreviations based on the term 'nucleoside monophosphate' are used. Thus AMP stands for adenosine monophosphate (adenylate), UMP for uridine monophosphate (uridylate), and NMP for any ribonucleoside monophosphate. Similarly, dAMP stands for deoxyadenosine monophosphate (deoxyadenylate), dUMP for deoxyuridine monophosphate (deoxyuridylate), and dNMP for any deoxyribonucleoside monophosphate. Thymidine monophosphate (thymidylate) often does not have the 'deoxy' prefix in its name, because thymidine is thymine deoxyriboside. However, the symbol including a 'd' is commonly used in biochemistry textbooks, so dTMP is adopted in this article.

1.3 Chemical Structures of Nucleotide-Related Compounds

Studies on purines and pyrimidines began in 1776 when the Swedish pharmacist Carl Wilhelm Scheele isolated uric acid from bladder stones. In 1846, Unger isolated guanine from the guano of Peruvian sea birds. At the end of the nineteenth century, several purines (adenine, xanthine, and hypoxanthine) and pyrimidines (thymine, cytosine, and uracil) were discovered by the German biochemist, Albrecht Kossel who believed they constituted the main part of cell nuclei. In 1874 Friedrich Miescher isolated nuclear material rich in phosphorus which he called 'nuclein'. In the same period, Emil Fischer (1884) elucidated the structures of caffeine and related compounds which he confirmed by chemical synthesis. Further information can be found in a historical survey by Burnstock and Verkhratsky (2012). The pyridine nucleotide, NAD was discovered by the British biochemists Arthur Harden and William John Young in the early twentieth century (Harden and Young 1906).

1.3.1 Purines

A purine is a heterocyclic compound that consists of a pyrimidine ring fused to an imidazole ring. The word, 'purine' ('Purum' + 'Uricum') was coined by Emil Fischer (1884).

1.3.1.1 Purine Bases

As shown in structure **1** the atoms of the purine ring are numbered in an anticlockwise manner. In plants, there are several naturally occurring purine bases. They include adenine (**2**) and guanine (**3**), which are constituents of nucleic acids, and hypoxanthine (**4**), xanthine (**5**), and uric acid (**6**), which are produced as catabolites of adenine and guanine. Purine alkaloids, such as theobromine (3,7-dimethylxanthine) (**7**), theophylline (1,3-dimethylxanthine) (**8**), caffeine (1,3,7-trimethylxanthine) (**9**), and theacrine (1,3,7,9-tetramethyluric acid) (**10**) are derived from purine nucleotides, as are the major cytokinin plant hormones isopentenyladenine (**11**), benzyladenine (**12**), and *trans*-zeatin (**13**) (Ashihara et al. 2013).

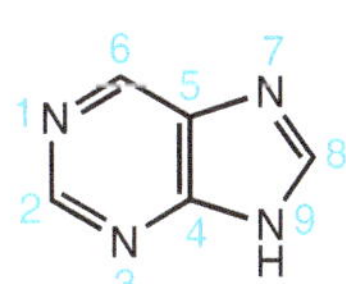

Purine: numbering system (**1**)

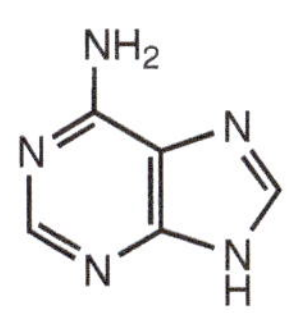

Adenine (**2**)

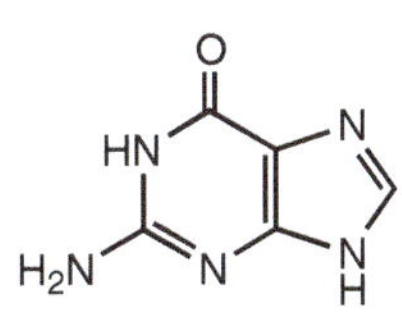

Guanine (**3**)

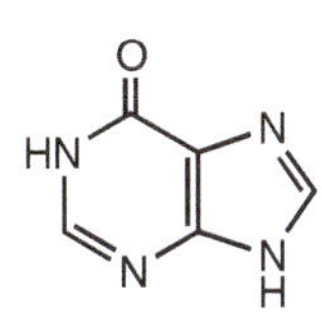

Hypoxanthine (**4**)

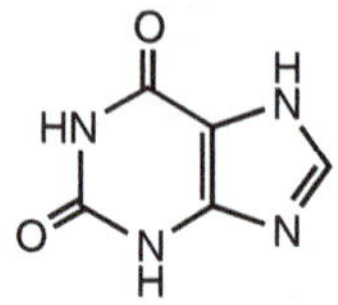
Xanthine (**5**)

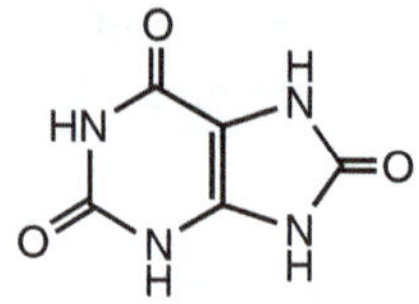
Uric acid (**6**)

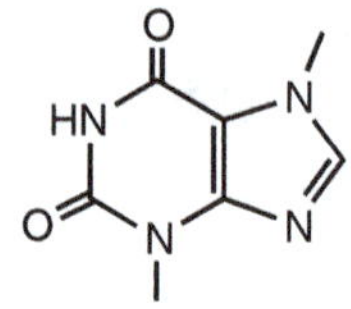
Theobromine (**7**)

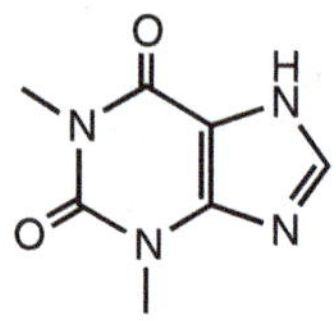
Theophylline (**8**)

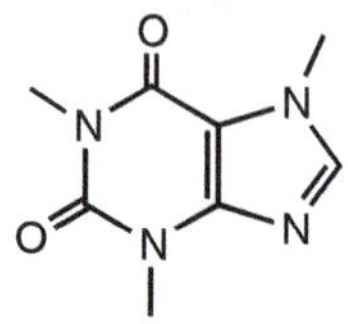
Caffeine (**9**)

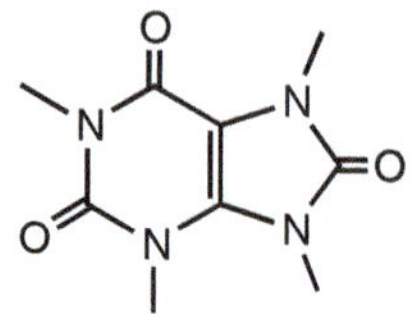
Theacrine (**10**)

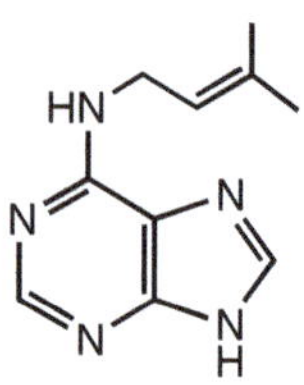
Isopentenyladenine (**11**)

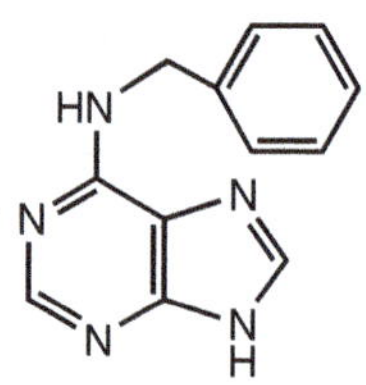
Benzyladenine (**12**)

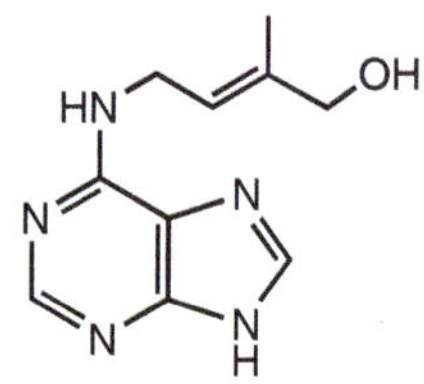
trans-Zeatin (**13**)

1.3.1.2 Purine Nucleosides

A nucleoside consists of a nucleobase and a five-carbon sugar (ribose or deoxyribose). Adenosine (**14**), guanosine (**15**), inosine (**16**), and xanthosine (**17**) are catabolites of purine ribonucleotides and RNA while deoxyadenosine (**18**) and deoxyguanosine (**19**) are catabolites of DNA. Cytokinins also occur as ribosides, namely, isopentenyladenine riboside (**20**), benzyladenine riboside (**21**), and *trans*-zeatin riboside (**22**) (Ashihara et al. 2013).

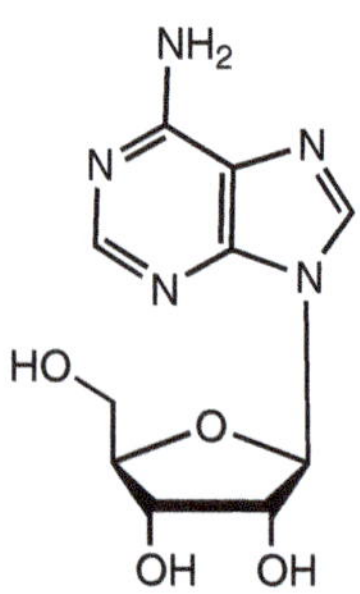
Adenosine (**14**)

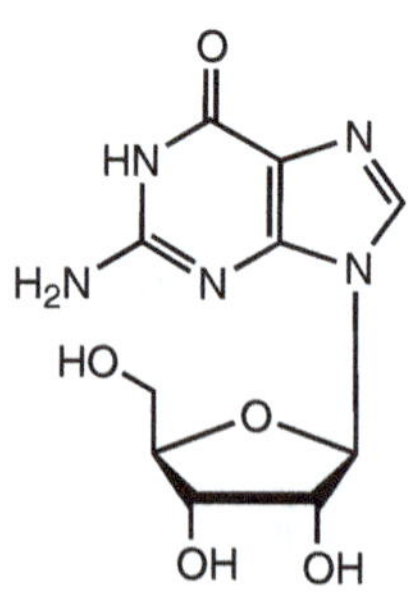
Guanosine (**15**)

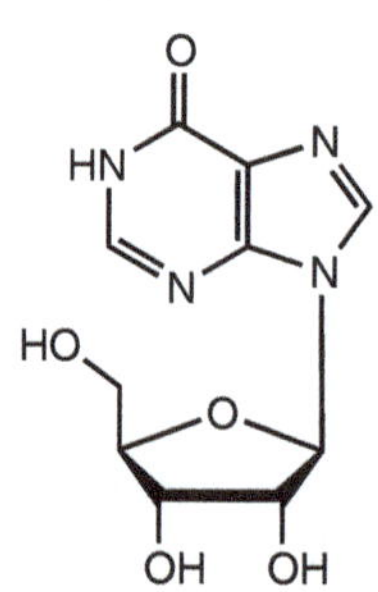
Inosine (**16**)

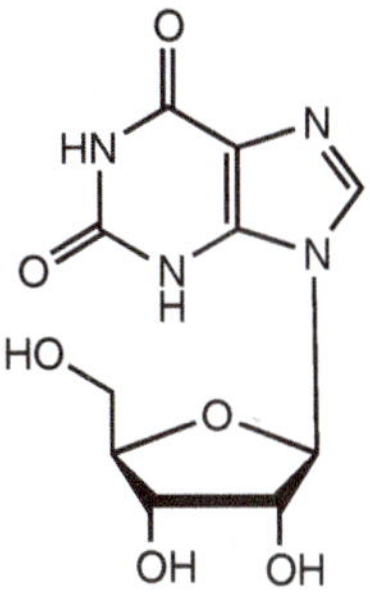

Xanthosine (**17**)

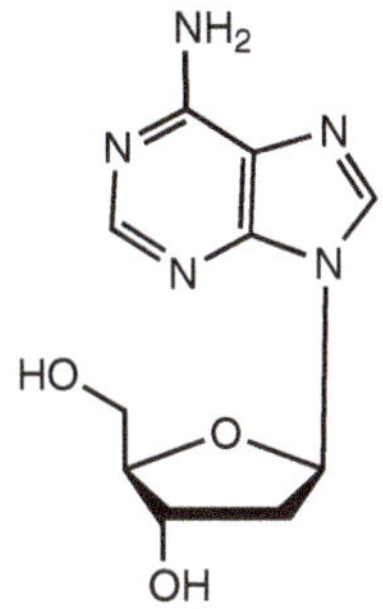

Deoxyadenosine (**18**)

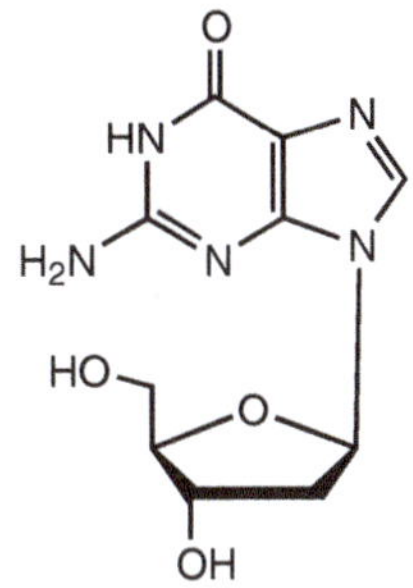

Deoxyguanosine (**19**)

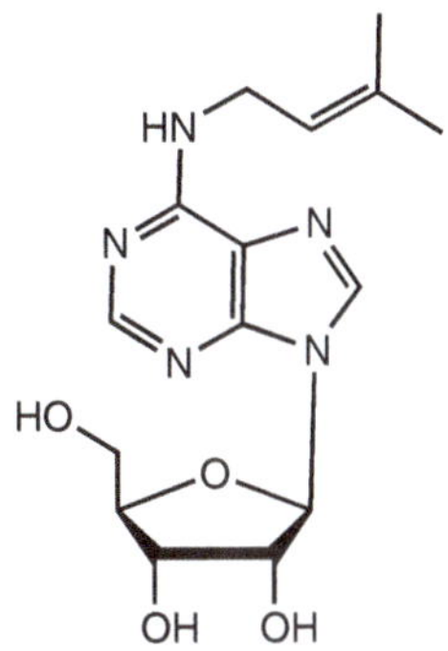

Isopentenyladenine riboside (**20**)

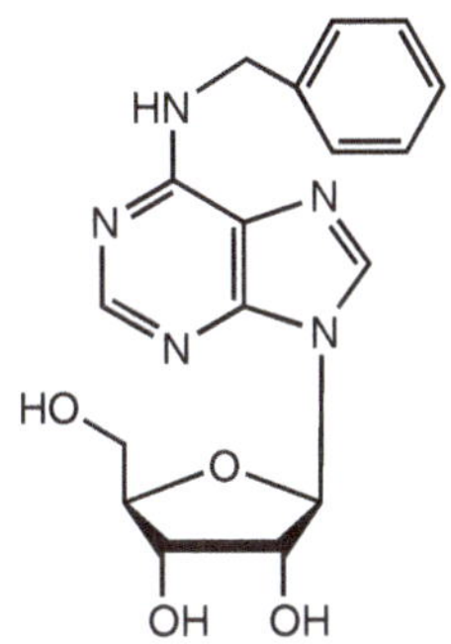

Benzyladenine riboside (**21**)

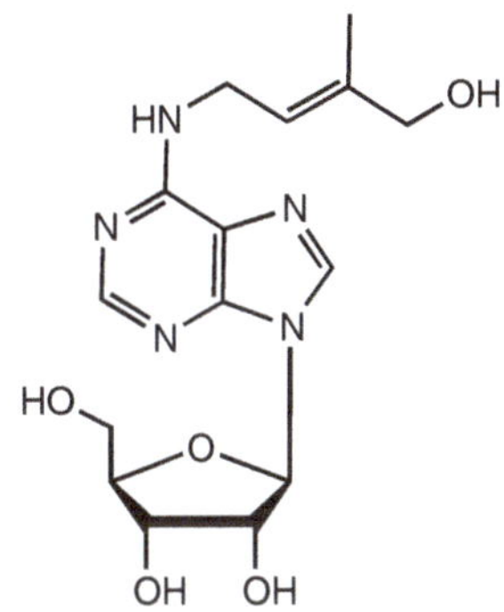

trans-Zeatin riboside (**22**)

1.3.1.3 Purine Nucleotides

A nucleotide is composed of a purine base, a sugar moiety (ribose or deoxyribose) and at least one phosphate group. The phosphate group is attached to either the C3′or C5′ position of the sugar. Nucleoside-5′-phosphates (5′-nucleotides) are the main purine nucleotides in plants as well as other organisms. Small nucleoside-3′-monophosphate (3′-nucleotides) pools are mainly produced as catabolites of nucleic acids. Examples of different forms of nucleotides are adenosine-5′-monophosphate (5′-AMP, usually abbreviated as AMP) (**23**), adenosine-3′-monophosphate (3′-AMP) (**24**), deoxyadenosine-5′-monophosphate (dAMP) (**25**), and isopentenyladenine ribotide (**26**).

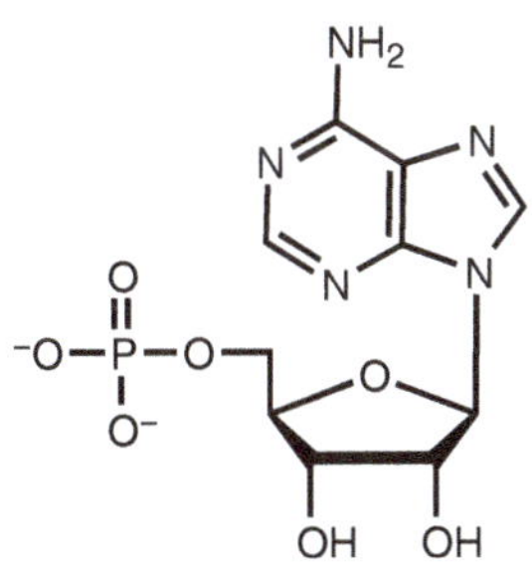

Adenosine-5′-monophosphate (**23**)

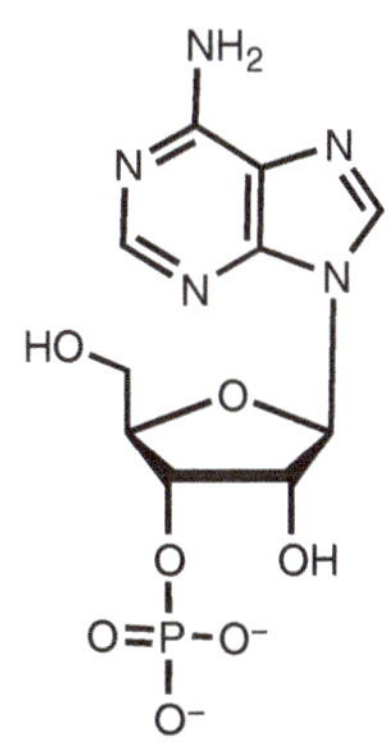

Adenosine-3′-monophosphate (**24**)

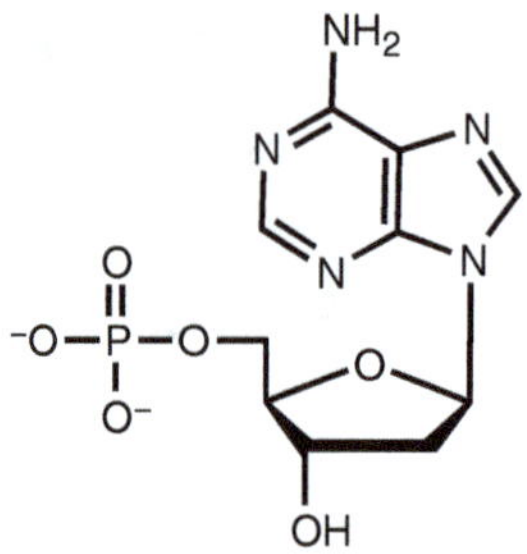

Deoyxadenosine-5′-monophosphate (**25**)

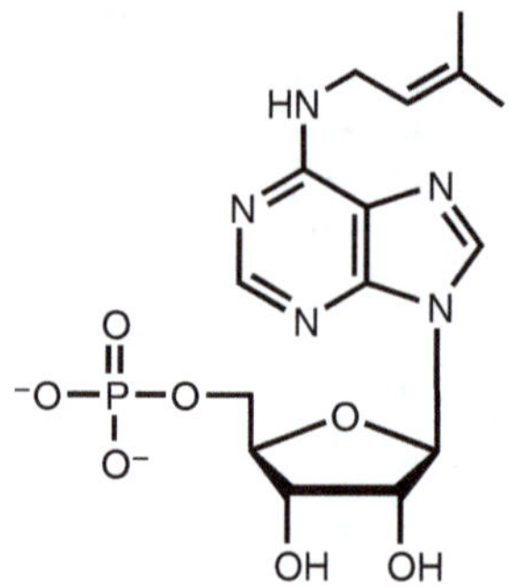

Isopentenyladenine ribotide (**26**)

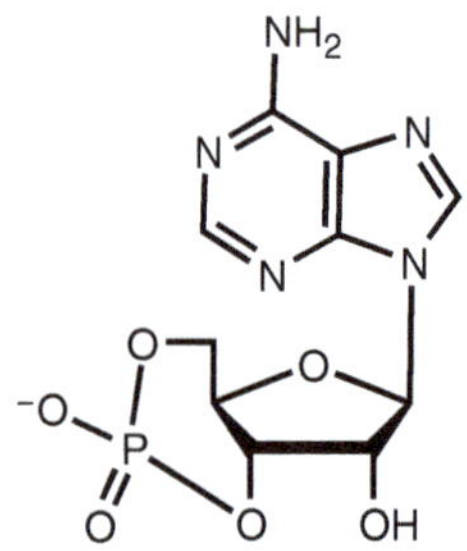

Cyclic adenosine-3′,5′-monophosphate (**27**)

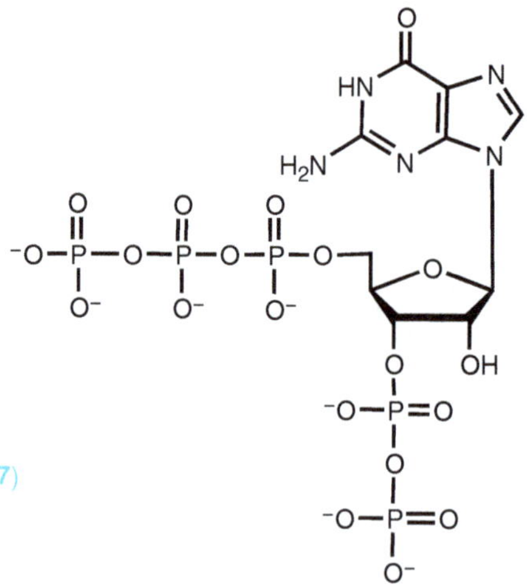

Guanosine pentaphosphate (**28**)

The cyclic monophosphates, cyclic adenosine-3′,5′-monophosphate (cAMP) (**27**) and cyclic guanosine-3′,5′-monophosphate (cGMP) occur in plants. These cyclic nucleotides are derived from ATP and GTP and act as second messengers. In addition, some unusual nucleotides, such as guanosine tetraphosphate (ppGpp), guanosine pentaphosphate (pppGpp) (**28**), diadenosine triphosphate (Ap_3A), and diadenosine tetraphosphate (Ap_4A), known as alarmones, which act as intracellular signal molecules, are produced in response to harsh environmental conditions (Boniecka et al. 2017; Pietrowska-Borek et al. 2011). Possible roles of these unusual nucleotides in plants are outlined in Part VIII.

1.3.2 Pyrimidines

Pyrimidine (**29**) is an aromatic heterocyclic organic compound similar to pyridine. The systematic study of pyrimidines was carried out and named 'pyrimidin' by a German chemist, Adolf Pinner (1885).

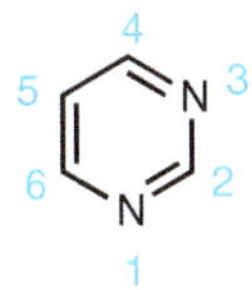

Pyrimidine: numbering system (**29**)

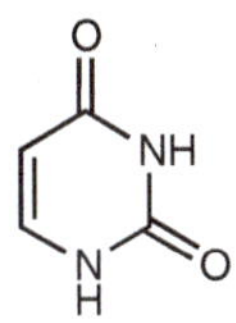

Uracil (**30**)

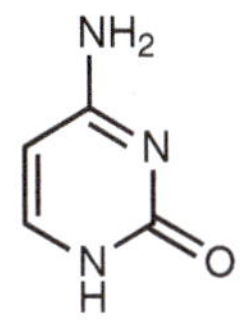

Cytosine (**31**)

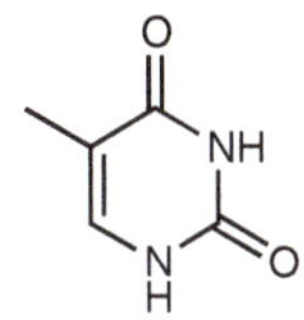

Thymine (**32**)

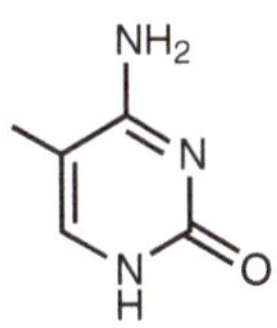

5-Methylcytosine (**33**)

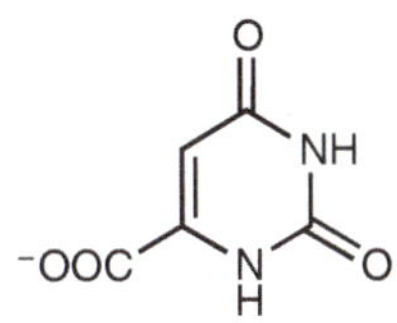

Orotate (**34**)

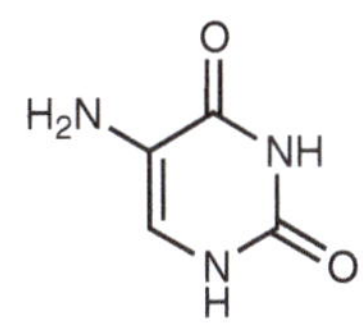

5-Aminouracil (**35**)

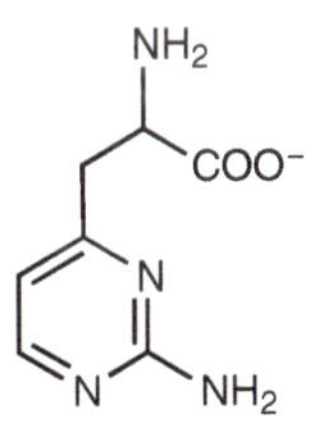

Lathyrine (**36**)

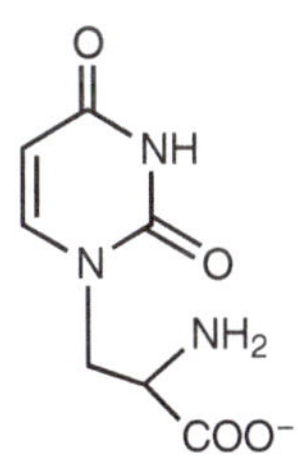

Willardine (**37**)

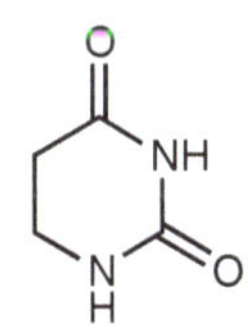

Dihydrouracil (**38**)

1.3.2.1 Pyrimidine Bases

Uracil (**30**), cytosine (**31**), and thymine (**32**) are the common nucleobases of nucleic acids. DNA and RNA also contain other bases that have been modified after formation of the nucleic acid chain. In the case of DNA, the most common modified base is 5-methylcytosine (**33**).

Orotate (pyrimidine carboxylic acid) (**34**) is an intermediate of the *de novo* pyrimidine biosynthesis. A number of secondary products, such as 5-aminouracil (**35**), lathyrine (**36**), and willardine (**37**) occur in plants (see Part VIII). Dihydrouracil (**38**) is an intermediate of uracil catabolism (see Part III).

1.3.2.2 Pyrimidine Nucleosides

A pyrimidine nucleoside consists of a pyrimidine base and a five-carbon sugar, either ribose or deoxyribose. Uridine (**39**) and cytidine (**40**) are produced as catabolites of pyrimidine ribonucleotides and RNA. Thymidine (**41**) is a catabolite of DNA. There are many modified bases in RNA, including those contained in the nucleosides, pseudouridine (5-ribosyluracil) (**42**), and dihydrouridine (**43**). Pseudouridine is an isomer of the nucleoside uridine (**39**) in which the uracil is attached via a carbon–carbon linkage instead of a nitrogen–carbon glycosidic bond. It is the most prevalent of the over 100 different modified nucleosides found in RNA.

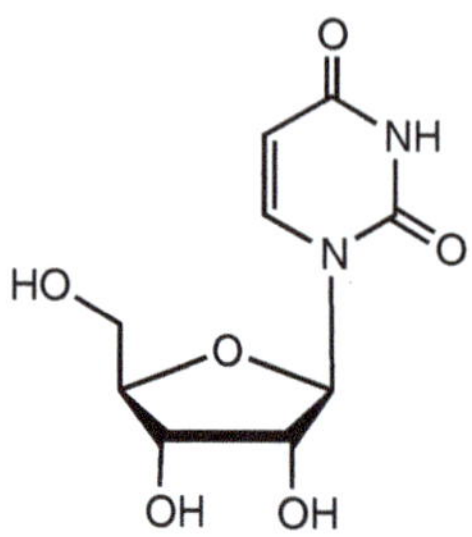

Uridine (**39**)

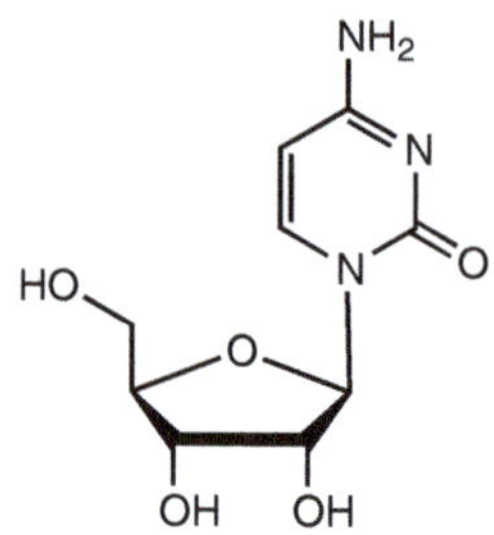

Cytidine (**40**)

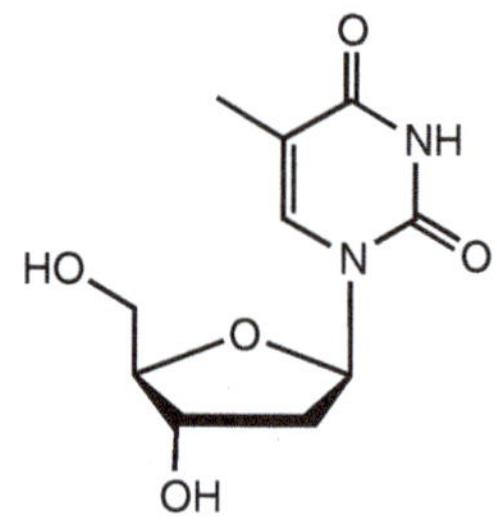

Thymidine (**41**)

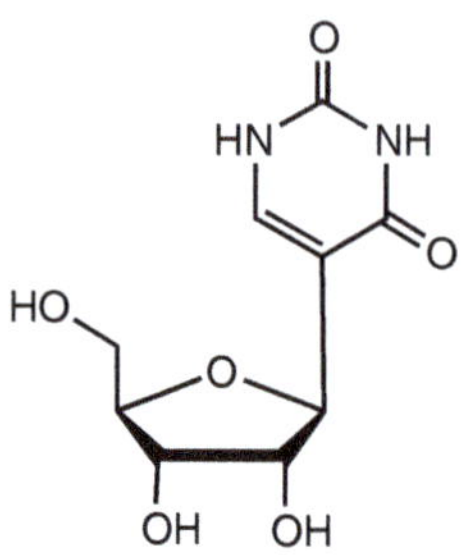

Pseudouridine (5-ribosyluracil) (**42**)

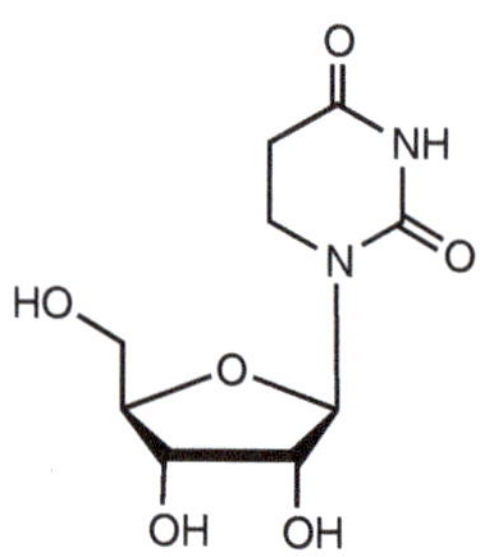

Dihydrouridine (**43**)

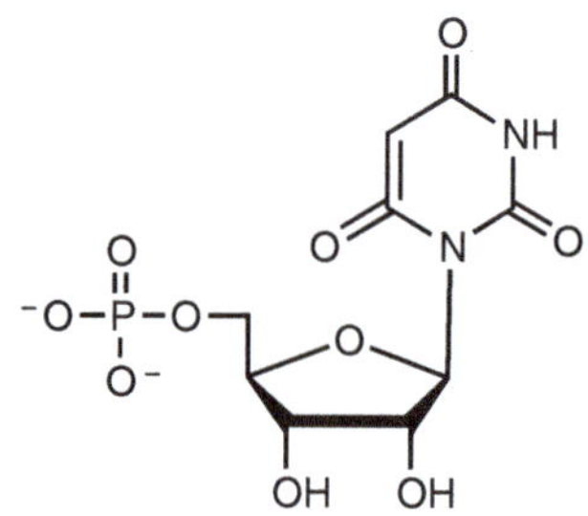

Uridine-5′-monophosphate (**44**)

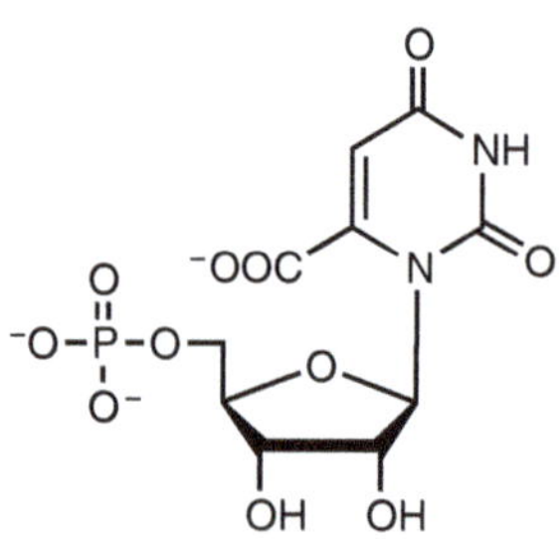

Orotidine-5-monophosphate (**45**)

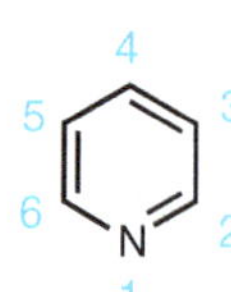

Pyridine numbering system (**46**)

1.3.2.3 Pyrimidine Nucleotides

A pyrimidine nucleotide is composed of a pyrimidine base, a sugar (ribose or deoxyribose) and at least a phosphate group. Examples of different forms of pyrimidine nucleotide structures are uridine-5′-monophosphate (5′-UMP usually abbreviated as UMP) (**44**) and orotidine-5′-monophosphate (5′-OMP) (**45**), an intermediate of the *de novo* pyrimidine biosynthesis.

1.3.3 Pyridines

Pyridine (**46**) is a basic heterocyclic organic compound. It is structurally related to benzene, with one methine group replaced by a nitrogen atom, and was discovered in 1849 by a Scottish chemist, Thomas Anderson, as one of the constituents of bone oil. Pyridine-related compounds include the catabolites nicotinate (**47**) and nicotinamide (**48**), and quinolinate (**49**) and NMN (**50**), which are intermediates of the biosynthesis of NAD (**51**) and NADP. These compounds act as common coenzymes involved in many redox reactions, carrying electrons from one reaction to another in all living cells. Each coenzyme consists of pyridine purine nucleotides joined through their phosphate groups. NAD(P) exists in two forms: an oxidized and reduced form abbreviated as $NAD(P)^+$ and NAD(P)H respectively.

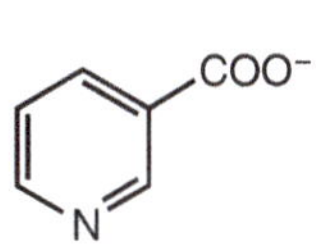

Nicotinate (**47**)

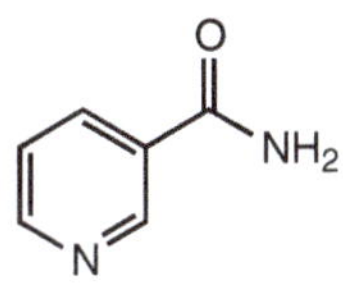

Nicotinamide (**48**)

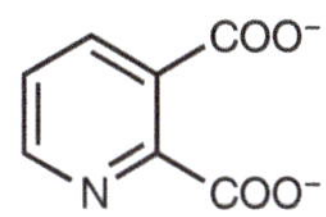

Quinolinate (**49**)

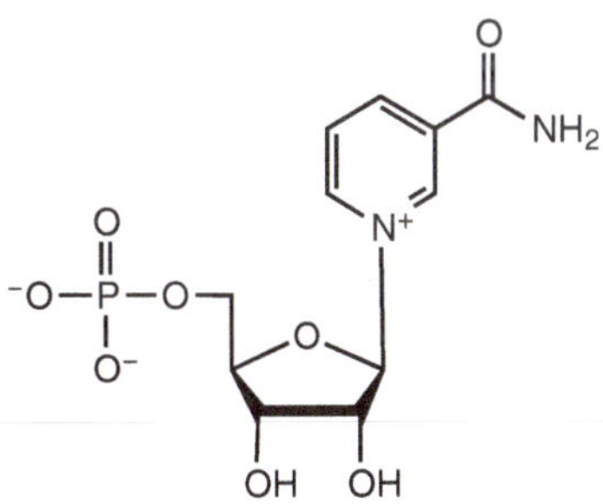

Nicotinamide mononucleotide (**50**)

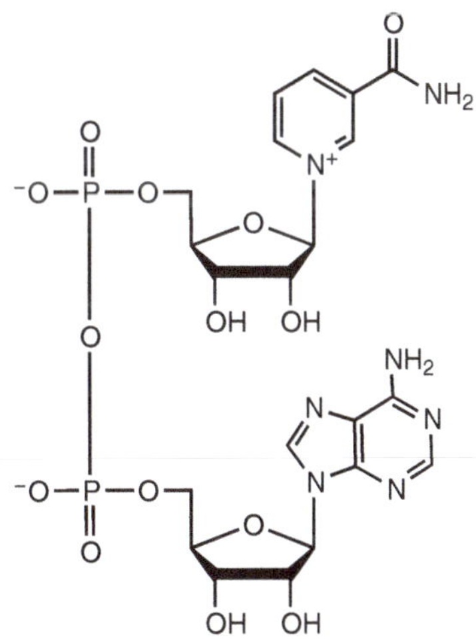

Nicotinamide adenine dinucleotide (**51**)

1.4 Summary

Nomenclature and abbreviations of nucleotide-related compounds and major chemical structures of purines, pyrimidine, and pyridine are described.

References

Ashihara, H., Yokota, T., and Crozier, A. (2013). Purine alkaloids, cytokinins, and purine-like neurotoxin alkaloids. In: *Natural Products: Phytochemistry, Botany and*

Metabolism of Alkaloids, Phenolics and Terpenes (eds. K.G. Ramawat and J.-M. Mérillon), 953–975. Berlin, Heidelberg: Springer.

Boniecka, J., Prusińska, J., Dąbrowska, G.B., and Goc, A. (2017). Within and beyond the stringent response-RSH and (p)ppGpp in plants. *Planta* 246: 817–842.

Burnstock, G. and Verkhratsky, A. (2012). Early history of purinergic signalling. In: *Purinergic Signalling and the Nervous System*, 7–66. Berlin, Heidelberg: Springer.

Fischer, E. (1884). Ueber die Harnsäure. I. *Ber. Dtsch. Chem. Ges.* 17: 328–338.

Harden, A. and Young, W.J. (1906). The alcoholic ferment of yeast-juice. Part II.—The coferment of yeast-juice. *Proc. R. Soc. London, Ser. B* 78: 369–375.

Henderson, J.F. and Paterson, A.R.P. (1973). *Nucleotide Metabolism - an Introduction*. New York: Academic Press.

IUPAC-IUB Commission on Biochemical Nomenclature (1970). Abbreviations and symbols for nucleic acids, polynucleotides and their constituents. Recommendations 1970. *J. Biol. Chem.* 245: 5171–5176.

Pietrowska-Borek, M., Nuc, K., Zielezińska, M., and Guranowski, A. (2011). Diadenosine polyphosphates (Ap3A and Ap4A) behave as alarmones triggering the synthesis of enzymes of the phenylpropanoid pathway in *Arabidopsis thaliana*. *FEBS Open Bio* 1: 1–6.

Pinner, A. (1885). Ueber die Einwirkung von Acetessigäther auf die Amidine Pyrimidin. *Ber. Dtsch. Chem. Ges.* A18: 759–760.

2

Occurrence of Nucleotides and Related Metabolites in Plants

2.1 Purines and Pyrimidines

Purines and pyrimidines are major chemical constituents of cells and occur primarily as components of DNA and RNA (polymerized nucleotides), and to a much lesser extent in the form of 'free' (aka 'soluble') nucleotides. Compared with nucleotides, free nucleosides and bases usually represent a very small fraction of the total purine and pyrimidine content of living cells, including plant cells. However, there are exceptions to this generalization, including the occurrence of substantial amounts caffeine (**1**) in plants such as coffee (*Coffea arabica* and *Coffea robusta*) and tea (*Camellia sinensis*), and theobromine (**2**) in cocoa (*Theobroma cocao*). Although not found in plants, the arabinosides of both thymine (**3**) and uracil (**4**) have been detected in the Caribbean sponge, *Cryptotethya crypta*. Guanine, in the silvery constituent of fish scales, and the excretion product, uric acid, are both free purines found in animal cells, but only occur in substantial amounts extracellularly (Henderson and Paterson 1973).

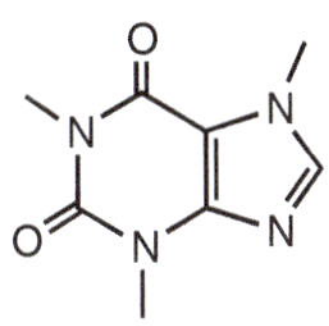
Caffeine (**1**)

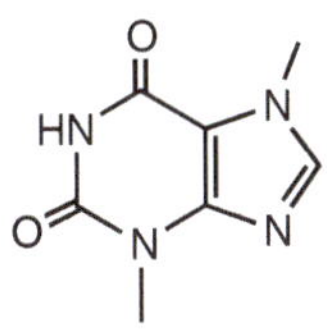
Theobromine (**2**)

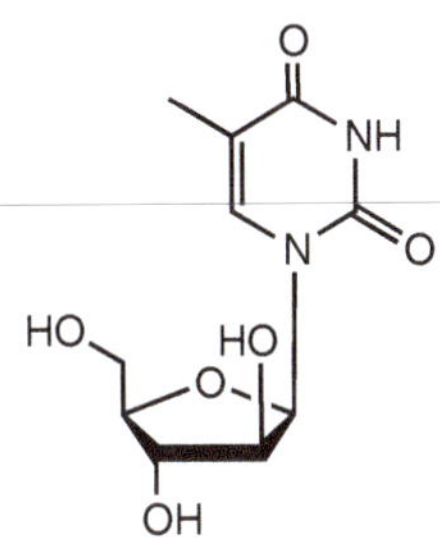
Thymine arabinoside (**3**)

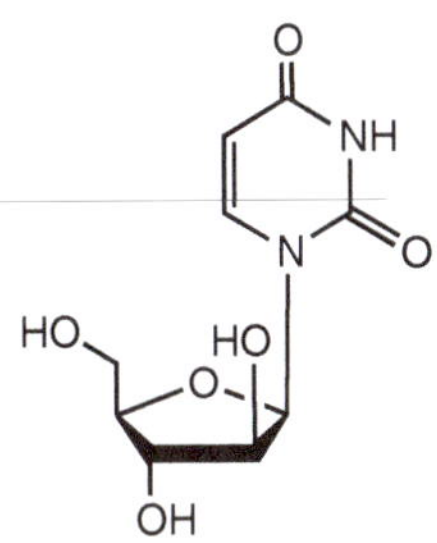
Uracil arabinoside (**4**)

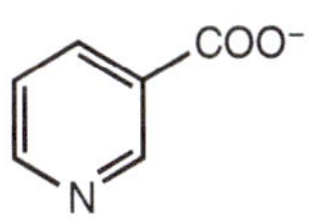
Nicotinate (**5**)

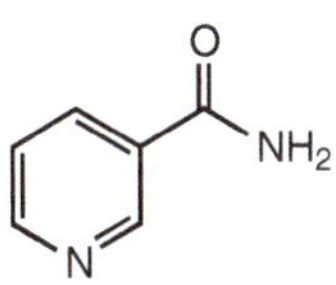
Nicotinamide (**6**)

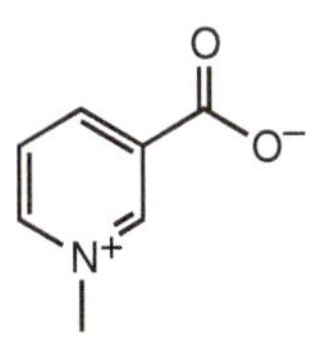
Trigonelline (**7**)

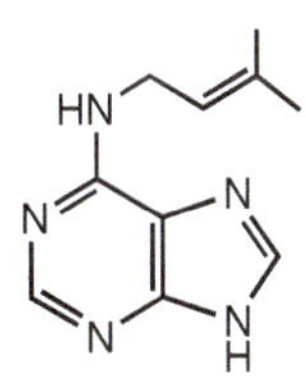
Isopentenyladenine (**8**)

Plant Nucleotide Metabolism – Biosynthesis, Degradation, and Alkaloid Formation, First Edition.
Hiroshi Ashihara, Iziar A. Ludwig and Alan Crozier.

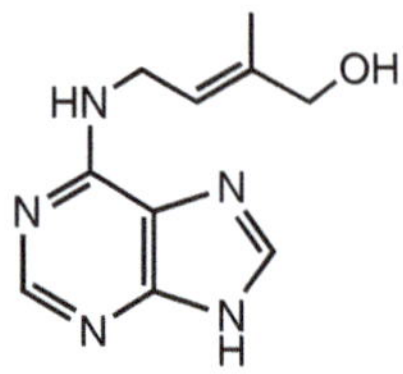

trans-Zeatin (**9**)

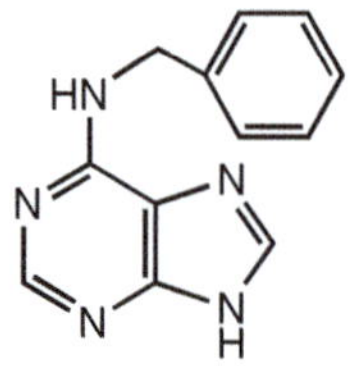

Benzyladenine (**10**)

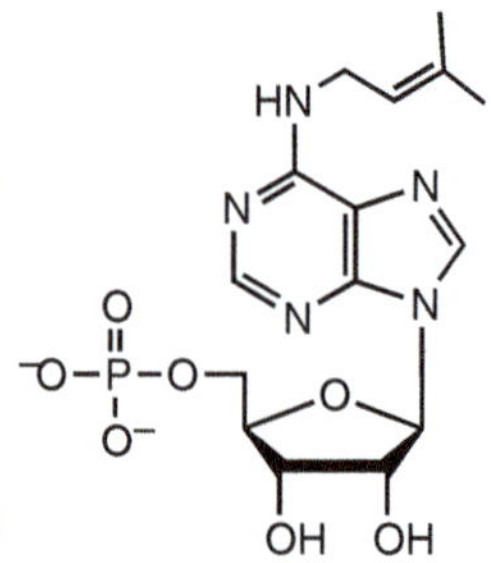

Isopentenyladenine ribotide (**11**)

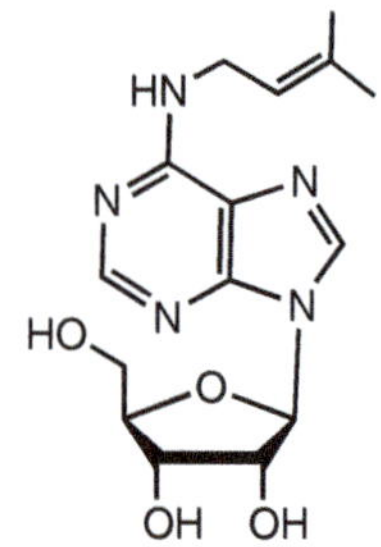

Isopentenyladenine riboside (**12**)

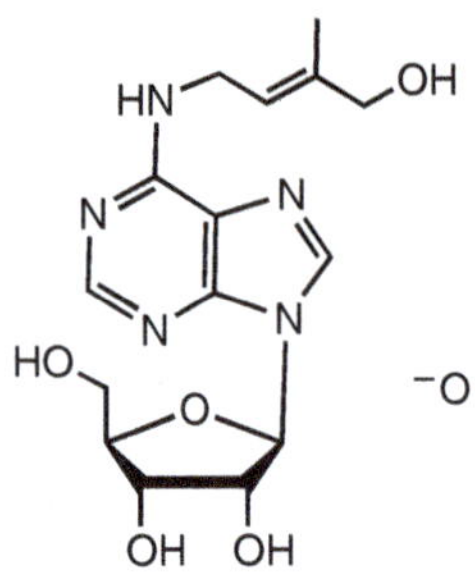

trans-Zeatin riboside (**13**)

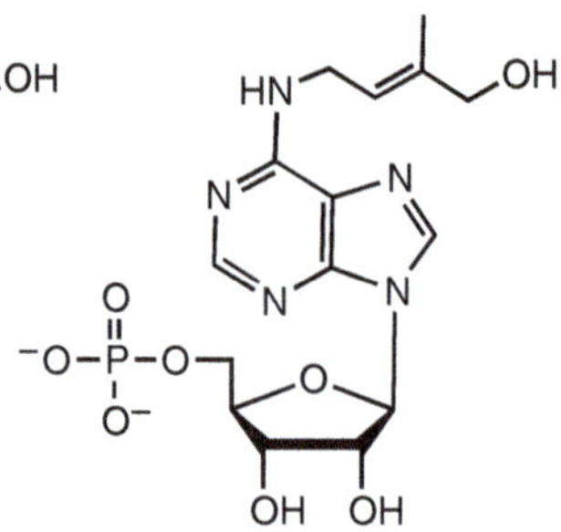

trans-Zeatin ribotide (**14**)

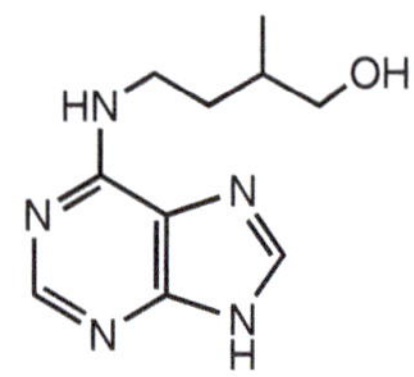

Dihydrozeatin (**15**)

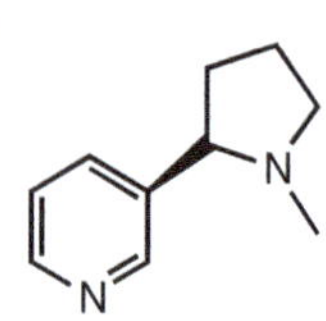

Nicotine (**16**)

2.1.1 Concentration of Purine and Pyrimidine Nucleotides

Plants contain complex mixtures of free nucleotides. The principal components of such mixtures can be isolated and determined using HPLC combined with absorbance or, preferably, mass spectrometric detection. However, the extraction of nucleotides from cells is a crucial step. With some plant tissues, the extraction procedure has a profound influence on the resultant analytical profile, and special precautions must be taken if the results are to truly represent the nucleotide composition of living cells. Because of the presence of enzymes that hydrolyse nucleotides and nucleosides, such as non-specific acid phosphatases and nucleosidases, extraction of nucleotide-related compounds from plant material is extremely challenging. These hydrolytic enzymes are not denatured during extraction and remain active even in cold methanolic extracts.

Free nucleotides are separated from cellular macromolecular components by cold extraction procedures in which the macromolecules are precipitated; the usual extraction solvents are dilute perchloric acid or trichloroacetic acid. Prior to analysis, perchloric acid extracts are neutralized with KOH, precipitating $KClO_4$, while those treated with trichloroacetic acid are partitioned against diethyl ether to remove the acid. Usually, the concentration of nucleotides is low, so the aqueous fraction is lyophilized, and redissolved in small amounts of an aqueous solvent prior to HPLC.

Some examples of the levels of purine and pyrimidine nucleotide levels in cultured plant cells and leaves are shown in Table 2.1. The results were obtained with cultured cells of *Arabidopsis thaliana* (thale cress), *Nicotiana tabacum* (tobacco), *Lycopersicon esculentum* (tomato), and *Catharanthus roseus* (Madagascar periwinkle) at the exponential stage of growth. For comparison, nucleotide levels of cultured *C. roseus* cells

Table 2.1 Concentration of nucleotides in plant cells and tissues.

	Arabidopsis thaliana		*Nicotiana tabacum*		*Lycopersicon esculentum*		*Camellia sinensis*	*Catharansus roseus*	
Nucleotide	**Cells**	**Leaves**	**Cells**	**Leaves**	**Cells**	**Leaves**	**Leaves**	**+Pi**	**−Pi**
ATP	145	40	124	48	190	50	177	100	22
ADP	19	77	8	23	38	30	35	18	16
AMP	17	59	9	33	—	55	14	16	12
GTP	26	tr	23	tr	24	tr	26	26	8
GDP	—	—	—	—	—	—	tr	3	3
GMP	tr	9	tr	—	10	—	tr	3	2
IMP	—	8	—	20	—	20	—	—	—
XMP	—	—	—	—	—	—	29	—	—
UTP	18	20	49	7	27	6	88	55	10
UDP	tr	18	—	23	59	16	46	29	22
UMP	178	81	187	74	174	50	55	33	24
CTP	9	tr	4	12	—	—	tr	7	2
CDP	tr	—	—	9	—	—	tr	4	2
CMP	—	tr	—	tr	—	tr	32	6	3
Relative rate									
ΣAN (%)	44	56	35	42	44	60	48	45	40
ΣGN (%)	6	3	6	0	7	0	6	11	10
ΣUN (%)	48	38	59	42	50	0	40	39	44
ΣCN (%)	2	0	1	8	0	32	7	6	6
Energy status									
EC	0.85	0.45	0.91	0.57	0.92	0.48	0.86	0.81	0.60
ATP/ADP	7.6	0.5	14.9	2.0	5.1	1.7	3.3	5.6	1.4

The values (nmol g^{-1} f.w.) are obtained from the suspension-cultured cells at the exponential phase of growth and young leaves. For comparison, the values from phosphate-deficient *Catharanthus roseus* cells are also shown. tr: trace and EC: energy charge.
Source: Based on Ashihara and Crozier (1999) and Riondet et al. (2005).

grown in a phosphate (Pi)-deficient medium are also shown. Table 2.1 also lists the levels in young leaves of *A. thaliana*, *N. tabacum*, *L. esculentum* and *Ca. sinensis*.

There are similar profiles of nucleotide pools in almost all plants. The pools of adenine nucleotides and uridine nucleotides (aka uracil nucleotides) always exceed the guanine nucleotide pool. The cytidine nucleotide pool is the smallest of these pools. The nucleoside triphosphate levels are usually much more substantial than those of nucleoside mono- and diphosphates.

The intracellular levels of nucleotides in plant cells are influenced by several environmental factors. For example, Pi starvation of cultured plant cells results in a marked decrease in the levels of nucleotides, especially nucleoside triphosphates (Table 2.1). The proportion of adenylates, i.e. the adenylate energy charge ([ATP] + ½[ADP])/(ATP + ADP + AMP) originally proposed by Atkinson (1977), is maintained at an almost constant ratio of 0.8–0.9 in actively growing plant cells. The energy charge in young tea leaves, 0.86, is within this range, but leaves of *A. thaliana*, tobacco, and tomato showed lower values, 0.5–0.6, because of the higher concentrations of ADP and AMP. While the energy charge fluctuates with diverse growth conditions, some reported changes might be ascribed to the presence of hydrolytic enzymes in extracts. The ATP/ADP ratio is also a major parameter of interest in the investigation of metabolic aspects of bioenergetics. The ratios are usually high in actively growing cells and tissues.

2.1.2 Concentration of Purine and Pyrimidine Bases and Nucleosides

Plant cells, as well as microorganisms and animal cells contain, at best, very low cellular pools of bases and nucleosides (Table 2.2). It is thought that this may be caused by the high activities of salvage enzymes resulting in a rapid conversion of purine bases into their respective nucleotides (Ashihara et al. 2018). In many plant cells, adenine and guanine pools are markedly smaller than those of their nucleotides (Ashihara et al. 1990; Brown 1963; Grzelczak and Buchowicz 1975). For example, the adenine and guanine content of wheat grain are 4 and 5 nmol g^{-1} d.w., respectively (Grzelczak and Buchowicz 1975). Shimano and Ashihara (2006) found only trace amounts of adenine and guanine in growing *C. roseus* cells, but the purine bases accumulated after prolonged Pi starvation, when adenine nucleotide contents were reduced (Table 2.2). A much higher content of adenine, up to 300 nmol g^{-1} d.w., has been reported in the youngest, upper leaves of barley (Sawert et al. 1988), and in soybean seeds (Yokozawa and Oura 1986).

Compared with purine bases, sizeable amounts of purine nucleosides have been detected in plants (Ashihara et al. 1990; Grzelczak and Buchowicz 1975; Sawert et al. 1988; Shimano and Ashihara 2006). For example, the adenosine, guanosine, and

Table 2.2 Concentration of nucleosides and nucleobases in plant cells and tissues.

Nucleosides and bases	*Nicotiana tabacum*		*Datura innoxia*	*Catharanthus roseus*		Peanut	Barley	Wheat	Rye
	Cells	Leaves	Cells	+Pi	−Pi[a]	Leaves	Leaves	Leaves	Leaves
Adenosine	3.8	—	1.0	—	8.1	9.0	69.0	41.9	40.6
Guanosine	—	—	2.3	1.3	13.6	—	27.6	22.3	4.6
Inosine	—	—	—	—	—	—	2.0	—	—
Uridine	4.2	2.0	4.7	1.0	4.1	—	57.1	15.8	4.1
Cytidine	—	—	—	0.9	17.5	—	—	—	—
Adenine	—	—	—	—	1.5	—	41.4	13.0	8.7
Guanine	—	—	—	—	7.2	—	—	—	—
Uracil	—	—	—	—	—	—	—	—	—
Cytosine	—	—	—	—	—	—	—	—	—

The values (nmol g^{-1} f.w.) were obtained from the suspension-cultured cells at the exponential phase of growth phase and young leaves. For comparison, the values from phosphate-deficient *Catharanthus roseus* cells are also shown.

a) Two week Pi starvation.

Source: Based on data of Wagner and Backer (1992) and Shimano and Ashihara (2006).

xanthosine content of mature pea seeds are 152, 67, and 91 nmol g^{-1} d.w., respectively (Brown 1963). Accumulation of nucleosides also occurs in some cereal leaves. For instance, in young wheat leaves, the adenosine pool (300 nmol g^{-1} d.w.) is similar to that of ATP (280 nmol g^{-1} d.w.) (Sawert et al. 1988). Nucleosides, such as adenosine, may be stored in vacuoles of plant cells as excess cytosolic nucleosides can disturb the cellular energy charge and inhibit the activity of several metabolic pathways associated with the *S*-adenosyl-L-methionine cycle (Ashihara et al. 2018; Oikawa et al. 2011).

2.2 Pyridine Nucleotides

2.2.1 Concentration of Pyridine Nucleotides

Pyridine nucleotides, NAD and NADP occur in oxidized (NAD^+ and $NADP^+$) and reduced forms (NADH and NADPH). While HPLC techniques exist for separating and quantifying NAD and NADP, both can be readily quantified through well-established and sensitive enzymatic cycling assays. It is a simple matter to distinguish between NAD and NADP either through the use of specific enzymes or by HPLC. More problematical is discrimination between the reduced and oxidized forms. Usually, this is achieved by exploiting the different stability of the two forms in acid and alkaline extraction buffers. High-throughput methods have been developed that can measure the two nucleotides at the same time as other redox compounds (Queval and Noctor 2007).

Concentrations of NAD^+, NADH, $NADP^+$ and NADPH reported in young *A. thaliana* are shown in Table 2.3. The NADH/NAD^+ and NADPH/$NADP^+$ ratios vary according to the physiological conditions. However, the concentration of the oxidized form (NAD^+) in plants is higher than that of the reduced form (NADH) (Wang and Pichersky (2007).

A full understanding of pyridine nucleotide-related biological phenomena requires techniques that determine specific pool sizes in different intracellular compartments. The fluorescent properties of reduced pyridine nucleotides have been exploited in the analysis of plant mitochondrial metabolism (Kasimova et al. 2006). However, these measurements do not distinguish between the NADH and NADPH signals. The limitations

Table 2.3 Concentrations of NAD^+, NADH, $NADP^+$, and NADH in *Arabidopsis thaliana* (ecotype Col-0, wild-type) plants.

Tissue	NAD^+	NADH	NAD^+/NADH	$NADP^+$	NADPH	$NADP^+$/NADPH
Root	6.5	0.2	(31.4)	1.7	0.3	(5.7)
Roselle leaves	10.2	4.5	(2.8)	2.8	—	—
Stem	12.0	—	—	3.1	—	—
Cauline leaves	11.3	4.0	(3.4)	3.6	—	—
Flower	27.0	3.6	(9.1)	4.5	1.0	(4.0)
Silique	17.9	2.8	(4.5)	3.8	0.6	(6.4)
Seedlings	9.9	0.9	(11.9)	2.0	0.4	(6.0)

The values (nmol g^{-1} f.w.) and the NAD(P)$^+$/NAD(P)H ratios are obtained from different tissues.
Source: In part based on data of Wang and Pichersky (2007).

of the information generated by the different methods of measuring pyridine nucleotides have been discussed by Hagedorn et al. (2004).

2.2.2 Concentration of Nicotinate and Nicotinamide

In contrast to pyridine nucleotides, the pool sizes of nicotinate (**5**) and nicotinamide (**6**) are very small. These metabolites, produced by pyridine nucleotide degradation, are converted rapidly to trigonelline (**7**, *N*-methylnicotinate) or nicotinate glucosides (Ashihara et al. 2015). Typically, when the concentration of niacin, which comprises nicotinamide, nicotinate, and vitamin B_3, is analysed in foodstuffs, the contribution of these individual components is not considered. Common extraction treatments, such as alkaline or acid hydrolysis, mean that any NAD, NADP, nicotinamide, and other nicotinate-related metabolites present in the sample are converted to nicotinate, which is the form that is then measured as niacin (Noctor et al. 2011).

2.3 Concentration of Cytokinins

Cytokinins are a class of phytohormones (aka growth substances) that promote cell division in plants. They are involved primarily in cell growth and differentiation, but also affect apical dominance, axillary bud growth, and leaf senescence. Cytokinins, such as isopentenyladenine (**8**), *trans*-zeatin (**9**), and benzyladenine (**10**) are adenine-derived plant hormones. The major sites of cytokinin biosynthesis are roots, cambium, and other actively dividing tissues. Cytokinins play vital roles in the regulation of multiple physiological processes in plants including cell proliferation and differentiation, nutrient allocation, leaf senescence, and responses to external biotic and abiotic signals (Bishop et al. 2015; Crozier et al. 2000).

Compared to purine nucleotides, the concentration of cytokinins is extremely low. Cytokinins are now routinely analysed using ultra-performance HPLC–high resolution mass spectrometry. The concentration of isopentenyladenine cytokinins, isopentenyladenine (**8**), isopentenyladenine ribotide (**11**), and isopentenyladenine riboside (**12**) in *A. thaliana* seedlings has been reported to be 0.4, 2.4, and 5.7 pmol g^{-1} f.w., respectively. While those of *trans*-zeatin cytokinins, *trans*-zeatin (**9**), *trans*-zeatin riboside (**13**) and *trans*-zeatin ribotide (**14**) were 0.6, 2.8, and 1.6 pmol g^{-1} f.w., respectively. Levels of dihydrozeatin (**15**) and benzyladenine cytokinins were <20 fmol g^{-1} f.w. (Novák et al. 2008).

2.4 Alkaloids Derived from Nucleotides

Some alkaloids, derived from purine and pyridine nucleotides, accumulate in specific plant species. Caffeine and nicotine (**16**) are typical nucleotide-derived alkaloids. For example, high concentrations of caffeine (~2% d.w.) are found in tea (*Ca. sinensis*) and coffee (*C. arabica*) (Ashihara et al. 2017), while nicotine (~1% d.w.) accumulates in tobacco (*N. tabacum*) (Saitoh et al. 1985). These alkaloids occur in more than 100 plant species. Detailed distribution and concentration of these alkaloids are described in purine and pyridine alkaloids in Parts V and VI.

2.5 Summary

Occurrence of major purine, pyrimidine, and pyridine nucleotides in plants are summarized. For comparison, cytokinins and nucleotide alkaloids are also noted. In most plants, the concentration of nucleotides is higher than nucleosides and bases. Cytokinins occur in much lower amounts than purine nucleotides. A much higher concentration of caffeine is found in tea and coffee plants.

References

Ashihara, H. and Crozier, A. (1999). Biosynthesis and catabolism of caffeine in low-caffeine-containing species of *Coffea*. *J. Agric. Food. Chem.* 47: 3425–3431.

Ashihara, H., Yabuki, N., and Mitsui, K. (1990). A high-performance liquid chromatography method for separation of purine bases, nucleosides and ureides: application to studies on purine catabolism in higher plants. *J. Biochem. Biophys. Methods* 21: 59–63.

Ashihara, H., Ludwig, I., Katahira, R. et al. (2015). Trigonelline and related nicotinic acid metabolites: occurrence, biosynthesis, taxonomic considerations, and their roles in planta and in human health. *Phytochem. Rev.* 14: 765–798.

Ashihara, H., Mizuno, K., Yokota, T., and Crozier, A. (2017). Xanthine alkaloids: occurrence, biosynthesis, and function in plants. In: *Progress in the Chemistry of Organic Natural Products*, vol. 105 (eds. A.D. Kinghorn, H. Falk, S. Gibbons and J. Kobayashi), 1–88. Cham: Springer International Publishing.

Ashihara, H., Stasolla, C., Fujimura, T., and Crozier, A. (2018). Purine salvage in plants. *Phytochemistry* 147: 89–124.

Atkinson, D.E. (1977). *Cellular Energy Metabolism and its Regulation*. New York: Academic Press.

Bishop, G., Sakakibara, H., Seo, M., and Yamaguchi, S. (2015). Biosynthesis of hormones. In: *Biochemistry and Molecular Biology of Plants*, 2e (eds. B.B. Buchanan, W. Gruissem and R.L. Jones), 769–833. Chichester, UK: Wiley.

Brown, E. (1963). Purine and pyrimidine derivatives in mature pea seeds. *Biochem. J.* 88: 498–504.

Crozier, A., Kamita, Y., Bishop, G., and Yokota, T. (2000). Biosynthesis of hormones and elicitor molecules. In: *Biochemistry and Molecular Biology of Plants* (eds. B.B. Buchanan, W. Gruissen and R.L. Jones), 850–829. Rockville: American Society of Plant Physiologists.

Grzelczak, Z. and Buchowicz, J. (1975). Purine and pyrimidine bases and nucleosides of germinating *Triticum aestivum* seeds. *Phytochemistry* 14: 329–331.

Hagedorn, P.H., Flyvbjerg, H., and Møller, I.M. (2004). Modelling NADH turnover in plant mitochondria. *Physiol. Plant* 120: 370–385.

Henderson, J.F. and Paterson, A.R.P. (1973). *Nucleotide Metabolism – an Introduction*. New York: Academic Press.

Kasimova, M.R., Grigiene, J., Krab, K. et al. (2006). The free NADH concentration is kept constant in plant mitochondria under different metabolic conditions. *Plant Cell* 18: 688–698.

Noctor, G., Hager, J., and Li, S. (2011). Biosynthesis of NAD and its manipulation in plants. In: *Advances in Botanical Research*, vol. 58 (eds. R. Fabrice and D. Roland), 153–201. Academic Press.

Novák, O., Hauserová, E., Amakorová, P. et al. (2008). Cytokinin profiling in plant tissues using ultra-performance liquid chromatography–electrospray tandem mass spectrometry. *Phytochemistry* 69: 2214–2224.

Oikawa, A., Matsuda, F., Kikuyama, M. et al. (2011). Metabolomics of a single vacuole reveals metabolic dynamism in an alga *Chara australis*. *Plant Physiol.* 157: 544–551.

Queval, G. and Noctor, G. (2007). A plate reader method for the measurement of NAD, NADP, glutathione, and ascorbate in tissue extracts: application to redox profiling during Arabidopsis rosette development. *Anal. Biochem.* 363: 58–69.

Riondet, C., Morel, S., and Alcaraz, G. (2005). Determination of total ribonucleotide pool in plant materials by high-pH anion-exchange high-performance liquid chromatography following extraction with potassium hydroxide. *J. Chromatogr. A* 1077: 120–127.

Saitoh, F., Noma, M., and Kawashima, N. (1985). The alkaloid contents of sixty *Nicotiana* species. *Phytochemistry* 24: 477–480.

Sawert, A., Backer, A.I., and Wagner, K.G. (1988). Age-dependent decrease of nucleoside pools in cereal leaves. *Plant Cell Physiol.* 29: 61–65.

Shimano, F. and Ashihara, H. (2006). Effect of long-term phosphate starvation on the levels and metabolism of purine nucleotides in suspension-cultured *Catharanthus roseus* cells. *Phytochemistry* 67: 132–141.

Wagner, K.G. and Backer, A.I. (1992). Dynamics of nucleotides in plants studied on a cellular basis. In: *International Review of Cytology*, vol. 134 (eds. K.W. Jeon and M. Friedlander), 1–84. San Diego: Academic Press.

Wang, G. and Pichersky, E. (2007). Nicotinamidase participates in the salvage pathway of NAD biosynthesis in Arabidopsis. *Plant J.* 49: 1020–1029.

Yokozawa, T. and Oura, H. (1986). Contents of free adenine in soybeans from several countries. *Agric. Biol. Chem.* 50: 1317–1319.

3

General Aspects of Nucleotide Biosynthesis and Interconversions

3.1 Introduction

This chapter focuses on (i) an outline of the formation of purine, pyrimidine, and pyridine ribonucleoside monophosphates from small molecular weight precursors, the initial metabolites of *de novo* nucleotide biosynthesis, (ii) conversion of these initial metabolites to nucleoside di- and triphosphates, and (iii) synthesis of deoxyribonucleotides from ribonucleotides. The biosynthesis of purine, pyrimidine, and pyridine monophosphate utilizes three distinct pathways. The individual pathways are presented in more detail in Part II (purine), Part III (pyrimidine), and Part VI (pyridine). In contrast, pathways ii and iii both involve common interconversions of purine and pyrimidine nucleotides, and are therefore, described in this chapter.

3.2 *De Novo* Biosynthesis of Ribonucleoside Monophosphates

In many organisms including plants, ribonucleotides are synthesized from precursors derived from amino acids, sugar phosphates, and carbon dioxide. *De novo* synthesis of purine, pyrimidines, and the ring of pyridine (nicotinamide adenine) nucleotides are synthesized via three distinct pathways. It is noteworthy that the initial products are nucleotides, namely inosine-5′-monophosphate (IMP), orotidine-5′-monophosphate (OMP) and nicotinate mononucleotide (NaMN), but not nucleosides or nucleobases. The *de novo* pathways of nucleotide biosynthesis are summarized in Figure 3.1. In the case of purine nucleotides, IMP is synthesized from 5-phosphoribosyl-1-pyrophosphate (PRPP) (Figure 3.1a). The carbon and nitrogen atoms of the purine ring come from multiple sources. Glycine contributes all its carbon and nitrogen atoms (positions 4, 5, and 7 of a purine ring), with additional nitrogen atoms from glutamine (positions 3 and 9) and aspartate (position 1), and further carbon atoms from formyl groups (positions 2 and 8), which are transferred from the coenzyme tetrahydrofolate as 10-formyltetrahydrofolate, and a carbon atom from bicarbonate (position 6). A detailed biochemical and molecular analysis of purine biosynthesis in plants has yet to be carried out, but the enzymatic steps are similar to those occurring in microorganisms and animals. More details of the 14 enzymatic steps involved in adenosine monophosphate (AMP) and GMP production from PRPP are presented in Part II.

Plant Nucleotide Metabolism – Biosynthesis, Degradation, and Alkaloid Formation, First Edition.
Hiroshi Ashihara, Iziar A. Ludwig and Alan Crozier.

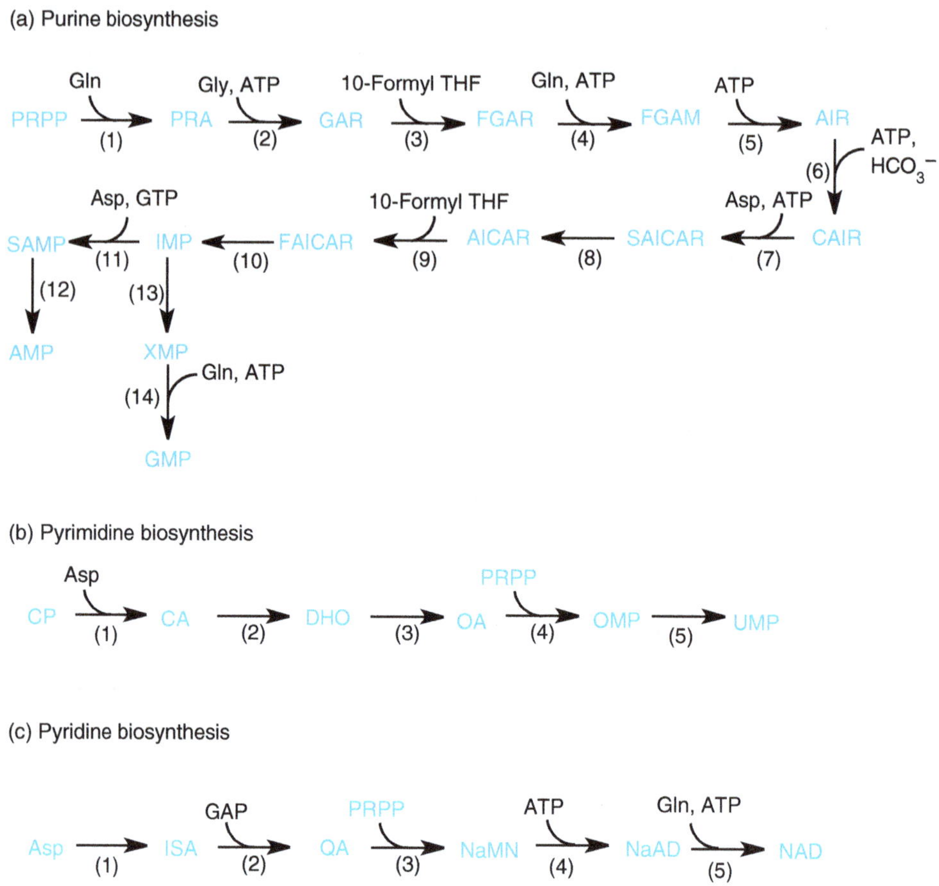

Figure 3.1 The major pathways of *de novo* biosynthesis of purine, pyrimidine and pyridine nucleotides. Detailed pathways are shown in Part II, III, and IV, respectively. **(a) Purine ribonucleotide biosynthesis:** abbreviations of metabolites are as follows: PRPP, 5-phosphoribosyl-1-pyrophosphate; PRA, 5-phosphoribosyl amine; GAR, glycineamide ribonucleotide; FGAR, formylglycineamide ribonucleotide; FGRAM, formylglycine amidine ribonucleotide; AIR, 5-aminoimidazole ribonucleotide; CAIR, 5-aminoimidazole 4-carboxylate ribonucleotide; SCAIR, 5-aminoimidazole-4-*N*-succinocar boxyamide ribonucleotide; AICAR, 5-aminoimidazole-4-carboxyamide ribonucleotide; FAICAR, 5-formamidoimidazole-4-carboxyamide ribonucleotide; SAMP, adenylosuccinate; XMP, xanthosine-5′-monophosphate. The participated enzymes are (1) PRPP amidotransferase (2.4.2.14); (2) GAR synthetase (6.3.4.13); (3) GAR formyl transferase (2.1.2.2); (4) FGAM synthetase (6.3.5.3); (5) AIR synthetase (6.3.3.1); (6) AIR carboxylase (4.1.1.21); (7) SAICAR synthetase (6.3.2.6); (8) adenylosuccinate lyase (4.3.2.2); (9) AICAR formyl transferase (2.1.2.3); (10) IMP cyclohydrolase (3.5.4.10); (11) SAMP synthetase (6.3.4.4); (12) adenylosuccinate lyase (4.3.2.2), the same enzyme for step 8; (13) IMP dehydrogenase (1.1.1.205); (14) GMP synthetase (6.3.5.2). **(b) Pyrimidine ribonucleotide biosynthesis:** CP, carbamoyl phosphate; CA, carbamoyl aspartate; DHO, dihydroorotate; OA, orotate; OMP, orotidine-5′-monophosphate. Enzymes: (1) aspartate transcarbamoylase (2.1.3.2); (2) dihydroorotase (3.5.2.3); (3) dihydroorotate dehydrogenase (1.3.99.11); (4)–(5) UMP synthase (orotate phos phoribosyltransferase [2.4.2.10] plus orotidine-5′-phosphate decarboxylase [4.1.1.23]). **(C) Pyridine ribonucleotide biosynthesis:** Asp, aspartate; ISA, α-iminosuccinate; QA, quinolinate; NaMN, nicotinate mononucleotide; NaAD; nicotinate adenine dinucleotide. Enzymes: (1) L-aspartate oxidase (1.4.3.16); (2) quinolinate synthase (2.5.1.72); (3) quinolinate phosphoribosyltransferase (decarboxylating) (2.4.2.19); (4) nicotinate mononucleotide adenylyltransferase (2.7.7.18); (5) NAD synthase (6.3.5.1).

Compared to purine biosynthesis, the pyrimidine biosynthetic pathway (aka the orotate pathway), namely the formation of uridine monophosphate (UMP) from carbamoyl phosphate, is more straight forward. The pathway consists of six reactions and the precursors are carbamoyl phosphate, aspartate, and PRPP (Figure 3.1b). Pyrimidine nucleotide biosynthesis, which is the same in plants as other organisms, is evolutionarily conserved in all species examined to date. Since nitrogen in carbamoyl phosphate is synthesized from glutamine, the nitrogen atoms of positions 1 and 3 of pyrimidine ring are derived from aspartate and glutamine, respectively. The carbon atoms at position 2 and positions 4–6 of the ring originate from carbon dioxide and aspartate (see Part III).

Pyridine nucleotides (nicotinamide adenine dinucleotide [NAD] and nicotinamide adenine dinucleotide phosphate [NADP]) consist of two mononucleotides, namely AMP and nicotinamide mononucleotide (NMN). Since the AMP moiety is a product of purine biosynthesis, in a narrow sense, pyridine nucleotide synthesis results in the formation of nicotinate monophosphate (NaMN). Two distinct pathways called the 'aspartate pathway' and the 'kynurenine pathway' occur in different organisms. The aspartate pathway, in which aspartate, glutamine, glyceraldehyde-3-phosphate and PRPP are used as precursors, operates in plants (Figure 3.1c). This route is also found in most bacteria, including *Escherichia coli*. In contrast, mammals, fungi, and some bacteria produce NAD by the kynurenine pathway. The steps leading from quinolinate to NAD are conserved among prokaryotes and eukaryotes. Further details of these biosynthetic pathways are covered in Part IV.

3.3 Interconversion of Nucleoside Monophosphates, Nucleoside Diphosphates, and Triphosphates

As illustrated in Figure 3.1, products of *de novo* biosynthesis pathways of purine, pyrimidine, and pyridine nucleotides are AMP and GMP, UMP and NaMN. Purine and pyrimidine ribonucleoside monophosphates are converted to ribonucleoside diphosphates and triphosphates as well as deoxyribonucleotides. These processes are common to both purine and pyrimidine nucleotides. The ribonucleoside mono- (NMP), di- (NDP) and triphosphates (NTP) are interconverted (see Figure 3.5), coupled with the various reactions shown in Sections 3.3.1, 3.3.2 and 3.4.

3.3.1 Nucleoside-Monophosphate Kinase

Nucleoside monophosphates are converted to nucleoside diphosphates by nucleoside-monophosphate kinase (EC 2.7.4.4) which catalyses reaction 1 as shown in Table 3.1. The substrate specificity of this enzyme is broad; many nucleoside monophosphates can act as acceptors and other nucleoside triphosphates can also serve as phosphate donors instead of ATP. However, the cellular concentration of ATP is usually much higher than that of other nucleoside triphosphates (Table 2.1), so ATP is the principal phosphate donor.

Table 3.1 Enzymes involved in the conversion of nucleoside mono-, di- and triphosphate in plants.

	Enzymes	EC	Reaction	References
1	Nucleoside-monophosphate kinase	2.7.4.4	ATP + Nucleoside monophosphate → ADP + Nucleoside diphosphate	Noda (1962)
2	Nucleoside-diphosphate kinase	2.7.4.6	ADP + Nucleoside triphosphate → ATP + Nucleoside diphosphate	Dorion and Rivoal (2015)
3	Adenylate kinase (= Myokinase)	2.7.4.3	AMP + ATP → 2 ADP	Johansson et al. (2008)
4	Guanylate kinase	2.7.4.8	GMP + ATP → GDP + ADP	Nomura et al. (2014)
5	UMP/CMP kinase (= Cytidylate kinase)	2.7.4.14	CMP + ATP → CDP + ADP UMP + ATP → UDP + ADP	Zhou et al. (1998)
6	ATP synthase (H^+-transporting two-sector ATPase)	3.6.3.14	ADP + Pi + H^+_{out} → ATP + H_2O + H^+_{in}	Li et al. (2012)
7	Phosphoglycerate kinase	2.7.2.3	ADP + 3-Phospho-D-glyceroyl phosphate → ATP + 3-Phospho-D-glycerate	Troncoso-Ponce et al. (2012)
8	Pyruvate kinase	2.7.1.40	ADP + Phosphoenolpyruvate → ATP + Pyruvate	Baysdorfer and Bassham (1984)
9	Succinate – CoA ligase (ADP-forming)	6.2.1.5	ADP + Pi + Succinyl-CoA → ATP + Succinate + CoA	Johnson et al. (1998)

3.3.2 Specific Nucleoside-Monophosphate Kinases

In addition to the non-specific nucleoside-monophosphate kinase mentioned above, there are at least three nucleoside-monophosphate kinases which contribute to the conversion of NMP to NDP. They are adenylate (AMP) kinase (EC 2.7.4.3), guanylate (GMP) kinase (EC 2.7.4.8) and cytidylate (UMP/cytidine-5′-monophosphate [CMP]) kinase (EC 2.7.4.14). It has been reported that genes encoding adenylate kinase show high sequence homology with those encoding UMP-CMP kinase (Fukami-Kobayashi et al. 1996). The equilibrium constants of these enzymes are ~1, so the reaction is reversible. Adenylate kinase is ubiquitous and is found in different subcellular locations, including the cytosol, mitochondria, and plastids of plants. The reactions of each enzyme, and associated references, are presented in Table 3.1.

3.4 Conversion of Nucleoside Diphosphates to Nucleoside Triphosphates

Phosphorylation of nucleoside diphosphates, especially ADP to ATP, is performed by two different types of reaction. One, ADP + Pi → ATP, is the electron transfer system of respiration and photosynthesis. The other is reactions involving substrate level phosphorylation, which is the formation of ATP or GTP by the direct transfer of a phosphoryl group to ADP or GDP from another phosphorylated metabolite, such as phosphoenolpyruvate.

3.4.1 ATP Synthesis by Electron Transfer Systems

The conversion of ADP to ATP is performed by phosphorylation linked to the electron transport chains of respiration and photosynthesis (Karp 2013; Niyogi et al. 2015). Briefly, in oxidative phosphorylation, electrons are derived from substrates that enter the TCA (tricarboxylic acid) cycle (aka the citrate or Krebs cycle) (Figure 3.2), whereas in photophosphorylation, they are furnished by chlorophyll in the presence of light. ATP synthesis is coupled to the subsequent successive oxidation–reduction of members of electron transport chains, which pass electron pairs through stages of successively lower potential energy until they reach oxygen, the terminal electron sink. ATP synthase (H^+-transporting two-sector ATPase, EC 3.6.3.14) is an enzyme that creates ATP. It is formed from ADP and Pi (Table 3.1).

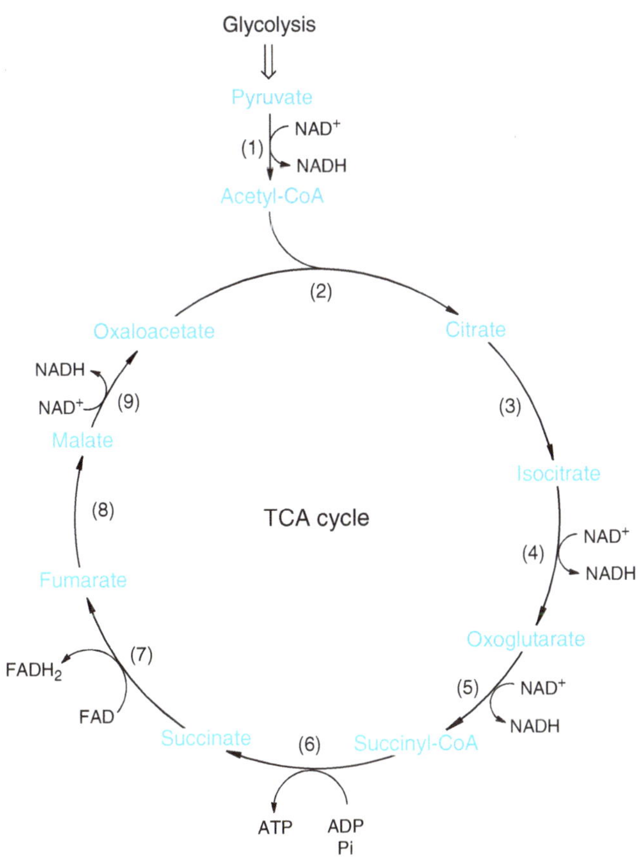

Figure 3.2 The tricarboxylic acid (Krebs) cycle in plants. (1) Pyruvate dehydrogenase; (2) citrate synthase; (3) aconitase; (4) isocitrate dehydrogenase; (5) 2-oxoglutarate dehydrogenase; (6) succinyl-CoA synthetase; (7) succinate dehydrogenase; (8) fumarase; (9) malate dehydrogenase.

3.4.2 Substrate-Level ATP Synthesis

Substrate-level phosphorylation of ADP is performed by the direct transfer of a phosphoryl group to ADP from another phosphorylated compound (Table 3.1). In the TCA cycle in plants, substrate-level phosphorylation occurs mainly in the cytosol or chloroplasts during glycolysis and in mitochondria.

Substrate-level phosphorylation in glycolysis occurs via two steps, 7 and 10, in Figure 3.3. After the conversion of glyceraldehyde-3-phosphate to glycerate-1,3-bisphos phate, catalysed by glyceraldehyde-3-phosphate dehydrogenase using Pi and NAD^+ (step 6), ATP is formed by the dephosphorylation of glycerate-1,3-bisphosphate (step 7). The second substrate-level phosphorylation is catalysed by pyruvate kinase (step 10) and ATP is produced from the phosphate group of phosphoenolpyruvate. The hexose molecule is split into two three-carbon molecules, via an aldolase reaction (step 4). Hexose is converted to fructose-1,6-bisphosphate by hexokinase (step 1) and phosphofructokinase (step 3a), in the process consuming 2 mol and producing 4 mol of ATP in steps 7 and 10. Thus, there is net production of 2 mol of ATP. The ATP-consuming and ATP-producing reactions of glycolysis are illustrated in Figure 3.4. If the conversion is performed by hexokinase (step 1) and pyrophosphate (PPi)-dependent phosphofructokinase (step 3b), net production of ATP is 3 mol from 1 mol of hexose. Two mol of NADH produced by glycolysis can be used in oxidative phosphorylation in mitochondria to generate more ATP (Zeeman 2015).

A further substrate-level phosphorylation occurs in the TCA cycle. In the matrix of the mitochondria, a substrate-level phosphorylation occurs at the succinate-CoA ligase reaction step (step 6 in Fig. 3.2). In contrast to animals, in plants ATP-specific succinyl-CoA synthetase (EC 6.2.1.5), but not GTP-succinyl-CoA synthetase (EC 6.2.1.4), acts as an enzyme in the TCA cycle (see Figure 3.4). This enzyme produces ATP from ADP accompanied by degradation of succinyl-CoA (Table 3.1 and Figure 3.4c). For further details of this topic, readers can refer to comprehensive plant biochemistry text books such as Bowsher et al. (2012) and Buchanan et al. (2015).

3.4.3 Nucleoside-Diphosphate Kinase

Using cellular ATP, purine and pyrimidine nucleoside triphosphate are produced by nucleoside-diphosphate kinase (EC 2.7.4.6). The enzyme catalyses the reaction: nucleoside diphosphate + ATP ↔ nucleoside triphosphate + ADP. Substrate specificity is broad. All purine and pyrimidine nucleoside diphosphates can act as acceptors, while not only ATP, but also any ribonucleoside- and deoxyribonucleoside triphosphate, can act as a donor. This enzyme contributes principally to the formation of nucleoside triphosphates, such as GTP, uridine-triphosphate (UTP) and cytidine-triphosphate

Figure 3.3 Glycolysis in plants. (1) Hexokinase; (2) phosphoglucoisomerase; (3a) phosphofructokinase; (3b) pyrophosphate dependent-fructose-6-phosphate 1-phosphotransferase; (4) aldolase; (5) triose phosphate isomerase; (6) glyceraldehyde-3-phosphate dehydrogenase; (7) phosphoglycerate kinase; (8) phosphoglycerate mutase; (9) enolase; (10) pyruvate kinase; (11) amylase etc.; (12) starch phosphorylase; (13) invertase; (14) sucrose synthase; (15) phosphoglucomutase; (16) UDP-glucose pyrophosphorylase; (17) fructokinase. Note: Fructose* produced from sucrose (reaction 13) is converted to fructose-6-P (reaction 17).

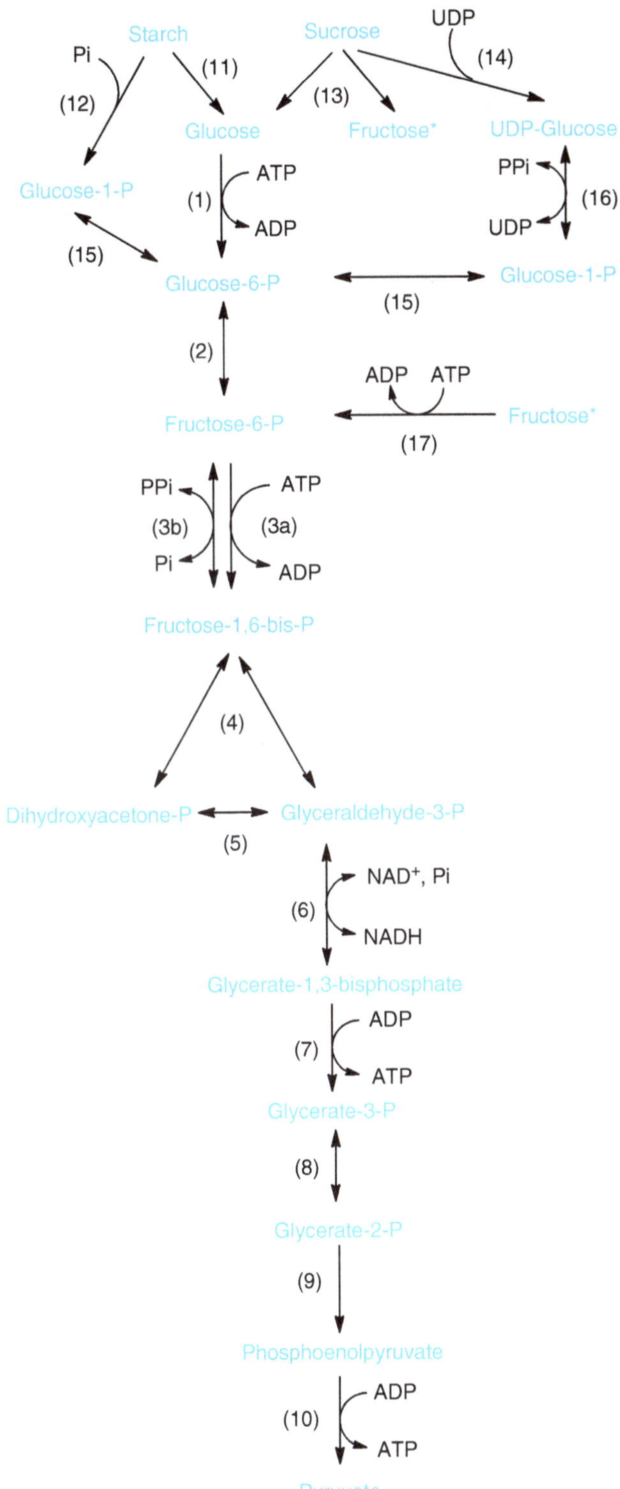
Starch
Sucrose
UDP
Pi
(11)
(14)
(12)
(13)
Glucose
Fructose*
UDP-Glucose
Glucose-1-P
ATP
PPi
(1)
(16)
ADP
UDP
(15)
Glucose-6-P
Glucose-1-P
(15)
(2)
ADP
ATP
Fructose-6-P
Fructose*
(17)
PPi
ATP
(3b)
(3a)
Pi
ADP
Fructose-1,6-bis-P
(4)
Dihydroxyacetone-P
Glyceraldehyde-3-P
(5)
NAD+, Pi
(6)
NADH
Glycerate-1,3-bisphosphate
ADP
(7)
ATP
Glycerate-3-P
(8)
Glycerate-2-P
(9)
Phosphoenolpyruvate
ADP
(10)
ATP
Pyruvate

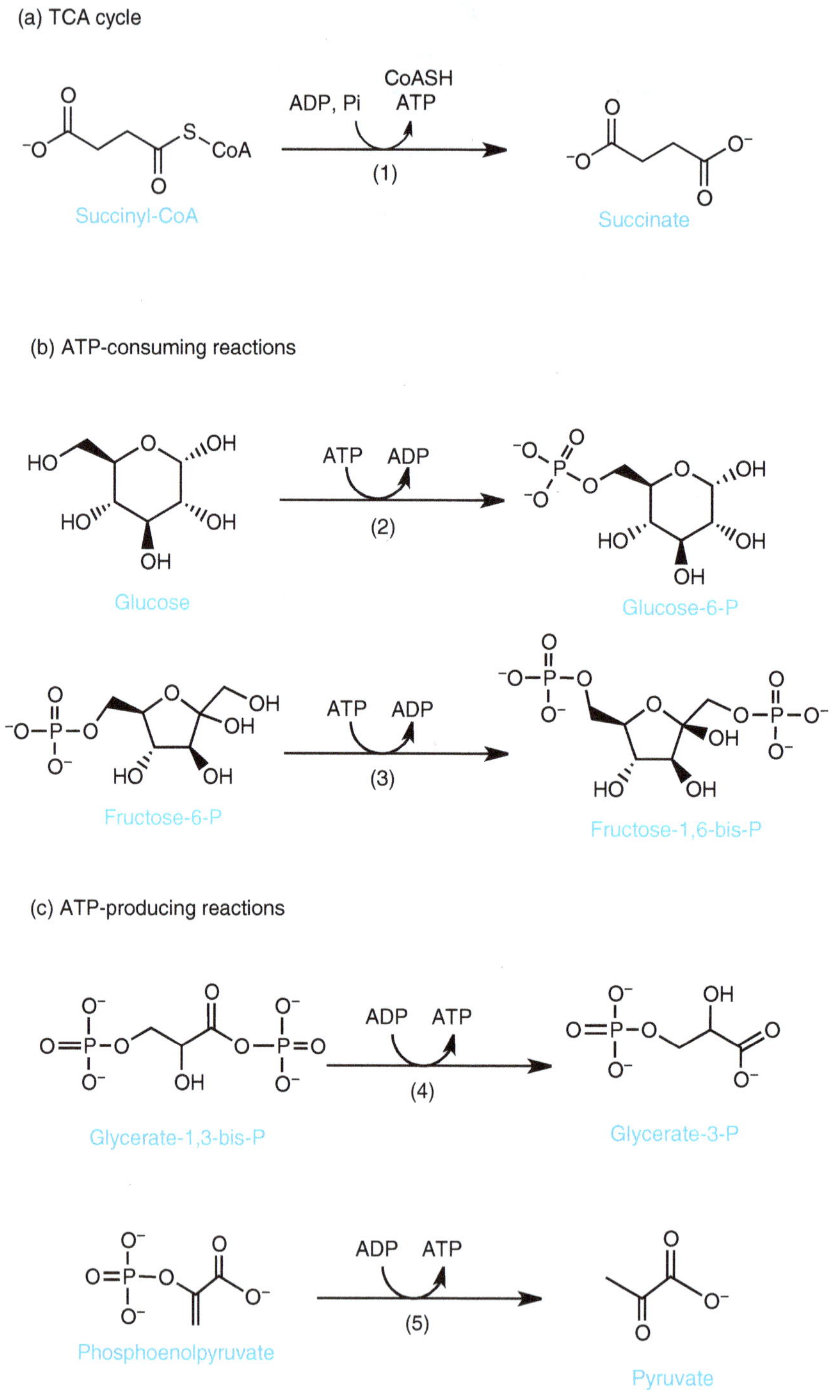

Figure 3.4 Reactions involved in the substrate level ATP production and consumption. (1) succinyl CoA synthetase; (2) hexokinase; (3) phosphofructokinase; (4) phosphoglycerate kinase; (5) pyruvate kinase.

(CTP), from the respective nucleoside-diphosphates. Nucleoside-diphosphate kinase occurs ubiquitously in animals, plants, and bacteria. Recent molecular genetic studies suggest that nucleoside-diphosphate kinases are major players in the synthesis of macromolecules since they provide the triphosphates used for cell anabolism, including nucleic acid synthesis, CTP for lipid synthesis, UTP for polysaccharide synthesis, and GTP for protein elongation, signal transduction, and microtubule polymerization. Five nucleoside-diphosphate kinase genes have been detected in *Arabidopsis thaliana* and rice (*Oryza sativa*). Cytosolic (type I) and chloroplast (type II) enzymes are involved in metabolism, growth, and stress responses and in photosynthetic development and oxidative stress management, respectively. Type III enzymes are located in mitochondria and chloroplasts and are involved principally in energy metabolism. The subcellular localization and precise function of the novel type IV enzyme has not as yet been determined (Dorion and Rivoal 2015).

3.5 Biosynthesis of Deoxyribonucleotides

Reduction of the ribose moiety of the ribonucleotide occurs at the ribonucleoside diphosphate level by ribonucleotide reductase (EC 1.17.4.1). The reaction is:

$$\text{ribonucleoside 5}'\text{-diphosphate (NDP)} + \text{thioredoxin}$$
$$\rightarrow \text{2}'\text{-deoxyribonucleoside 5}'\text{-diphosphate (dNDP)}$$
$$+ \text{ thioredoxin disulphide} + H_2O.$$

dNDP is phosphorylated to dNTP and used for replication and repair in DNA biosynthesis. It has been shown that ribonucleotide reductase, comprising two large (R1) and two small (R2) subunits, catalyses a rate-limiting step in the production of deoxyribonucleotides required for DNA synthesis. The large subunit (R1) contains the allosteric regulatory sites, while the small subunit (R2) encompasses a binuclear iron centre and a tyrosyl free radical (Elledge et al. 1992). In mammals, defective ribonucleotide reductase often leads to cell cycle arrest, growth retardation, and apoptosis, whereas abnormally increased levels result in higher mutation rates. Similar phenomena have been demonstrated in mutants of *A. thaliana*. The results suggest that ribonucleotide reductases are critical for cell cycle progression, DNA damage repair, and general development in plants (Wang and Liu 2006). The pathway for thymidine nucleotides, which are required for DNA synthesis, is described in Chapter 10

3.6 Nucleic Acid Biosynthesis

DNA polymerases (DNA-directed DNA polymerase, EC 2.7.7.7) synthesize DNA from deoxyribonucleotides. These enzymes are essential for DNA replication and usually work in pairs to create two identical DNA strands from a single original DNA molecule. During this process, DNA polymerase 'reads' the existing DNA strand to create two new strands that match the existing one.

These enzymes catalyse the following chemical reaction:

$$DNA_{(n)} + dNTP \leftrightarrow DNA_{(n+1)} + PPi$$

DNA polymerase adds nucleotides to the 3′ end of a DNA strand, one nucleotide at a time. At least 15 classes of DNA polymerase have been identified in animals and terminal deoxyribonucleotidyl transferases. Based on their properties, the polymerases (α to σ) are classified into four families, A, B, X, and Y (Burgers et al. 2001). There are few papers on plant DNA polymerases although there is one which reports that plant DNA polymerase-γ is a DNA repair enzyme which functions in plant meristematic and meiotic tissues, and that it can substitute for Pol β and terminal deoxyribonucleotidyl transferase (Uchiyama et al. 2004).

Biosynthesis of RNA is catalysed by RNA polymerase (DNA-directed RNA polymerase, EC 2.7.7.6). The reaction is:

$$RNA_{(n)} + \text{nucleoside triphosphate} \rightarrow RNA_{(n+1)} + PPi.$$

RNA polymerase, locally, opens the double-stranded DNA (usually about four turns of the double helix) so that one strand of the exposed nucleotides can be used as a template for the synthesis of RNA, namely transcription. A transcription factor and its associated transcription mediator complex must be attached to a DNA binding site, a promoter region, before RNA polymerase can initiate the DNA unwinding at that position. RNA polymerase has intrinsic helicase activity, therefore, no additional enzyme is required to unwind the DNA, in contrast to DNA polymerase. RNA polymerase, not only initiates RNA transcription, it also guides the nucleotides into position, facilitates attachment and elongation, has intrinsic proof reading and replacement capabilities, and a termination recognition function. In eukaryotes, RNA polymerase can build chains as long as 2.4 million nucleotides.

Plants have multiple types of nuclear RNA polymerase, each responsible for synthesis of a distinct subset of RNA. All are structurally and mechanistically related to each other and to bacterial RNA polymerase.

RNA polymerase I synthesizes a pre-rRNA 45S (35S in yeast), which matures into 28S, 18S, and 5.8S rRNAs that form the major RNA sections of the ribosome (Grummt 1998). RNA polymerase II synthesizes precursors of mRNAs and most small nuclear RNA (snRNA) and microRNA (miRNA) (Lee et al. 2004). RNA polymerase III synthesizes tRNAs, rRNA 5S and other small RNAs found in the nucleus and cytosol (Willis 1993). RNA polymerase IV synthesizes small interfering RNA (siRNA) in plants (Herr et al. 2005). RNA polymerase V synthesizes RNAs involved in siRNA-directed heterochromatin formation in plants (Wierzbicki et al. 2009).

Eukaryotic chloroplasts contain a plastid-encoded RNA polymerase that structurally is very similar to bacterial RNA polymerase. Chloroplasts also contain a second, structurally unrelated nucleus-encoded RNA polymerase. Mitochondria contain a structurally- and mechanistically-unrelated RNA polymerase that is a member of the single-subunit RNA polymerase protein family.

3.7 Supply of 5-Phosphoribosyl-1-Pyrophosphate

PRPP is an essential substrate for the biosynthesis of nucleotides. In the case of purine nucleotide synthesis, the first step is phosphoribosylamine (PRA) formation from PRPP (Figure 3.1a). In pyridine and pyridine nucleotide biosynthesis, PRPP is used in intermediate steps, namely the formation of nicotinate mononucleotide (NaMN) (step 3 in

Figure 3.1b) and OMP (step 4 in Figure 3.1b). In addition to its involvement in *de novo* nucleotide biosynthesis, PRPP also acts as a substrate in several pathways including salvage biosynthesis of purine, pyrimidine, and pyridine nucleotides as well as tryptophan and histidine biosynthesis (Becker 2001). The enzyme reactions which utilize PRPP as a substrate are summarized in Table 3.2.

PRPP is synthesized from ribose-5-phosphate and ATP by PRPP synthetase (EC 2.7.6.1, ATP: D-ribose-5-phosphate diphosphotransferase). In plants, ribose-5-phos phate is an intermediate in the photosynthetic reductive pentose phosphate cycle (Calvin–Benson–Bassham cycle), as well as the oxidative pentose phosphate pathway. The reductive cycle is the major CO_2 fixation pathway. CO_2, molecule at a time, is added to the acceptor molecule ribulose-1,5-bisphosphate (RuBP) generating two molecules of 3-phospho-glycerate (Figure 3.5). The oxidative pentose phosphate pathway is an alternative route for oxidizing glucose. As a result, the pentose phosphate pathway is a major source of reducing equivalents for biosynthesis reactions. The pentose phosphate pathway is important for the conversion of hexoses to pentoses such as ribose-5-phosphate (Figure 3.6). Therefore, ribose-5-phosphate is generated both in chloroplasts and the cytosol of plant cells.

Table 3.2 5-Phosphoribosyl-1-pyrophosphate PRPP-utilizing enzymes in plants.

Enzyme	EC	Reaction
Purine nucleotide biosynthesis		
Phosphoribosylamine synthetase	2.4.2.14	Glutamine + PRPP → PRA + PPi
Adenine phosphoribosyltransferase	2.4.2.7	A + PRPP → AMP + PPi
Hypoxanthine-Guanine phosphoribosyltransferase	2.4.2.8	HX + PRPP → IMP + PPi
		G + PRPP → GMP + PPi
Xanthine phosphoribosyltransferase	2.2.2.22	X + PRPP → XMP + PPi
Pyrimidine nucleotide biosynthesis		
Orotate phosphoribosyltransferase	2.4.2.10	O + PRPP → OMP + PPi
Uracil phosphoribosyltransferase	2.4.2.9	U + PRPP → UMP + PPi
Pyridine nucleotide biosynthesis		
Quinolinate phosphoribosyltransferase	2.4.2.19	Q + PRPP → NaMN + PPi + CO_2
Nicotinate phosphoribosyltransferase	2.4.2.11	Na + PRPP → NaMN + PPi
Nicotinamide phosphoribosyltransferase[a)]	2.4.2.12	N + PRPP → NMN + PPi
Tryptophan biosynthesis		
Anthranilate phosphoribosyltransferase	2.4.2.18	Anthranilate + PRPP → Phosphoribosylanthranilate + PPi
Histidine biosynthesis		
ATP phosphoribosyltransferase	2.4.2.17	ATP + PRPP → Phosphoribosyl-ATP + PPi

a) This activity is usually not detected in plants.

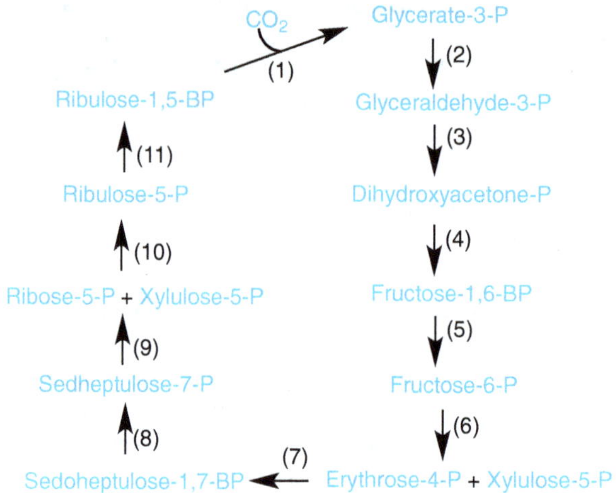

Figure 3.5 Photosynthetic Calvin–Benson–Bassham cycle. (1) Ribulose bisphosphate carboxylase/oxygenase; (2) phosphoglycerate kinase; (3) glyceraldehyde-3-phosphate dehydrogenase; (4) aldolase; (5) fructose-bisphosphatase; (6) transketolase; (7) aldolase; (8) sedoheptulose diphosphatase; (9) transketolase; (10) erythrose-4-phosphate isomerase; (11) phosphoribulokinase.

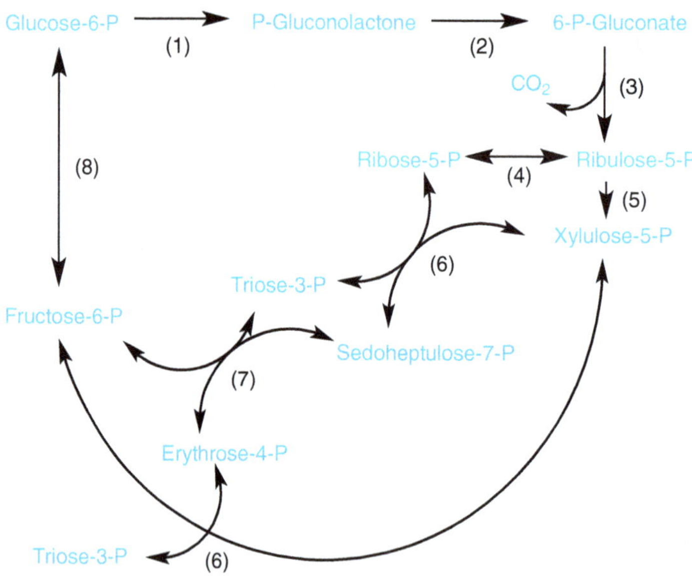

Figure 3.6 Oxidative pentose phosphate pathway. (1) Glucose-6-phosphate dehydrogenase; (2) 6-phosphogluconolactonase; (3) 6-phosphogluconate dehydrogenase (decarboxylating); (4) ribose-5-phosphate isomerase; (5) ribulose-5-phosphate 3-epimerase; (6) transketolase; (7) transaldolase; (8) glucose-6-phosphate isomerase.

Plant PRPP synthetase has been characterized using native enzymes and it has been shown that the Pi-independent properties are different from bacterial and mammalian PRPP synthetase for which Pi is essential for activity. Recent recombinant enzyme studies revealed that both Pi-dependent (class I) and Pi-independent PRPP synthetase (class II) occur in plants (see Ashihara 2016).

3.8 Supply of Amino Acids for Nucleotide Biosynthesis

For the nucleotide biosynthesis, certain amino acids, namely, glutamine, glycine, and aspartate are utilized for nucleotide biosynthesis. Interconversion of nucleoside monophosphates, nucleoside diphosphates, and triphosphates are summarized in Figure 3.7.

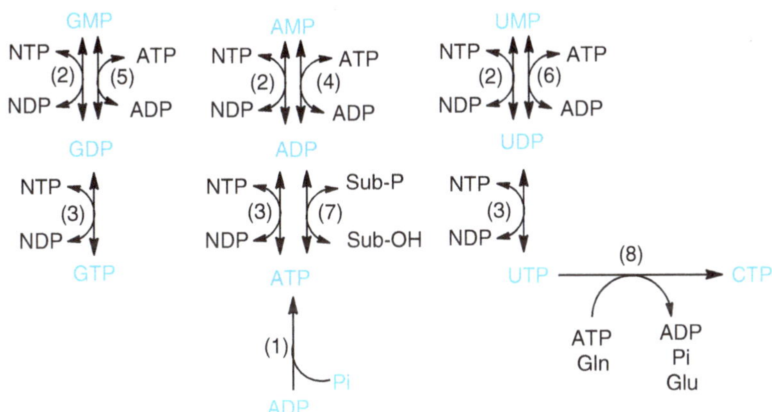

Figure 3.7 Conversion of nucleoside mono-, di- and triphosphates. (1) ATP synthase; (2) nucleoside-monophosphate kinase; (3) nucleoside-diphosphate kinase; (4) adenylate kinase; (5) guanylate kinase; (6) UMP/CMP kinase; (7) various kinases; [8] CTP synthetase (see Part III).

3.9 Nitrogen Metabolism and Amino Acid Biosynthesis in Plants

As described in Section 3.1, amino acids are the precursors of ribonucleotide monophosphates involved in the *de novo* biosynthesis of purine, pyrimidine, and pyridine nucleotides. Plants absorb nitrogen from the environment in the form of nitrate (NO_3^-) and ammonium (NH_4^+). Nitrate assimilation is performed by two enzymes: nitrate reductase (EC 1.7.1.1) and nitrite reductase (EC 1.7.7.1). In many plants, nitrate reductase occurs in the cytosol and catalyses the reaction: $\text{Nitrate} + \text{NADH} + H^+ \rightarrow \text{Nitrite} + \text{NAD}^+ + H_2O$ (step 1 in Figure 3.8). In contrast, nitrite reductase occurs in the chloroplast and other plastids. This reduction requires six electrons donated by reduced ferredoxin. The reaction catalysed is: nitrite + 6 reduced ferredoxin + $7H^+ \rightarrow NH_3 + 2H_2O + 6$ oxidized ferredoxin (step 2 in Figure 3.8).

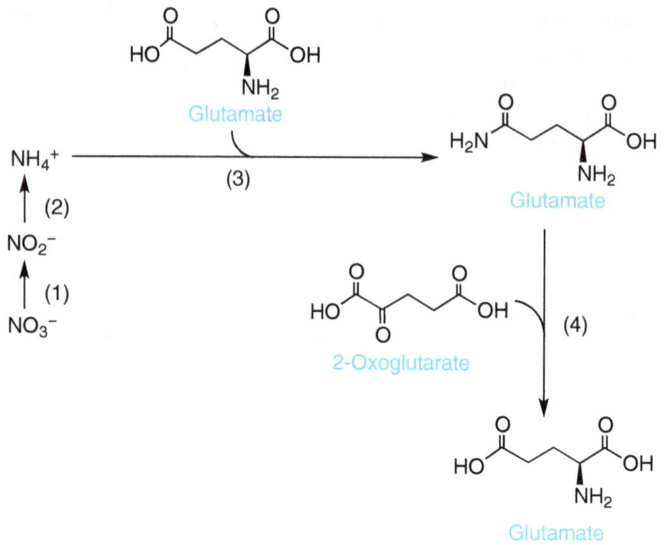

Figure 3.8 Nitrate reduction and assimilation of ammonia in plants. Enzymes shown are: (1) nitrate reductase (NR); (2) nitrite reductase (NiR); (3) glutamine synthetase (GS); (4) glutamate synthase (GOGAT).

Ammonium is assimilated by glutamine synthetase (GS, EC 6.3.1.2) and glutamate synthase (L-glutamine: 2-oxoglutarate aminotransferase, GOGAT, EC 1.4.1.13) and glutamate is formed (Figure 3.8). These two enzymes catalyse the following reactions:

$$\text{ATP} + \text{L-glutamate} + \text{NH}_3 \rightarrow \text{ADP} + \text{phosphate} + \text{L-glutamine (step 3 in Figure 3.8)}$$

and

$$\text{L-glutamine} + \text{2-oxoglutarate} + \text{NADPH} + \text{H}^+ \rightarrow 2\ \text{L-glutamate} + \text{NADP}^+ \text{ (step 4 in Figure 3.8).}$$

Plants have a high potential for nitrate assimilation in leaves and/or roots. In plants, unlike animals, all protein constituent amino acids are synthesized from the intermediates of the glycolysis, pentose phosphate pathway, and the TCA cycle (Figure 3.9).

Aspartate is used as a substrate in the biosynthesis of IMP, OMP, and NaMN nucleotides of the initial nucleotide products of purine, pyrimidine, and pyridine biosynthesis (see Figure 3.1). In addition, glutamine and glycine are used for purine ring synthesis. The genes and enzymes of nitrogen assimilation and amino acid biosynthesis have been well studied in plants (Coruzzi et al. 2015).

3.10 Summary

De novo biosynthesis of ribonucleoside monophosphate, interconversion of nucleoside phosphates, conversion of nucleoside diphosphates to triphosphates, biosynthesis of

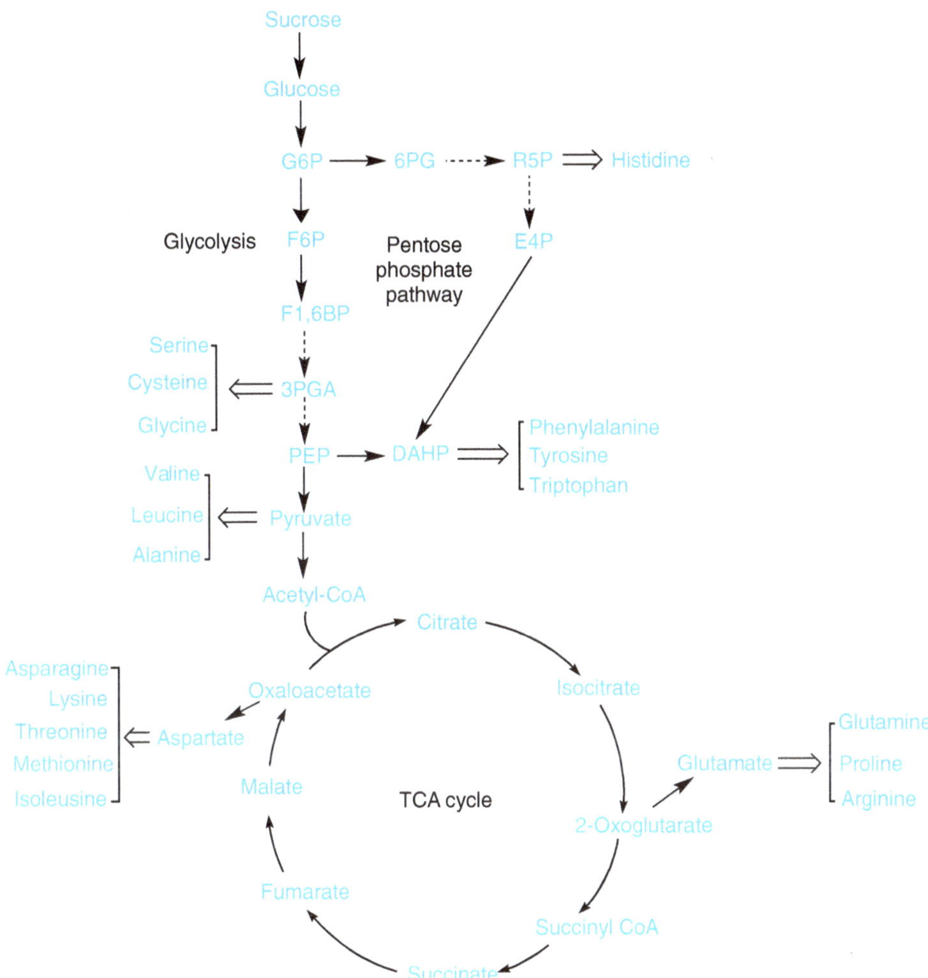

Figure 3.9 Outline of amino acids biosynthesis in plants. DAHP, 3-deoxy-D-arabino-heptulosonate-7-phosphate; E4P, erythrose-4-phosphate; F6P, fructose-6-phosphate; F1,6BP, fructose-1,6-bisphosphate; 6PG, 6-phosphogluconate; 3PGA, 3-phosphoglycerate; PEP, phosphoenolpyruvate.

deoxyribonucleotides, nucleic acid (DNA and RNA) biosynthesis, and supply of PRPP and amino acids for nucleotide biosynthesis are reviewed.

References

Ashihara, H. (2016). Biosynthesis of 5-phosphoribosyl-1-pyrophosphate in plants: a review. *Eur. Chem. Bull.* 5: 314–323.

Baysdorfer, C. and Bassham, J.A. (1984). Spinach pyruvate kinase isoforms: partial purification and regulatory properties. *Plant Physiol.* 74: 374–379.

Becker, M.A. (2001). Phosphoribosylpyrophosphate synthetase and the regulation of phosphoribosylpyrophosphate production in human cells. In: *Progress in Nucleic*

Acid Research and Molecular Biology., vol. 69 (ed. K. Moldave), 115–148. Academic Press.

Bowsher, C., Steer, M., and Tobin, A. (2012). *Plant Biochemistry.* New York: Garland Science.

Buchanan, B.B., Gruissen, W., and Jones, R.L. (2015). *Biochemistry and Molecular Biology of Plants.* Rockville: American Society of Plant Biologists.

Burgers, P.M.J., Koonin, E.V., Bruford, E. et al. (2001). Eukaryotic DNA polymerases: proposal for a revised nomenclature. *J. Biol. Chem.* 276: 43487–43490.

Coruzzi, G., Last, R., Dudareva, N., and Amrhein, N. (2015). Amino acids. In: *Biochemistry and Molecular Biology of Plants*, 2e (eds. B.B. Buchanan, W. Gruissem and R.L. Jones), 289–336. Wiley.

Dorion, S. and Rivoal, J. (2015). Clues to the functions of plant NDPK isoforms. *Naunyn Schmiedeberg's Arch. Pharmacol.* 388: 119–132.

Elledge, S.J., Zhou, Z., and Allen, J.B. (1992). Ribonucleotide reductase: regulation, regulation, regulation. *Trends Biochem. Sci.* 17: 119–123.

Fukami-Kobayashi, K., Nosaka, M., Nakazawa, A., and Gō, M. (1996). Ancient divergence of long and short isoforms of adenylate kinase molecular evolution of the nucleoside monophosphate kinase family. *FEBS Lett.* 385: 214–220.

Grummt, I. (1998). Regulation of mammalian ribosomal gene transcription by RNA polymerase I. In: *Progress in Nucleic Acid Research and Molecular Biology*, vol. 62 (ed. K. Moldave), 109–154. Academic Press.

Herr, A.J., Jensen, M.B., Dalmay, T., and Baulcombe, D.C. (2005). RNA polymerase IV directs silencing of endogenous DNA. *Science* 308: 118–120.

Johansson, M., Hammargren, J., Uppsäll, E. et al. (2008). The activities of nucleoside diphosphate kinase and adenylate kinase are influenced by their interaction. *Plant Sci.* 174: 192–199.

Johnson, J.D., Mehus, J.G., Tews, K. et al. (1998). Genetic evidence for the expression of ATP- and GTP-specific succinyl-CoA synthetases in multicellular eucaryotes. *J. Biol. Chem.* 273: 27580–27586.

Karp, G. (2013). *Cell and Molecular Biology: Concepts and Experiments*, 7e. Wiley.

Lee, Y., Kim, M., Han, J. et al. (2004). MicroRNA genes are transcribed by RNA polymerase II. *EMBO J.* 23: 4051–4060.

Li, L., Carrie, C., Nelson, C. et al. (2012). Accumulation of newly synthesized F(1) *in vivo* in Arabidopsis mitochondria provides evidence for modular assembly of the plant F(1)F(o) ATP synthase. *J. Biol. Chem.* 287: 25749–25757.

Niyogi, K.K., Wolosiuk, R.A., and Malkin, R. (2015). Photosynthesis. In: *Biochemistry & Molecular Biology of Plants*, 2e (eds. B.B. Buchanan, W. Gruissem and R.L. Jones), 508–566. Oxford: Wiley.

Noda, L. (1962). Nucleoside triphosphate-nucleoside monophosphokinases. In: *The Enzymes*, 2e, vol. 6 (eds. P.D. Boyer, H. Lardy and K. Myrbäck), 139–149. New York: Academic Press.

Nomura, Y., Izumi, A., Fukunaga, Y. et al. (2014). Diversity in guanosine 3′,5′-bisdiphosphate (ppGpp) sensitivity among guanylate kinases of bacteria and plants. *J. Biol. Chem.* 289: 15631–15641.

Troncoso-Ponce, M.A., Rivoal, J., Venegas-Calerón, M. et al. (2012). Molecular cloning and biochemical characterization of three phosphoglycerate kinase isoforms from developing sunflower (*Helianthus annuus* L.) seeds. *Phytochemistry* 79: 27–38.

Uchiyama, Y., Kimura, S., Yamamoto, T. et al. (2004). Plant DNA polymerase λ, a DNA repair enzyme that functions in plant meristematic and meiotic tissues. *Eur. J. Biochem.* 271: 2799–2807.

Wang, C. and Liu, Z. (2006). Arabidopsis ribonucleotide reductases are critical for cell cycle progression, DNA damage repair, and plant development. *Plant Cell* 18: 350–365.

Wierzbicki, A.T., Ream, T., Haag, J.R., and Pikaard, C.S. (2009). RNA polymerase V transcription guides ARGONAUTE4 to chromatin. *Nat. Genet.* 41: 630–634.

Willis, I.M. (1993). RNA polymerase III. *Eur. J. Biochem.* 212: 1–11.

Zeeman, S.C. (2015). Carbohydrate metabolism. In: *Biochemistry & Molecular Biology of Plants* (eds. B.B. Buchanan, W. Gruissem and R.L. Jones), 950–998. Chichester, UK: Wiley.

Zhou, L., Lacroute, F., and Thornburg, R. (1998). Cloning, expression in *Escherichia coli*, and characterization of *Arabidopsis thaliana* UMP/CMP kinase. *Plant Physiol.* 117: 245–254.

Part II

Purine Nucleotide Metabolism

4

Purine Nucleotide Biosynthesis *De Novo*

4.1 Introduction

The *de novo* biosynthesis of purine nucleotides was elucidated independently by Buchanan and Greenberg in the 1950s (see, Buchanan 1986). The purine ring is assembled from several small molecules, namely, glycine, glutamine, aspartate, 5-phosphoribosyl-1-pyrophosphate (PRPP), 10-formyl tetrahydrofolate, and carbon dioxide. In the early studies pigeon liver was used, because birds synthesize purines from excess amino nitrogen, the purines are then degraded to urate and excreted. Incorporation of individual precursors into different atoms of urate was used to reveal the purine biosynthesis pathway. Similar precursor feeding studies were performed with leaves of coffee and tea which accumulate the caffeine (Anderson and Gibbs 1962; Proiser and Serenkov 1963). In plant purine studies, N_2-fixing root nodules of tropical legumes have been used as experimental material to investigate *de novo* purine biosynthesis. Substantial amounts of ureides are produced as the major nitrogen-containing compounds derived from purine nucleotides (Schubert and Boland 1990).

Arabidopsis thaliana is a popular model organism in plant biology and genetics. It was the first plant genome to be sequenced, completed by the Arabidopsis-Genome-Initiative (2000). The most up-to-date version of the *A. thaliana* genome is maintained by the Arabidopsis Information Resource (TAIR). Although feeding experiments and detailed analysis of native enzymes of *A. thaliana* are difficult from a practical perspective, nucleotide metabolism research has developed aided by molecular genetic studies with the available sequence information from Arabidopsis and rice genomes, confirming the hypothesis that purine biosynthesis *de novo* in plants is essentially the same as that operating in microorganisms and animals (van der Graaff et al. 2004; Zrenner and Ashihara 2011). However, based on the genome sequence in *A. thaliana*, the genomic and pathway organization of purine biosynthesis in plants appears to be different from that occurring in other organisms. Plant organization seems to be somewhat similar to prokaryotes, but greater differences are found with higher eukaryotes such as *Drosophila*, chickens, and humans.

Prokaryotes, such as *Escherichia coli*, have single genes coding monofunctional proteins except for the *purH* gene which encodes the bifunctional protein

Plant Nucleotide Metabolism – Biosynthesis, Degradation, and Alkaloid Formation, First Edition.
Hiroshi Ashihara, Iziar A. Ludwig and Alan Crozier.

Table 4.1 Summary of enzymes involved in the purine biosynthesis *de novo* in plants.

Step	Enzyme name	EC	Abbreviation	Gene[a)]		TAIR locus
				Plant	*E. coli*	
1	5-Phosphoribosylpyrophosphate amidotransferase	2.4.2.14	PRAT1 PRAT2 PRAT3	*pur1*	*purF*	AT2G16570 AT4G34740 AT4G38880
2	Glycineamide ribonucleotide synthase	6.3.4.13	GARS	*pur2*	*purD*	AT1G09830
3	Glycineamide ribonucleotide formyl transferase	2.1.2.2	GARFT	*pur3*	*purN*	AT1G31220
4	Formylglycinamide ribonucleotide amidotransferase	6.3.5.3	FGAMS	*pur4*	*purL*	AT1G74260
5	5-Aminoimidazole ribonucleotide synthetase	6.3.3.1	AIRS	*pur5*	*purM*	AT3G55010
6	5-Aminoimidazole ribonucleotide carboxylase	4.1.1.21	AIRC	*pur6*	*purEK*	AT2G05140 AT2G37690
7	4-(*N*-Succinocarboxamide)-5-aminoimidazole synthetase	6.3.2.6	SAICARS	*pur7*	*purC*	AT3G21110
8	Adenylosuccinate lyase	4.3.2.2	ASL	*pur8/12*	*purB*	AT1G36280 AT4G18840
9	5-Aminoimidazole-4-carboxyamide ribonucleotide formyl transferase	2.1.2.3	AICARFT	*pur9/10*	*purH*	AT2G35040
10	IMP cyclohydrolase	3.5.4.10	IMPCH	*pur9/10*	*purH*	AT2G35040
11	Adenylosuccinate synthetase	6.3.4.4	ASS	*pur11*	*purA*	AT3G57610
12	Adenylosuccinate lyase	4.3.2.2	ASL	*pur8/12*	*purB*	AT1G36280 AT4G18840
13	IMP dehydrogenase	1.1.1.205	IMPDH	*gua13*	*guaB*	AT1G79470 AT1G16350
14	GMP synthetase	6.3.5.2	GMPS	*gua14*	*guaA*	AT1G63660

a) Gene designations used for plants and *E. coli* are shown. The plant gene names are according to Senecoff et al. (1996). The gene names used for *E. coli* are also often used in plant literature.

5-aminoimidazole-4-carboxyamide ribonucleotide formyl transferase/inosine-5′-monophosphate (IMP) cyclohydrolase (ATIC, steps 9 and 10, in Table 4.1). Eukaryotes contain single genes encoding mono-, bi- and trifunctional proteins (Smith and Atkins 2002; van der Graaff et al. 2004). The *de novo* purine biosynthesis pathway in plants consist of monofunctional proteins, except for the bifunctional enzymes AIRC (*pur6*) and ATIC (*pur9/10*) (see Table 4.1).

The *A. thaliana* genes involved in purine biosynthesis are organized either as single genes or small gene families. PRPP amidotransferase which catalyses the first step in this pathway is encoded by at least three genes of the same family. In addition a small gene family, *pur6*, *pur8/12,* and *gua13,* are each represented by two genes in the Arabidopsis genome (see Table 4.1).

4.2 Reactions and Enzymes

A summary of enzymes involved in purine nucleotide biosynthesis is presented in Table 4.1. Reaction steps, enzyme name, Enzyme Commission number (EC number), abbreviation of enzymes, gene names used in plants and *E. coli,* and TAIR locus are shown. The format of chromosomal-based nomenclature at TAIR is as follows: AT (*A. thaliana*) 1, 2, 3, 4, or 5 (chromosome number) or M for mitochondrial or C for chloroplast. G (gene), and five-digit code (numbered from top/north to bottom/south of chromosome).

The nomenclature of gene names in this field is not systematic, and considerable confusion exists in the genetic and biochemical literature. Although genes have often been called by the enzyme name, the genes of the *de novo* purine nucleotide synthesis are termed the *pur* genes. The bacterial genes are denoted by letters, e.g. *purF* for the step 1, but for ease of understanding, a numerical nomenclature proposed was by Smith and Atkins (2002) for plants, e.g. *pur1,* encodes the enzyme catalysing the first step of the pathway (Table 4.1). To distinguish genes from different plants the initials of species names are used as a prefix to the gene name, e.g. the *pur1* gene from *A. thaliana* is designated *Atpur1*.

The *de novo* purine biosynthesis pathway occurs in cytosol of bacteria and animals. However, in plants there is speculation about the localization of the enzymes in plastids and mitochondria (Zrenner et al. 2006), although this possibility is not mentioned in the latest comprehensive plant biochemistry textbook (Buchanan et al. 2015). The activity of the enzymes involved in *de novo* purine biosynthesis is extremely low and, as a consequence, measurement of activity in plants is difficult, as is the determination of enzyme activity in fractionated organelles. Plastids and mitochondria isolated from nodules of cowpea using Percoll gradients are capable of IMP synthesis from ribose-5-phosphate or PRPP (Atkins et al. 1997). Thus, the complete capacity for *de novo* purine biosynthesis has been found in these organelles when isolated by effective fractionation procedures.

Protein subcellular localization predictions have been attempted. In general, most proteins in chloroplasts, plastids, and mitochondria are encoded by the nuclear genome and synthesized in the cytosol as high molecular weight precursors with *N*-terminal extensions called transit peptides. It has been reported that import of these proteins into chloroplasts requires the peptides and a translocon complex to be in the chloroplast envelope (Schleiff and Soll 2000). Most of the *de novo* purine biosynthesis enzymes appear to be located in chloroplasts, plastids, and mitochondria. The deficiency of 5-phosphoribosyl-1-pyrophosphate amidotransferase (PRAT) has a detrimental effect on chloroplast biogenesis (Hung et al. 2004). Berthome et al. (2008) reported that transient expression of a translational fusion with the green fluorescent protein in plantlets of *A. thaliana* showed that the protein of formylglycinamidine ribonucleotide (FGAM) synthase, the fourth enzyme of *de novo* purine biosynthesis, is dually targeted to chloroplasts and mitochondria. This indicates that at least some steps of the *de novo* purine biosynthesis pathway can take place in both organelles in *A. thaliana,* as well as in the ureide-forming tropical legumes. However, there are many difficulties in predicting the actual localization of the enzymes. For example, a number of proteins that have been identified are directed to plastids and mitochondria by the same targeting sequence (Chow et al. 1997; Menand et al. 1998).

Information in the literature on purine metabolism, especially *de novo* biosynthesis in plants, is fragmentary. However, it is possible to follow progress using databases linked to the TAIR gene locus numbers of *A. thaliana* as well as the EC numbers (the International Union of Biochemistry and Molecular Biology) shown in Table 4.1. Useful databases for nucleotide metabolism are KEGG (Kyoto Encyclopedia of Genes and Genomes), BRENDA (The Comprehensive Enzyme Information System), and MetaCyc (BioCyc Database Collection).

4.2.1 Synthesis of Phosphoribosylamine

Purine biosynthesis *de novo* starts with the formation of phosphoribosylamine (PRA) from PRPP and glutamine (step 1, Figure 4.1, Reaction 1).

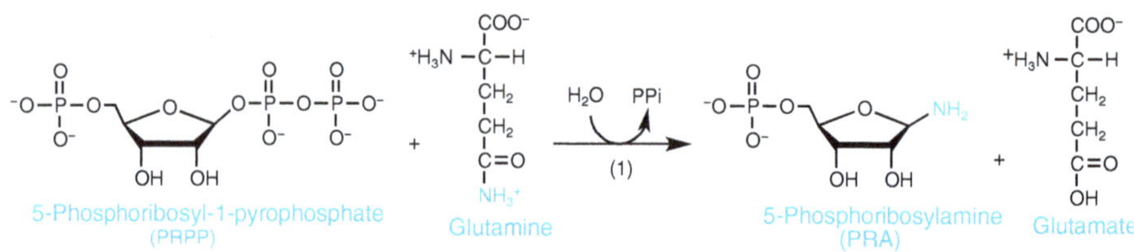

Reaction 1

This reaction is catalysed by 5-phosphoribosyl-1-pyrophosphate amidotransferase (PRAT, EC 2.4.2.14). PRAT has been purified from a number of sources, including *E. coli*, and *Bacillus subtilus*, avian liver, and human placenta. In plants, a highly purified native enzyme has been obtained from root nodules of soybean *(Glycine max)* (Reynolds et al. 1984). In tropical legumes, ammonia, a direct product of symbiotic nitrogen fixation in roots, is converted to ureides (allantoin and allantoic acid) which are translocated to the shoot where they serve as a nitrogen source. Ureides are generated through the oxidation of purines formed by the *de novo* purine biosynthetic pathway. PRAT catalyses the first committed step of ureide biosynthesis in soybean. Nodule PRAT has a pH optimum of 8.0, and Km values of 18 mM for glutamine and 0.4 mM for PRPP. Ammonia can replace glutamine as a substrate (Km ~14 mM) and the V_{max} with ammonia is double that obtained when glutamine is the substrate. Regulatory properties of the soybean root nodule enzyme are similar to those reported for the human placental enzyme (Holmes et al. 1973). Both are subject to regulation by end products of the purine biosynthetic pathway, IMP, xanthosine-5′-monophosphate (XMP), guanosine-5′-monophosphate (GMP), and adenosine-5′-monophosphate (AMP). Therefore, the first step of purine biosynthesis *de novo* is feedback controlled by the end products in both plants and animals.

In plants, there are three isoforms of PRAT (AtPRAT1–3) which are coded in a family of three isoforms of *pur1*. In *A. thaliana*, AtPRAT2 is the major isoform (Hung et al. 2004). One of the *pur1* genes which encodes AtPRAT1 protein is expressed in roots and flowers. The AtPRAT2-encoding gene is expressed in leaves as well as roots and flowers (Ito et al. 1994), while the gene that encodes AtPRAT3 is expressed at very low levels in silique, cauline leaves, and roots (Hung et al. 2004).

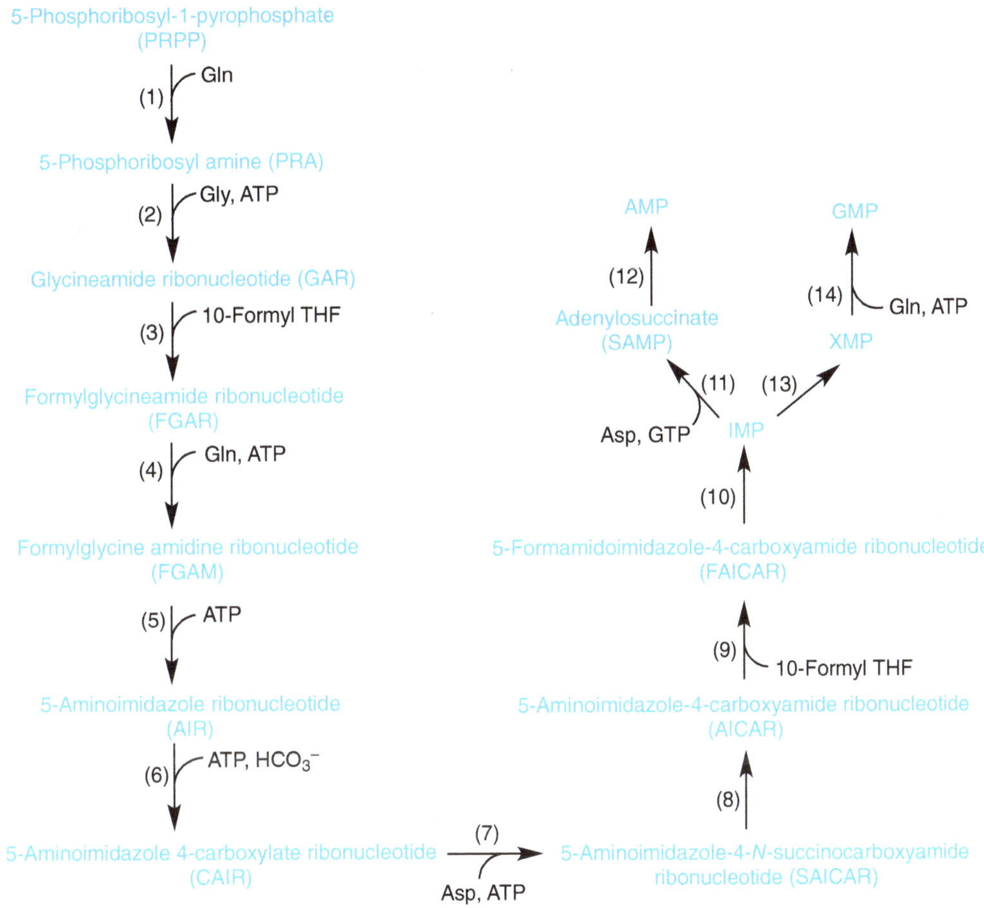

Figure 4.1 Purine ribonucleotide biosynthesis. Abbreviations of metabolites are as follows: PRPP, 5-phosphoribosyl-1-pyrophosphate; PRA, 5-phosphoribosyl amine; GAR, glycineamide ribonucleotide; FGAR, formylglycineamide ribonucleotide; FGAM, formylglycinamidine ribonucleotide; AIR, 5-aminoimidazole ribonucleotide; CAIR, 5-aminoimidazole 4-carboxylate ribonucleotide; SAICAR, 5-aminoimidazole-4-*N*-succinocarboxyamide ribonucleotide; AICAR, 5-aminoimidazole-4-carboxyamide ribonucleotide; FAICAR, 5-formamidoimidazole-4-carboxyamide ribonucleotide; SAMP, adenylosuccinate; XMP, xanthosine-5′-monophosphate; IMP, inosine-5′-monophosphate; GMP, guanosine-5′-monophosphate; AMP, adenosine-5′-monophosphate; Asp, aspartate; Gln, glutamine; Gly, glycine; 10-THF, 10-formyltetrahydrofolate. The participated enzymes are (1) PRPP amidotransferase (2.4.2.14); (2) GAR synthetase (6.3.4.13); (3) GAR formyl transferase (2.1.2.2); (4) FGAM synthetase (6.3.5.3); (5) AIR synthetase (6.3.3.1); (6) AIR carboxylase (4.1.1.21); (7) SAICAR synthetase (6.3.2.6); (8) adenylosuccinate lyase (4.3.2.2); (9) AICAR formyl transferase (2.1.2.3); (10) IMP cyclohydrolase (3.5.4.10); (11) SAMP synthetase (6.3.4.4); (12) adenylosuccinate lyase (4.3.2.2), the same enzyme for step 8; (13) IMP dehydrogenase (1.1.1.205); (14) GMP synthetase (6.3.5.2).

The nitrogen-fixing root nodules of *Phaseolus vulgaris* (common bean) also have three *pur1* genes coding for PRAT (Coleto et al. 2016). One of the three *pur1* genes which encodes PvPRAT3, is highly expressed in nodules and the enzymatic activity is correlated with nitrogen fixation activity. Some mutant studies suggest that PvPRAT3 is essential for the synthesis of ureides in nodules.

The formation of PRA from PRPP and glutamine occurs as two consecutive reactions that take place at separate active sites on PRAT.

(i) glutamine + $H_2O \rightarrow$ glutamate + NH_3
(ii) NH_3 + PRPP $\rightarrow$ PRA + PPi

Ammonia is released at the glutaminase domain and channelled to the PRPP binding site in the phosphoribosyltransferase domain (Zalkin and Dixon 1992). PRAT is structurally classified into two groups, namely, [4Fe-4S] cluster-dependent (type I) and independent-enzyme groups (type II). Plant PRAT displays the typical conserved protein structures of type I, containing a propeptide preceding the first cysteine and four conserved cysteine residues that are ligands to a [4Fe-4S] cluster and crucial for catalytic activity and feedback regulation (Ito et al. 1994).

4.2.2 Synthesis of Glycineamide Ribonucleotide

The second step of the *de novo* pathway is the ATP-dependent formation of glycineamide ribonucleotide (GAR) by the addition of glycine to PRA via an amide bond. This reaction is catalysed by glycineamide ribonucleotide synthetase (GARS, EC 6.3.4.13) (step 2, Figure 4.1, Reaction 2).

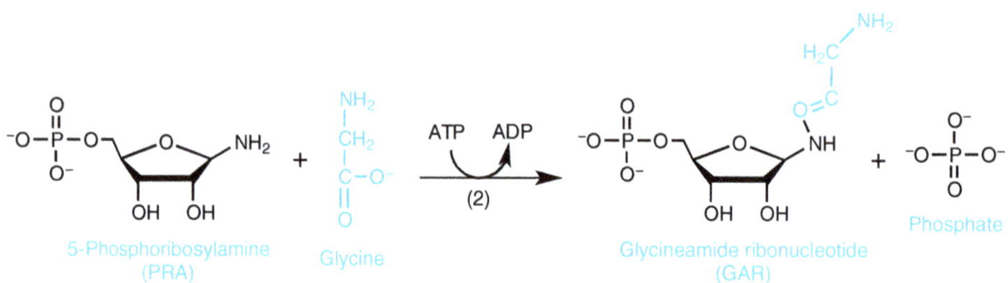

Reaction 2

The enzymatic properties of GARS are not well characterized. However, fluctuation in its activity during growth of cultured carrot cells has been investigated and the highest level was found in the cell division phase (Ashihara and Nygaard 1989).

4.2.3 Synthesis of Formylglycineamide Ribonucleotide

GAR is metabolised by the enzyme glycineamide ribonucleotide transformylase (GARFT, EC 2.1.2.2) using 10-formyltetrahydrofolate (10-Formyl-THF) to generate formylglycineamide ribonucleotide (FGAR) (step 3, Figure 4.1, Reaction 3).

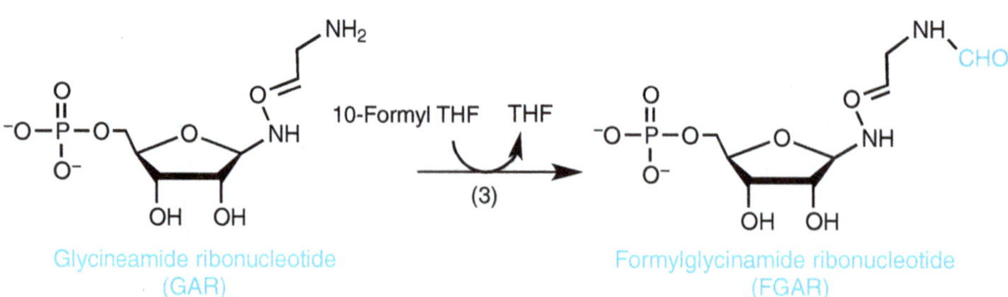

Reaction 3

Soybean nodule cDNA clones *pur2* and *pur3* encode enzymes that catalyse the second and the third steps of the *de novo* purine biosynthesis pathway, namely, GARS and GARFT, which were isolated by complementation of corresponding *E. coli* mutants. One class of the cDNA clone of GARS and the clones of three classes of GARFT cDNA were identified. These *pur2* and *pur3* genes are highly expressed in young and mature nodules but weakly expressed in roots and leaves. Expression levels of the mRNAs of these two genes were not enhanced even when ammonia was provided to non-nodulated roots. (Schnorr et al. 1996).

4.2.4 Synthesis of Formylglycinamidine Ribonucleotide

FGAM synthesis from FGAR and glutamine is catalysed by formylglycinamide ribonucleotide amidotransferase (FGAMS, EC 6.3.5.3), in a reaction that requires ATP (step 4 in Figure 4.1, Reaction 4).

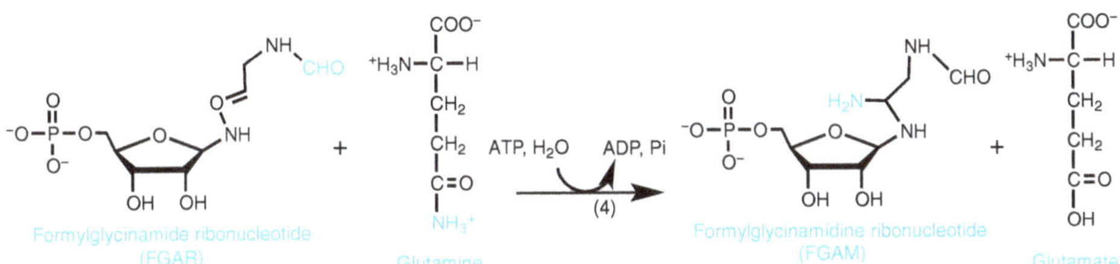

Reaction 4

The *pur4* gene encodes the *N*-terminal portion of the FGAMS protein which contains targeting sequences of the FGAMS protein of *A. thaliana* suggesting that this enzyme is located in mitochondria as well as chloroplasts (Berthome et al. 2008). It has also been postulated that the FGAMS-catalysed step is crucial for plant reproduction, in particular male gametophyte development, but it probably also occurs in the sporophytic tissues sustaining pollen and embryo sac developments (Berthome et al. 2008).

4.2.5 Synthesis of Aminoimidazole Ribonucleotide

FGAM undergoes ring closure to form 5-aminoimidazole ribonucleotide (AIR) in a reaction that requires ATP (step 5, Figure 4.1, Reaction 5). This step is catalysed by aminoimidazole ribonucleotide synthase (AIRS, EC 6.3.3.1).

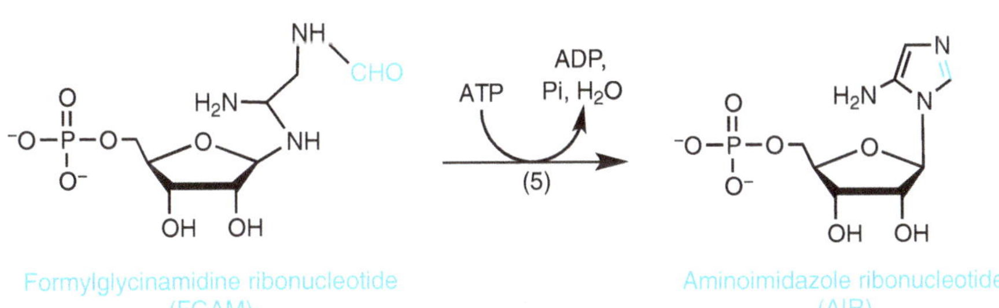

Reaction 5

The cDNA of *pur5* encoding AIRS was identified and isolated from *A. thaliana*. The functional confirmation of the enzymatic activity was achieved by functional suppression of *E. coli* auxotrophs using expressed *A. thaliana* leaf cDNAs (Schnorr et al. 1994; Senecoff and Meagher 1993).

4.2.6 Synthesis of Aminoimidazole Carboxylate Ribonucleotide

In contrast to animals, plants carboxylate AIR to 4-carboxy aminoimidazole ribonucleotide (CAIR) in a two-step reaction catalysed by aminoimidazole ribonucleotide carboxylase (AIRC, EC 4.1.1.21) (step 6, Figure 4.1, Reaction 6).

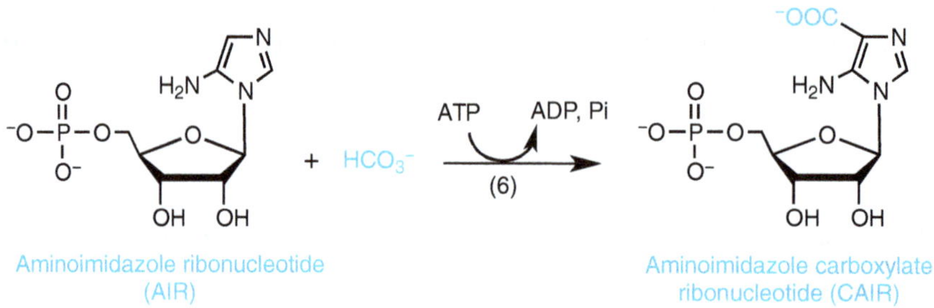

Reaction 6

This two-step reaction in plants is similar to that occurring in *E. coli,* where two separate enzymes, 5-(carboxyamino)imidazole ribonucleotide (N^5-CAIR) synthase (PurK, EC 6.3.4.18), and N^5-CAIR mutase (PurE, EC 5.4.99.18) are required to carry out the single reaction catalysed by AIRC (EC 4.1.1.21) in eukaryotes. In plants and yeast, PurK, and PurE proteins are fused and form an enzyme complex (Voet and Voet 2010). As in yeast, the moth bean (*Vigna aconitifolia*) AIRC has an *N*-terminal domain homologous to the bacterial *purK* gene product. This *purK*-like domain appears to facilitate the binding of HCO_3^- and is dispensable in the presence of high HCO_3^- concentrations (Chapman et al. 1994).

In animals, activities of AIRC and the enzyme catalysing the next step, 4-(*N*-succinocarboxamide)-5-aminoimidazole synthetase (SAICARS), are associated with a single bifunctional polypeptide. However, these two enzymes are distinct proteins in plants (Chapman et al. 1994).

4.2.7 Synthesis of Aminoimidazole Succinocarboxamide Ribonucleotide

Conversion of CAIR to SAICAR is catalysed by SAICARS (EC 6.3.2.6). In this reaction, aspartate is added at the α-amino group of CAIR with the consumption of ATP (step 7 in Figure 4.1, Reaction 7).

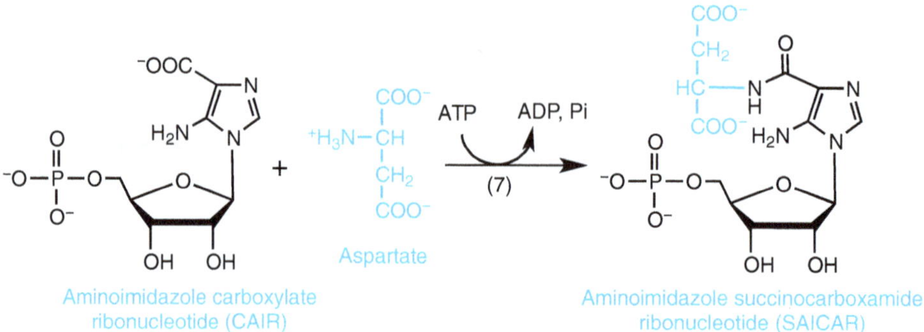

Reaction 7

The expression of the *pur7* gene is strongest in flowers, followed by stem, root, and leaves and is almost absent in siliques and pollen. It has been reported that the expression of the gene is associated with actively dividing tissues such as meristems, and is responsive to growth hormones (Senecoff et al. 1996).

4.2.8 Synthesis of Aminoimidazole Carboxamide Ribonucleotide

Fumarate is released from SAICAR in the formation of aminoimidazole carboxamide ribonucleotide (AICAR) (step 8, Figure 4.1, Reaction 8). The reaction is catalysed by adenylosuccinate lyase (ASL, EC 4.3.2.2). This enzyme also catalyses the cleavage of adenylosuccinate leading to the production of AMP and fumarate (step 12). The encoding *pur8/12* gene has been cloned from *A. thaliana*.

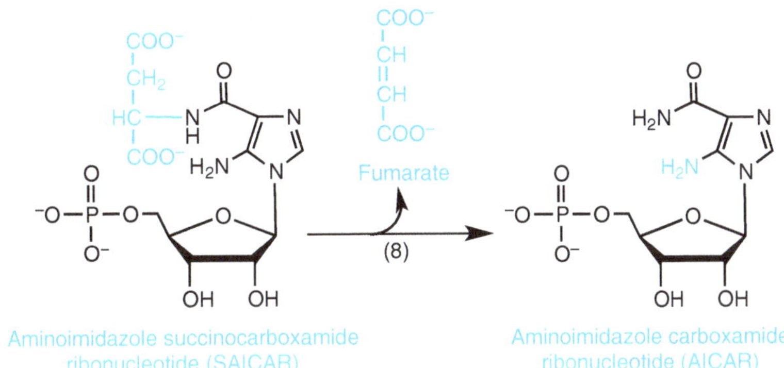

Reaction 8

4.2.9 Synthesis of IMP via Formamidoimidazole Carboxamide Ribonucleotide

The last two steps to form IMP, the first complete purine nucleotide, are catalysed by the bifunctional enzyme 5-aminoimidazole-4-carboxamide ribonucleotide formyltransferase (EC 2.1.2.3)/inosine monophosphate cyclohydrolase (EC 3.5.4.10) (abbreviated as ATIC). In the first part of this reaction the final carbon of the purine ring is provided by 10-formyl-THF to form 5-formamidoimidazole-4-carboxamide ribonucleotide (FAICAR) (step 9, Figure 4.1, Reaction 9). FAICAR then undergoes dehydration and ring closure to generate IMP (step 10, Figure 4.1, Reaction 10).

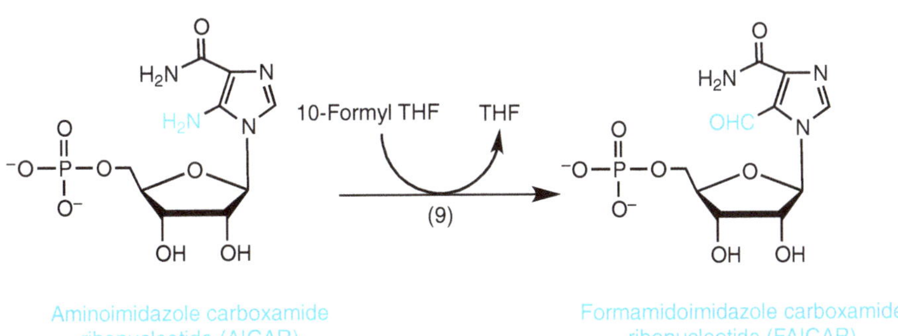

Reaction 9

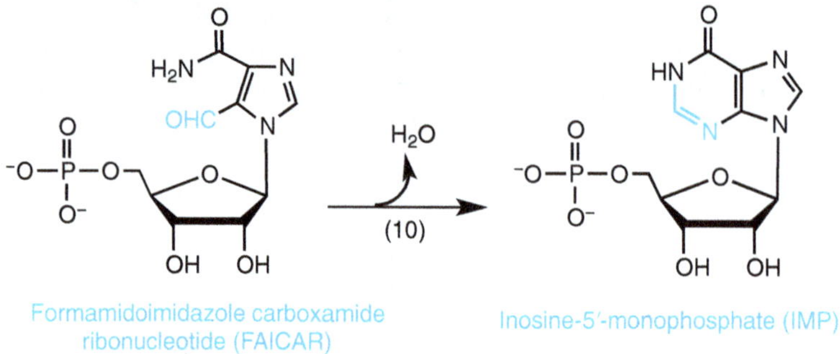

Reaction 10

4.2.10 Synthesis of AMP

AMP and GMP are synthesized from IMP. AMP is formed by replacing the carboxyl group at C6 with an amino group from aspartate. GTP is the donor for the energy-rich phosphate bond to form adenylosuccinate (SAMP) (step 11, Figure 4.1, Reaction 11). The reaction is catalysed by SAMP synthase (ASS, EC 6.3.4.4).

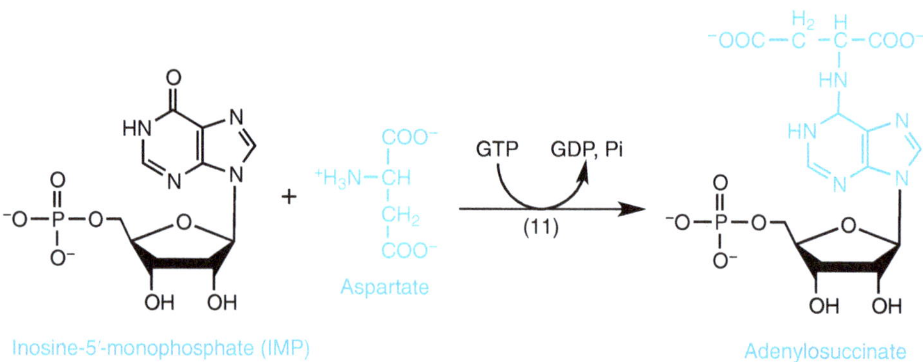

Reaction 11

The cDNA encoding ASS (*pur11*) has been isolated and characterized from *A. thaliana*. This enzyme is a known target for herbicides and antibiotics (Fonné-Pfister et al. 1996). The structure of ASS has been investigated using recombinant ASS proteins from *A. thaliana* and *Tritium aestivum* expressed in *E. coli*. Comparison with the known structures from *E. coli* revealed that the overall fold is very similar to that of the *E. coli* protein. The longer *N* terminus in the plant sequences is at the same place as the longer *C* terminus of the *E. coli* sequence in the 3D structure. The GDP-binding sites have one additional hydrogen-bonding partner, which is a plausible explanation for the lower Km value (Prade et al. 2000).

The removal of fumarate to form AMP is catalysed by ASL (step 12, Figure 4.1, Reaction 12). ASL also catalyses step 8 of the purine biosynthesis *de novo*.

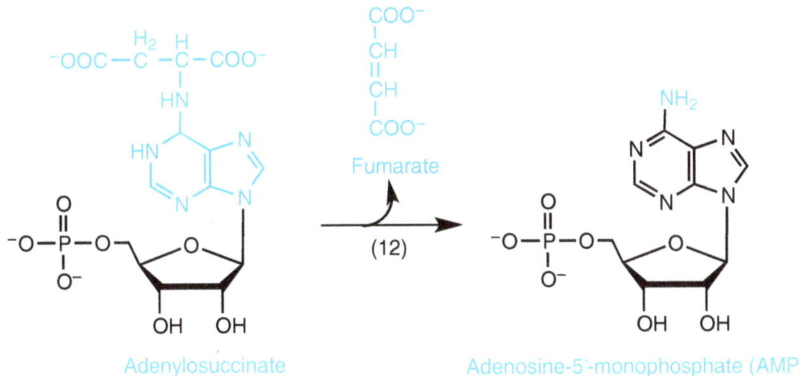

Reaction 12

4.2.11 Synthesis of GMP

Synthesis of GMP is initiated by the oxidation of IMP followed by the insertion of an amino group that is provided by glutamine. Xanthosine-5′-monophosphate (XMP) formation is catalysed by inosine-5′-monophosphate dehydrogenase (IMPDH, EC 1.1.1.205) using NAD^+ as the hydrogen acceptor (step 13, Figure 4.1, Reaction 13).

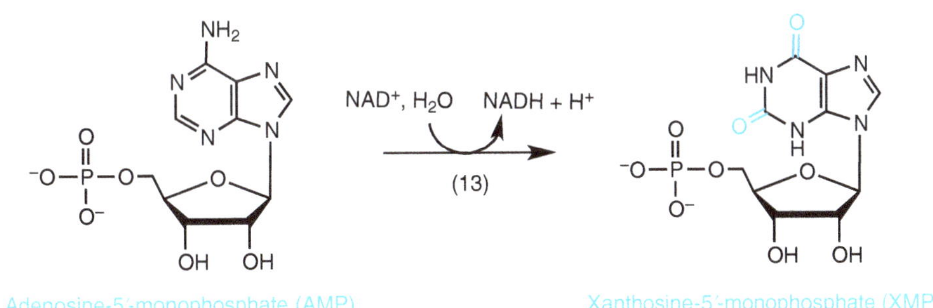

Reaction 13

The native IMPDH has been investigated in seeds of pea (*Pisum sativum*) (Turner and King 1961), *Catharanthus roseus* cells and young leaves of tea (*Camellia sinensis*) (Nishimura and Ashihara 1993) and the soluble proteins of the plant cell fractionation of nitrogen-fixing nodules of cowpea (*Vigna unguiculata*). The most detailed studies have been with the cowpea enzyme which has an alkaline pH optimum (pH 8.8) and a high affinity for IMP and nicotinamide adenine dinucleotide (NAD). Intermediates of ureide metabolism do not affect the enzyme, while AMP, GMP, and NADH are inhibitors (Atkins et al. 1985). Similar to the cowpea nodule enzymes, the Km values of *C. roseus* and tea enzymes for IMP are low (~20 μM) and inhibited by purine nucleotides, especially by GMP (Nishimura and Ashihara 1993). IMPDH is one of the key enzymes of caffeine biosynthesis (Keya et al. 2003). The full length cDNA (*TIDH*) encoding IMPDH was cloned from tea leaves and the expression of *TIDH* is highest in leaves where caffeine biosynthesis occurs than in other plant parts (Li et al. 2008). In addition to GMP and caffeine biosynthesis, the enzyme appears to contribute to the biosynthesis of ureides.

The final step in the formation of GMP is catalysed by GMP synthase (GMPS, EC 6.3.5.2) (step 14, Figure 4.1, Reaction 14). Properties of this enzyme have not yet been reported.

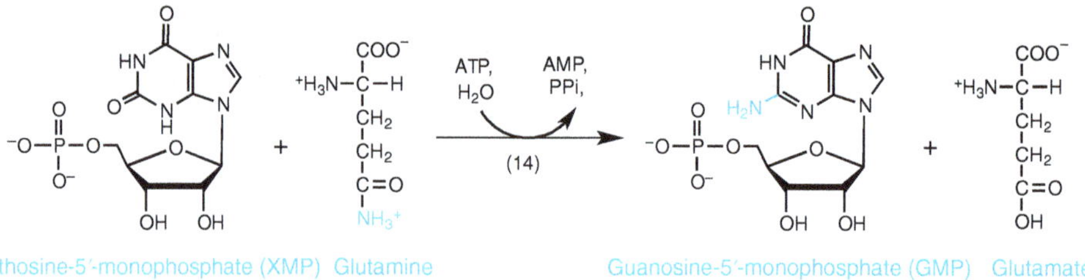

Reaction14

4.3 Summary

Reaction and enzymes of *de novo* purine nucleotide biosynthesis are described. There are 10 steps from PRPP to IMP. AMP and GMP are then produced by two steps from IMP.

References

Anderson, L. and Gibbs, M. (1962). The biosynthesis of caffeine in the coffee plant. *J. Biol. Chem.* 237: 1941–1944.

Arabidopsis-Genome-Initiative (2000). Analysis of the genome sequence of the flowering plant *Arabidopsis thaliana*. *Nature* 408: 796–815.

Ashihara, H. and Nygaard, P. (1989). Purine nucleotide synthesis and RNA synthesis in suspension cultured cells of carrots. *Physiol. Plant* 75: 31–36.

Atkins, C.A., Shelp, B.J., and Storer, P.J. (1985). Purification and properties of inosine monophosphate oxidoreductase from nitrogen-fixing nodules of cowpea (*Vigna unguiculata* L. Walp). *Arch. Biochem. Biophys.* 236: 807–814.

Atkins, C.A., Smith, P., and Storer, P.J. (1997). Reexamination of the intracellular localization of *de novo* purine synthesis in cowpea nodules. *Plant Physiol.* 113: 127–135.

Berthome, R., Thomasset, M., Maene, M. et al. (2008). *pur4* mutations are lethal to the male, but not the female, gametophyte and affect sporophyte development in Arabidopsis. *Plant Physiol.* 147: 650–660.

Buchanan, J.M. (1986). A backward glance. In: *Comprehensive Biochemistry*, vol. 36 (eds. A. Neuberger, L.L.M. Van Deenen and G. Semenza), 1–69. Amsterdam: Elsevier.

Buchanan, B.B., Gruissen, W., and Jones, R.L. (2015). *Biochemistry and Molecular Biology of Plants*. Rockville, USA: American Society of Plant Biologists.

Chapman, K.A., Delauney, A.J., Kim, J.H., and Verma, D.P.S. (1994). Structural organization of *de novo* purine biosynthesis enzymes in plants: 5-aminoimidazole ribonucleotide carboxylase and 5-aminoimidazole-4-*N*-succinocarboxamide ribonucleotide synthetase cDNAs from *Vigna aconitifolia*. *Plant Mol. Biol.* 24: 389–395.

Chow, K.-S., Singh, D.P., Roper, J.M., and Smith, A.G. (1997). A single precursor protein for ferrochelatase-I from Arabidopsis is imported *in vitro* into both chloroplasts and mitochondria. *J. Biol. Chem.* 272: 27565–27571.

Coleto, I., Trenas, A.T., Erban, A. et al. (2016). Functional specialization of one copy of glutamine phosphoribosyl pyrophosphate amidotransferase in ureide production from symbiotically fixed nitrogen in *Phaseolus vulgaris*. *Plant Cell Environ.* 39: 1767–1779.

Fonné-Pfister, R., Chemla, P., Ward, E. et al. (1996). The mode of action and the structure of a herbicide in complex with its target: binding of activated hydantocidin to the feedback regulation site of adenylosuccinate synthetase. *Proc. Natl. Acad. Sci. U.S.A.* 93: 9431–9436.

van der Graaff, E., Hooykaas, P., Lein, W. et al. (2004). Molecular analysis of "*de novo*" purine biosynthesis in solanaceous species and in *Arabidopsis thaliana*. *Front. Biosci.* 9: 1803–1816.

Holmes, E.W., McDonald, J.A., McCord, J.M. et al. (1973). Human glutamine phosphoribosylpyrophosphate amidotransferase: kinetic and regulatory properties. *J. Biol. Chem.* 248: 144–150.

Hung, W.-F., Chen, L.-J., Boldt, R. et al. (2004). Characterization of Arabidopsis glutamine phosphoribosyl pyrophosphate amidotransferase-deficient mutants. *Plant Physiol.* 135: 1314–1323.

Ito, T., Shiraishi, H., Okada, K., and Shimura, Y. (1994). Two amidophosphoribosyltransferase genes of *Arabidopsis thaliana* expressed in different organs. *Plant Mol. Biol.* 26: 529–533.

Keya, C.A., Crozier, A., and Ashihara, H. (2003). Inhibition of caffeine biosynthesis in tea (*Camellia sinensis*) and coffee (*Coffea arabica*) plants by ribavirin. *FEBS Lett.* 554: 473–477.

Li, Y., Ogita, S., Keya, C.A., and Ashihara, H. (2008). Expression of caffeine biosynthesis genes in tea (*Camellia sinensis*). *Z. Naturforsch.* 63C: 267–270.

Menand, B., Maréchal-Drouard, L., Sakamoto, W. et al. (1998). A single gene of chloroplast origin codes for mitochondrial and chloroplastic methionyl–tRNA synthetase in *Arabidopsis thaliana*. *Proc. Natl. Acad. Sci. U.S.A.* 95: 11014–11019.

Nishimura, K. and Ashihara, H. (1993). IMP dehydrogenase from tea leaves and suspension-cultured *Catharanthus roseus* cells. *Phytochemistry* 34: 613–615.

Prade, L., Cowan-Jacob, S.W., Chemla, P. et al. (2000). Structures of adenylosuccinate synthetase from *Triticum aestivum* and *Arabidopsis thaliana*. *J. Mol. Biol.* 296: 569–577.

Proiser, E. and Serenkov, G.P. (1963). Caffeine biosynthesis in tea leaves. *Biokhimiya (Moscow)* 28: 857–861.

Reynolds, P.H.S., Blevins, D.G., and Randall, D.D. (1984). 5-Phosphoribosylpyrophosphate amidotransferase from soybean root nodules: kinetic and regulatory properties. *Arch. Biochem. Biophys.* 229: 623–631.

Schleiff, E. and Soll, J. (2000). Travelling of proteins through membranes: translocation into chloroplasts. *Planta* 211: 449–456.

Schnorr, K.M., Nygaard, P., and Laloue, M. (1994). Molecular characterization of *Arabidopsis thaliana* cDNAs encoding three purine biosynthetic enzymes. *Plant J.* 6: 113–121.

Schnorr, K.M., Laloue, M., and Hirel, B. (1996). Isolation of cDNAs encoding two purine biosynthetic enzymes of soybean and expression of the corresponding transcripts in roots and root nodules. *Plant Mol. Biol.* 32: 751–757.

Schubert, K.R. and Boland, M.J. (1990). The ureides. In: *The Biochemistry of Plants*, vol. 16 (eds. P.K. Stump and E.E. Conn), 197–282. San Diego: Academic Press.

Senecoff, J.F. and Meagher, R.B. (1993). Isolating the *Arabidopsis thaliana* genes for *de novo* purine synthesis by suppression of *Escherichia coli* mutants. I. 5′-Phosphoribosyl-5-aminoimidazole synthetase. *Plant Physiol.* 102: 387–399.

Senecoff, J.F., McKinney, E.C., and Meagher, R.B. (1996). *De Novo* purine synthesis in *Arabidopsis thaliana* II. The PUR7 gene encoding 5′-phosphoribosyl-4-(*N*-succinocarboxamide)-5-aminoimidazole synthetase is expressed in rapidly dividing tissues. *Plant Physiol.* 112: 905–917.

Smith, P.M.C. and Atkins, C.A. (2002). Purine biosynthesis. Big in cell division, even bigger in nitrogen assimilation. *Plant Physiol.* 128: 793–802.

Turner, J.F. and King, J.E. (1961). Inosine 5′-phosphate dehydrogenase of pea seeds. *Biochem. J* 79: 147–151.

Voet, D. and Voet, J.G. (2010). *Biochemistry*. Hoboken, NJ: Wiley.

Zalkin, H. and Dixon, J.E. (1992). De novo purine nucleotide biosynthesis. In: *Progress in Nucleic Acid Research and Molecular Biology*, vol. 42 (eds. W.E. Cohn and K. Moldave), 259–287. Academic Press.

Zrenner, R. and Ashihara, H. (2011). Nucleotide metabolism. In: *Plant Metabolism and Biotechnology* (eds. H. Ashihara, A. Crozier and A. Komamine), 135–162. Chichester, UK: Wiley.

Zrenner, R., Stitt, M., Sonnewald, U., and Boldt, R. (2006). Pyrimidine and purine biosynthesis and degradation in plants. *Annu. Rev. Plant Biol.* 57: 805–836.

5

Salvage Pathways of Purine Nucleotide Biosynthesis

5.1 Introduction

Traditionally the term 'salvage' refers to the rescue of a wrecked or disabled ship or its cargo, but it also is used in a much narrower sense in biochemistry for the recovery of preformed bases and nucleosides for the regeneration of nucleotides and nucleic acids in nucleotide metabolism. Some salvage pathways, which recycle purine bases and nucleosides for nucleotide synthesis, function in plants and other organisms (Ashihara et al. 2018; Moffatt and Ashihara 2002). The net formation of purine nucleotides is performed by the *de novo* pathway, but rapid turnover of nucleic acids, especially RNA, is required for nucleotide production by the salvage pathways.

In contrast to *de novo* purine nucleotide synthesis, which consists of energy consuming multistep reactions when compounds with a purine skeleton are available, purine nucleotides can be synthesized by simpler savage pathways. In salvage pathways, 1 mol of 5-phosphoribosyl-1-pyrophosphate (PRPP) or adenosine-5′-monophosphate (AMP) is utilized for the synthesis of 1 mol of purine nucleotides from a purine base or a purine nucleoside, respectively, by purine phosphoribosyltransferases or purine nucleoside kinases.

In plant cells, purine bases and nucleosides originate from the intercellular breakdown of nucleic acids and nucleotides, as well as other reactions which release purine bases and nucleosides. These substrates can be transported from other tissues, such as storage organs and senescing leaves. It is also possible that they are taken up by the roots from the soil where the purines are breakdown products of fallen leaves, dead insects and bacteria. Salvage of purine bases and nucleosides is performed by phosphoribosyltransferases and nucleoside kinases. Reduced salvage activity inhibits the normal growth of plants and other organisms (Ashihara et al. 2018).

In mammals, genetic deficiency of the purine base salvage enzyme, hypoxanthine/guanine phosphoribosyltransferase (HGPRT, EC 2.4.2.8), causes Lesch–Nyhan syndrome which is characterized by neurological and behavioural abnormalities and the overproduction of uric acid (Nyhan 1997). In plants, adenine phosphoribosyltransferase (APRT, EC 2.4.2.7) appears to be a more important enzyme in purine salvage than HGPRT. In addition to producing adenine nucleotides for energy metabolism and nucleic acid biosynthesis, APRT and adenosine kinase (AK) are involved in the removal of adenine and adenosine, which act as inhibitors of several reactions. These salvage enzymes are also involved in caffeine biosynthesis and the activation and inactivation of cytokinins (see Part VI). In addition, deficiency of APRT causes male sterility in

Plant Nucleotide Metabolism – Biosynthesis, Degradation, and Alkaloid Formation, First Edition.
Hiroshi Ashihara, Iziar A. Ludwig and Alan Crozier.

plants (Gaillard et al. 1998). Thus, purine salvage is not only the energy-saving route of nucleotide biosynthesis, but is an essential pathway in plants. For a comprehensive review on purine salvage in plants, see Ashihara et al. (2018).

5.2 Characteristics of Purine Salvage in Plants

The purine salvage mechanism in plants is essentially the same in microorganisms and animals but some of the participating enzymes are different. Profiles of *in vitro* activity of enzymes involved in purine salvage has been reported in cell-free preparations from young tuber tissues of potato (*Solanum tuberosum*) (Katahira and Ashihara 2006) and young leaves of tea (*Camellia sinensis*) (Deng and Ashihara 2010), see Table 5.1.

In plants, as shown in Figure 5.1, purine bases are salvaged by phosphoribosyltransferases (reactions 1, 2, and/or 2a), while purine nucleosides are salvaged by nucleoside kinases (reaction 3 and/or 4) and nucleoside phosphotransferase (NPT) (reaction 7). In some cases, purine nucleosides are hydrolysed to purine bases by nucleosidases (reaction 8 and/or 9) and then salvaged by the phosphoribosyltransferases mentioned above.

Possible salvage routes of four purine bases (adenine, hypoxanthine, guanine, and xanthine) and six purine nucleosides (adenosine, inosine, guanosine, xanthosine, deoxyadenosine, and deoxyguanosine) are illustrated in Figure 5.1. Activity profiles of enzymes involved in purine salvage in potato tubers and tea leaves are shown in Table 5.1.

Purine base salvage in plants is carried out by two distinct phosphoribosyltransferases, APRT and HGPRT. In mammalian tissues HGPRT has a central role in the purine salvage (Adams and Harkness 1976). However, in plant tissues activity of this enzyme is much lower (~6%) than that of APRT (Table 5.1). In bacteria and animals purine bases can also be converted to their respective purine nucleosides, utilizing ribose-1-phosphate, by purine nucleoside phosphorylase (EC 2.4.2.1) (Bzowska et al. 2000). However, this phosphorylase was not detected in enzyme extracts from yellow lupin seeds and seedlings (Guranowski 1982), potato tubers (Katahira and Ashihara 2006), or tea leaves (Deng and Ashihara 2010). Hence, the production of ribonucleotides by purine nucleoside phosphorylase in plants would appear to have a very minor role in purine salvage.

Purine nucleosides, namely, adenosine, inosine, guanosine, deoxyadenosine, and deoxyguanosine are converted to their respective nucleoside monophosphates. In plants, as illustrated in Figure 5.1, there are three possible routes: (i) direct formation by purine nucleoside kinases (steps 3–6); (ii) interconversion by a non-specific NPT (step 7); and (iii) a two-step reaction, hydrolysis of nucleosides to bases (steps 8 or 9) and salvage by phosphoribosyltransferases (steps 1 or 2).

There are two distinct purine ribonucleoside kinases in plants, AK (EC 2.7.1.20) and inosine/guanosine kinase (IGK) (EC 2.7.1.73). Deoxyadenosine kinase (dAK) (EC 2.7.1.76) and deoxyguanosine kinase (dGK) (EC 2.7.1.113) activities have been detected in plant extracts, and they participate in phosphorylation of deoxyribonucleoside. However, a non-specific deoxyribonucleoside kinase (EC 2.7.1.145) may be the main contributor to deoxyribonucleoside salvage, at least in *Arabidopsis thaliana*.

In addition to nucleoside kinases, NPT (aka non-specific NPT, EC 2.7.1.77) also participates in purine nucleoside salvage. High activity is found with adenosine, inosine,

Table 5.1 Profile of activity of enzymes involved in purine salvage in growing potato tubers (Katahira and Ashihara 2006) and young tea leaves (Deng and Ashihara 2010).

Enzyme (EC number)	Reaction (step number in Figure 5.1) (*^{14}C-Labelled substrates and products)	Potato tubers	Tea leaves
1) Phosphoribosyltransferases			
Adenine phosphoribosyltransferase (2.4.2.7)	Adenine* + PRPP → AMP* + PPi (**1**)	101.4	96.9
Hypoxanthine/guanine phosphoribosyltransferase (2.4.2.8)	Hypoxanthine* + PRPP → IMP* + PPi (**2**)	6.4	0.79
	Guanine* + PRPP → GMP* + PPi (**2**)	4.8	1.26
Xanthine phosphoribosyltransferase (2.4.2.22)	Xanthine* + PRPP → XMP* + PPi		0.22
2) Nucleoside kinase			
Adenosine kinase (2.7.1.20)	Adenosine* + ATP → AMP* + ADP (**3**)	76.3	34.7
Inosine/guanosine kinase (2.7.1.73)	Inosine* + ATP → IMP* + ADP (**4**)	16.8	0.37
	Guanosine* + ATP → GMP* + ADP (**4**)	15.5	0.13
	Xanthosine* + ATP → XMP* + ADP	a)	a)
Deoxyadenosine kinase (2.7.1.76)	Deoxyadenosine* + ATP → dAMP* + ADP (**5**)	75.9	
Deoxyguanosine kinase (2.7.1.113)	Deoxyguanosine* + ATP → dGMP* + ADP (**6**)	53.8	
3) Nucleoside phosphotransferase			
Nucleoside phosphotransferase (2.7.1.77)	Adenosine* + AMP → AMP* + Adenosine (**1**)	39.7	
	Inosine* + AMP → IMP* + Adenosine (**6**)	19.8	
	Guanosine* + AMP → GMP* + Adenosine (7)	20.9	
	Xanthosine* + AMP → XMP* + Adenosine	a)	
	Xanthosine* + IMP → XMP* + Inosine	a)	
	Deoxyadenosine* + AMP → dAMP* + Adenosine (**4**)	89.8	
	Deoxyguanosine* + AMP → dGMP* + Adenosine (**5**)	55.0	
4) Nucleosidases			
Adenosine nucleosidase (3.2.2.7)	Adenosine* → Adenine* + Ribose (**8**)	9.5	84.9
	Deoxyadenosine* → Adenine* + Deoxyribose (**8**)	13.2	
Inosine/guanosine nucleosidase (3.2.2.2)	Inosine* → Hypoxanthine* + Ribose (**9**)	1.1	9.21
	Guanosine* → Guanine* + Ribose (**9**)	0.1	8.53
	Deoxyguanosine* → Guanine* + Deoxyribose (**9**)	<0.1	
Purine nucleosidase (3.2.2.1)	Xanthosine* → Xanthine* + Ribose (**9**)	8.7	29.1

Enzyme activity is expressed as pkat mg^{-1} protein; a) no activity detected.

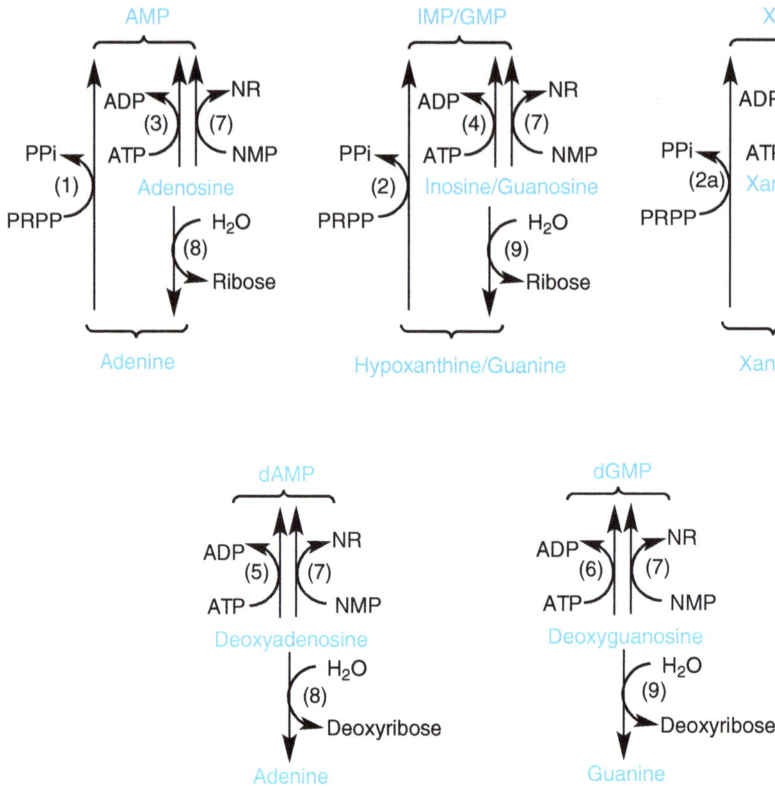

Figure 5.1 Outline of purine salvage reactions and related enzymes in plants. **Phosphoribosyltransferases:** adenine phosphoribosyltransferase (1), hypoxanthine/guanine phosphoribosyltransferase (2), xanthine phosphoribosyltransferase or side reaction of hypoxanthine/guanine phosphoribosyltransferase (2a). **Nucleoside kinases:** adenosine kinase (3), inosine/guanosine kinase (4), deoxyadenosine kinase (5), deoxyguanosine kinase (6). Steps 5 and 6 may be catalysed by deoxynucleoside kinase (see Section 5.3.3). Nucleoside monophosphate phosphotransferase: non-specific nucleoside monophosphate phosphotransferase (7). **Nucleosidases:** adenosine nucleosidase (8), inosine/guanosine nucleosidase (9). Steps 8 and 9 are also catalysed by non-specific purine nucleosidase. Nucleosides are sometimes hydrolysed to purine bases and salvaged by purine phosphoribosyltransferase.

guanosine, deoxyadenosine, and deoxyguanosine, but there is an absence of activity with xanthosine and xanthine (Table 5.1).

Since activity of adenosine nucleosidase (EC 3.2.2.7), inosine/guanosine nucleosidase (EC 3.2.2.2), and/or purine nucleosidase (EC 3.2.2.1) occur in potato tubers and tea leaves (Table 5.1), simple hydrolysis of purine nucleosides to purine bases would appear to be the main route in plants. However, purine base formation from purine nucleosides catalysed by purine phosphorylase, or by the reverse reaction of phosphoribosyltransferases, appears not to participate in this hydrolysis in plants. Properties of nucleosidases are described with the interconversion of purines in Chapter 6.

5.3 Properties of Purine Phosphoribosyltransferases

Phosphoribosyltransferases are enzymes which mediate nucleotide formation through the transfer of nucleobases to PRPP with the release of PPi as shown in reaction 1 in Figure 5.2.

5.3.1 Adenine Phosphoribosyltransferase

Adenine phosphoribosyltransferase (APRT) (EC 2.4.2.7) catalyses the formation of AMP from adenine and PRPP (reaction 1 in Figure 5.1). APRT, and PRPP, the substrate of this enzyme, was first isolated from yeast by Arthur Kornberg, the American Nobel Prize Laureate, and subsequently APRT was found in bacteria, animals and plants (Kornberg et al. 1955a,b).

Kinetic properties of the native and recombinant APRT and HGPRT with the Arabidopsis Information Resource (TAIR) locus of the genes encoding these phosphoribosyltransferases in *A. thaliana* are listed in Table 5.2. Studies on APRT from several plants indicate high affinity with adenine; the Km values are usually low (1–10 μM). In contrast, the Km values for PRPP can fluctuate between 1 and 300 μM. APRT activity is inhibited by AMP, but no significant effect has been observed with other nucleotides (Hirose and Ashihara 1983).

The Arabidopsis genome has five *APRT* sequences (Table 5.2), three of which, *APT1*, *APT2*, and *APT3*, have been cloned, overexpressed, and their catalytic properties characterized (Moffatt et al. 1992). All three isozymes bind adenine (Km: <3 μM). One of the isoenzymes (APT1) is 40–80 times more efficient, judged by the Vmax/Km ratio, in metabolizing adenine than the other isozymes. Since none of the predicted amino acid sequences for the adenine phosphoribosyltransferases (APTs) of *A. thaliana* appear to contain transit signalling peptides, APTs are assumed to be located in the cytosol. However, the findings of Zybailov et al. (2008) suggest that an isoform of APT1 has a transit peptide for chloroplasts. Traditional biochemical analysis indicates that in addition to the cytosol, APRT also occurs in plastids and mitochondria in some plants, (see Ashihara et al. 2018). Biochemical analysis indicates that large quantities of APRT activity are located in cytosol, but substantial amounts are also found in chloroplasts of spinach (Ashihara and Ukaji 1985) and tea leaves (Koshiishi et al. 2001), as well as mitochondria of *Catharanthus roseus* cells (Hirose and Ashihara 1982) and tubers of Jerusalem artichoke (*Helianthus tuberosus*) (Le Floc'h and Lafleuriel 1983). Although the occurrence of APRT in mitochondria has not yet been confirmed at the molecular level, purified intact mitochondria from *C. roseus* are capable of synthesizing adenine nucleotides from ^{14}C-labelled adenine (Ukaji et al. 1986).

5.3.2 Hypoxanthine/Guanine Phosphoribosyltransferase

HGPRT (IMP/GMP): diphosphate phospho-D-ribosyltransferase, EC 2.4.2.8) catalyses the formation of IMP (or GMP) from inosine (or guanine) and PRPP. Although HGPRT activity is generally high in mammalian tissues (Adams and Harkness 1976), in plants it is usually much lower than that of APRT (Table 5.1).

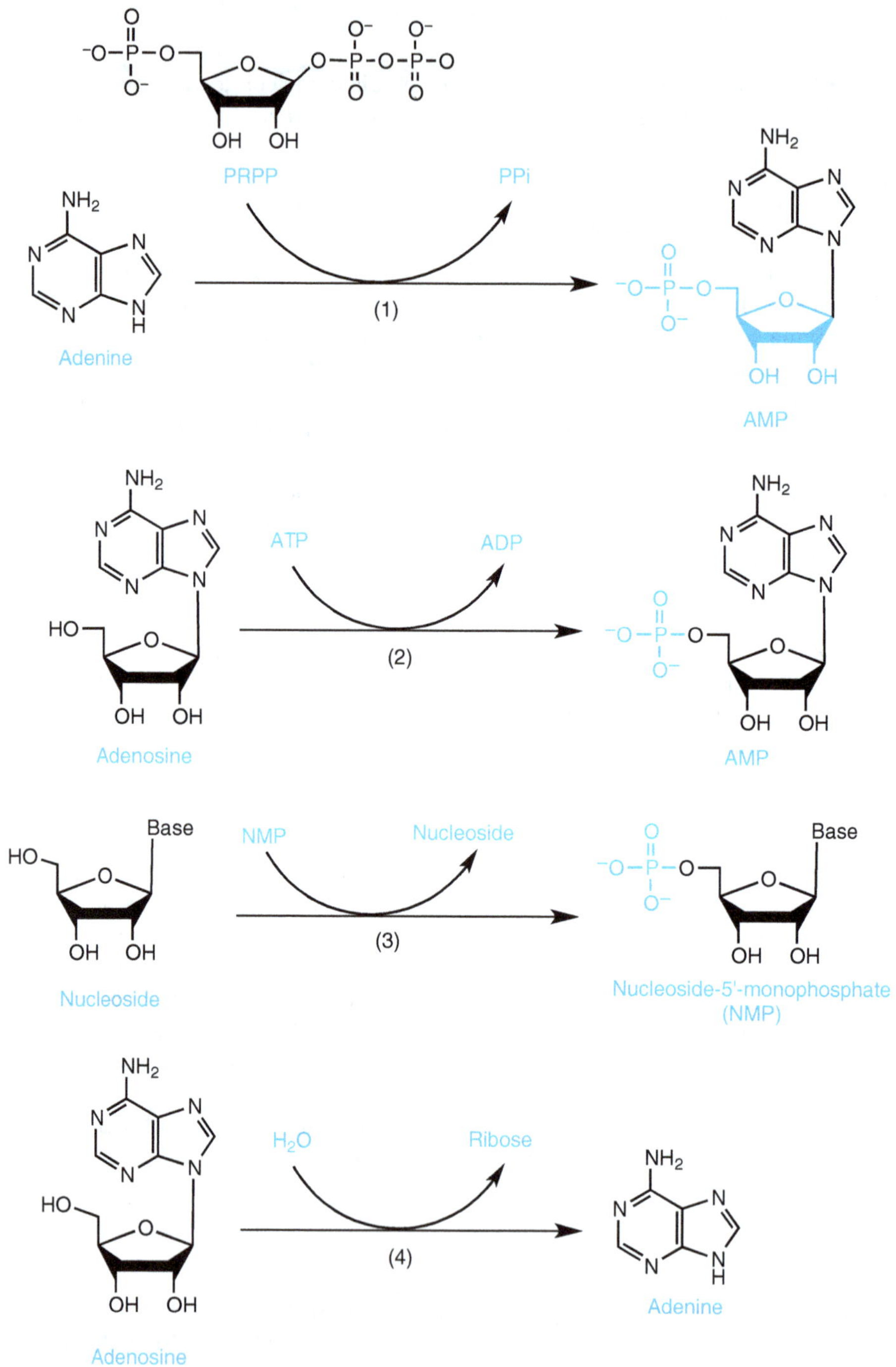

Figure 5.2 Reactions involved in purine salvage in plants. **Reaction 1:** phosphoribosyltransfer reaction: this reaction transfers the 5-phosphoribosyl group from PRPP to the purine (adenine phosphoribosyltransferase reaction is shown as an example). **Reaction 2:** kinase reaction: the transfer of γ-phosphate from ATP to purine nucleoside leading to formation of nucleoside monophosphate (example, adenosine kinase reaction). **Reaction 3:** non-specific nucleoside phosphotransferase reaction. Nucleoside 5′-monophosphate is used for this transfer reaction. **Reaction 4:** Purine nucleosidase enzyme reaction. Purine nucleoside is hydrolysed forming purine base and ribose.

Table 5.2 Properties of the native and recombinant adenine phosphoribosyltransferase (APRT) and hypoxanthine/guanine phosphoribosyltransferase (HGPRT) from plants.

Enzyme	Enzyme source	Isozyme	Optimum pH	Km values (μM)					Gene	TAIR locus	Reference
				A	PRPP	*t*Z	iP	BA			
APRT	Jerusalem artichoke[a]		5.5–6.5	5.5	64						Le Floc'h and Lafleuriel (1978)
	Strawberry[a]		7.5	10	130						Robert and Pétel (2000)
	Catharansus roseus[a]		7.6–8.0	7.0	9.6						Hirose and Ashihara (1983)
	Brassica juncea[a]		9.2	3.8	15						Moffatt and Somerville (1990)
	Wheat germ[a]		7.5	74	[c]		130	154			Chen et al. (1982)
	Tomato roots[a]		~8	4.0	2.0			50			Burch and Stuchbury (1986)
	Tomato leaves[a]		~9	2.0	1.0			125			
	Hevea brasiliensis[a]		7.0	3.8	250						Gallois et al. (1996)
	Prunus persica		8.5	0.7	25	300		[c]			Lecomte and Le Floc'h (1999)
	Arabidopsis thaliana[a]		8.8	4.5	290		[c]	[c]			Lee and Moffatt (1993)
	A. thaliana[b]	APT1	8.8	1.0	55	1800	2500	2400	*apt1*	AT1G27450	Allen et al. (2002)
		APT2	7.4–7.5	2.6		330	110	15	*apt2*	AT1G80050	Zybailov et al. (2008)
		APT3	7.4–7.5	0.8		76	440	370	*apt3*	AT4G22570	
		APT4							*apt4*	AT4G12440	
		APT5							*apt5*	AT5g11160	
HGPRT				**HX**	**G**	**X**	**PRPP**				
	Jerusalem artichoke[a]		7.5–8.5	6.1	5.5		8.8				Le Floc'h and Lafleuriel (1981)
	A. thaliana[b]			176	29		[c]		*hpt/gpt*	AT1G71750	Liu et al. (2007)

A, Adenine; PRPP, 5-phosphoribosyl-1-pyrophosphate. Cytokinin bases: BA, benzyladenine; iP, isopentenyladenine; *t*Z, *trans*-zeatin. HX, Hypoxanthine; G, guanine; X, xanthine.

—, not determined.

a) Native enzyme.

b) Recombinant enzyme.

c) Activity was detected but no data was published on the Km value.

Compared with APRT, only a few kinetic analyses of HGPRT activity have been carried out with plant enzymes (Table 5.2). A native HGPRT, partially purified from Jerusalem artichoke shoots, exhibited the low Km values for hypoxanthine, guanine, and PRPP (<10 μM), which have high affinities for the enzyme. The rate of IMP formation from hypoxanthine was greater than GMP production from guanine. Enzyme activity was inhibited by the products, IMP and GMP (Le Floc'h and Lafleuriel 1981).

Molecular and functional analysis of HGPRT of *A. thaliana* was carried out by Liu et al. (2007). The kinetic analysis of the recombinant HGPRT revealed that the enzyme catalyses the conversion of guanine and hypoxanthine to their respective nucleoside monophosphates, but that xanthine is not a substrate. The Km values of the recombinant enzyme for hypoxanthine and guanine were 176 and 29 μM, respectively. The relatively low affinity of the *A. thaliana* recombinant enzyme for hypoxanthine is different to that from other sources, including humans (Keough et al. 1999) and yeast (Ali and Sloan 1982), as well as the native plant enzyme (Le Floc'h and Lafleuriel 1981), which displays similar Km values (<10 μM) towards guanine and hypoxanthine. The reason why the recombinant *A. thaliana* HGPRT possesses a substantially higher affinity for guanine than for hypoxanthine remains to be resolved. Like the enzymes in humans and yeast, *A. thaliana* HGPRT does not act on xanthine (Liu et al. 2007).

5.3.3 Xanthine Phosphoribosyltransferase

As noted above, in most prokaryotes and nearly all eukaryotes, HGPRT catalyses the salvage of hypoxanthine and guanine. However, in a few species, xanthine salvage enzyme activity is also detected. Guanine and xanthine phosphoribosyltransferase (GXPRT, EC 2.4.2.22) activity has been observed in *Lactobacillus casei* and *E. coli* (Krenitsky et al. 1970). More specific xanthine phosphoribosyltransferase (XPRT) is found in *Leishmania donovani*, a protozoan parasite. This enzyme preferentially uses xanthine; the Km value for xanthine (7 μM) is much lower than for hypoxanthine and guanine (100–450 μM) (Jardim et al. 1999). In plants, low phosphoribosylation of xanthine was detected in tea leaf extracts *in vitro* (Deng and Ashihara 2010). However, it is unclear whether this activity is due to a distinct XPRT or if it is a side reaction of HGPRT (Ashihara et al. 2018).

5.4 Properties of Nucleoside Kinases

Nucleoside kinases are enzymes which catalyse the transfer of γ-phosphate from ATP to nucleosides leading to formation of nucleoside-5′-monophosphates.

5.4.1 Adenosine Kinase

Adenosine kinase (AK, ATP: adenosine-5′-phosphotransferase, EC 2.7.1.20) catalyses the phosphorylation of adenosine to AMP using ATP as the phosphate donor (reaction 2 in Figure 5.2). The enzyme was first isolated from yeast and animal tissues (Caputto, 1951; Kornberg and Pricer, 1951). AK is widely distributed in plant cells and tissues and there are reports on both purified native and recombinant enzymes (Table 5.3).

Table 5.3 Properties of the native and recombinant adenosine kinase (AK) and inosine/guanosine kinase (IGK) from plants.

Enzyme	Enzyme source	Isozyme	Optimum pH	Km values (μM)					Gene	TAIR Locus	References
				AR	ATP	*t*ZR	iPR	DHZR			
AK	Wheat germ		6.8–7.2	8.7			31				Chen and Eckert (1977)
	Yellow lupin seeds[a]		7.0–7.5	1.5	300						Guranowski (1979b)
	Peach flower buds[a]		9.5	1.8	200						Faye and Le Floc'h (1997)
	Tobacco BY-2 cells[a]										Laukens et al. (2003)
	Arabidopsis thaliana[b]	ADK1	8.0	0.5	350		3.2		*adk1*	AT3G09820	Moffatt et al. (2000)
		ADK2	9.5	0.3	370		4.8		*adk2*	AT5G03300	
	Tobacco BY-2 cells[b]	ADK1T	7.5	4.0		2.4	0.3	3.9	*adk1t*		Kwade et al. (2005)
		ADK1S	7.0	1.9		2.5	6.1	3.9	*adk1s*		
		ADK2T	8.5	7.3		3.2	8.7	2.6	*adk2t*		
		ADK2S	8.8	3.9		2.2	1.7	3.8	*adk2s*		
	Physcomitrella (moss)[b]						[c]				von Schwartzenberg et al. (1998)
IGK				**IR**	**GR**	**XR**	**ATP**				
	Jerusalem artichoke[a]		7.6	70	14		1450				Combés et al. (1989)
dNK				**AdR**	**GdR**						
	A. thaliana[b]			14	27				*dnk*	AT1G72040	Clausen et al. (2012)

AR, Adenosine; ATP, adenosine triphosphate. Cytokinin ribosides: DHZR, dihydrozeatin riboside; iPR, isopentenyladenosine; *t*ZR, *trans*-zeatin riboside. IR, Inosine; GR, guanosine; XR, xanthosine; AdR, deoxyadenosine; GdR, deoxyguanosine.

a) Native enzyme.

b) Recombinant enzyme.

c) Activity was detected but no data was published on the Km value.

Almost homogeneous AK was isolated from seeds of yellow lupin (*Lupinus luteus*) by Guranowski (1979b). The Km value for adenosine was low (1.5 μM), but was higher (300 μM) for ATP. This trend is also found with AK from other plant sources (Table 5.3). ATP is the main phosphate donor. dATP and guanosine triphosphate (GTP) can substitute for ATP, but less effectively. AK has been investigated in the context of phosphorylation of cytokinins (see Part V).

Cloning and characterization of AK was first performed with the moss *Physcomitrella patens* (von Schwartzenberg et al. 1998). A gene *adk* was cloned from a cDNA library by functional complementation of an *E. coli* purine auxotrophic strain. The deduced amino acid sequence had a 52% homology with the human *adk*. Subsequently, cDNAs and genes encoding two isoforms of AK were isolated from *A. thaliana* (Moffatt et al. 2000). The *adk1*- and *adk2*-coding sequences were very similar sharing, respectively, 92% and 89% amino acid and nucleotide identity. Each cDNA was overexpressed in *E. coli*, and the catalytic activity of the two isoforms determined. Both AK isozymes have similar catalytic properties (Table 5.3). Four AK isoforms, designated 1S, 2S, 1T, and 2T have been identified in BY2 tobacco cells (Kwade et al. 2005). In contrast to AK from other plant sources, all four tobacco AK isoforms displayed a high affinity for adenosine and three cytokinin ribosides (see Part V).

5.4.2 Inosine/Guanosine Kinase

Inosine/guanosine kinase (IGK, ATP: inosine/guanosine 5′-phosphotransferase, EC 2.7.1.73) catalyses the phosphorylation of inosine to IMP and that of guanosine to GMP using ATP as a phosphate donor. This enzyme activity was first found in cell-free extracts of *Ehrlich ascites* tumour cells (Pierre and LePage 1968). In contrast to AK, the occurrence of IGK is limited. Unusually, only a few prokaryotes, including *E. coli* and *Salmonella typhimurium*, contain IGK as opposed to AK activity (Nygaard 1983).

IGK has been detected in plants (Deng and Ashihara 2010; Katahira and Ashihara 2006) (Table 5.1). Partial purification of IGK from mitochondria of Jerusalem artichoke was reported by Combés et al. (1989) (Table 5.3). The enzyme appears to be located in the intermembrane space of mitochondria. The Km values for guanosine (14 μM) are lower than for inosine (70 μM). There are no reports of the plant IGK being cloned, but molecular studies with bacteria and animals indicate that AK and IGK belong to the ribokinase family of proteins that share a number of unique primary and tertiary structural elements (Park and Gupta 2008).

5.4.3 Deoxyribonucleoside Kinases

Deoxyribonucleoside kinases salvage deoxyribonucleosides by transfer of a phosphate group to the 5′ position of a deoxyribonucleoside. This salvage pathway is well characterized in mammals, but little is known about deoxyribonucleoside salvage enzymes in plants.

Relatively high activity of deoxyadenosine kinase (dAK, EC 2.7.1.76) and deoxyguanosine kinase (dGK, EC 2.7.1.113) was detected in extracts of potato tubers (Katahira and Ashihara 2006) (see Table 5.1) and *A. thaliana* (Clausen et al. 2012). There are two types of genes that encode deoxyribonucleoside salvage enzymes in the *A. thaliana* genome. One is a single *AtdNK* gene which codes deoxynucleoside kinase (dNK, EC 2.7.1.145).

The enzyme is able to use deoxyadenosine, deoxyguanosine, and deoxycytidine as substrates. Another is a thymidine kinase-like enzyme which participates in pyrimidine salvage (Clausen et al. 2012; Clausen et al. 2008) (see Part III). These findings suggest that the activity of dAK and dGK detected in plant extracts may be due to dNK which possesses broad substrate specificity.

5.5 Properties of Nucleoside Phosphotransferase

In contrast to nucleoside kinases (see Section 5.4, reaction 2 in Figure 5.2), NPTs use nucleoside monophosphates as phosphate donors (reaction 3 in Figure 5.2). The non-specific nucleoside phosphotransferase (NPT, nucleotide: nucleoside 5′-phosphotransferase, EC 2.7.1.77) catalyses the conversion of nucleosides to nucleoside monophosphate using 5′-nucleoside monophosphate. For example, guanosine + AMP → GMP + adenosine. In potato tubers, *in vitro* activity of NPT is substantial and at a similar or slightly lower level than the respective nucleoside kinases. Neither NPT nor IGK can convert xanthosine to xanthosine monophosphate (XMP) (see Table 5.1).

NPT is widely distributed in plants (Brawerman and Chargaff 1955). Highly purified, native NPT has been obtained from carrot roots (Brunngraber and Chargaff 1967, 1970), cotyledons of yellow lupin seedlings (Guranowski 1979a), and barley seedlings (Prasher et al. 1982) (Table 5.4). The lupin enzyme is similar to the enzyme purified from the bacterium, *Erwinia herbicola* (Chao 1976). With the yellow lupin NPT, purine and pyrimidine nucleosides are good phosphate acceptors and 5′-nucleotides are more effective phosphate donors than adenosine-3′-monophosphate (3′-AMP). The high Km values (>0.4 mM) indicate that NPT binds purine nucleosides but with much lower affinity than purine nucleoside kinases (Km: <10 μM) (Table 5.3). Therefore, if the cellular concentration of purine nucleosides is low, purine nucleoside kinases (AK and IGK) are preferentially involved in purine nucleoside salvage. In the *A. thaliana* database, no

Table 5.4 Properties of the native nucleoside phosphotransferase (NPT) from plants.

Enzyme	Enzyme source	Isozyme	Optimum pH	Km values (μM) AR	IR	GR	AdR	AMP	Gene	TAIR Locus	References
NPT	Carrot root[a]		5.0	[b]	[b]	[b]	[b]	[b]			Brunngraber and Chargaff (1967)
	Yellow lupin cotyledons[a]		8.0	400	400			400		—	Guranowski (1979a)
	Barley seedlings[a]			540		2130	87	[b]		—	Prasher et al. (1982)

AR, Adenosine; IR, inosine; GR, guanosine; AdR, deoxyadenosine; AMP, adenosine-5′-monophosphate.

a) Native enzyme.

b) Activity was found but the Km value is not available.

matches have been found for search string non-specific nucleotide triphosphate (NTP) (EC 2.7.1.77).

5.6 Role of Purine Salvage in Plants

The importance of purine salvage has been demonstrated in plants as well as other organisms. Purine salvage enzyme-deficient mutants do not grow normally. This recycling pathway is not only energetically advantageous for nucleotide formation, but also the rapid turnover of nucleotides by the salvage reactions is essential for normal cellular growth and homeostasis. A further role of salvage enzymes is to catalyse the removal of purine bases and nucleosides which inhibit the flow of metabolites and cause toxicity. The salvage activity is closely associated with physiological events in plants. Onset of seed germination, cell division and growth, breaking dormancy, and recovery from several stresses all require purine nucleotides produced by these purine salvage reactions. Purine salvage activity associated with plant developmental and physiological events has been discussed in detail by Ashihara et al. (2018).

5.7 Summary

In plants purine bases are salvaged by phosphoribosyltransferases while purine nucleosides are salvaged by nucleoside kinases and NPT. Possible salvage routes of four purine bases, adenine, hypoxanthine, guanine, and xanthine, and six purine nucleosides, adenosine, inosine, guanosine, xanthosine, deoxyadenosine, and deoxyguanosine, are present in plants. Adenine and adenosine are the best substrates. The purine salvage mechanism in plants is different from what occurs in other organisms.

References

Adams, A. and Harkness, R.A. (1976). Developmental changes in purine phosphoribosyltransferases in human and rat tissues. *Biochem. J.* 160: 565–576.

Ali, L.Z. and Sloan, D.L. (1982). Studies of the kinetic mechanism of hypoxanthine-guanine phosphoribosyltransferase from yeast. *J. Biol. Chem.* 257: 1149–1155.

Allen, M., Qin, W., Moreau, F., and Moffatt, B. (2002). Adenine phosphoribosyltransferase isoforms of Arabidopsis and their potential contributions to adenine and cytokinin metabolism. *Physiol. Plant.* 115: 56–68.

Ashihara, H. and Ukaji, T. (1985). Presence of adenine phosphoribosyltransferase and adenosine kinase in chloroplasts of spinach leaves. *Int. J. Biochem.* 17: 1275–1277.

Ashihara, H., Stasolla, C., Fujimura, T., and Crozier, A. (2018). Purine salvage in plants. *Phytochemistry* 147: 89–124.

Brawerman, G. and Chargaff, E. (1955). On the distribution and biological significance of the nucleoside phosphotransferases. *Biochim. Biophys. Acta* 16: 524–532.

Brunngraber, E.F. and Chargaff, E. (1967). Purification and properties of a nucleoside phosphotransferase from carrot. *J. Biol. Chem.* 242: 4834–4840.

Brunngraber, E.F. and Chargaff, E. (1970). Nucleoside phosphotransferase from carrot: kinetic studies and exploration of active sites. *J. Biol. Chem.* 245: 4825–4831.

Burch, L.R. and Stuchbury, T. (1986). Purification and properties of adenosine nucleosidases from tomato (*Lycopersicon esculentum*) roots and leaves. *J. Plant Physiol.* 125: 267–273.

Bzowska, A., Kulikowska, E., and Shugar, D. (2000). Purine nucleoside phosphorylases: properties, functions, and clinical aspects. *Pharmacol. Ther.* 88: 349–425.

Caputto, R. (1951). The enzymatic synthesis of adenylic acid; adenosinekinase. *J. Biol. Chem.* 189: 801–814.

Chao, H.M. (1976). Nucleoside phosphotransferase from *Erwinia herbicola*, a new membrane-bound enzyme. *J. Biol. Chem.* 251: 2330–2333.

Chen, C.-M. and Eckert, R.L. (1977). Phosphorylation of cytokinin by adenosine kinase from wheat germ. *Plant Physiol.* 59: 443–447.

Chen, C.-M., Melitz, D.K., and Clough, F.W. (1982). Metabolism of cytokinin: Phosphoribosylation of cytokinin bases by adenine phosphoribosyltransferase from wheat germ. *Arch. Biochem. Biophys.* 214: 634–641.

Clausen, A.R., Girandon, L., Knecht, W. et al. (2008). A multisubstrate deoxyribonucleoside kinase from plants. *Nucleic Acids Symp. Ser.* 52: 489–490.

Clausen, A.R., Girandon, L., Ali, A. et al. (2012). Two thymidine kinases and one multisubstrate deoxyribonucleoside kinase salvage DNA precursors in *Arabidopsis thaliana*. *FEBS J.* 279: 3889–3897.

Combés, A., Lafleuriel, J., and Le Floc'h, F. (1989). The inosine-guanosine kinase activity of mitochondria in tubers of *Jersalem artichoke*. *Plant Physiol. Biochem.* 27: 729–736.

Deng, W.-W. and Ashihara, H. (2010). Profiles of purine metabolism in leaves and roots of *Camellia sinensis* seedlings. *Plant Cell Physiol.* 51: 2105–2118.

Faye, F. and Le Floc'h, F. (1997). Adenosine kinase of peach tree flower buds: Purification and properties. *Plant Physiol. Biochem.* 35: 15–22.

Gaillard, C., Moffatt, B.A., Blacker, M., and Laloue, M. (1998). Male sterility associated with APRT deficiency in *Arabidopsis thaliana* results from a mutation in the gene *APT1*. *Mol. Gen. Genet.* 257: 348–353.

Gallois, R., Prevot, J.C., Clement, A., and Jacob, J.L. (1996). Purification and characterization of an adenine phosphoribosyl- transferase from rubber tree latex. Physiological implications. *Plant Physiol. Biochem.* 34: 527–537.

Guranowski, A. (1979a). Nucleoside phosphotransferase from yellow lupin seedling cotyledons. *Biochim. Biophys. Acta* 569: 13–22.

Guranowski, A. (1979b). Plant adenosine kinase: purification and some properties of the enzyme from Lupinus luteus seeds. *Arch. Biochem. Biophys.* 196: 220–226.

Guranowski, A. (1982). Purine catabolism in plants. Purification and some properties of inosine nucleosidase from yellow lupin (*Lupinus leteus* L.) seeds. *Plant Physiol.* 70: 344–349.

Hirose, F. and Ashihara, H. (1982). Adenine phosphoribosyltransferase activity in mitochondria of *Catharanthus roseus* cells. *Z. Naturforsh.* 37c: 1288–1289.

Hirose, F. and Ashihara, H. (1983). Adenine phosphoribosyltransferase of *Catharanthus roseus* cells: purification, properties and regulation. *Z. Pflanzenphysiol.* 110: 135–145.

Jardim, A., Bergeson, S.E., Shih, S. et al. (1999). Xanthine phosphoribosyltransferase from *Leishmania donovani* : molecular cloning, biochemical characterization, and genetic analysis. *J. Biol. Chem.* 274: 34403–34410.

Katahira, R. and Ashihara, H. (2006). Profiles of purine biosynthesis, salvage and degradation in disks of potato (*Solanum tuberosum* L.) tubers. *Planta* 225: 115–126.

Keough, D., Ng, A.-L., Winzor, D. et al. (1999). Purification and characterization of *Plasmodium falciparum* hypoxanthine–guanine–xanthine phosphoribosyltransferase and comparison with the human enzyme. *Mol. Biochem. Parasitol.* 98: 29–41.

Kornberg, A., Lieberman, I., and Simms, E.S. (1955a). Enzymatic synthesis and properties of 5-phosphoribosylpyrophosphate. *J. Biol. Chem.* 215: 389–402.

Kornberg, A., Lieberman, I., and Simms, E.S. (1955b). Enzymatic synthesis of purine nucleotides. *J. Biol. Chem.* 215: 417–427.

Kornberg, A. and Pricer, W.E. (1951). Enzymatic phosphorylation of adenosine and 2,6-diaminopurine riboside. *J. Biol. Chem.* 193: 481–496.

Koshiishi, C., Kato, A., Yama, S. et al. (2001). A new caffeine biosynthetic pathway in tea leaves: utilisation of adenosine released from the *S*-adenosyl-L-methionine cycle. *FEBS Lett.* 499: 50–54.

Krenitsky, T.A., Neil, S.M., and Miller, R.L. (1970). Guanine and xanthine phosphoribosyltransfer activities of *Lactobacillus casei* and *Escherichia coli*. *J. Biol. Chem.* 245: 2605–2611.

Kwade, Z., Świątek, A., Azmi, A. et al. (2005). Identification of four adenosine kinase isoforms in tobacco By-2 cells and their putative role in the cell cycle-regulated cytokinin metabolism. *J. Biol. Chem.* 280: 17512–17519.

Laukens, K., Lenobel, R., Strnad, M. et al. (2003). Cytokinin affinity purification and identification of a tobacco BY-2 adenosine kinase. *FEBS Lett.* 533: 63–66.

Lecomte, I. and Le Floc'h, F. (1999). Adenine phosphoribosyltransferase of peach tree leaves: Purification and properties. *J. Plant Physiol.* 154: 289–295.

Lee, D. and Moffatt, B.A. (1993). Purification and characterization of adenine phosphoribosyl transferase from *Arabidopsis thaliana*. *Physiol. Plant.* 87: 483–492.

Le Floc'h, F. and Lafleuriel, J. (1978). L'adeninephosphoribosyltransferase des pousses de topinambour Helianthus tuberosus. *Phytochemistry* 17: 643–646.

Le Floc'h, F. and Lafleuriel, J. (1981). L'hypoxanthine-guanine phosphoribosyltransferase des pousses de topinambour *Helianthus tuberosus*. *Z. Pflanzenphysiol.* 104: 331–340.

Le Floc'h, F. and Lafleuriel, J. (1983). The role of mitochondria in the recycling of adenine into purine nucleotides in the Jerusalem artichoke (*Helianthus tuberosus* L.). *Z. Pflanzenphysiol.* 113: 61–71.

Liu, X., Qian, W., Liu, X. et al. (2007). Molecular and functional analysis of hypoxanthine-guanine phosphoribosyltransferase from *Arabidopsis thaliana*. *New Phytol.* 175: 448–461.

Moffatt, B.A. and Ashihara, H. (2002). Purine and pyrimidine nucleotide synthesis and metabolism. In: *The Arabiopsis Book*, vol. 1, 1–20. Rockville, USA: American Society of Plant Biologists https://doi.org/10.1199/tab.0018.

Moffatt, B.A., McWhinnie, E.A., Burkhart, W.E. et al. (1992). A complete cDNA for adenine phosphoribosyltransferase from *Arabidopsis thaliana*. *Plant Mol. Biol.* 18: 653–662.

Moffatt, B.A. and Somerville, C.R. (1990). Purification of adenine phosphoribosyltransferase from *Brassica juncea*. *Arch. Biochem. Biophys.* 283: 484–490.

Moffatt, B.A., Wang, L., Allen, M.S. et al. (2000). Adenosine kinase of Arabidopsis. Kinetic properties and gene expression. *Plant Physiol.* 124: 1775–1785.

Nygaard, P. (1983). Utilization of preformed purine bases and nucleosides. In: *Metabolism of Nucleotides, Nucleosides and Nucleobases in Microorganisms* (ed. A. Munch-Petersen), 27–93. London: Academic Press.

Nyhan, W.L. (1997). The recognition of Lesch-Nyhan syndrome as an inborn error of purine metabolism. *J. Inherited Metab. Dis.* 20: 171–178.

Park, J. and Gupta, R.S. (2008). Adenosine kinase and ribokinase – the RK family of proteins. *Cell. Mol. Life Sci.* 65: 2875–2896.

Pierre, K.J. and LePage, G.A. (1968). Formation of inosine-5′-monophosphate by a kinase in cell-free extracts of *Ehrlich ascites* cells in vitro. *Proc. Soc. Exp. Biol. Med.* 127: 432–440.

Prasher, D.C., Carr, M.C., Ives, D.H. et al. (1982). Nucleoside phosphotransferase from barley. Characterization and evidence for ping pong kinetics involving phosphoryl enzyme. *J. Biol. Chem.* 257: 4931–4939.

Robert, F. and Pétel, G. (2000). Seasonal changes in adenine phosphoribosyltransferase and adenosine kinase activities as markers of the dormancy and the growth of *Fragaria x ananassa*. *Plant Physiol. Biochem.* 38: 395–402.

von Schwartzenberg, K., Kruse, S., Reski, R. et al. (1998). Cloning and characterization of an adenosine kinase from *Physcomitrella* involved in cytokinin metabolism. *Plant J.* 13: 249–257.

Ukaji, T., Hirose, F., and Ashihara, H. (1986). Purine salvage in mitochondria of cultured *Catharanthus roseus* cells. *J. Plant Physiol.* 125: 191–197.

Zybailov, B., Rutschow, H., Friso, G. et al. (2008). Sorting signals, *N*-terminal modifications and abundance of the chloroplast proteome. *PLOS One* 3: e1994.

6

Interconversion of Purine Nucleotides

6.1 Introduction

Four purine skeletons, namely, adenine, guanine, hypoxanthine, and xanthine occur as bases, nucleosides, and nucleotides. In the mid-twentieth century, it was realized that adenine and guanine are interconverted, as ^{15}N-adenine fed to rats were found in both adenine and guanine in visceral nucleic acids (Brown et al. 1948). Subsequently, a similar conversion of adenine to guanine was shown to take place in plants in studies using ^{14}C-labelled adenine and adenosine, which were converted to adenine nucleotides and then incorporated into both adenine and guanine residues of nucleic acids (Ashihara et al. 2000; Deng and Ashihara 2010; Katahira and Ashihara 2006; Yin et al. 2014). The interconversion of purines is often classified as purine catabolism (Stasolla et al. 2003). However, as noted above, some of the interconversion reactions are involved in nucleic acid biosynthesis. The initial step of purine degradation involves xanthine. Therefore, reactions from purine nucleotides to xanthine are discussed in this chapter while catabolism of xanthine is described in Chapter 7.

The conversion of inosine-5′-monophosphate (IMP), formed by the *de novo* and salvage pathways (see Chapter 4), to adenosine-5′-monophosphate (AMP) or guanosine-5′-monophosphate (GMP) involves, respectively, steps 2–3, and steps 5–6 as illustrated in Figure 6.1. Since adenine and guanine possess an NH_2 group, conversion to hypoxanthine and xanthine requires deamination reactions. This entails steps 1, 10, or 11 for deamination of adenine and steps 4, 12, or 13 for guanine. Conversion of nucleotides (AMP, IMP, GMP, and xanthosine-5′-monophosphate [XMP]) to their respective nucleosides requires dephosphorylation (step 7) while metabolism of nucleosides to purine bases involves glycosidic bond cleavage (step 8 or 9). The enzymes participating in these processes in plants are listed with EC numbers and The Arabidopsis Information Resource (TAIR) Locus in Table 6.1. The interconversion of purine is performed by deamination (see Section 6.2), dephosphorylation (see Section 6.3), and glycosidic bond cleavage reactions (see Section 6.4).

6.2 Deamination Reactions

There is diversity in the deamination reactions in different species as they can occur at the nucleotide, nucleoside, or nucleobase levels (Henderson et al. 2008; Henderson and Paterson 1973). In plants, deamination adenine and guanine structures are catalysed,

Plant Nucleotide Metabolism – Biosynthesis, Degradation, and Alkaloid Formation, First Edition.
Hiroshi Ashihara, Iziar A. Ludwig and Alan Crozier.

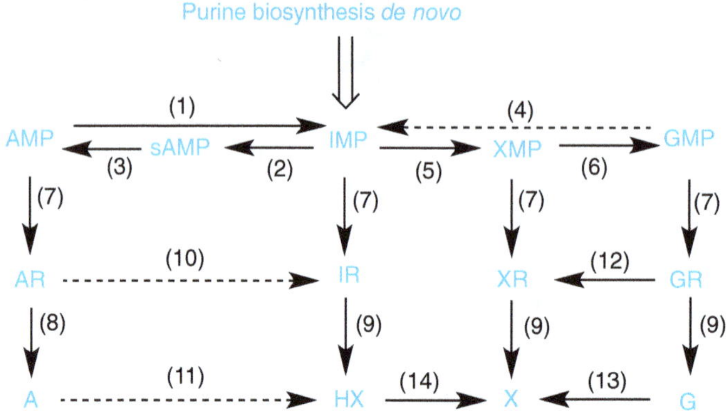

Figure 6.1 The interconversion of purines in plants. Abbreviations: A, adenine; AR, adenosine; G, guanine; GR, guanosine; HX, hypoxanthine; IR, inosine; X, xanthine; XR, xanthosine; sAMP, adenylosuccinate. Dashed arrows show non-active reactions in plants. Enzymes which catalyse steps (1)–(14) are shown in Table 6.1.

Table 6.1 Summary of enzymes involved in the purine interconversion in plants.

Step[a)]	Enzyme name	EC	Enzyme abbreviation	TAIR locus
1	AMP deaminase	3.5.4.6	AMPD	AT2G38280
2	Adenylosuccinate synthetase	6.3.4.4	ASS	AT3G57610
3	Adenylosuccinate lyase	4.3.2.2	ASL	AT1G36280 AT4G18840
4	GMP reductase	1.7.1.7	GMPR	NM
5	IMP dehydrogenase	1.1.1.205	IMPDH	AT1G79470 AT1G16350
6	GMP synthetase	6.3.5.2	GMPS	AT1G63660
7	5′-Nucleotidase	3.1.3.5		AT1G72880 etc.
8	Adenosine nucleosidase	3.2.2.7	AN	AT5G18860
9	Inosine/guanosine nucleosidase	3.2.2.2	IGN	AT5G18890
10	Adenosine deaminase	3.5.4.4	ADA	AT4G04880
11	Adenine deaminase	3.5.4.2		NM
12	Guanosine deaminase	3.5.4.15		NM
13	Guanine deaminase	3.5.4.3		AT3G05300 AT5G28050
14	Xanthine dehydrogenase	1.17.1.4	XDH	AT4G34900 etc.

NM: no matches found for search string of this EC number.

a) Steps in Figure 6.1.

respectively, by AMP deaminase (Table 6.1; step 1 in Figure 6.1) and guanosine deaminase (Table 6.1; step 12 in Figure 6.1).

6.2.1 Routes of Deamination of Adenine Ring

Adenine deaminase (EC 3.5.4.2) catalyses the reaction: adenine + $H_2O \rightarrow$ hypoxanthine + NH_3. The enzyme was discovered in *Azotobacter vinelandii* (Heppel et al. 1957) and occurs in various bacteria including *Escherichia coli* (Kamat et al. 2011). However, to date, no adenine deaminase activity has been detected in plants, including potato tubers (Katahira and Ashihara 2006; Wagner and Backer 1992).

Adenosine deaminase (EC 3.5.4.4) catalysed the reaction: adenosine + $H_2O \rightarrow$ inosine + NH_3. This enzyme was discovered in spores of *Bacillus cereus* (Powell and Hunter 1956) and is widely distributed in microorganisms and animals including humans. It is an important enzyme in humans as it functions in the development and maintenance of the immune system (Wilson et al. 1991).

Interestingly, adenosine deaminase is usually not detected in plant cells (Stasolla et al. 2003). Very low adenosine deaminase activity (<5 pkat mg^{-1} protein) has been detected in leaves and roots of alfalfa (*Medicago sativa*) (Edwards 1996). A putative adenosine deaminase from *Arabidopsis thaliana* was obtained as a recombinant protein by molecular cloning of the corresponding gene (*AT4G04880*) and overexpression in *E. coli*. The recombinant enzyme showed low level deaminase activity without obvious preference for adenine, adenosine, AMP, or ATP (PospíŠilová et al. 2008). These results strongly suggest that adenosine deaminase, if it occurs in plants, has only a minor function.

6.2.2 AMP Deaminase

AMP deaminase (EC 3.5.4.6) is an enzyme that converts AMP to IMP, releasing ammonia in the process. The reaction is shown Figure 6.2. This enzyme activity was originally found in skeletal rabbit muscle (Kalckar and Shafran 1947) and was purified by Lee (1957). Plant AMP deaminase was first characterized in pea seeds by Turner and Turner (1961).

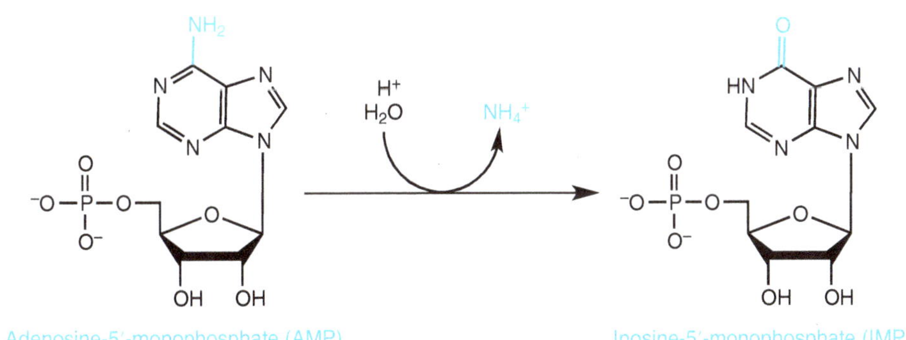

Figure 6.2 The AMP deaminase-catalysed conversion of adenosine-5′-monophosphate to inosine-5′-monophosphate (reaction 1 in Figure 6.1).

In the light of these findings it has been concluded that in plants AMP deaminase contributes mainly to the deamination of the adenine ring (Wagner and Backer 1992; Yabuki and Ashihara 1992; Yoshino and Murakami 1980). The kinetic properties of AMP deaminase in pea shoots (Dancer et al. 1997), spinach leaves (Yoshino and Murakami 1980), Jerusalem artichoke tubers (Le Floc'h and Lafleuriel 1983), and *Catharanthus roseus* cells (Yabuki and Ashihara 1991, 1992) have been reported and are summarized in Table 6.2.

The activity of partially purified AMP deaminase from cultured *C. roseus* cells has an absolute requirement for ATP. The Ka (50% activation) for ATP is 0.6 mM. GTP is a potent inhibitor as 50% inhibition was observed with 0.25 and 0.7 mM GTP in the presence of 2 and 5 mM ATP, respectively. The enzyme is highly specific for AMP as neither GMP nor ATP are metabolized (Yabuki and Ashihara 1992). The catabolism of AMP is strictly controlled by the cellular level of ATP. In addition, AMP deaminase participates in the conversion of adenylates to guanylates. ATP activates AMP deaminase and, furthermore, may stimulate the conversion of XMP to GMP. When the optimal level of GTP is reached, GTP inhibits the activity of AMP deaminase and, as a result, conversion of adenine nucleotides to guanine nucleotides may be prevented (Yabuki and Ashihara 1992).

6.2.3 Routes of Deamination of Guanine Ring

Guanine deaminase (EC 3.5.4.3), also known as guanase, catalyses the reaction: guanine + $H_2O \rightarrow$ xanthine + NH_3. Guanine is a constituent of fish scales and guanine deaminase activity was first observed in *Girella nigricans*, commonly known as rudder-fish (Hitchings and Falco 1944). This enzyme is distributed from bacteria (Rabinowitz and Barker 1956) to humans (Gupta and Glantz 1985). In plants, low activity has been detected in tea leaves (Negishi et al. 1994). In microorganisms and animals, guanine deaminase, but not guanosine deaminase, occurs as the principal deamination enzyme (Maynes et al. 2000; Nygaard et al. 2000; Yuan et al. 1999).

Table 6.2 Properties of plant AMP deaminase (AMPD), guanosine deaminase (GRD), and guanine deaminase.

Enzyme	Enzyme source	Optimum pH	Km values (mM)			References
			AMP	GR	G	
AMPD	Pea seeds	~6	13			Turner and Turner (1961)
	Pea shoots		0.4			Dancer et al. (1997)
	Spinach leaves		10			Yoshino and Murakami (1980)
	Jerusalem artichoke tubers		a)			Le Floc'h and Lafleuriel (1983)
	Catharanthus roseus cells	7.5–9.5	0.6			Yabuki and Ashihara (1992)
GRD	Tea leaves	7.5		0.01		Negishi et al. (1994)
	Arabidopsis thaliana			a)		Dahncke and Witte (2013)
GD	Tea leaves	~7.2 and 8.0			0.04	Negishi et al. (1994)

a) Activity was detected but no information was published on the Km value.

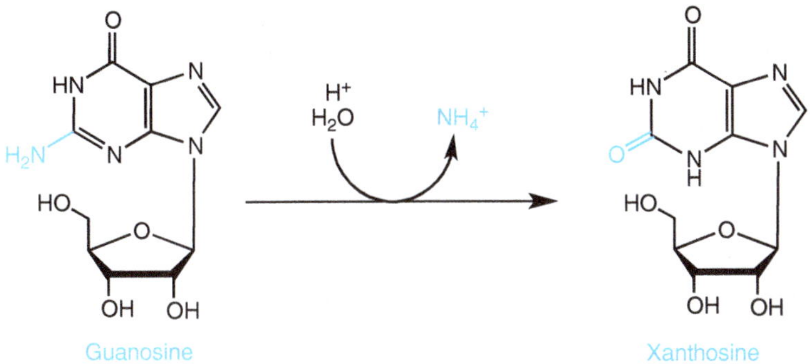

Figure 6.3 The guanosine deaminase-catalysed conversion guanosine to xanthosine (reaction 12 in Figure 6.1).

6.2.4 Guanosine Deaminase

Guanosine deaminase (EC 3.5.4.15) catalyses the conversion of guanosine to xanthosine as shown in Figure 6.3. It was first purified from *Pseudomonas convexa* and characterized by Ishida et al. (1969). Guanosine deaminase is the predominant deamination enzyme of the guanine ring in plants (Dahncke and Witte 2013; Negishi et al. 1994). As noted above, guanine deaminase has been detected in tea leaves, but its activity was ~30 times lower than that of guanosine deaminase (Negishi et al. 1994).

GMP reductase (EC 1.7.1.7, GMP oxidoreductase) (Table 6.1), which catalyses the conversion of GMP to IMP, has been detected in bacteria and animals (Andrews and Guest 1988; Mager and Magasanik 1960; Renart and Sillero 1974; Stephens and Whittaker 1973). It catalyses: $GMP + NADPH + H^+ \rightarrow IMP + NH_3 + NADP^+$. However, GMP reductase activity has not been detected in plants (Ashihara et al. 1991; Le Floc'h et al. 1982). These findings indicate that guanosine deaminase has the predominant role in deamination of guanine, and that participation of guanine deaminase, may be possible, but to a lesser degree.

Native plant guanosine deaminase from tea leaves has been partially purified and characterized (Negishi et al. 1994). Substrate specificity studies indicate that this enzyme is also able to deaminate 2′-deoxyguanosine and cytidine. However, the relative catalytic activity with these substrates are, respectively, 54% and 30% of that of guanosine. Recombinant guanosine deaminase of *A. thaliana* was obtained by Dahncke and Witte (2013). The kinetic properties of these enzyme are shown in Table 6.2. Like the tea enzyme, recombinant guanosine deaminase also deaminates guanosine and 2′-deoxyguanosine. Molecular genetic studies suggest that guanosine deaminase belongs to the cytidine/dCMP deaminase family of proteins in *A. thaliana*. Guanosine deaminase is conserved in plants, including the moss *Physcomitrella patens*, but is absent in the algae and species outwith the plant kingdom (Dahncke and Witte 2013).

6.3 Dephosphorylation Reactions

Dephosphorylation of purine nucleotides in plant extracts is catalysed by various enzymes including 5′-nucleotidase and some phosphatases. However, the identity of the enzyme(s) that dephosphorylate adenine and guanine nucleotides *in planta* have not as yet been clearly established. Plant extracts do, however, readily hydrolyse AMP, GMP, IMP, and XMP *in vitro* to their respective purine nucleosides, i.e. adenosine, guanosine, inosine, and xanthosine (Hirose and Ashihara 1984; Shimano and Ashihara 2006) (step 7 in Figure 6.1).

5′-Nucleotidase (EC 3.1.3.5) is widely distributed in animals, plants, and bacteria (Sharma et al. 1986). Soluble and membrane-bound 5′-nucleotidase have been found in plants (Polya and Ashton 1973; Sharma et al. 1986). The soluble wheat enzyme has a Km of 1.4 mM for 5′-AMP. It also hydrolyses 3′-AMP, but the Km value is higher at 6.1 mM (Polya and Ashton 1973). Membrane-bound 5′-nucleotidase has been detected in the plasma membrane (Sharma et al. 1986) and Golgi apparatus of peanut cotyledons (Mittal et al. 1988). The plasma membrane enzyme is specific for 5′-AMP. 5′-GMP is not hydrolysed (Sharma et al. 1986). Acid phosphatases (EC 3.1.3.2) are enzymes which catalyse the breakdown of a wide variety of phosphate esters and usually exhibit a pH optima below 6.0 (Ullah and Gibson 1988). The Km values for an artificial substrate, *p*-nitropheny phosphate, is 70 μM and for ATP is 125 μM, but no values for AMP and GMP are available. The enzymes mentioned above are examples of enzymes potentially involved in the dephosphorylation of purine nucleotides *in vivo*.

6.4 Glycosidic Bond Cleavage Reactions

Glycosidic bond cleavage, which releases ribose from ribonucleosides, is accomplished either hydrolytically or phosphorolytically. In plants, hydrolytic cleavage by nucleosidases is common, while phosphorolytic cleavage of nucleosides by nucleoside phosphorylases is the major reaction in mammals (Ashihara et al. 2018).

Nucleosidases (nucleoside hydrolases) are a class of enzymes that hydrolyse the *N*-glycosidic bond between the base and sugar of selected nucleosides. They have been isolated from a number of sources including bacteria, parasitic protozoans, plants, marine invertebrates, and baker's yeast, but not mammals (Arivett et al. 2014). Following traditional enzyme purification and characterization studies using various plants sources, two distinct adenosine nucleosidase (ARN, adenosine ribohydrolase, EC3.2.2.7) and inosine/guanosine nucleosidase (IR/GRN, inosine ribohydrolase, EC 3.2.2.2) have been reported (Table 6.3).

Arabidopsis thaliana recombinant nucleosidases, obtained using the homologs of the well-characterized nucleosidase of *Crithidia fasciculate*, a species of parasitic excavates, were used to generate recombinant proteins, which then had their enzymatic capacity assessed. Recombinant enzymes expressed in *E. coli* and yeast have multiple substrate specificities and are called uridine ribohydrolase (URH) or nucleoside ribohydrolase (NRH) (Table 6.4). These recombinant enzymes have as yet not been detected as native proteins from plants.

6.4.1 Adenosine Nucleosidase

Adenosine nucleosidase (adenosine ribohydrolase, EC 3.2.2.7) catalyses the hydrolysis of adenosine to adenine (Figure 6.4). This enzyme was first discovered in Brussels sprouts

Table 6.3 Properties of the native adenosine nucleosidase (ARN), inosine/guanosine nucleosidase (IR/GRN), and non-specific purine nucleosidase (PRN) from plants.

Enzyme	Enzyme source	Optimum pH	Km values (μM)				References
			AR	IR	GR	XR	
ARN	Brussels sprouts	3.5–4.5	2400				Mazelis and Creveling (1963)
	Spinach beet leaves	4.5	11				Poulton and Butt (1976)
	Barley leaves	4.7–5.4	0.8–2.3				Guranowski and Schneider (1977)
	Tea leaves	4.0–4.5	[a]				Imagawa et al. (1979)
	Wheat germ	4.7	1.4				Chen and Kristopeit (1981)
	Yellow lupin seeds	7.5	4.8				Abusamhadneh et al. (2000)
	Coffee leaves	6.0	6.3				Campos et al. (2005)
	Jerusalem artichoke tubers	5–7	17				Le Floc'h and Lafleuriel (1981b)
	Potato						Riewe et al. (2008)
IR/GRN	Jerusalem artichoke tubers	5–7		2.5	8.5		Le Floc'h and Lafleuriel (1981b)
	Yellow lupin seeds	8		65			Guranowski (1982)
	Yellow lupin seeds	4.7–5.5		2.7	2.7		Szuwart et al. (2006)
PRN	Cowpea nodules	8.0	[a]	0.8	[a]	0.8	Atkins et al. (1989)

a) Activity was detected using this nucleoside, but the Km was not measured.

Table 6.4 Properties of plant nucleosidases.

Enzyme	Gene source	Optimum pH	Km values (μM)						References
			AR	IR	GR	XR	UR	iPAR	
URH1	*Arabidopsis thaliana*		700	1400			800		Jung et al. (2009)
ZmNRH3	Maize	7–9	39	201	104	396	1030	196	Kopečná et al. (2013)
PpNHR1	*Physcomitrella patens*	7–9	113	78	61	116	890		Kopečná et al. (2013)

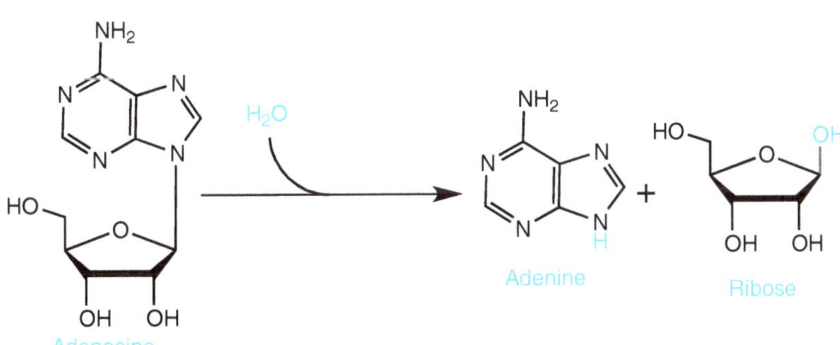

Figure 6.4 The adenosine nucleosidase-catalysed conversion of adenosine to adenine (reaction 8 in Figure 6.1).

(*Brassica oleracea*) by Mazelis and Creveling (1963). It has subsequently been purified from wheat germ, yellow lupin seed, Jerusalem artichoke tubers, and leaves of spinach beet, barley, tea, and coffee (Table 6.3). Cell wall-bound adenosine nucleosidase, which may be involved in the salvage of extracellular ATP, has also been detected potato tubers (Riewe et al. 2008). The Km values and optimum pH are presented in Table 6.3. Most enzymes are specific for adenosine, and do not hydrolyse other purine and pyrimidine nucleosides. The lack of activity with inosine, suggests that a 6-amino group is a prerequisite for activity and, possibly, is required for one site of attachment to the enzyme (Mazelis and Creveling 1963).

6.4.2 Inosine/Guanosine Nucleosidase

Inosine/guanosine nucleosidase (inosine/guanosine ribohydrolase, EC 3.2.2.2) catalyses the reaction: inosine/guanosine + $H_2O \rightarrow$ hypoxanthine/guanine + D-ribose. The activity has been found in *E. coli* (Koch and Lamont 1956), fish (Tarr 1955), and plants (Deng and Ashihara 2010; Katahira and Ashihara 2006). An enzyme, with a slightly different substrate specificity, has been detected in Jerusalem tubers (Le Floc'h and Lafleuriel 1981a) and yellow lupin seeds (Guranowski 1982; Szuwart et al. 2006) (Table 6.3). Guranowski and his co-workers reported two distinct enzymes which hydrolyse inosine and guanosine. Guranowski (1982) partially purified inosine nucleosidase from yellow lupin seeds that catalysed the hydrolysis of adenosine and guanosine but at lower rates than inosine. Subsequently, this group reported on a highly purified guanosine/inosine-preferring nucleosidase that was also isolated from yellow lupin seeds (Szuwart et al. 2006).

6.4.3 Non-specific Purine Nucleosidases

An enzyme with broad substrate specificity has been purified to homogeneity from the soluble cell fraction of N_2-fixing nodules of the tropical legume cowpea (*Vigna unguiculata*) (Atkins et al. 1989). Among the purine nucleosides, the Vmax values are in the ratio 28 : 7 : 1 : 0.4, respectively, for xanthosine, inosine, adenosine, and guanosine. The enzyme can, therefore, be classified as a purine nucleosidase (PRN) (nucleoside hydrolase; EC 3.2.2.1), however, it also had significant activity with xanthosine and inosine, so it can be referred to as a xanthine/inosine deaminase.

6.4.4 Recombinant Non-Specific Nucleosidases

Some studies on plant nucleosidases have been carried out using genes of *A. thaliana* with recombinant enzymes being expressed in *E. coli* or yeast (Jung et al. 2009; Kopečná et al. 2013). Jung et al. (2009) describe the cloning of a nucleosidase from *A. thaliana*, URH1 which was characterized by complementation of a yeast mutant. Furthermore, URH1 was synthesized as a recombinant protein in *E. coli*. The purified recombinant protein exhibited hydrolase activity for both pyrimidine and purine nucleosides (Table 6.4).

A much higher Km value of 2.1 mM was found for uridine with the recombinant enzyme expressed in yeast. These values are ~1000 times higher than those of most native adenosine nucleosidase and inosine/guanosine nucleosidase (see Table 6.3). In a subsequent paper, it was reported that xanthosine, as well as uridine, were better substrates for NSH1 (formerly designation URH1) than inosine and adenosine (Jung et al. 2011). The apparent Km value for NSH1 for xanthosine cleavage (1.7 mM) is high, and the catalytic activity with xanthosine is sixfold lower than with uridine. However, extremely low affinity of substrates and broad substrate specificity observed with this recombinant enzyme may reflect differences in the post-translational modification of the protein expressed in *E. coli*, yeast, and in plants.

Characterization of recombinant nucleosidases from *Physcomitrella patens* (PpNRH) and maize (ZmNRH) has been reported (Kopečná et al. 2013). The substrate specificity of these enzymes are broad, as they can hydrolyse xanthosine, inosine, adenosine, guanosine, uridine, and cytokinin-ribosides with varying efficiencies. Two isoforms of nucleosidases were found; one isoform preferentially used inosine and xanthosine, and the other hydrolysed uridine and xanthosine. The Km values of typical recombinant enzymes (0.1–0.4 mM) are much higher than those observed with native nucleosidases (1–10 μM) in plants. At present, there are big gaps between the information obtained from these recombinant enzymes and the nucleosidases *in planta*.

6.5 *In Situ* Metabolism of ^{14}C-Labelled Purine Nucleotides

In addition to enzyme studies, information from feeding experiments is useful to reveal the interconversion of purine metabolites *in planta*.

6.5.1 Metabolism of Adenine Nucleotides

To investigate the *in situ* metabolic fate of purine nucleotides, especially adenylates, cell cultures are often used. Since nucleotides are not taken up efficiently by cells, the cultures were pre-labelled with ^{14}C-labelled nucleosides or nucleobases for a short period (~1 hour). The labelled cells are then transferred to a ^{14}C-precursor-free medium and the radioactivity in metabolites chased. This pre-incubation method, using labelled adenosine or adenine, has often been used for studies on the interconversion and catabolism of adenine nucleotides in mammalian cells (Altschuld et al. 1987; Zoref-Shani et al. 1987). In plants, for example, *C. roseus* cells, ~90% of radioactivity was distributed in nucleotides and nucleic acid during the pre-incubation period and the predominant labelled product was ATP (Yabuki and Ashihara 1991). When the cells were incubated with non-labelled 50 μM adenosine, ~70% of label disappeared from nucleotides within one hour, and concomitantly, radioactivity increased in catabolites, most notably in CO_2. The release of $^{14}CO_2$ from [8-^{14}C]adenosine was inhibited almost completely when the cells were incubated with allopurinol, an inhibitor of xanthine oxidoreductase, and instead radioactivity accumulated in xanthine (Yabuki and Ashihara 1991). These observations strongly suggest that turnover of adenine nucleotides is rapid and if there is a continuing supply of sufficient substrate, adenine nucleotides are rapidly catabolized.

The catabolism of adenine nucleotides is initiated by deamination and dephosphorylation reactions. Several pathways have been proposed for different organs in mammals (Henderson and Paterson 1973). Theoretically, the following pathways may operate in plant cells:

1. AMP → IMP → inosine → hypoxanthine
2. AMP → IMP → XMP → xanthosine → xanthine
3. AMP → adenosine → inosine → hypoxanthine
4. AMP → adenosine → adenine → hypoxanthine

Treatment of cultured *C. roseus* cells with coformycin, an inhibitor of adenosine deaminase and AMP deaminase (Klenow and Østergaard 1988), results in the accumulation of AMP but not adenosine (Yabuki and Ashihara 1991). Therefore, reactions 3 and 4 can be eliminated. On the other hand, tiazofurin, an inhibitor of IMP dehydrogenase which catalyses conversion of IMP to XMP (Lui et al. 1984), has little or no effect. This eliminates reaction 2 and indicates that reaction 1 is the most plausible pathway in plants (Yabuki and Ashihara 1991).

Adenine nucleotides are also utilized in nucleic acid synthesis. The radioactivity from [8-^{14}C]adenosine incubated with *C. roseus* cells is incorporated into both adenine and guanine residues of nucleic acids. This suggests that an AMP → IMP → XMP → GMP → GDP → GTP → RNA route functions in plant cells (Ashihara et al. 2000).

6.5.2 Metabolism of Guanine Nucleotides

To investigate the metabolic fate of guanine nucleotides in plants, *in situ* metabolism of ^{14}C-guanosine was studied (Ashihara et al. 1997). The pulse-chase experiments using liquid-cultured *C. roseus* cells revealed that some guanosine was converted to guanine nucleotides (pathway 1) while the remainder was catabolized (pathway 2). This is due to the activity of guanosine deaminase.

1. Guanosine → GMP → GDP → GTP → RNA
2. Guanosine → xanthosine → xanthine → uric acid → purine catabolites

These findings indicate that guanosine is an intermediate of both salvage and deamination reactions. Guanosine catabolism is described in Section 6.6.2.

In contrast to adenine nucleotides, no radioactivity from guanosine was incorporated into the adenine residue of nucleic acids. *In situ* experiments indicate that no conversion of GMP → AMP occurs *in planta* (Katahira and Ashihara 2006).

6.6 *In Situ* Metabolism of Purine Nucleosides and Bases

In situ metabolism of purine nucleosides and bases has been investigated in many plant cells and tissues (Table 6.5). The metabolic fate of ^{14}C-labelled compounds varies in different organs and plant species, while the metabolic profiles of individual purine nucleosides and nucleobases tend to be similar. In purine alkaloid-forming plants,

Table 6.5 Plant materials used for *in situ* ^{14}C-purine tracer experiments. Used tracers are of adenine (A), adenosine (AR), guanine (G), guanosine (GR), hypoxanthine (HX), inosine (IR), xanthine (X), xanthosine (XR), deoxyadenosine (AdR), deoxyguanosine (GdR), and aminoimidazole carboxamide ribonucleoside (AICAR).

Plant species	Plant name	Material	^{14}C-Purine	References
Ginkgo biloba	Ginkgo tree	Leaves	A, HX	Ashihara and Nobusawa (1981)
Picea glauca	White spruce	Cell culture	A, AR, IR	Ashihara et al. (2000)
Arabidopsis thaliana		Cell culture	A, AR, HX, IR, G, GR, X, XR	Yin et al. (2014)
Brassica napus	Canola	Embryos	A, IR	Ashihara et al. (2008)
Daucus carota	Carrot	Roots	A, HX	Ashihara and Nobusawa (1981)
Acer buergerianum	Trident maple	Leaves	A, HX	Ashihara and Nobusawa (1981)
Solanum tuberosum	Potato	Tubers	A, AR, HX, IR, XR, X, AdR, GdR	Katahira and Ashihara (2006)
Nicotiana tabacum	Tobacco	Leaves, Callus	A, HX	Ashihara and Nobusawa (1981)
Pisum sativum	Pea	Seedlings	A, HX	Ashihara and Nobusawa (1981)
Vigna (=Phaseolus) mungo	Black gram	Seedlings	A, AR, HX, IR, G, GR	Ashihara (1983), Ashihara et al. (1997), Nobusawa and Ashihara (1983)
Vigna radiate (=Phaseolus aureus)	Mungbean	Seedlings	XR, X	Ashihara (2012)
Catharansus roseus	Madagascar periwinkle	Cell culture, leaves	A, AR, HX, IR, G, GR	Ashihara et al. (1997), Hirose and Ashihara (1983)
Rhizophora stylosa	Spotted mangrove	Leaves	AR	Ashihara et al. (2003)
Bruguiera sexangula	Upriver orange mangrove	Cell culture	AR, HX, GR, AICAR	Suzuki-Yamamoto et al. (2006)
Bruguiera gymnorrhiza	Black mangrove	Leaves, propagules	A, AR	Ashihara et al. (2003), Yin et al. (2012)
Kandelia obovata (=K. candel)	Mehirugi mangrove	Leaves	AR	Ashihara et al. (2003)
Avicennia marina	Grey mangrove	Leaves	AR	Ashihara et al. (2003)
Populus alba	Silver poplar	Leaves	AR	Ashihara et al. (2003)
Prunus persica	Peach	Leaves	A, AR	Le Floc'h and Faye (1995)
Camellia sinensis	Tea	Leaves, roots, flowers, fruits Tissue culture	A, AR, HX, IR, G, GR, X, XR, AdR, GdR, AICAR	Deng and Ashihara (2010), Deng et al. (2008), Fujimori and Ashihara (1990, 1993), Ito and Ashihara (1999), Negishi et al. (1992), Suzuki and Takahashi (1975), Terrasaki et al. (1994)
Camellia ptilophylla	Cocoa tea	Leaves	A, X	Ashihara et al. (1998)
Ilex paraguariensis	Maté	Leaves	A, GR, HX, X	Ashihara (1993), Yin et al. (2015)
Theobroma cacao	Cacao	Leaves	A, AR, HX, IR, G, GR, X, XR	Koyama et al. (2003)
Coffea arabica	Arabica coffee	Leaves, fruits	A, G, IR, XR	Ashihara et al. (1996), Keya et al. (2003), Koshiro et al. (2006)
Coffea canephora	Robusta coffee	Fruits	A	Koshiro et al. (2006)
Coffea eugenioides		Leaves	A	Ashihara and Crozier (1999)
Coffea millotii	Mascarocoffea	Leaves	A, IR, GR, XR	Deng et al. (2017)

theobromine and caffeine are synthesized mainly from xanthosine. As examples, the metabolic fate of four nucleosides and four nucleobases in growing tubers of potato and theobromine-forming leaves of cacao are shown in Tables 6.6 and 6.7, respectively.

6.6.1 Metabolism of Adenine and Adenosine

The metabolic fate of ^{14}C-adenine and adenosine is similar. ^{14}C-Labelled adenosine and adenine are both converted to AMP, and enter the adenylate pool which consists of ATP, ADP, and AMP. Significant amounts of radioactivity are then incorporated into nucleic acids, mainly RNA. Turnover of adenine nucleotides in plant cells is rapid, and after longer incubation periods, typically 18 hours, radioactivity is incorporated into degradation products of purine catabolism, such as allantoin, allantoic acid, and CO_2 (Table 6.7). Conversion of adenine nucleotides to guanine nucleotides also occurs. Nearly 30% of radioactivity from adenine and adenosine is incorporated into the guanine residues of RNA (Table 6.6). In purine alkaloid-forming plants, adenine and adenosine are good precursors for theobromine and caffeine. Methylxanthine biosynthesis starts from xanthosine, with AMP being converted to xanthosine via IMP. Adenine and adenine are

Table 6.6 Metabolic fate of [8-^{14}C]adenine (A), adenosine (AR), guanine (G), guanosine (GR), hypoxanthine (HX), inosine (IR), xanthine (X), and xanthosine (XR) in growing potato tubers.

^{14}C-Metabolites (%)	A	AR	G	GR	HX	IR	X	XR
Salvage product								
Nucleotides	74.3	64.0	26.0	29.0	2.9	19.0	nd	nd
RNA								
A-residue	8.5	5.7	nd	nd	nd	nd	nd	nd
G-residue	3.1	2.2	24.2	29.1	4.6	14.8	nd	nd
Nucleosides and bases								
IR	nd	nd	nd	nd	nd	6.3	nd	nd
HX	nd	nd	nd	nd	0.8	4.7	nd	nd
X + XR	nd	nd	nd	nd	1.5	nd	8.4	33.8
A + AR	13.5	27.3	nd	nd	nd	nd	nd	nd
G	nd	nd	47.9	nd	nd	nd	nd	nd
GR	nd	nd	nd	37.7	nd	nd	nd	nd
Catabolites								
Allantoin	nd	0.1	nd	1.9	41.1	13.2	32.4	17.0
Allantoic acid	nd	0.1	nd	nd	31.4	35.8	47.6	25.5
Urea	nd	nd	nd	nd	5.0	4.7	5.3	2.7
Carbon dioxide	0.2	0.1	0.5	1.4	1.6	0.5	2.5	7.0
Others	0.5	0.5	1.4	0.9	1.0	1.0	3.8	14.0

nd, not detected. Incubation time was 4 hours. Values are expressed as % of radioactivity and total uptake.
Source: Adapted from Katahira and Ashihara (2006).

Table 6.7 Metabolic fate of [8-^{14}C]adenine (A), adenosine (AR), guanine (G), guanosine (GR), hypoxanthine (HX), inosine (IR), xanthine (X), and xanthosine (XR) in young leaves of cacao plants.

^{14}C-Metabolites (%)	A	AR	G	GR	HX	IR	X	XR
Salvage products								
Nucleotides	16.1	17.5	5.9	8.7	5.7	7.7	nd	nd
RNA	28.7	29.8	24.4	30.6	29.3	33.3	nd	nd
Nucleosides and bases								
IR	1.1	1.1	nd	nd	nd	nd	nd	nd
HX	0.5	0.5	nd	nd	0.8	1.1	nd	nd
X + XR	0.7	0.7	nd	nd	1.5	2.1	6.7	4.9
A	4.9	5.1	nd	nd	nd	nd	nd	nd
AR	9.0	10.3	nd	nd	nd	nd	nd	nd
G + GR	nd	nd	1.9	3.7	nd	nd	nd	nd
Purine alkaloids								
3mX	0.3	0.3	0.6	1.7	1.0	1.0	0.9	2.5
7mX + 7mXR	1.9	1.6	1.3	0.8	0.6	0.6	nd	nd
Theobromine	6.7	7.8	7.6	6.9	12.3	8.8	8.7	16.2
Caffeine	nd	nd	nd	nd	nd	nd	nd	nd
Catabolites								
Ureides	15.9	11.8	31.6	33.2	1.7	2.9	59.7	37.8
Carbon dioxide	13.1	10.3	26.8	13.4	46.1	41.3	22.2	35.4
Others	0.3	0.2	0.0	0.0	0.5	0.8	0.6	1.1

nd, not detected. 3mX, 3-methylxanthine; 7mX, 7-methylxanthine; 7mXR, 7-methylxanthosine. Incubation time was 18 hours. Values are expressed as % of radioactivity.
Source: Adapted from Koyama et al. (2003).

not deaminated directly which is in keeping with *in vitro* enzyme profiles showing an absence of adenine deaminase and adenosine deaminase activity (see Section 6.2.1). An outline of adenine and adenosine metabolism in plants (Table 6.5) is presented in Figure 6.5.

6.6.2 Metabolism of Guanine and Guanosine

Compared with adenine and adenosine, salvage velocity of guanine and guanosine is slower. The rates of incorporation of radioactivity from ^{14}C-labelled guanine and guanosine into RNA are, however, often higher than from ^{14}C-labelled adenine and adenosine. This is a consequence of the smaller size of the guanine nucleotide pool (see Table 2.1). Radioactivity is incorporated predominantly into the guanine residues of RNA. This is in agreement with the absence of GMP dehydrogenase in plants (see Section 6.2.1). Catabolic activity of guanosine and guanine varies with plant species and different organs. As shown in Tables 6.6 and 6.7, these compounds are catabolized in

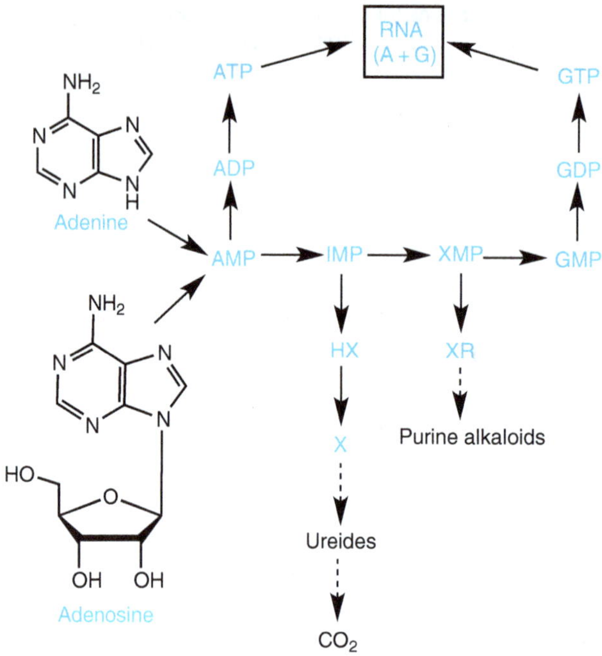

Figure 6.5 *In situ* metabolism of adenine and adenosine in plants. Abbreviations: A, adenine; G, guanine; HX, hypoxanthine; X, xanthine; XR, xanthosine.

young cacao leaves but not in young developing potato tubers (Katahira and Ashihara 2006; Koyama et al. 2003). Outlines of guanine and guanosine metabolism are presented in Figure 6.6.

6.6.3 Metabolism of Hypoxanthine and Inosine

Hypoxanthine and inosine are precursors for purine salvage as well as purine catabolism. In general, salvage activity for these compounds are much lower than those of adenine, adenosine, guanine, and guanosine (Table 6.6), but in some plant materials, considerably higher salvage activity is observed (Table 6.7). Little or no radioactivity from ^{14}C-labelled hypoxanthine and inosine is incorporated into the adenine residues of RNA. Conversion of IMP formed from hypoxanthine and inosine does not occur *in planta* (Table 6.6). In most plant materials, hypoxanthine and inosine appear to be converted into xanthine which is further metabolized by the conventional purine catabolism pathway (see Chapter 7). An outline of metabolism of hypoxanthine and inosine is shown in Figure 6.7.

6.6.4 Metabolism of Xanthine and Xanthosine

Xanthine and xanthosine are not utilized for nucleotide and nucleic acid biosynthesis, and the radioactivity from ^{14}C-xanthine and ^{14}C-xanthosine is incorporated into the

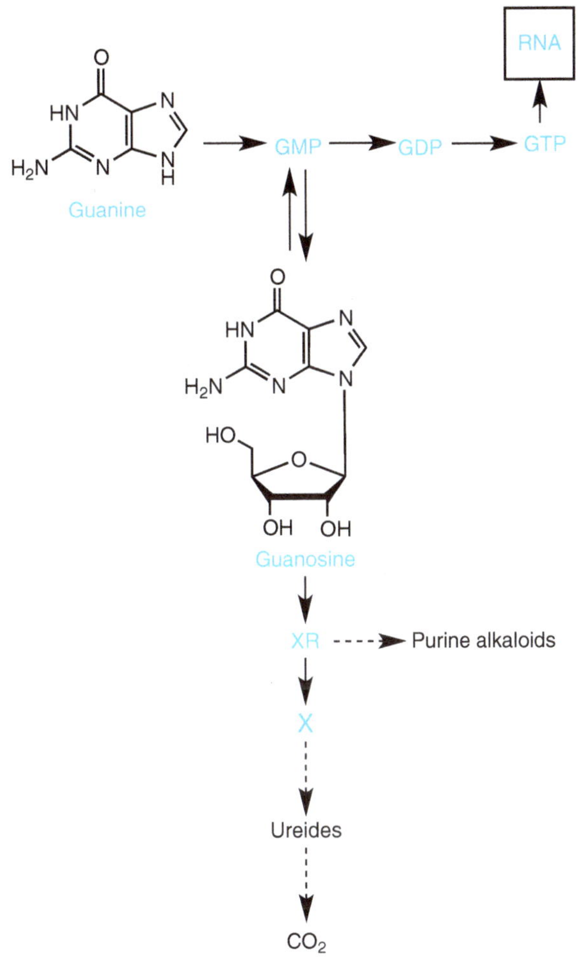

Figure 6.6 *In situ* metabolism of guanine and guanosine in plants. Abbreviations: X, xanthine; XR, xanthosine.

ureides, allantoin, and allantoic acid. After longer incubation periods, >24 hours, most of radioactivity is released as $^{14}CO_2$. Although very low activity of xanthine phosphoribosyltransferase have been demonstrated in tea leaf extracts (Deng and Ashihara 2010). An outline of xanthine and xanthosine metabolism is shown in Figure 6.8.

6.6.5 Metabolism of Deoxyadenosine and Deoxyguanosine

Only limited investigations have been carried out on purine deoxynucleotide metabolism in plants. Katahira and Ashihara (2006) reported the metabolic fate of ^{14}C-labelled deoxyadenosine and deoxyguanosine incubated with young potato tubers. Deoxyadenosine was salvaged to nucleotides and nucleic acids. Very little radioactivity was found in DNA, with the vast majority of radioactivity being recovered in RNA (Table 6.8).

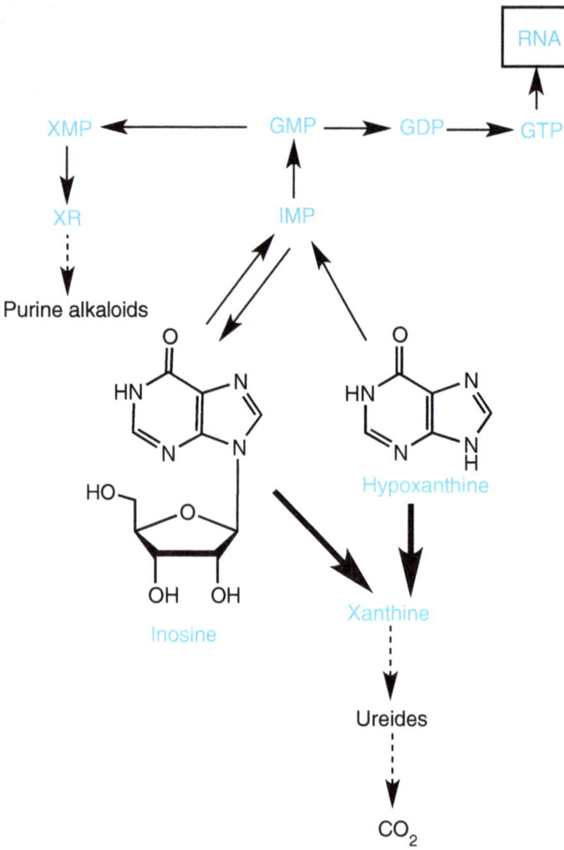

Figure 6.7 *In situ* metabolism of hypoxanthine and inosine in plants. Abbreviation: XR, xanthosine.

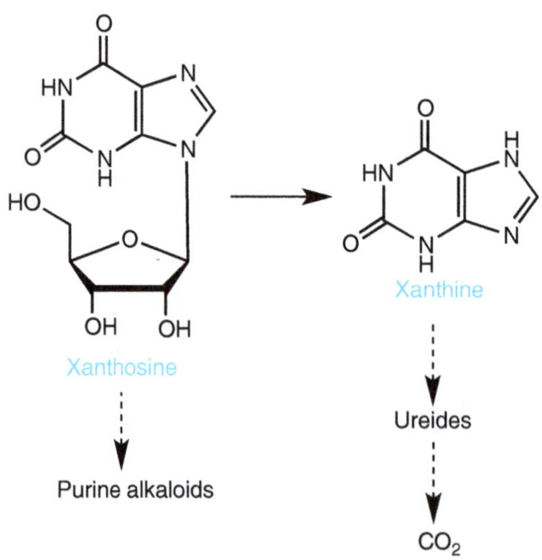

Figure 6.8 *In situ* metabolism of xanthine and xanthosine in plants.

Table 6.8 Metabolic fate of [8-^{14}C]deoxyladenosine (AdR) and deoxyguanosine (GdR) in growing potato tubers.

^{14}C-Metabolites (%)	AdR	GdR
Purine metabolites		
ATP + ADP	45.5	nd
GTP + GDP	nd	24.2
AMP	1.9	nd
dAMP	0.8	nd
GMP	nd	1.0
dGMP	nd	3.0
RNA		
A-residue	36.2	nd
G-residue	3.1	46.1
DNA	0.4	1.8
A + AR	5.1	nd
AdR	5.6	nd
G	nd	4.0
GdR	nd	16.5
Catabolites		
Allantoin	1.3	2.4
Allantoic acid	nd	nd
Urea	nd	nd
Carbon dioxide	nd	0.6
Others	0.1	0.4

nd, not detected. A, adenine; AR, adenosine; G, guanine.
Incubation time – four hours. Values are expressed as % of radioactivity and total uptake.
Source: Adapted from Katahira and Ashihara (2006).

Only a limited amount of deoxyadenosine appears to be directly salvaged to dAMP, as it was mainly salvaged to AMP by adenine phosphoribosyltransferase, after conversion to adenine. In a similar manner, ^{14}C-deoxyguanosine was salvaged to GMP and ~50% of radioactivity was found in guanine residues of RNA. Only limited radioactivity, <2%, was incorporated into DNA. Catabolism of both deoxyadenosine and deoxyguanosine was negligible during the four hours incubation period. *In vitro* enzyme assay indicates that deoxyadenosine hydration activity by adenosine nucleosidase is high, but deoxyguanosine hydration activity is very low. Deoxyguanosine degradation enzymes have not been studied in detail in any organism. Outlines of deoxyadenosine and deoxyguanosine metabolism are shown in Figure 6.9.

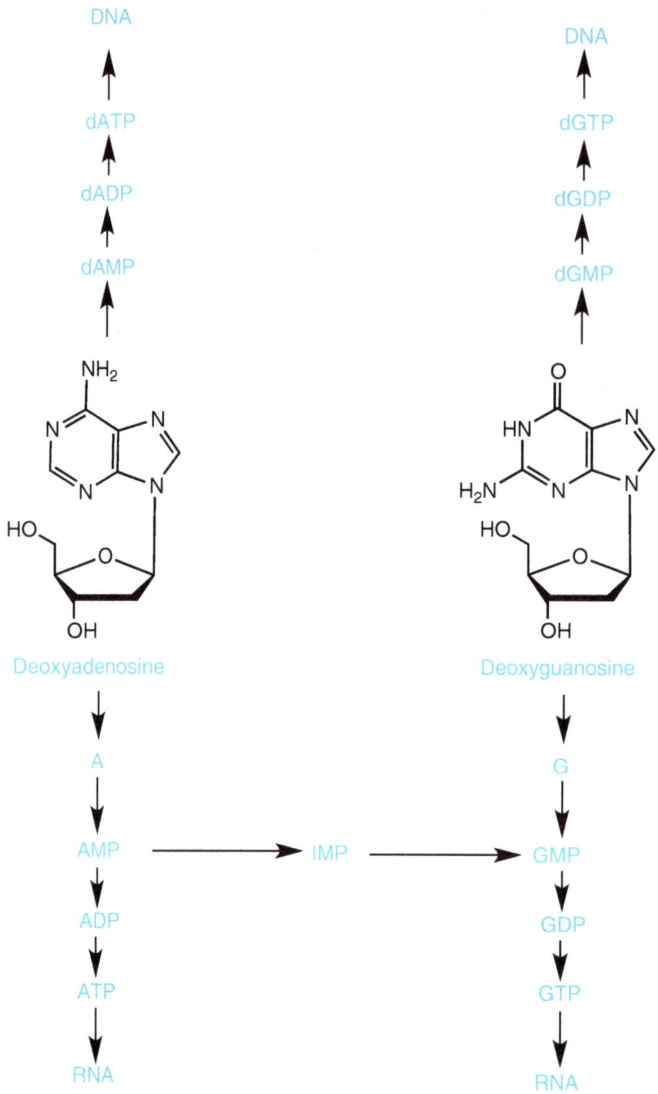

Figure 6.9 *In situ* metabolism of deoxyadenosine and deoxyguanosine in plants. Abbreviations: A, adenine; G, guanine.

6.7 Summary

A diversity of purine interconversions occur in different organisms. In plants, adenine nucleotides are converted to guanine nucleotides by an AMP → IMP → XMP → GMP pathway. The reverse conversion of guanine nucleotides to adenine nucleotides, does not occur in plants. A major route of xanthine formation from AMP is AMP → IMP → inosine → hypoxanthine → xanthine, while a major route from GMP is GMP → guanosine → xanthosine → xanthine. In plants, key enzymes for purine interconversions are AMP deaminase and guanosine deaminase.

References

Abusamhadneh, E., McDonald, N.E., and Kline, P.C. (2000). Isolation and characterization of adenosine nucleosidase from yellow lupin (*Lupinus luteus*). *Plant Sci.* 153: 25–32.

Altschuld, R.A., Gamelin, L.M., Kelley, R.E. et al. (1987). Degradation and resynthesis of adenine nucleotides in adult rat heart myocytes. *J. Biol. Chem.* 262: 13527–13533.

Andrews, S.C. and Guest, J.R. (1988). Nucleotide sequence of the gene encoding the GMP reductase of *Escherichia coli* K12. *Biochem. J.* 255: 35–43.

Arivett, B., Farone, M., Masiragani, R. et al. (2014). Characterization of inosine-uridine nucleoside hydrolase (*RihC*) from *Escherichia coli*. *Biochim. Biophys. Acta* 1844: 656–662.

Ashihara, H. (1983). Changes in activities of purine salvage and ureide synthesis during germination of black gram (*Phaseolus mungo*) seeds. *Z. Pflanzenphysiol.* 113: 47–60.

Ashihara, H. (1993). Purine metabolism and the biosynthesis of caffeine in maté leaves. *Phytochemistry* 33: 1427–1430.

Ashihara, H. (2012). Xanthosine metabolism in plants: metabolic fate of exogenously supplied ^{14}C-labelled xanthosine and xanthine in intact mungbean seedlings. *Phytochem. Lett.* 5: 100–103.

Ashihara, H. and Crozier, A. (1999). Biosynthesis and catabolism of caffeine in low-caffeine-containing species of *Coffea*. *J. Agric. Food. Chem.* 47: 3425–3431.

Ashihara, H., Kato, M., and Ye, C.X. (1998). Biosynthesis and metabolism of purine alkaloids in leaves of cocoa tea (*Camellia ptilophylla*). *J. Plant Res.* 111: 599–604.

Ashihara, H., Luit, B., Belmonte, M., and Stasolla, C. (2008). Metabolism of nicotinamide, adenine and inosine in developing microspore-derived canola (*Brassica napus*) embryos. *Plant Physiol. Biochem.* 46: 752–759.

Ashihara, H., Mitsui, K., Yabuki, N., and Nygaard, P. (1991). Adenosine metabolism and growth of adenosine-requiring mutant cells of *Datura innoxia*. *Int. J. Purine Pyrim. Res.* 2: 129–135.

Ashihara, H., Monteiro, A.M., Gillies, F.M., and Crozier, A. (1996). Biosynthesis of caffeine in leaves of coffee. *Plant Physiol.* 111: 747–753.

Ashihara, H. and Nobusawa, E. (1981). Metabolic fate of [8-^{14}C]adenine and [8-^{14}C]hypoxanthine in higher plants. *Z. Pflanzenphysiol.* 104: 443–458.

Ashihara, H., Stasolla, C., Fujimura, T., and Crozier, A. (2018). Purine salvage in plants. *Phytochemistry* 147: 89–124.

Ashihara, H., Stasolla, C., Loukanina, N., and Thorpe, T.A. (2000). Purine and pyrimidine metabolism in cultured white spruce (*Picea glauca*) cells: metabolic fate of ^{14}C-labeled precursors and activity of key enzymes. *Physiol. Plant.* 108: 25–33.

Ashihara, H., Takasawa, Y., and Suzuki, T. (1997). Metabolic fate of guanosine in higher plants. *Physiol. Plant.* 100: 909–916.

Ashihara, H., Wakahara, S., Suzuki, M. et al. (2003). Comparison of adenosine metabolism in leaves of several mangrove plants and a poplar species. *Plant Physiol. Biochem.* 41: 133–139.

Atkins, C.A., Storer, P.J., and Shelp, B.J. (1989). Purification and properties of purine nucleosidase from N_2 -fixing nodules of cowpea (*Vigna unguiculata*). *J. Plant Physiol.* 134: 447–452.

Brown, G.B., Roll, P.M., Plentl, A.A., and Cavalieri, L.F. (1948). The utilization of adenine for nucleic acid synthesis and as a precursor of guanine. *J. Biol. Chem.* 172: 469–484.

Campos, A., Rijo-Johansen, M.J., Carneiro, M.F., and Fevereiro, P. (2005). Purification and characterisation of adenosine nucleosidase from *Coffea arabica* young leaves. *Phytochemistry* 66: 147–151.

Chen, C.M. and Kristopeit, S.M. (1981). Metabolism of cytokinin: Deribosylation of cytokinin ribonucleoside by adenosine nucleosidase from wheat germ cells. *Plant Physiol.* 68: 1020–1023.

Dahncke, K. and Witte, C.P. (2013). Plant purine nucleoside catabolism employs a guanosine deaminase required for the generation of xanthosine in Arabidopsis. *Plant Cell* 25: 4101–4109.

Dancer, J.E., Hughes, R.G., and Lindell, S.D. (1997). Adenosine-5′-phosphate deaminase. A novel herbicide target. *Plant Physiol.* 114: 119–129.

Deng, W.-W. and Ashihara, H. (2010). Profiles of purine metabolism in leaves and roots of *Camellia sinensis* seedlings. *Plant Cell Physiol.* 51: 2105–2118.

Deng, W.W., Li, Y., Ogita, S., and Ashihara, H. (2008). Fine control of caffeine biosynthesis in tissue cultures of *Camellia sinensis*. *Phytochem. Lett.* 1: 195–198.

Deng, W.-W., Rakotomalala, J.J., Nagai, C., and Ashihara, H. (2017). Caffeine biosynthesis and purine metabolism in leaves of *Mascarocoffea* species. *Eur. Chem. Bull.* 6: 223–228.

Edwards, R. (1996). *S*-adenosyl-L-methionine metabolism in alfalfa cell cultures following treatment with fungal elicitors. *Phytochemistry* 43: 1163–1169.

Fujimori, N. and Ashihara, H. (1990). Adenine metabolism and the synthesis of purine alkaloids in flowers of *Camellia* plants. *Phytochemistry* 29: 3513–3516.

Fujimori, N. and Ashihara, H. (1993). Biosynthesis of caffeine in flower buds of *Camellia sinensis*. *Ann. Bot.* 71: 279–284.

Gupta, N.K. and Glantz, M.D. (1985). Isolation and characterization of human liver guanine deaminase. *Arch. Biochem. Biophys.* 236: 266–276.

Guranowski, A. (1982). Purine catabolism in plants. Purification and some properties of inosine nucleosidase from yellow lupin (*Lupinus leteus* L.) seeds. *Plant Physiol.* 70: 344–349.

Guranowski, A. and Schneider, Z. (1977). Purification and characterization of adenosine nucleosidase from barley leaves. *Biochim. Biophys. Acta* 482: 145–158.

Henderson, J.F., Lowe, J.K., and Barankiewicz, J. (2008). Purine and pyrimidine metabolism: pathways, pitfalls and perturbations. In: *Ciba Foundation Symposium 48 – Purine and Pyrimidine Metabolism* (eds. K. Elliott and D.W. Fitzsimons), 3–21. Wiley.

Henderson, J.F. and Paterson, A.R.P. (1973). *Nucleotide Metabolism – An Introduction*. New York: Academic Press.

Heppel, L.A., Hurwitz, J., and Horecker, B.L. (1957). Adenine deaminase of *Azotobacter vinelandii*. *J. Am. Chem. Soc.* 79: 630–633.

Hirose, F. and Ashihara, H. (1983). Comparison of purine metabolism in suspension cultured cells of different growth phases and stem tissue of *Catharanthus roseus*. *Z. Naturforsch.* 38c: 375–381.

Hirose, F. and Ashihara, H. (1984). Changes in activity of enzymes involved in purine "salvage" and nucleic acid degradation during growth of *Catharanthus roseus* cells in suspension culture. *Physiol. Plant.* 60: 532–538.

Hitchings, G.H. and Falco, E.A. (1944). The identification of guanine in extracts of *Girella Nigricans*: the specificity of guanase. *Proc. Natl. Acad. Sci. U. S. A.* 30: 294–297.

Imagawa, H., Yamano, H., Inoue, K., and Takino, Y. (1979). Purification and properties of adenosine nucleosidases from tea leaves. *Agric. Biol. Chem.* 43: 2337–2343.

Ishida, Y., Shirafuji, H., Kida, M., and Yoneda, M. (1969). Studies on the guanosine degrading system in bacterial cell. Part III. Purification and properties of guanosine deaminase. *Agric. Biol. Chem.* 33: 384–390.

Ito, E. and Ashihara, H. (1999). Contribution of purine nucleotide biosynthesis *de novo* to the formation of caffeine in young tea (*Camellia sinensis*) leaves. *J. Plant Physiol.* 254: 145–151.

Jung, B., Flörchinger, M., Kunz, H.-H. et al. (2009). Uridine-ribohydrolase is a key regulator in the uridine degradation pathway of *Arabidopsis. Plant Cell* 21: 876–891.

Jung, B., Hoffmann, C., and Möhlmann, T. (2011). Arabidopsis nucleoside hydrolases involved in intracellular and extracellular degradation of purines. *Plant J.* 65: 703–711.

Kalckar, H.M. and Shafran, W. (1947). Differential spectrophotometry of purine compounds by means of specific enzymes: III. Studies of the enzymes of purine metabolism. *J. Biol. Chem.* 167: 461–475.

Kamat, S.S., Bagaria, A., Kumaran, D. et al. (2011). Catalytic mechanism and three-dimensional structure of adenine deaminase. *Biochemistry* 50: 1917–1927.

Katahira, R. and Ashihara, H. (2006). Profiles of purine biosynthesis, salvage and degradation in disks of potato (*Solanum tuberosum* L.) tubers. *Planta* 225: 115–126.

Keya, C.A., Crozier, A., and Ashihara, H. (2003). Inhibition of caffeine biosynthesis in tea (*Camellia sinensis*) and coffee (*Coffea arabica*) plants by ribavirin. *FEBS Lett.* 554: 473–477.

Klenow, H. and Østergaard, E. (1988). Adenosine induction of rapid catabolism of adenine ribonucleotides and independent elevation of the ATP content in quiescent mouse fibroblasts. *J. Cell. Physiol.* 137: 565–570.

Koch, A.L. and Lamont, W.A. (1956). Some enzymes of nucleoside metabolism of *Escherichia coli. J. Biol. Chem.* 223: 535–549.

Kopečná, M., Blaschke, H., Kopečný, D. et al. (2013). Structure and function of nucleoside hydrolases from *Physcomitrella patens* and maize catalyzing the hydrolysis of purine, pyrimidine, and cytokinin ribosides. *Plant Physiol.* 163: 1568–1583.

Koshiro, Y., Zheng, X.Q., Wang, M. et al. (2006). Changes in content and biosynthetic activity of caffeine and trigonelline during growth and ripening of *Coffea arabica* and *Coffea canephora* fruits. *Plant Sci.* 171: 242–250.

Koyama, Y., Tomoda, Y., Kato, M., and Ashihara, H. (2003). Metabolism of purine bases, nucleosides and alkaloids in theobromine-forming *Theobroma cacao* leaves. *Plant Physiol. Biochem.* 41: 977–984.

Le Floc'h, F. and Faye, F. (1995). Metabolic fate of adenosine and purine metabolism enzymes in young plants of peach tree. *J. Plant Physiol.* 145: 398–404.

Le Floc'h, F. and Lafleuriel, J. (1981a). L'hypoxanthine-guanine phosphoribosyltransferase des pousses de topinambour *Helianthus tuberosus. Z. Pflanzenphysiol.* 104: 331–340.

Le Floc'h, F. and Lafleuriel, J. (1981b). The purine nucleosidases of Jerusalem artichoke shoots. *Phytochemistry* 20: 2127–2129.

Le Floc'h, F. and Lafleuriel, J. (1983). The particulate AMP aminohydrolase of Jerusalem artichoke tubers: partial purification and properties. *Physiol. Veg.* 21: 15–27.

Le Floc'h, F., Lafleuriel, J., and Guillot, A. (1982). Interconversion of purine nucleotides in Jerusalem artichoke shoots. *Plant Sci. Lett.* 27: 309–316.

Lee, Y.-P. (1957). Adenylic acid deaminase: I. Isolation of the crystalline enzyme from rabbit skeletal muscle. *J. Biol. Chem.* 5, 227: 987–992.

Lui, M.S., Faderan, M.A., Liepnieks, J.J. et al. (1984). Modulation of IMP dehydrogenase activity and guanylate metabolism by tiazofurin (2-β-D-ribofuranosylthiazole-4-carboxamide). *J. Biol. Chem.* 259: 5078–5082.

Mager, J. and Magasanik, B. (1960). Guanosine 5′-phosphate reductase and its role in the interconversion of purine nucleotides. *J. Biol. Chem.* 235: 1474–1478.

Maynes, J.T., Yuan, R.G., and Snyder, F.F. (2000). Identification, expression, and characterization of *Escherichia coli* guanine deaminase. *J. Bacteriol.* 182: 4658–4660.

Mazelis, M. and Creveling, R.K. (1963). An adenosine hydrolase from Brussels sprouts. *J. Biol. Chem.* 238: 3358–3361.

Mittal, R., Das, J., and Sharma, C.B. (1988). Purification and characterization of 5′-nucleotidase from the Golgi apparatus of peanut cotyledons. *Plant Sci.* 55: 93–101.

Negishi, O., Ozawa, T., and Imagawa, H. (1992). Biosynthesis of caffeine from purine nucleotides in tea plant. *Biosci. Biotechnol. Biochem.* 56: 499–503.

Negishi, O., Ozawa, T., and Imagawa, H. (1994). Guanosine deaminase and guanine deaminase from tea leaves. *Biosci. Biotechnol. Biochem.* 58: 1277–1281.

Nobusawa, E. and Ashihara, H. (1983). Purine metabolism in cotyledons and embryonic axes of black gram (*Phaseolus mungo* L.) seedlings. *Int. J. Biochem.* 15: 1059–1065.

Nygaard, P., Bested, S.M., Andersen, K.A.K., and Saxild, H.H. (2000). *Bacillus subtilis* guanine deaminase is encoded by the *yknA* gene and is induced during growth with purines as the nitrogen source. *Microbiology* 146: 3061–3069.

Polya, G.M. and Ashton, A.R. (1973). Inhibition of wheat seedling 5′(3′)-ribonucleotide phosphohydrolase by adenosine 3′,5′-cyclic monophosphate. *Plant Sci. Lett.* 1: 349–357.

PospíŠilová, H., Šebela, M., Novák, O., and Frébort, I. (2008). Hydrolytic cleavage of N^6-substituted adenine derivatives by eukaryotic adenine and adenosine deaminases. *Biosci. Rep.* 28: 335–347.

Poulton, J.E. and Butt, V.S. (1976). Partial purification and properties of adenosine nucleosidase from leaves of spinach beet (*Beta vulgaris* L.). *Planta* 131: 179–185.

Powell, J.F. and Hunter, J.R. (1956). Adenosine deaminase and ribosidase in spores of *Bacillus cereus*. *Biochem. J.* 62: 381–387.

Rabinowitz, J.C. and Barker, H.A. (1956). Purine fermentation by *Clostridium cylindrosporum*. II. Purine transformations. *J. Biol. Chem.* 218: 161–173.

Renart, M.F. and Sillero, A. (1974). GMP reductase in *Artemia salina*. *Biochim. Biophys. Acta* 341: 178–186.

Riewe, D., Grosman, L., Fernie, A.R. et al. (2008). A cell wall-bound adenosine nucleosidase is involved in the salvage of extracellular ATP in *Solanum tuberosum*. *Plant Cell Physiol.* 49: 1572–1579.

Sharma, C.B., Mittal, R., and Tanner, W. (1986). Purification and properties of a glycoprotein adenosine 5′-monophosphatase from the plasma membrane fraction of *Arachis hypogaea* cotyledons. *Biochim. Biophys. Acta* 884: 567–577.

Shimano, F. and Ashihara, H. (2006). Effect of long-term phosphate starvation on the levels and metabolism of purine nucleotides in suspension-cultured *Catharanthus roseus* cells. *Phytochemistry* 67: 132–141.

Stasolla, C., Katahira, R., Thorpe, T.A., and Ashihara, H. (2003). Purine and pyrimidine nucleotide metabolism in higher plants. *J. Plant Physiol.* 160: 1271–1295.

Stephens, R.W. and Whittaker, V.K. (1973). Calf thymus GMP reductase: control by XMP. *Biochem. Biophys. Res. Commun.* 53: 975–981.

Suzuki, T. and Takahashi, E. (1975). Metabolism of xanthine and hypoxanthine in the tea plant (*Thea sinensis* L.). *Biochem. J.* 146: 79–85.

Suzuki-Yamamoto, M., Mimura, T., and Ashihara, H. (2006). Effect of short-term salt stress on the metabolic profiles of pyrimidine, purine and pyridine nucleotides in cultured cells of the mangrove tree, *Bruguiera sexangula*. *Physiol. Plant.* 128: 405–414.

Szuwart, M., Starzynska, E., Pietrowska-Borek, M., and Guranowski, A. (2006). Calcium-stimulated guanosine-inosine nucleosidase from yellow lupin (*Lupinus luteus*). *Phytochemistry* 67: 1476–1485.

Tarr, H.L.A. (1955). Fish muscle riboside hydrolases. *Biochem. J.* 59: 386–391.

Terrasaki, Y., Suzuki, T., and Ashihara, H. (1994). Purine metabolism and the biosynthesis of purine alkaloids in tea fruits during development. *Plant Physiol. (Life Sci. Adv.)* 13: 135–142.

Turner, D.H. and Turner, J.F. (1961). Adenylic deaminase of pea seeds. *Biochem. J.* 79: 143–147.

Ullah, A.H.J. and Gibson, D.M. (1988). Purification and characterization of acid phosphatase from cotyledons of germinating soybean seeds. *Arch. Biochem. Biophys.* 260: 514–520.

Wagner, K.G. and Backer, A.I. (1992). Dynamics of nucleotides in plants studied on a cellular basis. In: *International Review of Cytology*, vol. 134 (eds. K.W. Jeon and M. Friedlander), 1–84. San Diego: Academic Press.

Wilson, D., Rudolph, F., and Quiocho, F. (1991). Atomic structure of adenosine deaminase complexed with a transition-state analog: understanding catalysis and immunodeficiency mutations. *Science* 252: 1278–1284.

Yabuki, N. and Ashihara, H. (1991). Catabolism of adenine nucleotides in suspension-cultured plant cells. *Biochim. Biophys. Acta* 1073: 474–480.

Yabuki, N. and Ashihara, H. (1992). AMP deaminase and the control of adenylate catabolism in suspension-cultured *Catharanthus roseus* cells. *Phytochemistry* 31: 1905–1909.

Yin, Y., Ashihara, H., Watanabe, S., and Yin, Y. (2012). Metabolic fate of ^{14}C-labelled nicotinamide and adenine in germinating propagules of the mangrove *Bruguiera gymnorrhiza*. *Z. Naturforsch.* 67c: 319–326.

Yin, Y., Katahira, R., and Ashihara, H. (2014). Metabolism of purine nucleosides and bases in suspension-cultured *Arabidopsis thaliana* cells. *Eur. Chem. Bull.* 3: 925–934.

Yin, Y., Katahira, R., and Ashihara, H. (2015). Metabolism of purine alkaloids and xanthine in leaves of mate (*Ilex paraguariensis*). *Nat. Prod. Commun.* 10: 707–712.

Yoshino, M. and Murakami, K. (1980). AMP deaminase from spinach leaves. Purification and some regulatory properties. *Z. Pflanzenphysiol.* 99: 331–338.

Yuan, G., Bin, J.C., McKay, D.J., and Snyder, F.F. (1999). Cloning and characterization of human guanine deaminase: purification and partial amino acid sequence of the mouse protein. *J. Biol. Chem.* 274: 8175–8180.

Zoref-Shani, E., Shainberg, A., and Sperling, O. (1987). Pathways of adenine nucleotide catabolism in primary rat muscle cultures. *Biochim. Biophys. Acta* 926: 287–295.

7

Degradation of Purine Nucleotides

7.1 Introduction

In contrast to most animals, plants possess a complete oxidative purine catabolism pathway, with purine structures being fully degraded via urate and allantoin to CO_2, glyoxylate, and NH_3. Xanthine is the key starting compound of purine catabolism. All purine nucleotides are converted to xanthine before purine ring cleavage is initiated, as discussed in Chapter 6 and summarized in Figure 7.1. There are two possible routes from inosine-5′-monophosphate (IMP) to urate via xanthine. The classic IMP → inosine → hypoxanthine → xanthine → urate pathway (steps 1, 3, 4, and 7 in Figure 7.1) was suggested from studies with ^{14}C-labelled inosine and hypoxanthine (Ashihara and Nobusawa 1981). An alternative IMP → XMP → xanthosine → xanthine → urate route (steps 1, 2, 3, 4, and 7 in Figure 7.1) occurs, most notably in nodules of tropical legumes. In nodules that produce the ureides, allantoin, and allantoate, IMP dehydrogenase which participates in step 2 is found in proplastids, and the kinetic properties suggest that oxidation of IMP facilitates the channelling of IMP into purine oxidation. Thus, ureides are produced by effective scavenging of IMP (Boland and Schubert 1983; Shelp and Atkins 1983). Therefore, the latter route appears to be involved in the pathway for the biogenesis of ureides by *de novo* purine nucleotide biosynthesis (Schubert and Boland 1990). In plants, urate is degraded to CO_2, glyoxylate, and NH_3 via allantoin (Figure 7.2).

Animals do not possess the wide-ranging oxidative purine catabolic pathways found in plants. Instead a variety of selective conversions occur (Figure 7.2). This was one of the twentieth century's fascinating topics of evolutionary and comparative biochemistry (Baldwin 1970; Florkin 1949). Urate is the principal end product of purine metabolism in higher primates, including humans, and it is excreted in the urine. Birds, terrestrial reptiles, and many insects also produce urate which, as it is only sparingly soluble in water, is excreted as a paste of crystals that contains very little water. Mammals, other than primates, oxidize urate to allantoin which is excreted. A further degradation product, allantoate, is excreted by teleost (bony) fish. Cartilaginous fish and amphibia further degrade allantoate to urea prior to excretion while marine invertebrates decompose urea

Plant Nucleotide Metabolism – Biosynthesis, Degradation, and Alkaloid Formation, First Edition.
Hiroshi Ashihara, Iziar A. Ludwig and Alan Crozier.

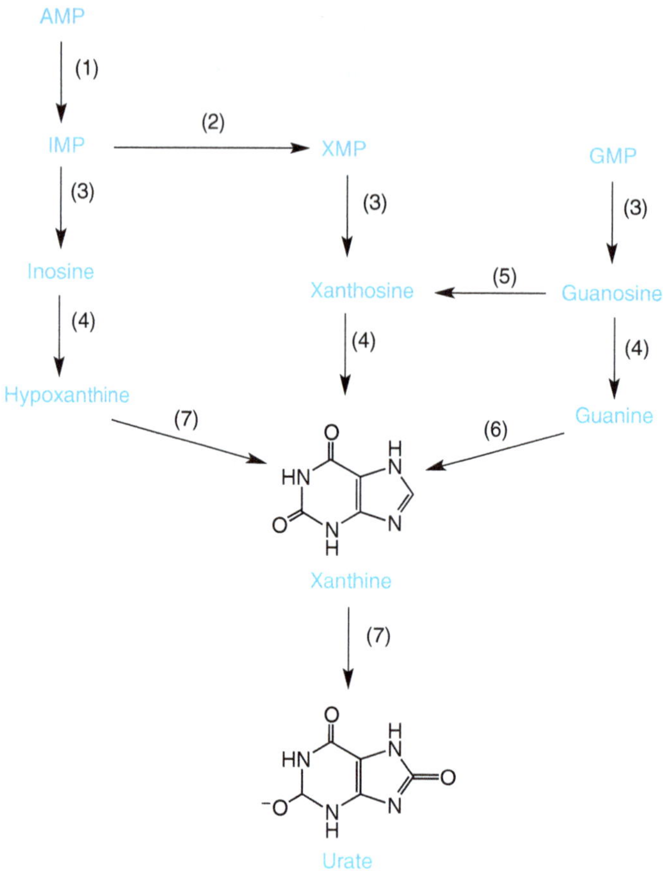

Figure 7.1 The routes of urate synthesis from purine nucleotides in plants. Two possible routes occur from IMP to xanthine. One route is an IMP → inosine → hypoxanthine → xanthine pathway (steps 3, 4 and 7) and another route is IMP → XMP → xanthosine → xanthine pathway (steps 2, 3 and 4). The latter is the primary pathway in ureide-forming nodules (Schubert and Boland 1990). From GMP, the pathway via xanthosine (steps 3, 5 and 4) is the major route (see Chapter 6). Enzymes: (1) AMP deaminase; (2) IMP dehydrogenase; (3) 5′-nucleotidase; (4) inosine/guanosine nucleosidase or purine nucleosidase; (5) guanosine deaminase; (6) guanine deaminase; (7) xanthine dehydrogenase.

to ammonia (Figure 7.2). This pattern is dependent upon the phylogenetic occurrence of the various enzymes involved in urate catabolism. There are exceptions, such as the excretion of hypoxanthine by leeches, and guanine by spiders, which present interesting patterns of enzyme specialization (Henderson and Paterson 1973; Voet and Voet 2010).

Although all plant species have a complete catabolic pathway of purine nucleotides which produces NH_3, glyoxylate, and CO_2 (Ashihara et al. 1997; Yabuki and Ashihara 1991), some plant organs, such as roots of the tropical legumes and maple (*Acer* sp.) trees, accumulate allantoin and/or allantoate, which play an important role in the storage and transport of nitrogen. Allantoin and allantoate are the major nitrogenous solutes exported from the nodulated roots of some taxonomically-related N_2-fixing legumes, including soybean (*Glycine max*) and cowpea (*Vigna unguiculata*) (Schubert and Boland 1990).

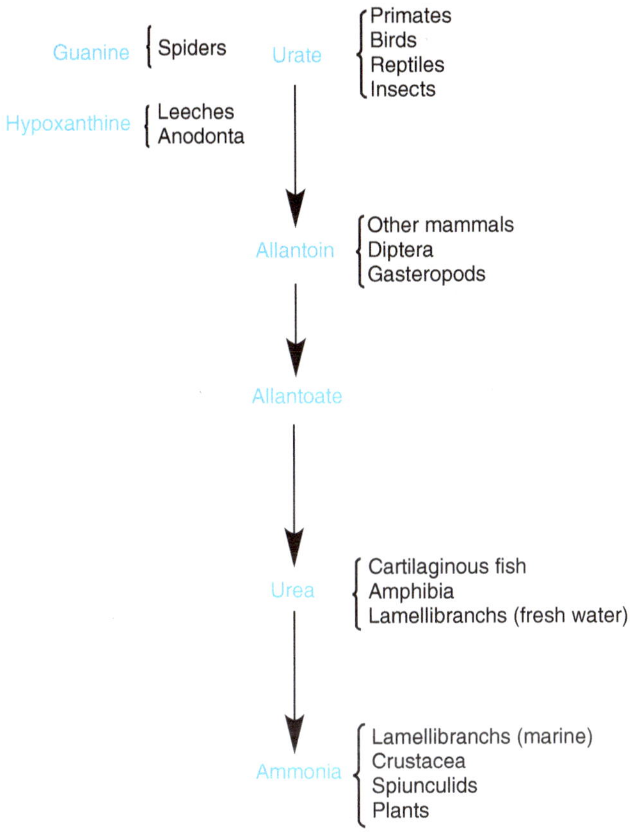

Figure 7.2 End products of purine metabolism in animals and plants. Based on Henderson and Paterson (1973) and Voet and Voet (2010).

The main pathway for the conversion of xanthine to allantoin is: xanthine → 5-hydroxyisourate (HIU) → 2-oxo-4-hydroxy-4-carboxy-5-ureidoimidazoline (OHCU) → (*S*)-allantoin (Figure 7.3). The naturally occurring optical isomer of allantoin is (*S*)-allantoin which is named according to the nomenclature of Cahn–Ingold–Prelog (see, Smith and March 2007). Allantoate produced from (*S*)-allantoin is further catabolized to ammonia, glyoxylate, and CO_2. The allantoate amidohydrolase pathway, namely allantoate → (*S*)-ureidoglycine → (*S*)-ureidoglycolate → glyoxylate operates in most plants (Werner et al. 2013). An alternative urea-producing allantoicase pathway, namely allantoate → (*S*)-ureidoglycolate → glyoxylate, occurs in *Chlamydomonas reinhardtii* (Piedras et al. 2000) (see Figure 7.6 in Section 7.3).

7.2 (*S*)-Allantoin Biosynthesis from Xanthine

A summary of the enzymes involved in purine nucleotide biosynthesis in plants is presented in Table 7.1 and Figure 7.3. It was originally thought that conversion of urate to allantoin was catalysed in a single step by urate oxidase (Schubert and Boland 1990).

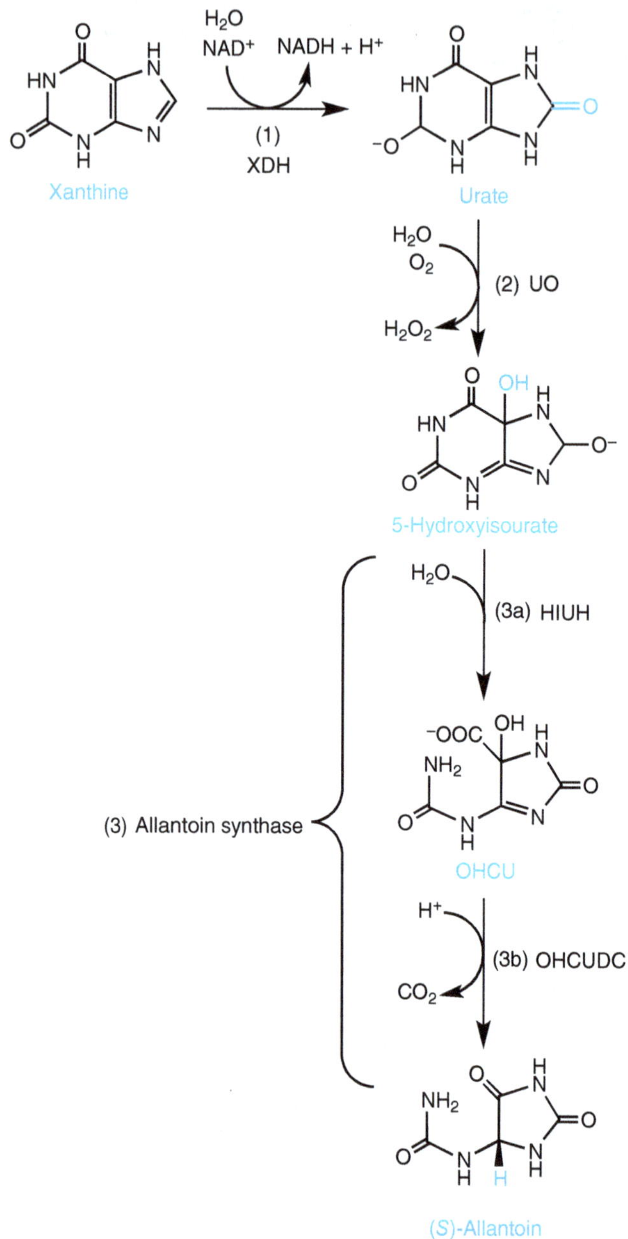

Figure 7.3 (*S*)-Allantoin biosynthesis from xanthine in plants. Enzyme names with EC numbers and the Arabidopsis Information Resource (TAIR) locus of *Arabidopsis thaliana* genes are shown in Table 7.1. Abbreviations: XDH, xanthine dehydrogenase; UO, urate oxidase; HIU, hydroxyisourate; HIUH, hydroxyisourate hydrolase; OHCU, 2-oxo-4-hydroxy-4-carboxy-5-ureidoimidazoline; OHCUDC, 2-oxo-4-hydroxy-4-carboxy-5-ureidoimidazoline decarboxylase.

Table 7.1 Summary of enzymes involved in allantoin formation from xanthine in plants.

Step[a)]	Enzyme name	EC	Enzyme abbreviation	Isozyme	TAIR locus
1	Xanthine dehydrogenase	1.17.1.4	XDH	AtXDH1	AT4G34890
				AtXDH2	AT4G34900
2	Urate oxidase (uricase)	1.7.3.3	UO		AT2G26230
3	Allantoin synthase[b)]				AT5G58220
3a	Hydroxyisourate (HIU) hydrolase	3.5.2.17	HIUH		AT5G58220
3b	2-Oxo-4-hydroxy-4-carboxy-5-ureidoimidazoline (OHCU) decarboxylase	4.1.1.97	OHCUDC		AT5G58220

a) Steps in Figure 7.3.
b) A bifunctional enzyme that catalyses the last two steps (3a and 3b) in the pathway.

However, it was subsequently shown that the process involves three different enzymes (Ramazzina et al. 2006).

7.2.1 Xanthine Dehydrogenase

Xanthine is converted to urate by xanthine dehydrogenase (EC 1.17.1.4). Initially, this enzyme was classified as acting on a CH-OH group (EC 1.1.1.204), but this was corrected in 2004 (IUBMB Enzyme Nomenclature). The enzyme can utilize either xanthine or hypoxanthine as a substrate: the reaction is: hypoxanthine + NAD^+ + H_2O → xanthine + NADH + H^+, and xanthine + NAD^+ + H_2O → urate + NADH + H^+. *In planta*, hypoxanthine is continuously converted to urate by this enzyme (Figure 7.4).

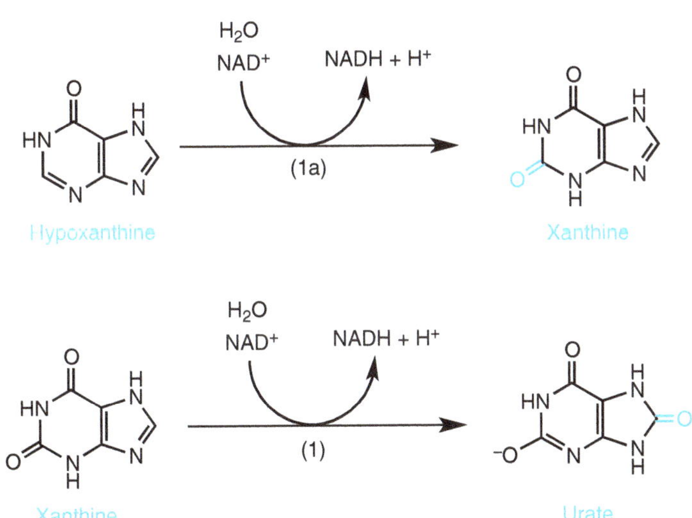

Figure 7.4 Reactions catalysed by xanthine dehydrogenase.

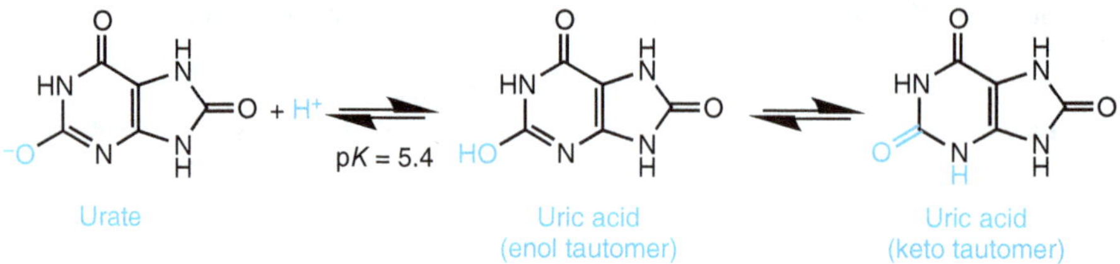

Figure 7.5 Tautomerization of uric acid.

Xanthine and hypoxanthine are respectively hydroxylated at the C8 and C2 positions yielding uric acid in its enol form which tautomerizes to the more stable keto form (Voet and Voet 2010). Uric acid is a diprotic acid with $pKa_1 = 5.4$ and $pKa_2 = 10.3$. At biological pH, it forms the singly-charged hydrogen urate or acid urate ion. Since its second ionization is weak, the full urate salts hydrolyse back to hydrogen urate salts at neutral pH values (Figure 7.5).

Xanthine dehydrogenases from eukaryotes contains [2Fe-2S], flavin adenine dinucleotide (FAD), and a molybdenum centre. The plant, as well as the mammalian enzyme, exists predominantly as the NAD-dependent dehydrogenase. However, during purification it is converted mainly to an O_2-dependent form, xanthine oxidase (EC 1.17.3.2) (Corte and Stirpe 1972).

Xanthine dehydrogenase has been purified from nodules of soybean *(Glycine max)*. (Boland et al. 1983; Triplett et al. 1982). The Km values for xanthine and NAD^+ are 5.0 and 12.5 μM, respectively. The pattern of substrate dependence suggested a Ping-Pong mechanism. The reaction with hypoxanthine has Km values of 52 and 20 μM for hypoxanthine and NAD^+, respectively. The Vmax for this reaction was twice that of the xanthine-dependent reaction. Furthermore, the Km for xanthine had a pKa of 7.5, consistent with the protonated form of xanthine being the true substrate.

In *Arabidopsis thaliana*, there are two highly homologous xanthine dehydrogenase genes, *AtXDH1* and *AtXDH2* which are duplicated in tandem. Both genes are widely expressed. *AtXDH1* shows altered transcript levels in response to ageing, senescence, salt and cold stress, and abscisic acid treatment. Heterologous expression and characterization of AtXDH1 indicated it is a strict dehydrogenase, not an oxidase. NAD^+, but not molecular oxygen, acts as the electron acceptor. The enzyme efficiently catalyses the oxidation of both hypoxanthine and xanthine (Hesberg et al. 2004).

7.2.2 Urate Oxidase

In the first step of the allantoin synthesis pathway, urate is oxidized to (*S*)-5-hydroxyisourate by urate oxidase (aka uricase). The enzyme catalyses the reaction: urate + O_2 + H_2O → (*S*)-5-hydroxyisourate + H_2O_2 (step 2 in Figure 7.3). Historically, glyoxysome-associated uricase activity was reported in fat-degrading castor bean endosperm and kinetic studies revealed that the Km for urate was 7.4 μM (Theimer and Beevers 1971). A high level of uricase activity was detected in soybean nodules (Tajima and Yamamoto 1975), located in intact peroxisomes (Hanks et al. 1981). The native soybean nodule uricase is a monomer at pH 7.5 which forms a dimer at pH 8.8. The pH optimum was 9.5. The Km values for urate and oxygen were 10 and 31 μM, respectively (Lucas et al. 1983). Uricase was purified to electrophoretic homogeneity from leaves of chickpea (*Cicer arietimum*), broad bean (*Vicia faba*), and wheat (*Triticum aestivum*). These three plant uricases were found by sodium dodecyl sulphate-polyacrylamide gel electrophoresis to be tetramers with a similar molecular mass and to have identical or similar sized subunits. They also had a similar optimum pH (9.0–9.5) and showed a hyperbolic kinetics with Km values from 9 to 24 μM. These properties suggest that plant uricase is a unique enzyme, (see Montalbini et al. 1997).

The product of urate oxidase, 5-hydroxyisourate (HIU), is relatively unstable and decomposes *in vitro* to yield allantoin. However, this spontaneous reaction is unlikely to be physiologically relevant, since spontaneous decomposition yields racemic allantoin, whereas organisms contain mostly the (*S*)-(+)-allantoin enantiomer (Kahn et al. 1997).

7.2.3 Allantoin Synthase

The oxidation of HIU can be carried out by two distinct enzymes. Eukaryotes and some bacteria use HIU hydrolase (factor-independent urate hydroxylase) (Kahn and Tipton 1997) which catalyses the hydrolysis of HIU yielding (*S*)-2-oxo-4-hydroxy-4-carboxy-5-ureidoimidazoline (OHCU) (step 3a in Figure 7.3). HIU hydrolase from soybean root nodules has been purified to apparent homogeneity. The native enzyme consists of a single subunit (~68 kD) and the Km value for HIU is 15 μM (Sarma et al. 1999).

The second enzyme, OHCU decarboxylase, which catalyses the conversion of OHCU to (*S*)-allantoin (step 3b in Figure 7.3), was first discovered in mice by Ramazzina et al. (2006) and subsequently purified from *A. thaliana* (Lamberto et al. 2010; Pessoa et al. 2010). Based on *in vitro* assays using a recombinant enzyme, the HIU hydrolase and OHCU decarboxylase from *A. thaliana* are present as a bifunctional enzyme (Lamberto et al. 2010; Pessoa et al. 2010). *A. thaliana* genes have fused to encode a single polypeptide, but most organisms have two separate genes for the two enzymes (Kim et al. 2007; Lamberto et al. 2010; Pessoa et al. 2010). Since allantoin is the final product of two successive reactions, Lamberto et al. (2010) proposed the name *S*-allantoin synthase for the plant enzyme.

7.3 Catabolism of (*S*)-Allantoin

Allantoin, an intermediate of purine nucleotide catabolism, is further degraded to NH_3, glyoxylate, and CO_2 (Figure 7.6). A summary of the enzymes involved in these conversions is shown in Table 7.2. Ammonia released from allantoin is used mainly for reassimilation into amino acids. In tropical legumes, such as soybean, allantoin and/or allantoate are used for nodule-to-shoot translocation of fixed-nitrogen (see Section 7.4).

In *A. thaliana*, soybean and rice *(Oryza sativa)*, four enzymes, (1) allantoinase (ALN, EC 3.5.2.5), (2) allantoate amidohydrolase (AAH, EC 3.5.3.9), (3) ureidoglycine aminohydrolase (UGlyAH, EC 3.5.3.26), and (4) ureidoglycolate amidohydrolase (UAH, EC 3.5.1.116), catalyse the complete hydrolysis of allantoin (Werner et al. 2013) (steps 1–3 and 5 in Figure 7.6).

There are alternative routes for allantoin catabolism. Allantoate is also converted to (*S*)-ureidoglycolate releasing urea, by allantoicase (allantoate amidinohydrolase, EC 3.5.3.4) (step 4 in Figure 7.6). This enzyme has been detected in the green algae,

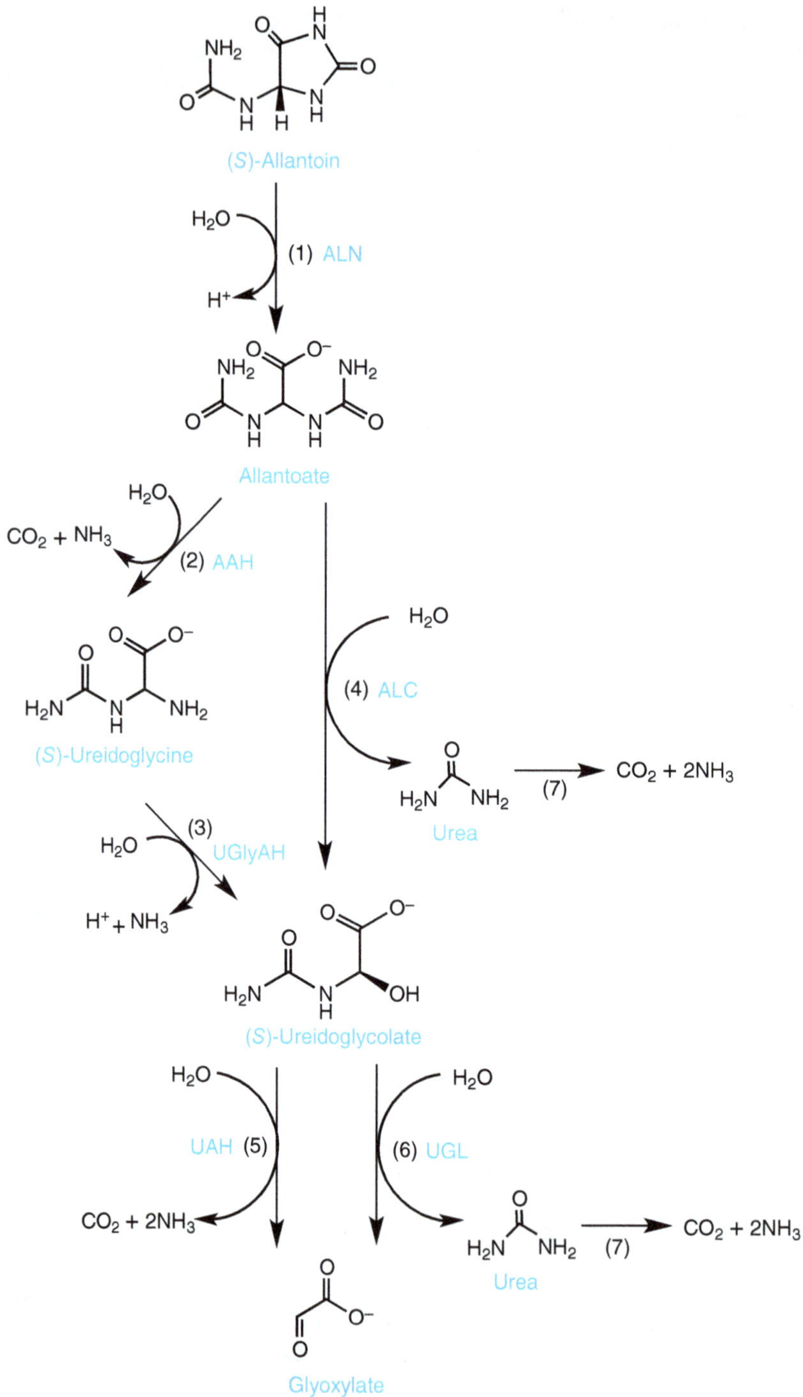

Figure 7.6 Degradation pathway of (*S*)-allantoin in plants. Enzyme names with EC numbers and gene locus of *A. thaliana* and *Glycine max* are shown in Table 7.2. Abbreviations: ALN, allantoinase: AAH, allantoate amidohydrolase; UGlyAH, (*S*)-ureidoglycine aminohydrolase; UAH, ureidoglycolate amidohydrolase; ALC, allantoicase; UGL, ureidoglycolate lyase. Some commonly used abbreviations shown in this figure are different from the IUBMB nomenclature. See Table 7.2 for details.

Table 7.2 Summary of enzymes involved in the allantoate catabolism in plants.

Step[a)]	Enzyme name	EC	Enzyme abbreviation	TAIR locus	Isozyme (GM)[b)]	Soybean locus
1	Allantoin amidohydrolase (Allantoinase)	3.5.2.5	ALN	AT4G04955	ALN1	GM15G07910
					ALN2	GM13G31430
					ALN3	GM15G07920
					ALN4	GM13G31420
2	Allantoate deiminase (Allantoate amidohydrolase)	3.5.3.9	AAH	AT4G20070	AAH1	GM09G05600
					AAH2	GM15G16870
3	(*S*)-Ureidoglycine aminohydrolase	3.5.3.26	UGlyAH	AT4G17050	UGlyAH1	GM17G15820
					UGlyAH2	GM05G05490
4	Allantoate amidinohydrolase (Allantoicase)	3.5.3.4	ALC			
5	Ureidoglycolate amidohydrolase	3.5.1.116	UAH	AT5G43600	UAH1	GM10G32850
					UAH2	GM20G34790
6	Ureidoglycolate lyase	4.3.2.3	UGL			
7	Urease	3.5.1.5		AT2G34470		GM05G27840
						GM11G37250

a) Steps in Figure 7.6.
b) Isozymes in soybean (*Glycine max*).

Chlamydomonas reinhardtii (Piedras et al. 2000). The next step is the conversion of (*S*)-ureidoglycolate to glyoxylate. This conversion can proceed with the liberation of either urea, by ureidoglycolate-urea-lyase (EC 4.3.2.3) (step 6 in Figure 7.6), or ammonia by ureidoglycolate amidohydrolase (EC 3.5.1.116) (step 5 in Figure 7.6). Ureidoglycolate-urea-lyase has been found in chickpea *(Cicer arietinum)* (Muñoz et al. 2001) and French bean *(Phaseolus vulgaris)* (Muñoz et al. 2006).

7.3.1 Allantoinase

Allantoinase [(*S*)-allantoin amidohydrolase, EC 3.5.2.5] catalyses the reaction: (*S*)-allantoin + H_2O → allantoate (step 1 in Figure 7.6). This enzyme occurs in all organisms that can degrade allantoin (Vogels et al. 1966). Purification and characterization of allantoinase have been performed with seedlings of mung bean (*Vigna radiata*) (Nagai and Funahashi 1961), *Lathyrus sativus* (grass pea) (Nirmala and Sastry 1975), and *Dolichos biflorus* (lablab-bean) (Mary and Sivarama Sastry 1978); nodules of *Cajanus cajan* (pigeon pea) (Amarjit and Singh 1985) and *Arachis hypogaea* (peanut) (Venkateswara Rao et al. 1988); shoot tissues and seeds of soybean (Thomas et al. 1983; Webb and Lindell 1993); and various tissues of *Phaseolus vulgaris* (Raso et al. 2007b).

The Km values of allantoinase for allantoin are high, for example, 2.6 and 10 mM for the enzymes from pea grass (Nirmala and Sastry 1975) and soybean shoot, respectively

(Thomas et al. 1983). Two allantoinase isozymes were purified homogeneity from developing *P. vulgaris* fruits. The two isozymes have similar biochemical properties and the Km values are ~60 mM (Raso et al. 2007b).

Allantoinase resides in the endoplasmic reticulum (ER) of plants (Hanks et al. 1981; Werner et al. 2008). Therefore, allantoin formed in peroxisomes might be translocated into the ER. Peroxisome-to-ER vesicle traffic or peroxisome–ER connectivity have been proposed, but the hypothesis remains rather speculative (Werner and Witte 2011).

7.3.2 Allantoate Amidohydrolase

Allantoate amidohydrolase (AAH, allantoate deiminase, EC 3.5.3.9) catalyses the reaction: allantoate + H_2O → (*S*)-ureidoglycine + NH_3 + CO_2 (step 2 in Figure 7.6). Mn^{2+}-dependent AAH activity was detected in partially purified extracts of developing soybeans. The main products were CO_2, NH_3, and glyoxylate. A broad optimal pH range of 8–9 was found and a Km of 1.0 μM for allantoate (Winkler et al. 1985). The activity of partially-purified AAH was completely dependent on phenylhydrazine and the optimal pH was 7.0 (Raso et al. 2007a). Purification and characterization of native AAHs have been difficult, seemingly because of their instability.

Todd and Polacco (2006) first reported the identification and cloning of an AAH encoded cDNA, *AtAAH*, from *A. thaliana*. The *AtAAH* genes of *A. thaliana* and *GmAAH* from soybean were cloned, expressed *in planta* and purified from leaf extracts. Both AAH proteins form homodimers and release two mol of NH_3 per mol of allantoate. Km values for allantoate are, respectively, 30 and 81 μM for AtAAH and GmAAH (Werner et al. 2008).

It has been proposed that one of the final products of AAH is (*S*)-ureidoglycolate (Werner et al. 2008) with ureidoglycine acting as a reactive intermediate (Vogels and Van der Drift 1976). However, it has been shown that (*S*)-ureidoglycine, CO_2 and NH_3 are the final products of AAH (step 2 in Figure 7.6) and that a further enzyme, (*S*)-ureidoglycine aminohydrolase (UGlyAH) is required to generate (*S*)-ureidoglycolate (step 3 in Figure 7.6) (Serventi et al. 2010; Werner et al. 2010; Werner and Witte 2011).

7.3.3 (*S*)-Ureidoglycine Aminohydrolase

(*S*)-Ureidoglycine aminohydrolase (UGlyAH, EC 3.5.3.26) catalyses the reaction, (*S*)-2-ureidoglycine + H_2O → (*S*)-ureidoglycolate + NH_3 (step 3 in Figure 7.6) The substrate for this reaction, (*S*)-ureidoglycine is unstable and reactive *in vitro* (Van Der Drift et al. 1970; Winkler et al. 1985), and decays rapidly to glyoxylate, urea, and NH_3 (Serventi et al. 2010; Werner et al. 2010). In contrast, (*S*)-ureidoglycolate is formed *in vivo*, which suggests that (*S*)-ureidoglycine is hydrolysed enzymatically releasing NH_3 (Werner and Witte 2011).

A gene encoding UGlyAH was identified in cross-genome comparisons with the enzymes from *A. thaliana* and *E. coli* (Serventi et al. 2010; Werner et al. 2010). The *UGlyAH* possesses a predicted ER targeting sequence. The purified UGlyAH protein shows no activity with allantoate as substrate, but when combined with AAH, NH_3 release is accelerated, and stoichiometric amounts of (*S*)-ureidoglycolate are generated (Werner et al. 2010).

7.3.4 Allantoate Amidinohydrolase

Allantoate amidinohydrolase (allantoicase, EC 3.5.3.4) catalyses the reaction: allantoate + H_2O → (*S*)-ureidoglycolate + urea (step 4 in Figure 7.6). Allantoate amidinohydrolase from *Chlamydomonas reinhardtii* was purified as an allantoate-degrading enzyme (Piedras et al. 2000). The enzyme uses both allantoate and (*S*)-ureidoglycolate as substrates, producing urea in both cases. Vmax with allantoate was nine times higher than with (*S*)-ureidoglycolate. The Km values for allantoate and (*S*)-ureidoglycolate are 2 and 0.7 mM, respectively.

7.3.5 Ureidoglycolate Amidohydrolase

Ureidoglycolate amidohydrolase (UAH, EC 3.5.1.116) catalysis the reaction: (*S*)-ureidoglycolate + H_2O → glyoxylate +2 NH_3 + CO_2 (step 5 in Figure 7.6). An enzyme complex consisting of AAH and UAH was isolated from soybean seed coat (Winkler et al. 1988). Since no urea was produced by the enzyme, UAH is a new type NH_3-releasing enzyme, like AAH. The Km value for (*S*)-ureidoglycolate is 85 μM. UAH was partially purified from developing seeds of *Phaseolus vulgaris* (Wells and Lees 1991). The Vmax increased in the presence of Mn^{2+}. The Km values for ureidoglycolate in the presence and the absence of Mn^{2+} were 2.0 and 5.4 mM, respectively, and the pH optimum ranged from pH 7.0–8.5. The activity was associated with peroxisomes.

A candidate gene encoding an UAH in the *A. thaliana* genome was identified as a consequence of its similarity to the *AAH* gene and its possession of an ER signal peptide. The recombinant UAH can hydrolyse (*S*)-ureidoglycolate but not allantoate. The Km value is 39 μM (Werner et al. 2010).

7.3.6 (S)-Ureidoglycolate-urea Lyase

(*S*)-Ureidoglycolate-urea-lyase (ureidoglycolate lyase, EC 4.3.2.3) catalyses the reaction: (*S*)-ureidoglycolate → glyoxylate + urea (step 6 in Figure 7.6). The gene encoding this urea-lyase enzyme is not found in Arabidopsis database although the enzyme has been reported to occur in leguminous plants (Muñoz et al. 2001; Muñoz et al. 2006; Todd and Polacco 2004).

(*S*)-Ureidoglycolate-urea-lyase was purified 1500-fold from pods of chickpea. The native enzyme consists of six identical subunits. It exhibited hyperbolic kinetics for (*S*)-ureidoglycolate with Km values of 6 μM in the presence of Mn^{2+} (Muñoz et al. 2001). An (*S*)-ureidoglycolate-urea-lyase was purified to homogeneity from developing seeds of *Phaseolus vulgaris*. Its properties were similar to the chickpea enzyme mentioned above, but the activity of the French bean enzyme is dependent on Mn^{2+} and asparagine, and it has a much higher Km value (3.9 mM) for ureidoglycolate than the chickpea enzyme (Muñoz et al. 2006).

7.3.7 Urease

Urease (urea amidohydrolase, EC 3.5.1.5) catalyses the reaction: urea + H_2O → CO_2 + 2 NH_3. The one-step enzymatic hydrolysis of urea by urease yields NH_3 and carbamate.

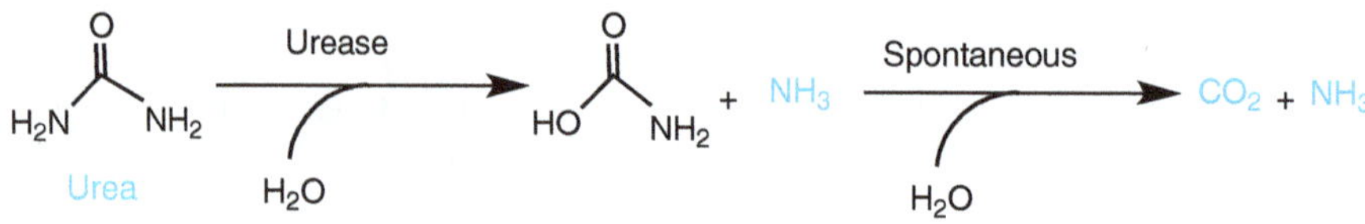

Figure 7.7 The reaction of urease. Enzymatic hydrolysis of urea to carbamate catalysed by urease and subsequent non-enzymatic decay of carbamate. Source: Adapted from Witte (2011).

The carbamate rapidly decays, non-enzymatically forming a second molecule of NH_3 and CO_2 (Figure 7.7.) see Witte (2011).

Since the isolation of urease from jack bean (*Canavalia ensiformis*) by Sumner (1926), there have been numerous investigations on characterization of this enzyme in a variety of organisms. Plant ureases are similar in sequence and structure to bacterial ureases and form a basic trimeric structure (Follmer 2008). Soybean has two urease isozymes; a highly expressed embryo-specific isozyme and a ubiquitous isozyme which is found in all tissues. In contrast, only a single urease gene occurs in *A. thaliana* (Witte 2011).

There are many kinetic reports on urease in plants. The Km values are usually high. For example, 0.45–0.9 mM for leaves of soybean and maize (*Zea mays*) (Davies and Shih 1984), 0.2 mM for leaves of mulberry (*Morus alba*) (Hirayama et al. 2000) and 3.0 mM for seeds of pigeon pea (*Cajanus cajan*) (Das et al. 2002). It has been reported that urease is located in the cytosol of jack bean seeds (Faye et al. 1986).

7.4 Purine Nucleotide Catabolism in Plants

Profiles of the catabolism of adenine nucleotides have been investigated using pulse-chase ^{14}C-feeding experiments with intact cultured plant cells (Yabuki and Ashihara 1991). Adenine nucleotides, pre-labelled by incubation of *Catharanthus roseus* cells with [8-^{14}C]adenosine, were catabolized rapidly and most of the radioactivity appeared in $^{14}CO_2$ (Figure 7.8). Allantoin and allantoate, intermediates of the oxidative catabolic pathway of purines, were labelled transiently. Incorporation into catabolites was inhibited completely by allopurinol, an inhibitor of xanthine hydrogenase. The rate of AMP catabolism is dependent upon the cellular concentration of nucleotides. To maintain the nucleotide level, excess nucleotides are removed by catabolism. Accumulation of radioactivity from [8-^{14}C]adenosine was not incorporated into other possible catabolites, including xanthine, urate, and urea (Figure 7.8). Therefore, the xanthine dehydrogenase- and uricase-catalysing steps are not rate limiting. On the basis of these findings, it is difficult to conclude whether or not the catabolic pathway uses the urea releasing reaction. It has been reported that in *Coffea arabica* radioactivity from [8-^{14}C]-labelled adenine, hypoxanthine, and xanthine was recovered as allantoin and urea (Nazario and Lovatt 1993).

The nitrogen atoms of positions 1, 3, 7, and 9 of purine skeletons are incorporated into NH_3 or urea. While position 6 carbon is released as CO_2 in the formation of ureide,

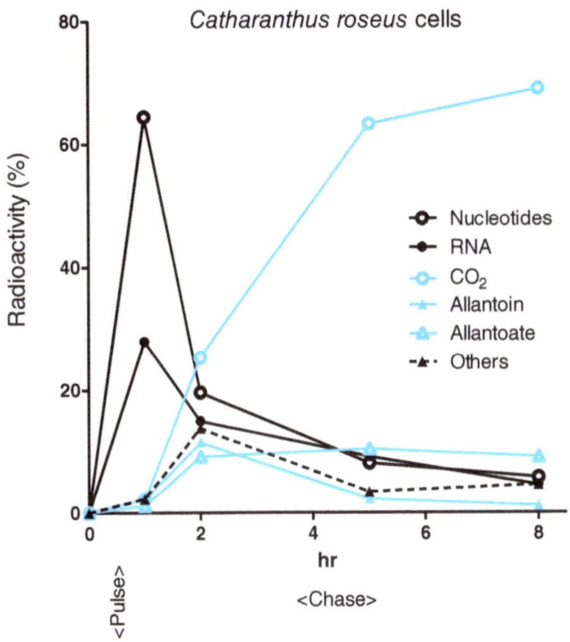

Figure 7.8 Metabolic fate of ^{14}C-labelled adenine nucleotides in intact *Catharanthus roseus* cells. After incubation with 5 μM [8-^{14}C]adenosine for 1 hour (pulse), ~90% of the radioactivity in the cells was distributed in free purine nucleotides and purine residues of RNA. The radioactivity was chased by replacement of the incubation medium with one containing 50 μM non-labelled adenosine for 7 hour. Most of radioactivity was transferred from nucleotides to allantoin, allantoate, and CO_2 (shown in red). The others fraction shown in the dotted line contained degradation metabolites, such as adenosine and adenine. When allopurinol, an inhibitor of xanthine dehydrogenase, was added, radioactivity accumulated in xanthine (data not shown). Source: Adapted from Yabuki and Ashihara (1991).

positions 2 and 8 are released as CO_2 during ureide degradation and positions 4 and 5 are retained in glyoxylate (Figure 7.9).

7.5 Accumulation and Utilization of Ureides in Plants

7.5.1 Ureides in Plant Tissues and Xylem Sap

The ureides, allantoin and allantoate have long been recognized as the major nitrogenous compounds in a variety of plant species (Schubert and Boland 1990; Thomas and Schrader 1981). Historically, a high concentration of allantoin was reported in the roots of *Platanus orientalis* (old world sycamore) by Schulze and Barbieri (1881). Since then allantoin has been detected in tissues of a number of plant genera including sycamore/maple (*Acer pseudoplatanus*), other woody species, and leguminous plants. Allantoate was discovered in the immature fruits of *Phaseolus vulgaris* by Fosse in 1926. Details of the occurrence of ureides in plants from a historical perspective have been described by Schubert and Boland (1990).

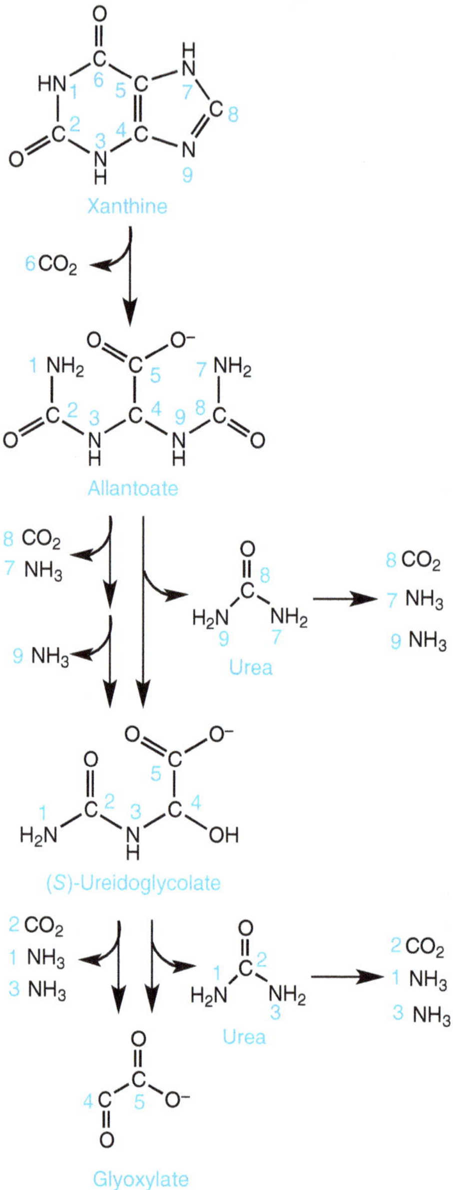

Figure 7.9 Fate of N and C atoms in xanthine structure during degradation of purine ring.

Fosse et al. (1931) reported that the allantoin level in seeds of 50 species from 11 families ranged from 0.01% to 0.33% d.w. with the highest levels occurring in seeds of *Vigna mungo* (black gram) and cowpea. Ureides have also been detected in the xylem sap of *A. pseudoplatanus* in which during the spring almost all nitrogen is in the form of allantoin and allantoate.

7.5.2 Role of Ureides in Nitrogen Storage and Transport

High levels of allantoin and allantoate occur in tissues and in the transport system of some plant species. In perennial species, such as maple and comfrey (*Symphytum officinale*), the concentration of the ureides is very high in dormant roots. In the spring, the ureides are exported from the roots, ascending through the xylem to the emerging leaves. Within the leaves, ureides are metabolized and the nitrogen released is reassimilated and utilized for protein and nucleic acid synthesis. As a consequence, the reserves of ureides in storage tissues are depleted. In the autumn this process is reversed. Protein and nucleic acids in senescing leaves are degraded and the nitrogen released is used to synthesize ureides, which are translocated to the roots for storage. Thus, allantoin and allantoate are central metabolites in the seasonal nitrogen cycle of these plants (Schubert and Boland 1990). The ureide-related nitrogen cycle in ureide-accumulating plants is summarized in Figure 7.10.

7.5.3 Role of Ureides in Germination and Development of Seeds

In some annual species, most notably tropical legumes, allantoin and allantoate are exported from the root system and accumulate in stems, petioles, leaves, pods, and seeds. Using a feeding experiment with [1, 3-^{15}N]allantoin, it was confirmed that ureide nitrogen was utilized as a source of nitrogen for the synthesis of seed protein in cowpea (Atkins et al. 1982). During seed germination the process is reversed. The levels of ureides increase in germinating seedlings accumulating in cotyledons, roots, and axes. The concentration of the ureides increases dramatically during soybean seedling

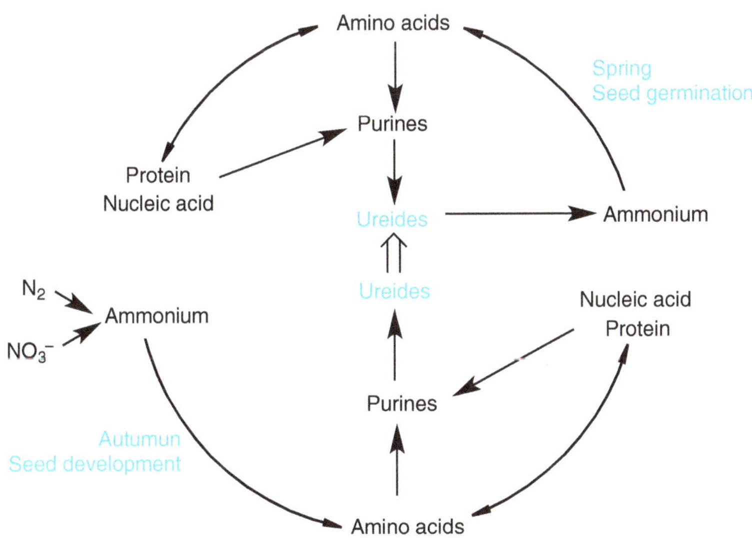

Figure 7.10 Ureide cycle in ureide-accumulating plants. Activity of the reactions in the cycle are altered by different circumstances, such as seasonal change (autumn and spring) and growth stages (developing seeds and germinating seeds). Source: adapted from Schubert and Boland (1990).

development (Matsumoto et al. 1977; Polayes and Schubert 1984). In soybean seedlings at the peak of ureide accumulation, ureides account for 0.7% and 6% of the total nitrogen of cotyledons and axes, respectively, (Polayes and Schubert 1984). Allantoin and allantoate are apparently formed as a result of turnover of seed storage reserves, transported throughout the plant, and used as a nitrogen source by the growing seedling (Polayes and Schubert 1984).

In dry seeds of black gram (*Vig. mungo*), the ureide pool is low, but during germination levels increase as they are formed from stored purine derivatives (Ashihara 1983). In cotyledons, the purine salvage activity was extremely high 12–48 hour after imbibition. [8-^{14}C]adenine absorbed into the cotyledons with imbibition (0–6 hour) was converted to ureides via AMP. The ureides formed in cotyledons are translocated to the embryonic axes. In the developing plants, ureides are degraded and the end products used for the growth of seedlings (Ashihara 1983).

NH_3 can be reassimilated by glutamine synthetase. Another end product, glyoxylate can be converted to malate by the glyoxylate cycle and further metabolized by the tricarboxylic acid cycle. Glyoxylate is also utilized directly for glycine formation. CO_2 may be released to the atmosphere, but HCO_3^- in cells can be fixed by phosphoenolpyruvate carboxylase and other carboxylases, such as carbamoyl phosphate synthetase (see Chapter 8) (Ashihara 1983; Bowsher et al. 2008).

7.5.4 Ureide Formation in Nodules of Tropical Legumes

In symbiotic nitrogen fixation, prokaryotes supply host plants with reduced nitrogen in the form of NH_3. The major nitrogen metabolites of symbiotically-fixed N_2 which are exported from root nodules, are different in temperate and the tropical legume species that exhibit either amide- or a ureide-type of nitrogen metabolism. In the amide-type species, such as alfalfa (*Medicago sativa*), pea (*Pisum sativum*), broad bean (*Vic. faba*), lupin (*Lupinus sp.*), and trefoil (*Lotus sp.*), asparagine and/or glutamine are the major metabolites exported from the nodules. In contrast, allantoin and allantoate are the principal metabolites in the tropical legumes of the tribe Phaseoleae, such as *Glycine, Vigna,* and *Phaseolus.* Around 70–90% of symbiotically-fixed root nitrogen is transported into the shoots as the ureides, allantoin, and allantoate (Reynolds et al. 1982; Schubert 1986).

Therefore, ureide synthesis and catabolism are carried out in the nodules of a limited number of plant species, some of which, such as soybean, are of economic importance.

The synthesis of ureides from NH_3 produced by N_2 fixation in nodules, and catabolism of ureides in shoots of tropical ureide-type legumes are summarized in Figure 7.11. Ammonia released by bacteroids in the nodules is assimilated and used to generate purines, which are then partially degraded to allantoin and allantoate. These ureides are exported from the nodules via the xylem into the shoot where they are degraded by the action of catabolic enzymes.

7.5.5 Other Role of Ureides in Plants

Ureides appear to be involved in protecting plants from abiotic stress. In *A. thaliana* seedlings, allantoin and allantoate accumulate as a result of stresses induced by darkness and natural senescence. The ureides may protect the leaves from damage induced by reactive oxygen species. See a review by Werner and Witte (2011).

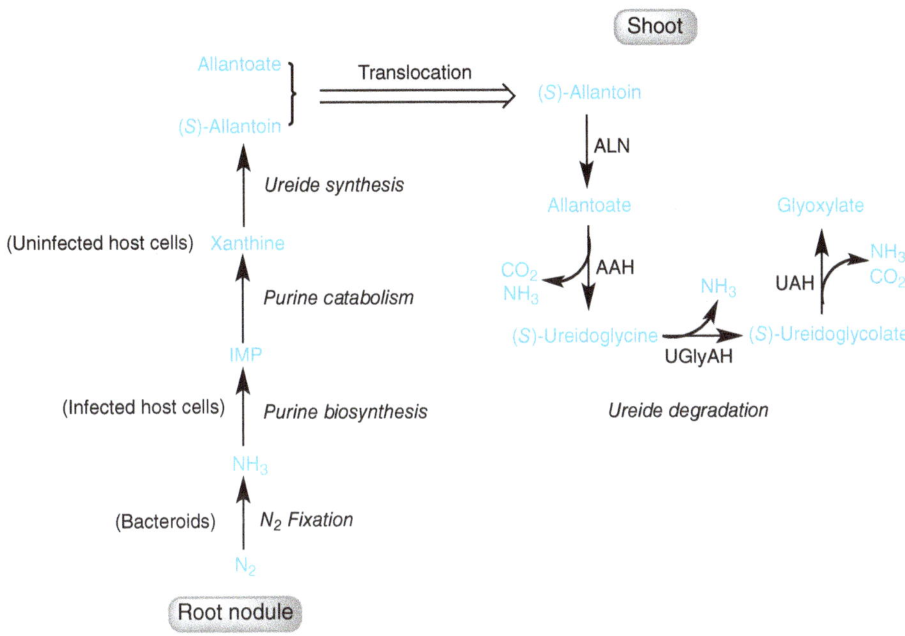

Figure 7.11 Ureide synthesis in nodules and catabolism in shoots in soybean (*Glycine max*). Source: Adapted from Werner et al. (2013).

7.6 Summary

The end products of purine catabolism in various animals differ from those found in plants. Xanthine is the starting material of purine degradation and end products in plants are NH_3, CO_2, and glyoxylate. In some plant species, there is a transient accumulation of allantoin and allantoate, intermediates of purine catabolism, which function as storage reserves or as transporters of nitrogen.

References

Amarjit and Singh, R. (1985). Allantoinase from nodules of pigeonpea (*Cajanus cajan*). *Phytochemistry* 24: 415–418.

Ashihara, H. (1983). Changes in activities of purine salvage and ureide synthesis during germination of black gram (*Phaseolus mungo*) seeds. *Z. Pflanzenphysiol.* 113: 47–60.

Ashihara, H. and Nobusawa, E. (1981). Metabolic fate of [8-^{14}C]adenine and [8-^{14}C]hypoxanthine in higher plants. *Z. Pflanzenphysiol.* 104: 443–458.

Ashihara, H., Takasawa, Y., and Suzuki, T. (1997). Metabolic fate of guanosine in higher plants. *Physiol. Plant.* 100: 909–916.

Atkins, C.A., Pate, J.S., Ritchie, A., and Peoples, M.B. (1982). Metabolism and translocation of allantoin in ureide-producing grain legumes. *Plant Physiol.* 70: 476–482.

Baldwin, E. (1970). *An Introduction to Comparative Biochemistry*, 4e. Cambridge University Press.

Boland, M.J. and Schubert, K.R. (1983). Biosynthesis of purines by a proplastid fraction from soybean nodules. *Arch. Biochem. Biophys.* 220: 179–187.

Boland, M.J., Blevins, D.G., and Randall, D.D. (1983). Soybean nodule xanthine dehydrogenase: A kinetic study. *Arch. Biochem. Biophys.* 222: 435–441.

Bowsher, C., Steer, M., and Tobin, A. (2008). *Plant Biochemistry*. New York: Garland Science.

Corte, E.D. and Stirpe, F. (1972). The regulation of rat liver xanthine oxidase. Involvement of thiol groups in the conversion of the enzyme activity from dehydrogenase (type D) into oxidase (type O) and purification of the enzyme. *Biochem. J.* 126: 739–745.

Das, N., Kayastha, A.M., and Srivastava, P.K. (2002). Purification and characterization of urease from dehusked pigeonpea (*Cajanus cajan* L.) seeds. *Phytochemistry* 61: 513–521.

Davies, H.M. and Shih, L.-M. (1984). Urease from leaves of *Glycine max* and *Zea mays*. *Phytochemistry* 23: 2741–2745.

Faye, L., Greenwood, J.S., and Chrispeels, M.J. (1986). Urease in jack-bean (*Canavalia ensiformis* (L.) DC) seeds is a cytosolic protein. *Planta* 168: 579–585.

Florkin, M. (1949). *Biochemical Evolution*. New York: Academic Press.

Follmer, C. (2008). Insights into the role and structure of plant ureases. *Phytochemistry* 69: 18–28.

Fosse, R., Brunel, A., and Thomas, P.-E. (1931). Application de l'analyse quantitative spectrophotométrique de l'allantoïne au sang de quel- ques mammifères et à la graine de nombreux végétaux. *C. R. Acad. Sci. Paris* 193: 7–11.

Hanks, J.F., Tolbert, N.E., and Schubert, K.R. (1981). Localization of enzymes of ureide biosynthesis in peroxisomes and microsomes of nodules. *Plant Physiol.* 68: 65–69.

Henderson, J.F. and Paterson, A.R.P. (1973). *Nucleotide Metabolism – An Introduction*. New York: Academic Press.

Hesberg, C., Hänsch, R., Mendel, R.R., and Bittner, F. (2004). Tandem orientation of duplicated xanthine dehydrogenase genes from *Arabidopsis thaliana*: Differential gene expression and enzyme activities. *J. Biol. Chem.* 279: 13547–13554.

Hirayama, C., Sugimura, M., Saito, H., and Nakamura, M. (2000). Purification and properties of urease from the leaf of mulberry, *Morus alba*. *Phytochemistry* 53: 325–330.

Kahn, K. and Tipton, P.A. (1997). Kinetic mechanism and cofactor content of soybean root nodule urate oxidase. *Biochemistry* 36: 4731–4738.

Kahn, K., Serfozo, P., and Tipton, P.A. (1997). Identification of the true product of the urate oxidase reaction. *J. Am. Chem. Soc.* 119: 5435–5442.

Kim, K., Park, J., and Rhee, S. (2007). Structural and functional basis for (*S*)-allantoin formation in the ureide pathway. *J. Biol. Chem.* 282: 23457–23464.

Lamberto, I., Percudani, R., Gatti, R. et al. (2010). Conserved alternative splicing of Arabidopsis transthyretin-like determines protein localization and *s*-allantoin synthesis in peroxisomes. *Plant Cell* 22: 1564–1574.

Lucas, K., Boland, M.J., and Schubert, K.R. (1983). Uricase from soybean root nodules: purification, properties, and comparison with the enzyme from cowpea. *Arch. Biochem. Biophys.* 226: 190–197.

Mary, A. and Sivarama Sastry, K. (1978). An unusual allantoinase from *Dolichos biflorus*. *Phytochemistry* 17: 397–399.

Matsumoto, T., Yatazawa, M., and Yamamoto, Y. (1977). Distribution and change in the contents of allantoin and allantoic acid1 in developing nodulating and non-nodulatng soybean plants. *Plant Cell Physiol.* 18: 353–359.

Montalbini, P., Redondo, J., Caballero, J.L. et al. (1997). Uricase from leaves: its purification and characterization from three different higher plants. *Planta* 202: 277–283.

Muñoz, A., Piedras, P., Aguilar, M., and Pineda, M. (2001). Urea is a product of ureidoglycolate degradation in chickpea. Purification and characterization of the ureidoglycolate urea-lyase. *Plant Physiol.* 125: 828–834.

Muñoz, A., Raso, M., Pineda, M., and Piedras, P. (2006). Degradation of ureidoglycolate in French bean (*Phaseolus vulgaris*) is catalysed by a ubiquitous ureidoglycolate urea-lyase. *Planta* 224: 175–184.

Nagai, Y. and Funahashi, S. (1961). Allantoinase from mung bean seedlings. *Agric. Biol. Chem.* 25: 265–268.

Nazario, G.M. and Lovatt, C.J. (1993). Regulation of purine metabolism in intact leaves of *Coffea arabica*. *Plant Physiol.* 103: 1195–1201.

Nirmala, J. and Sastry, K.S. (1975). The allantoinase of *Lathyrus sativus*. *Phytochemistry* 14: 1971–1973.

Pessoa, J., Sárkány, Z., Ferreira-da-Silva, F. et al. (2010). Functional characterization of *Arabidopsis thaliana* transthyretin-like protein. *BMC Plant Biol.* 10: 30.

Piedras, P., Muñoz, A., Aguilar, M., and Pineda, M. (2000). Allantoate amidinohydrolase (allantoicase) from *Chlamydomonas reinhardtii*: Its purification and catalytic and molecular characterization. *Arch. Biochem. Biophys.* 378: 340–348.

Polayes, D.A. and Schubert, K.R. (1984). Purine synthesis and catabolism in soybean seedlings. The Biogenesis of Ureides. *Plant Physiol.* 75: 1104–1110.

Ramazzina, I., Folli, C., Secchi, A. et al. (2006). Completing the uric acid degradation pathway through phylogenetic comparison of whole genomes. *Nat. Chem. Biol.* 2: 144–148.

Raso, M.J., Muñoz, A., Pineda, M., and Piedras, P. (2007a). Biochemical characterisation of an allantoate-degrading enzyme from French bean (*Phaseolus vulgaris*): the requirement of phenylhydrazine. *Planta* 226: 1333–1342.

Raso, M.J., Pineda, M., and Piedras, P. (2007b). Tissue abundance and characterization of two purified proteins with allantoinase activity from French bean (*Phaseolus vulgaris*). *Physiol. Plant.* 131: 355–366.

Reynolds, P.H.S., Blevins, D.G., Boland, M.J. et al. (1982). Enzymes of ammonia assimilation in legume nodules: A comparison between ureide- and amide-transporting plants. *Physiol. Plant.* 55: 255–260.

Sarma, A.D., Serfozo, P., Kahn, K., and Tipton, P.A. (1999). Identification and purification of hydroxyisourate hydrolase, a novel ureide-metabolizing enzyme. *J. Biol. Chem.* 274: 33863–33865.

Schubert, K.R. (1986). Products of biological nitrogen fixation in higher plants: Synthesis, transport, and metabolism. *Annu. Rev. Plant Physiol.* 37: 539–574.

Schubert, K.R. and Boland, M.J. (1990). The ureides. In: *The Biochemistry of Plants* (eds. P.K. Stump and E.E. Conn), 197–282. San Diego, CA: Academic Press.

Schulze, E. and Barbieri, J. (1881). Ueber das Vorkommen von Allantoin im Pflanzenorganismus. *Berichte der deutschen chemischen Gesellschaft* 14: 1602–1605.

Serventi, F., Ramazzina, I., Lamberto, I. et al. (2010). Chemical basis of nitrogen recovery through the ureide pathway: Formation and hydrolysis of *S*-ureidoglycine in plants and bacteria. *ACS Chem. Biol.* 5: 203–214.

Shelp, B. and Atkins, C.A. (1983). Role of inosine monophosphate oxidoreductase in the formation of ureides in nitrogen-fixing nodules of cowpea (*Vigna unguiculata* L. Walp.). *Plant Physiol.* 72: 1029–1034.

Smith, M.B. and March, J. (2007). *Chapter 4: Stereochemistry. March's Advanced Organic Chemistry: Reactions, Mechanisms, and Structure*, 6e, 136–233. Hoboken, NJ: Wiley.

Sumner, J.B. (1926). The isolation and crystallization of the enzyme urease: Preliminary paper. *J. Biol. Chem.* 69: 435–441.

Tajima, S. and Yamamoto, Y. (1975). Enzymes of purine catabolism in soybean plants. *Plant Cell Physiol.* 16: 271–282.

Theimer, R.R. and Beevers, H. (1971). Uricase and allantoinase in glyoxysomes. *Plant Physiol.* 47: 246–251.

Thomas, R.J. and Schrader, L.E. (1981). Ureide metabolism in higher plants. *Phytochemistry* 20: 361–371.

Thomas, R.J., Meyers, S., and Schrader, L.E. (1983). Allantoinase from shoot tissues of soybeans. *Phytochemistry* 22: 1117–1120.

Todd, C.D. and Polacco, J.C. (2004). Soybean cultivars 'Williams 82' and 'Maple Arrow' produce both urea and ammonia during ureide degradation. *J. Exp. Bot.* 55: 867–877.

Todd, C. and Polacco, J. (2006). *AtAAH* encodes a protein with allantoate amidohydrolase activity from *Arabidopsis thaliana*. *Planta* 223: 1108–1113.

Triplett, E.W., Blevins, D.G., and Randall, D.D. (1982). Purification and properties of soybean nodule xanthine dehydrogenase. *Arch. Biochem. Biophys.* 219: 39–46.

Van Der Drift, C., De Windt, F.E., and Vogels, G.D. (1970). Allantoate hydrolysis by allantoate amidohydrolase. *Arch. Biochem. Biophys.* 136: 273–279.

Venkateswara Rao, N., Subhash Reddy, R., and Sivarama Sastry, K. (1988). Allantoinases of nodulated *Arachis hypogaea*. *Phytochemistry* 27: 693–695.

Voet, D. and Voet, J.G. (2010). *Biochemistry*. Hoboken, NJ: Wiley.

Vogels, G.D. and Van der Drift, C. (1976). Degradation of purines and pyrimidines by microorganisms. *Bacteriol. Rev.* 40: 403–468.

Vogels, G.D., Trijbels, F., and Uffink, A. (1966). Allantoinases from bacterial, plant and animal sources I. Purification and enzymic properties. *Biochim. Biophys. Acta* 122: 482–496.

Webb, M.A. and Lindell, J.S. (1993). Purification of allantoinase from soybean seeds and production and characterization of anti-allantoinase antibodies. *Plant Physiol.* 103: 1235–1241.

Wells, X.E. and Lees, E.M. (1991). Ureidoglycolate amidohydrolase from developing french bean fruits (*Phaseolus vulgaris* [L.].). *Arch. Biochem. Biophys.* 287: 151–159.

Werner, A.K. and Witte, C.P. (2011). The biochemistry of nitrogen mobilization: Purine ring catabolism. *Trend. Plant Sci.* 16: 381–387.

Werner, A.K., Sparkes, I.A., Romeis, T., and Witte, C.-P. (2008). Identification, biochemical characterization, and subcellular localization of allantoate amidohydrolases from Arabidopsis and soybean. *Plant Physiol.* 146: 418–430.

Werner, A.K., Romeis, T., and Witte, C.-P. (2010). Ureide catabolism in *Arabidopsis thaliana* and *Escherichia coli*. *Nat. Chem. Biol.* 6: 19–21.

Werner, A.K., Medina-Escobar, N., Zulawski, M. et al. (2013). The ureide-degrading reactions of purine ring catabolism employ three amidohydrolases and one aminohydrolase in arabidopsis, soybean, and rice. *Plant Physiol.* 163: 672–681.

Winkler, R.G., Polacco, J.C., Blevins, D.G., and Randall, D.D. (1985). Enzymic degradation of allantoate in developing soybeans. *Plant Physiol.* 79: 787–793.

Winkler, R.G., Blevins, D.G., and Randall, D.D. (1988). Ureide catabolism in soybeans III. Ureidoglycolate amidohydrolase and allantoate amidohydrolase are activities of an allantoate degrading enzyme complex. *Plant Physiol.* 86: 1084–1088.

Witte, C.-P. (2011). Urea metabolism in plants. *Plant Sci.* 180: 431–438.

Yabuki, N. and Ashihara, H. (1991). Catabolism of adenine nucleotides in suspension-cultured plant cells. *Biochim. Biophys. Acta* 1073: 474–480.

Part III

Pyrimidine Nucleotide Metabolism

8

Pyrimidine Nucleotide Biosynthesis *De Novo*

8.1 Introduction

The *de novo* pyrimidine biosynthetic pathway (aka the orotate pathway), defined as the synthesis of uridine-5′-monophosphate (UMP) from carbamoyl-phosphate (carbamoyl-P) (steps 1–5b in Figure 8.1), is evolutionarily conserved from prokaryotes to eukaryotes (Henderson and Paterson 1973). The production of the nucleotide cytidine-triphosphate (CTP) from UMP is performed at the nucleoside triphosphate level with uridine-triphosphate (UTP) acting as the immediate precursor (steps 6–8 in Figure 8.1). Thymidine nucleotides are produced from deoxyuridine-5′-monophosphate (dUMP). The pathway from UMP to thymine deoxyribonucleoside triphosphate, a precursor of DNA synthesis (Figure 8.10) is described in this chapter.

As illustrated in Figure 8.1, in *de novo* pyrimidine nucleotide biosynthesis glutamine, aspartate, ATP, and HCO_3^- are used as substrates. The pathway consists of (i) the synthesis of carbamoyl-aspartate, (ii) formation of orotate, and (iii) the production of UMP. The pathway was first investigated in bacteria and mammals. Step ii, namely the carbamoyl-aspartate → dihydroorotate → orotate route was elucidated by Lieberman and Kornberg (1954). In the 1950s, the small molecule precursors of the pyrimidine ring atoms were identified. The biogenesis of orotate from different precursors was investigated using rat liver slices. Carbon dioxide was found to be a specific precursor for the ureide carbon (C-2) of nucleic acid pyrimidines. Using [^{15}N]-, [1,4-^{13}C]-, and [2,3-^{14}C]L-aspartic acid, it was demonstrated that N-3 plus C-4, C-5, C-6, and C-7 of orotic acid were derived from aspartate (Reichard and Lagerkvist 1953). This finding was confirmed using radiolabelled precursors which demonstrated the rapid and very specific incorporation of ^{14}C-bicarbonate into C-2 of the soluble uridine nucleotides in Ehrlich ascites tumour cells (Hager and Jones 1965; Kusama and Roberts 1963). A detailed historical review on this topic can be found in Henderson and Paterson (1973).

Evidence of the existence of various parts of the orotate pathway in plants, based on ^{14}C-feeding experiments, enzyme assays and inhibitor studies, began to accumulate in the 1960s. The entire *de novo* pyrimidine pathway was demonstrated in *Cucurbita pepo* cells (Lovatt 1983). Papers on pyrimidine biosynthesis in plants published up to 1980 were reviewed by Ross (1981). Molecular analysis of pyrimidine biosynthesis in plants began after the unravelling of the Arabidopsis genome (Arabidopsis-Genome-Initiative 2000). In *Arabidopsis thaliana*, there is only one gene locus for each enzyme of the *de novo* pathway (Boldt and Zrenner 2003; Moffatt and Ashihara 2002).

Plant Nucleotide Metabolism – Biosynthesis, Degradation, and Alkaloid Formation, First Edition.
Hiroshi Ashihara, Iziar A. Ludwig and Alan Crozier.

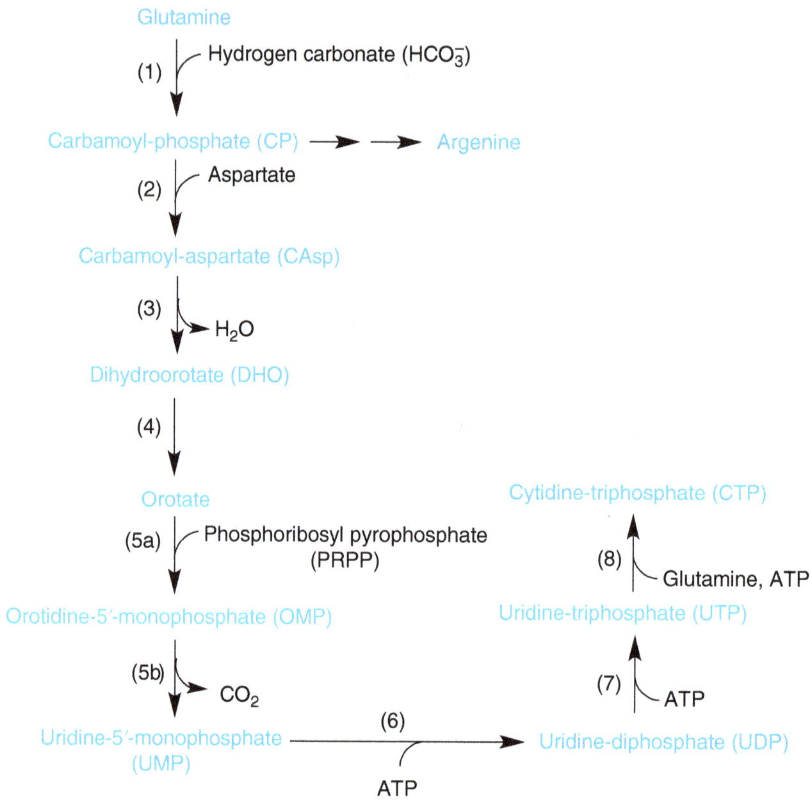

Figure 8.1 Outline of pyrimidine nucleotide biosynthesis *de novo*. Usually formation of UMP (steps 1 to 5b) is defined as the *de novo* pathway of pyrimidine nucleotides or the orotate pathway. Cytidine nucleotides biosynthesis *de novo* (steps 6–8) is also shown. Enzymes: (1) carbamoyl-phosphate synthetase, (2) aspartate carbamoyltransferase, (3) dihydroorotase, (4) Dihydroorotate dehydrogenase, (5a) orotate phosphoribosyltransferase, (5b) orotidine-5′-monophosphate decarboxylase, (6) UMP kinase, (7) nucleoside diphosphate kinase, (8) cytidine triphosphate synthase.

Although chemical reactions of pyrimidine *de novo* synthesis themselves are conserved from prokaryotes to eukaryotes, there are distinct differences in the enzymes in bacteria, yeast, and higher eukaryotes (Nara et al. 2000). In *Escherichia coli* and other bacteria, each enzymatic step is catalysed by a different protein that is encoded by a specific gene (Neuhard and Nygaard 1987). In plants, the first three steps are catalysed by three separate enzymes (Doremus 1986), whereas in mammals all three steps are carried out by a single multi-functional protein namely carbamoyl-phosphate synthase II aspartate carbamoyltransferase dihydroorotase (Christopherson and Szabados 1997; Iwahana et al. 1996).

In mammals and plants, the last two steps are catalysed by distinct domains of a bi-functional protein called UMP synthase (Nasr et al. 1994). From a phylogenetic viewpoint, the organization of the pathway in plants is chimeric: carbamoyl-P synthetase small- and large subunit clusters in the cyanobacteria, dihydroorotase in proteobacteria, aspartate carbamoyltransferase in eukaryotes, and dihydroorotate dehydrogenase and UMP synthase in higher eukaryotes (Nara et al. 2000). A number of publications have reviewed *de novo* plant pyrimidine nucleotide biosynthesis in the context of nucleotide metabolism (Moffatt and Ashihara 2002; Ross 1981; Stasolla et al. 2003; Zrenner and Ashihara 2011; Zrenner et al. 2006).

8.2 Reactions and Enzymes of the *De Novo* Biosynthesis

An outline of the *de novo* biosynthesis of UMP, its conversion to CTP, and the enzymes involved are presented in Figure 8.1 and Table 8.1. Table 8.1 includes the reaction steps shown in Figure 8.1, the Enzyme Commission (EC) number, the abbreviations of the enzymes and gene names used in *E. coli,* and the Arabidopsis Information Resource (TAIR) loci.

8.2.1 Synthesis of Carbamoyl-phosphate

Production of carbamoyl-P is the first step in the biosynthesis of both pyrimidine nucleotides and arginine (Sasamoto and Ashihara 1988). Three forms of carbamoyl-P

Table 8.1 Summary of enzymes involved in the pyrimidine biosynthesis *de novo* in plants illustrated in Figure 8.1.

				Gene[a]	
Step	**Enzyme name**	**EC**	**Abbreviation**	***E. coli***	**TAIR locus**
1	Carbamoyl phosphate synthetase	6.3.5.5	CPS		
1a	(Small subunit)			*carA*	AT3G27740
1b	(Large subunit)			*carB*	AT1G29900
2	Aspartate carbamoyltransferase	2.1.3.2	ACT	*pyrB*	AT3G20330
3	Dihydroorotase	3.5.2.3	DHO	*pyrC*	AT4G22930
4	Dihydroorotate dehydrogenase	1.3.5.2	DHODH	*pyrD*	AT5G23300
5	UMP synthase		UMPS	*pyrEF*	AT3G54470
5a	Orotate phosphoribosyltransferase	2.4.2.10	OPRT		
5b	Orotidine-5′-monophosphate decarboxylase	4.1.1.23	ODC		
6	UMP/CMP kinase	2.7.4.14	UMPK	*pyr6*	AT5G26667
7	Nucleoside diphosphate kinase	2.7.4.6	NDPK	*ndp*	AT4G09320[b]
8	Cytidine triphosphate synthase	6.3.4.4	CTPS	*ntps*	AT4G20320[b]

a) Gene designations used for *E. coli* are shown. These gene names used are also often used in plant literature.

b) Only a typical locus is shown.

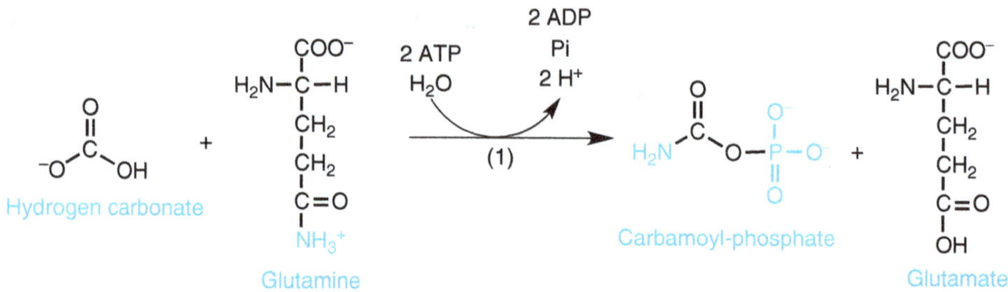

Figure 8.2 Reaction 1 in Figure 8.1. (1) Carbamoyl-phosphate synthetase.

synthetase have been reported, categorized as CPSI, CPSII, and CPSIII. In mammals, *N*-acetyl-L-glutamate-dependent carbamoyl-P synthetase (CPSI, EC 6.3.4.16), localized in mitochondria of liver cells, serves as the first enzyme for arginine and urea synthesis. Glutamine-dependent carbamoyl-P synthetase (CPSII, EC 6.3.5.5) is present in the cytosol of growing prokaryote and eukaryote cells and provides carbamoyl-P for pyrimidine synthesis (Tatibana 1978). CPSIII occurs only in fish (Saha et al. 2007).

CPSII occurs only in higher plants, and generates carbamoyl-phosphate for the biosynthesis of pyrimidines as well as arginine (Ross 1981; Sasamoto and Ashihara 1988; Stasolla et al. 2003). Pyrimidine biosynthesis *de novo* starts with the formation of carbamoyl-P from glutamine (Figure 8.2). This reaction (step 1 in Figure 8.1) is catalysed by carbamoyl-P synthetase (CPS, EC 6.3.5.5). The full IUBMB name for this enzyme is carbamoyl-phosphate synthase (glutamine-hydrolysing).

Native CPSII was partially purified from shoots of pea (*Pisum sativum*) (O'Neal and Naylor 1969) and mung bean (*Phaseolus aureus)* seedlings (Ong and Jackson 1972b). Both enzymes were more active with glutamine than ammonia. The Km value for glutamine was 0.1–0.2 mM, but for ammonia was >6 mM. UMP is a potent feedback inhibitor (Ki value: <2 μM). Ornithine stimulates the activity and counteracts inhibition by UMP (O'Neal and Naylor 1976; Ong and Jackson 1972b). Other activators were inosine-5′-monophosphate (IMP) and some purine nucleotides. The activators ornithine and IMP completely reversed inhibition caused by pyrimidine nucleotides (O'Neal and Naylor 1976). This is in agreement with the hypothesis that a cell with sufficient purine nucleotides diverts carbon, nitrogen, and phosphate into pyrimidine nucleotides (Ross 1981). Feedback control mechanism of pyrimidine nucleotide biosynthesis is discussed in Section 8.3.1.

Plant CPSII consists of two subunits. The small subunit exhibits a glutamine-amido-transferase activity and provides ammonia to the large subunit which catalyses the CPS reaction and possesses regulatory properties. Cloning and characterization of the sub-unit cDNAs from *A. thaliana*, *Nicotiana tabacum*, and *Medicago sativa (*alfalfa aka lucerne) revealed that they are encoded by individual genes (Giermann et al. 2002). The amino acid sequences deduced from these cDNAs suggest that the plant proteins are synthesized as precursors that contain chloroplast-transit peptides.

The localisation of CPSII has been demonstrated biochemically in isolated chloro-plasts from spinach (*Spinacia okracca*), soybean (*Glycine max*), and maize (*Zea mays*) leaves, and radish (*Raphanus sativus*). Light induces a marked increase in CPSII activity in dark-grown radish seedlings (Shibata et al. 1986). The chloroplast CPSII comprises about >65% of the total CPSII activity in radish cotyledon cells with the remainder

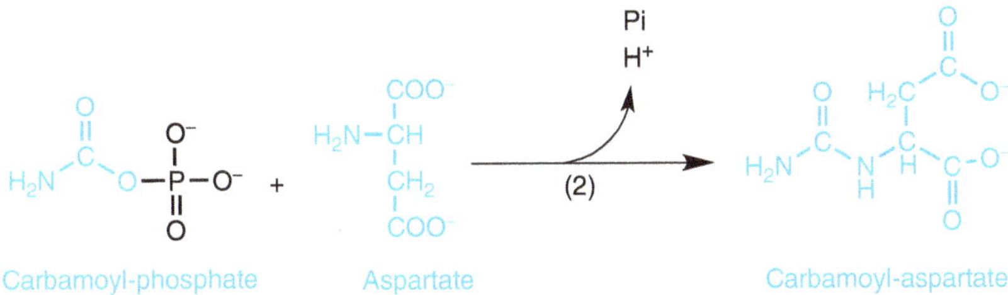

Figure 8.3 Reaction 2 in Figure 8.1. (2) Aspartate transcarbamoylase.

potentially in the cytosol (Shibata et al. 1986). Major activity of CPSII was found in the cytosol of dark grown *Catharanthus roseus* cells (Kanamori et al. 1980). Pyrimidine *de novo* pathway has been detected in non-photosynthetic tissues, such as roots of summer squash (*C. pepo*) (Lovatt et al. 1979). Therefore, in addition to chloroplasts, CPSII may be localized in the cytosol and/or proplastids of these cells.

8.2.2 Formation of Carbamoyl-aspartate

Aspartate transcarbamoylase (ATC) (EC 2.1.3.2) (aka aspartate transcarbamoylase) catalyses the reaction (step 2 in Figure 8.1), carbamoyl-P + L-aspartate → *N*-carbamoyl-L-aspartate + Pi (Figure 8.3).

The enzymatic formation of carbamoyl-aspartate is reversible, but the equilibrium strongly favours the formation of carbamoyl-aspartate. Compared with the first enzyme, CPSII, ATC activity is 20–1000 times higher and it is stable in plant extracts (Kanamori et al. 1980; Mazuś and Buchowicz 1972). It is, therefore, easy to purify and characterize and consequently there are many reports on native ATC from several plant species (Achar et al. 1974; Jagannatha Rao et al. 1979; Khan et al. 1999; Mazuś and Buchowicz 1972; Ong and Jackson 1972a; Yon 1972, 1973). In general, pH optima are ~10. The Km values for carbamoyl-P and aspartate are 0.1 and 1 mM, respectively. The end product, UMP, in the range of 0.1–1.0 mM, inhibits the activity of this enzyme.

The cDNA encoding ATC in *A. thaliana* has been isolated and characterized functionally (Nasr et al. 1994). Three cDNAs encoding different pea (*P. sativum*) ATCs were obtained by complementation of an *ApyrS* mutant of *E. coli*. They exhibit typical chloroplast transit peptide sequences. Subcellular localisation of ATC in chloroplasts has also been reported in leaves of pea seedlings (Doremus and Jagendorf 1985). In the recombinant pea ATC, the carbamoyl-P and aspartate binding domains are highly homologous with those of other prokaryotic and eukaryotic ATCs and critical active-site residues are fully conserved. The pea ATC also exhibits a putative pyrimidine-binding site (Kafer et al. 2004; Williamson and Slocum 1994). Feed-back control of the pyrimidine biosynthesis is exerted on the activity of CPSII and ATC (see Section 8.3.1).

8.2.3 Formation of Dihydroorotase from Carbamoyl-aspartate

Conversion of carbamoyl-aspartate to dihydroorotate (step 3 in Figure 8.1) is catalysed by dihydroorotase (DHO), also known as dihydroorotate hydrolase (EC 3.5.2.3), which catalyses *N*-carbamoyl-L-aspartate ↔ dihydroorotate + H_2O (Figure 8.4). The native DHO from pea leaves has been purified and characterized (Mazus and Buchowicz

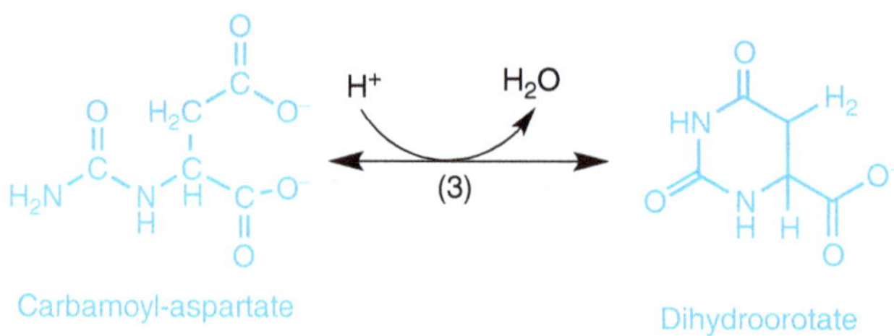

Figure 8.4 Reaction 3 in Figure 8.1. (3) Dihydroorotase.

1968). The reaction is reversible. The optimum pH for dihydroorotate formation is ~6 and the Km value for carbamoyl-aspartate is high at 6.2 mM.

Sequences of the gene encoding proteins with DHO activity have been characterized from potato (*Solanum tuberosum*) and *A. thaliana* (Giermann et al. 2002). The amino acid sequences deduced from these cDNAs suggest that DHO is synthesized via a precursor containing an *N*-terminal transit peptide, which indicates localisation in plastids or mitochondria. Chloroplastic localisation of DHO activity in pea leaves has been demonstrated biochemically (Doremus and Jagendorf 1985).

8.2.4 Formation of Orotate from Dihydroorotate

Dihydroorotate is converted to orotate by dihydroorotate dehydrogenase (DHODH, EC 1.3.5.2) (step 4 in Figure 8.1). This 'class 2'-DHODH enzyme contains flavin mononucleotide (FMN) and is located in the mitochondrial membrane of eukaryotes (Fagan et al. 2006). The catalysed reaction is dihydroorotate + quinone → orotate + quinol (Figure 8.5).

NAD-dependent DHODH has been demonstrated in extracts of wheat embryos (Kapoor and Waygood 1965). However, this is involved in 'class 1'-DHODH (EC 1.3.1.14) which uses NAD^+ as the electron acceptor which is not common in plants. In biochemical studies, occurrence of DHODH in mitochondria has been reported in ruptured pea leaf protoplasts (Doremus and Jagendorf 1985) and in cultured tomato (*Solanum lycopersicum*) cells (Miersch et al. 1986).

The gene encoding proteins with DHODH activity have been characterized from *A. thaliana* and tobacco and they belong to the membrane bound 'class 2'-DHODH (Giermann et al. 2002). The deduced amino acid sequences suggest that the plant protein is synthesized via a precursor containing an *N*-terminal transit peptide for mitochondrial targeting. The substrate specificity of the recombinant *A. thaliana* DHODH has been investigated. The Km values are 40 μM for dihydroorotate, and

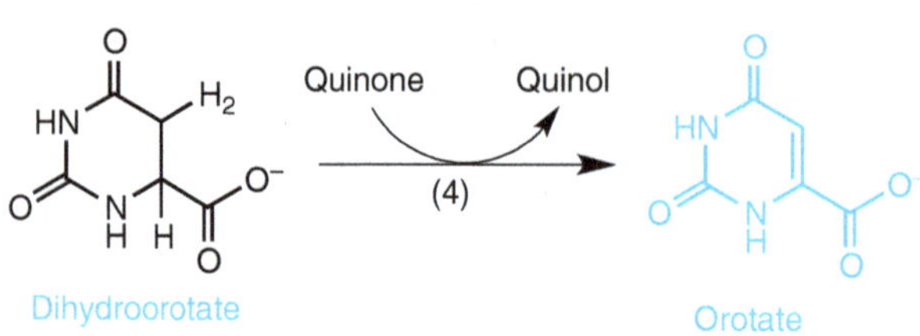

Figure 8.5 Reaction 4 in Figure 8.1. (4) Dihydroorotate dehydrogenase.

112 μM for decylubiquinone (Q_D). The endogenous electron acceptor of plant DHODH is not known. Coenzyme Q_{10} (ubiquinone) is present in all parts of the plant whereas plastoquinone and menaquinones (vitamins K1 and K2) appear to occur only in green tissues. Decylplastoquinone, Q_D, and 2,5-dimethyl-*p*-benzoquinone can also be used as co-substrates for recombinant *A. thaliana* DHODH (Ullrich et al. 2002).

8.2.5 Synthesis of UMP from Orotate

The conversion of orotate to UMP includes two different reactions (steps 5a and 5b in Figure 8.1). The first step (5a) involves orotate phosphoribosyltransferase (OPRT, EC 2.4.2.10) which catalyses the reaction: orotate + 5-phosphoribosyl-1-pyrophosphate (PRPP) → orotidine-5′-monophosphate (OMP) + PPi (Figure 8.6).

The second step (5b) is catalysed by orotidine-5′-monophosphate decarboxylase (ODC) (EC 4.1.1.23). The reaction is: OMP → UMP + CO_2 (Figure 8.7).

In higher eukaryotes, including plants, OPRT and ODC occur as an enzyme complex known as UMP synthase. Purification and characterization of the native OPRT from several plants have been reported including wheat embryos (Kapoor and Waygood 1965), cotyledons and shoot apices of pea seedlings (Murray and Ross 1971), and *Phaseolus mungo* (black gram) seedlings (Ashihara 1978). ODC has been detected in leaves of several angiosperms (Wolcott and Ross 1966) including etiolated *P. mungo* seedlings (Ashihara 1978). Activities of plant OPRT and ODC were not separable during purification (Ashihara 1978; Walther et al. 1984). This suggests that, like their mammalian counterparts, OPRT and ODC exist as a complex in plant cells.

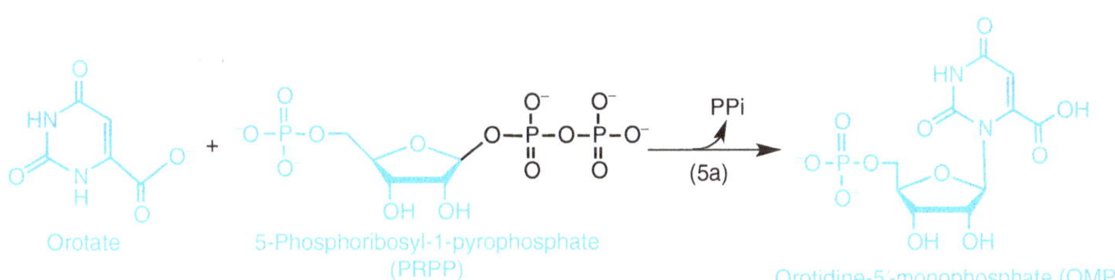

Figure 8.6 Reaction 5a in Figure 8.1. (5a) Orotate phosphoribosyltransferase.

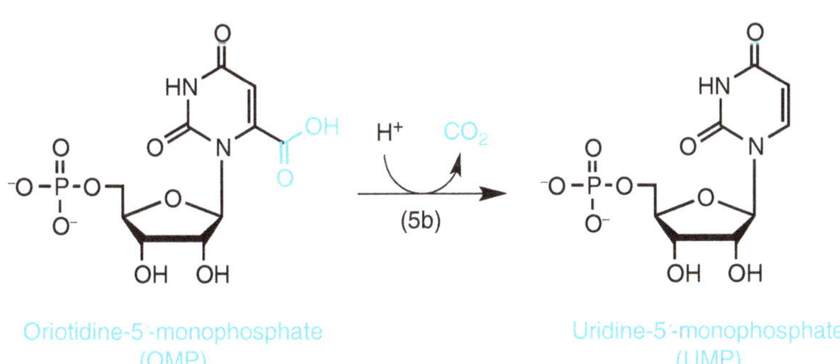

Figure 8.7 Reaction 5b in Figure 8.1. (5b) Orotidine-5′-monophosphate decarboxylase.

The conversion of orotate to UMP requires PRPP and Mg^{2+}, and is stimulated by dithiothreitol. The Km values of OPRT for PRPP and orotate at pH 7.2 are 1.6 and 4.5 μM, respectively. The pH optima for OPRT and ODC are, respectively, 8.0 and 6.2–9.2. The conversion of orotate to UMP by plant OPRT and ODC is inhibited by xanthosine-5′-monophosphate (XMP), UMP, and uridine diphosphate (UDP). Since the structures of the inhibitors, XMP and UMP, resemble that of OMP in having two carbonyl groups in a six membered ring structure, these inhibitors would appear to compete for the catalytic site (Ashihara 1978). The importance of the product inhibition of ODC by UMP in animal tissues has been discussed by Reid (1962). The level of UMP in some plant cells is appreciably higher than that of other pyrimidine nucleotides (Table 2.1). Hence, the inhibition by UMP may be one of the fine control mechanisms of pyrimidine biosynthesis in plants.

The *UMPS* gene encoding UMP synthase which catalyses the conversion of orotate to UMP in plants has been characterized (Nasr et al. 1994; Santoso and Thornburg 1992). Similar to other higher eukaryotes, the plant UMP synthase has both OPRT and ODC activities. The *A. thaliana UMPS* gene has an interesting structure. The 5′-half of the gene encodes the OPRT function while the 3′-half encodes the ODC function. Like other genes in the *de novo* pyrimidine pathway, *UMPS* is single gene in *A. thaliana*. However, three genes encoding UMP synthase has been found in rice (Kafer et al. 2004; Zrenner et al. 2006).

8.2.6 Synthesis of CTP from UTP

UMP produced by the orotate pathway is converted to UTP by two sequential phosphorylation reactions (steps 6 and 7 in Figure 8.1). UDP synthesis is catalysed by thymidine monophosphate kinase (EC 2.7.4.9) and conversion of UDP to UTP is mediated by a non-specific nucleoside diphosphate kinase (EC 2.7.4.6) as described in Chapter 3.2.

Conversion of UTP to CTP is catalysed by CTP synthase (glutamine hydrolysing) (EC 6.3.4.2). The overall reaction is UTP + L-glutamine + ATP → CTP + L-glutamate + ADP + Pi (Figure 8.8).

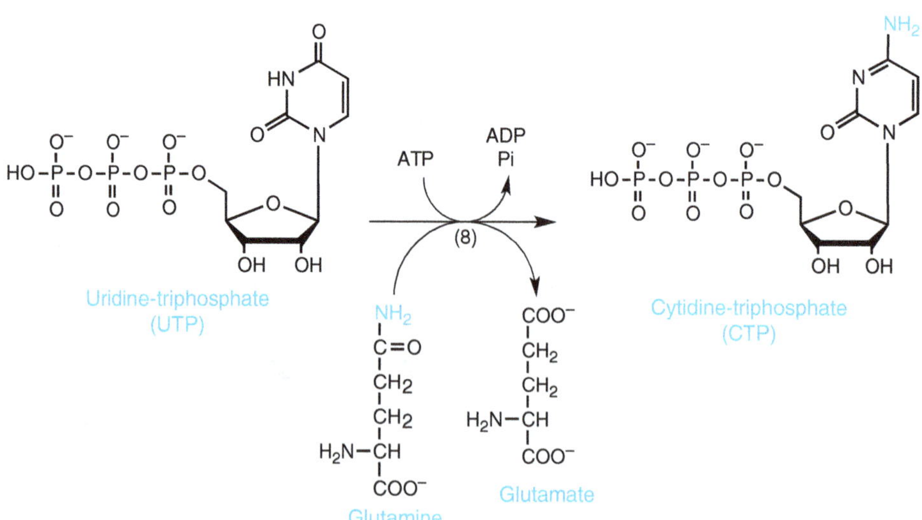

Figure 8.8 Reaction 8 in Figure 8.1. (8) Cytidine triphosphate synthase.

The reaction mechanism of CTP synthase is complex consisting of two reactions: (i) L-glutamine + $H_2O \rightarrow$ L-glutamate + NH_3; and (ii) ATP + UTP + $NH_3 \rightarrow$ ADP + Pi + CTP. The enzyme contains three functionally distinct sites: an allosteric GTP-binding site, a glutaminase site where glutamine hydrolysis occurs, and the active site where CTP synthesis takes place. The reaction proceeds via phosphorylation of UTP by ATP to give an activated intermediate 4-phosphoryl UTP and ADP. Ammonia then reacts with this intermediate generating CTP and phosphate. The enzyme can also use ammonia from the surrounding medium (Levitzki and Koshland 1972; Lewis and Villafranca 1989).

Although the existence of the CTP synthase reaction in plants was suggested from the results obtained in feeding experiments and inhibitor studies (Ross 1965; Ross and Cole 1968), no purification or characterization of native CTP synthase has been reported. The sequence of the gene encoding this enzyme is in the *A. thaliana* database, but there are no reports on the properties of recombinant proteins.

8.3 Control Mechanism of *De Novo* Pyrimidine Ribonucleotide Biosynthesis

8.3.1 Fine Control of the *De Novo* Pathway

Enzyme kinetics indicate that the activities of the first two enzymes, CPSII and ACT are efficiently controlled by both end products and co-substrates. However, since carbamoyl-P is a substrate for UMP as well as arginine synthesis (Sasamoto and Ashihara 1988) these *in vitro* kinetics are not straight forward as there are potential problems when relating the data to what might operate in cells *in vivo* (Ross 1981). Mechanisms for the fine control of the biosynthesis of pyrimidine nucleotides and arginine have, however, been postulated, and it has been proposed that the feedback inhibition of CPSII and ACT by UMP and the activation of CPSII by ornithine are the most important regulators of the biosynthesis of pyrimidine nucleotides and arginine in plant cells (O'Neal and Naylor 1969; Ong and Jackson 1972a, 1972b) (Figure 8.9).

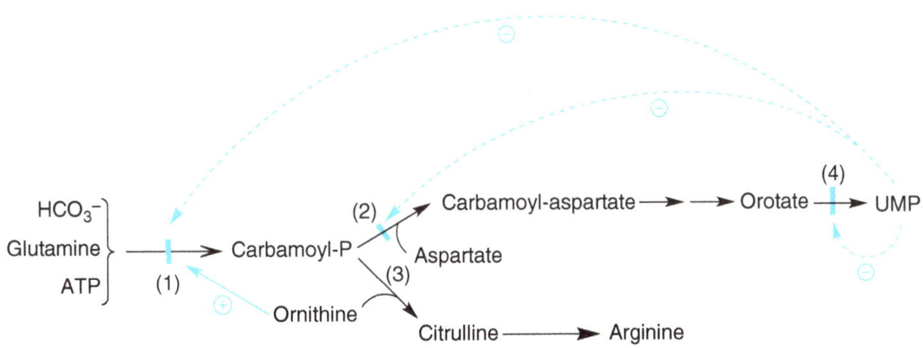

Figure 8.9 Fine control mechanism of UMP synthesis. CPSII (1), ATC (2) and UMPS (3) activity is inhibited by UMP. CPSII (1) is activated by ornithine, a substrate of ornithine carbamoyltransferase (EC 2.1.3.3) (4) for arginine biosynthesis. Source: Adapted from Ross (1981).

From the results of experiments which involved addition of exogenous uridine to intact plant cells, Lovatt and Cheng (1984) and Lin and Lovatt (1986) argued that inhibition of aspartate transcarbamoylase by UMP was the most plausible mechanism for the feedback control of synthesis of pyrimidine nucleotides in *C. pepo*. The level of UMP in *C. roseus* cells in culture varies from 19 to 82 nmol g^{-1} f.w. (Sasamoto and Ashihara 1988). Although the localization of enzymes which participate in the biosynthesis of pyrimidine nucleotides and arginine in heterotrophic plant cells are not well known, these enzymes may be present in the cytosol and/or plastids (Doremus and Jagendorf 1985; Kanamori et al. 1980; Shibata et al. 1986). If both UMP and the enzymes for pyrimidine biosynthesis are located in the cytoplasm/plastids of the cells, and if the volume of the cytoplasm/plastids is assumed to be 10% of the total cell volume, the concentration of UMP can be calculated to be 0.2–0.8 mM. This would be sufficient for significant inhibition of the activity of both CPSII and ATC (Kanamori et al. 1980; Sasamoto and Ashihara 1988). Therefore, feedback control of the biosynthesis of pyrimidine nucleotides is possible *in planta*. The feeding experiments using $NaH^{14}CO_3$ demonstrated that a decrease in pyrimidine nucleotide synthesis was observed with the increase of UMP level at the exponential phase of cell growth, at which point, an increase in the rate of arginine biosynthesis occurs.

Ross (1981) stated in his review that the low Km of ATC for carbamoyl-P and the irreversible binding between enzyme and substrate cause preferential channelling of carbamoyl-P into the pyrimidine pathway. As the concentration of UMP rises, the levels of both carbamoyl-P and carbamoyl-aspartate decline because of feedback inhibition of the first two enzymes of the *de novo* pyrimidine pathway (steps 1 and 2 in Figure 8.9). Another effective mechanism, activation of CPSII and its stabilization by ornithine, also takes place. Ornithine also removes the inhibition of carbamoyl-P synthesis caused by UMP (step 3 in Figure 8.9). Thus, if UMP inhibition lowers the carbamoyl-P level too far for effective arginine synthesis, the ornithine level rises, allowing formation of carbamoyl-P and arginine. In addition, as shown in the Section 8.2.5, the UMP synthase step (step 4 in Figure 8.10) is inhibited by UMP (Ashihara 1978). However, this has not been observed when orotate and/or OMP accumulate in plant cells.

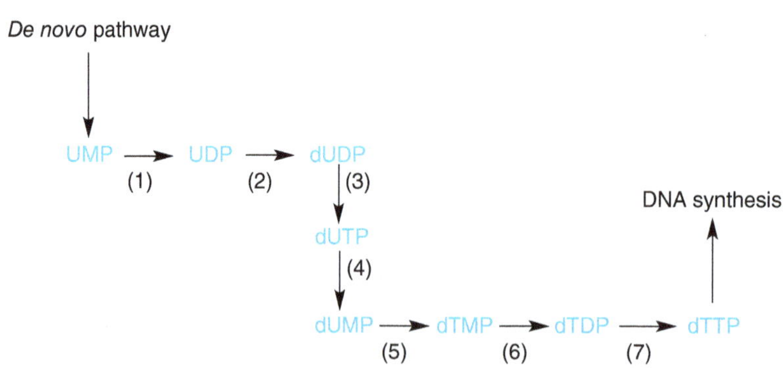

Figure 8.10 Outline of thymidine nucleotide biosynthesis *de novo*. UMP produced by the pyrimidine nucleotide biosynthesis *de novo* is converted to dTTP by steps 1–7. Enzymes: (1) UMP/CMP kinase, (2) ribonucleotide reductase, (3) nucleoside diphosphate kinase, (4) dUTP pyrophosphatase, (5) thymidylate synthase/dihydrofolate reductase, (6) thymidine monophosphate kinase, (7) nucleoside diphosphate kinase.

8.3.2 Coarse Control of the *De Novo* Pathway

In addition to the fine control, biosynthesis of pyrimidine nucleotides and arginine appears to be regulated by the levels of enzymes in the respective pathways, i.e. by so-called coarse control. The *in vitro* activity of these enzymes and biosynthetic rates change with the growth of *C. roseus* (Kanamori-Fukuda et al. 1981). Transcriptional control of genes encoding the enzymes involved in the *de novo* pyrimidine biosynthesis appears not to be especially pronounced in plants, except for some changes of expression along developmental gradients (Giermann et al. 2002; Zrenner et al. 2006). Transcriptional control during phosphate limitation indicated that transcripts for CPSII, ATC, and UMPS in roots of *A. thaliana* increased together with transcripts of genes of the salvage pathway. This may reflect the coordinate transcriptional control of UMP biosynthesis (Hewitt et al. 2005).

8.4 Biosynthesis of Thymidine Nucleotide

8.4.1 Formation of dUMP

Reduction of the ribose moiety of uridine ribonucleotide occurs at the level of UDP (see Section 3.4). For DNA synthesis, thymine deoxyribonucleoside triphosphate (dTTP) is required. It is produced by a pathway leading from UMP, the end product of *de novo* pyrimidine biosynthesis. The route in plants is summarized in Figure 8.10 and Table 8.2.

Table 8.2 Summary of enzymes involved in thymidine nucleotide biosynthesis pathway in plants illustrated in Figure 8.10.

Step	Enzyme name	EC	Abbreviation	TAIR locus
1	UMP/CMP kinase	2.7.4.14	UMPK	AT3G10030[a]
2	Ribonucleotide reductase	1.17.4.1	RNR	AT2G21790[a]
3	Nucleoside diphosphate kinase	2.7.4.6	NDPK	AT4G09320
4	Deoxyuridine triphosphate pyrophosphatase	3.6.1.23	DUT[b]	AT3G46940
5	Thymidylate synthase/dihydrofolate reductase	2.1.1.45/1.5.1.3	TS/DHFR	AT2G16370 AT2G21550 AT4G34570
6	Thymidine monophosphate kinase	2.7.4.9	TMPK	AT5G59440
7	Nucleoside diphosphate kinase	2.7.4.6	NDPK	AT4G09320

a) Only a typical locus is shown.
b) Abbreviation used for gene only.

UMP is phosphorylated to UDP (step 1 in Figure 8.10) which ribonucleotide reductase transforms to deoxyurinidine diphosphate (dUDP) (step 2). In plants, dUDP is then converted to deoxyuridine triphosphate (dUTP) by nucleoside diphosphate kinase (step 3) before being hydrolysed to deoxyuridine-5′-monophosphate (dUMP) by dUTP pyrophosphatase (step 4). Deoxythymidine-5′-monophosphate (dTMP) is formed from dUMP (step 5) by thymidylate synthase. Once dTMP is formed, two phosphorylations follow yielding dTTP. The initial step is catalysed by the highly specific thymidylate kinase (step 6), while the second step is catalysed by a nucleoside diphosphate kinase (step 7) which has a broad substrate specificity. The end product, dTTP is used for DNA synthesis (Stasolla et al. 2003).

8.4.2 Conversion of UMP to dUMP via dUTP

Conversion of UMP to dUTP, namely, UMP → UDP → dUDP → dUTP (steps 1–3) has already been described in Chapter 3 and the enzymes involved in plants are summarized in Table 8.2. Formation of dUMP from dUTP (step 4 in Figure 8.10) is catalysed by a specific enzyme, dUTP pyrophosphatase (dUTP diphosphatase, EC 3.6.1.23). The reaction is dUTP + H_2O → dUMP + PPi. dUMP is a substrate for thymidylate synthase and it is essential for cell survival due to its role in reducing the effective ratio of dUTP to dTTP, thereby restricting the possible incorporation of dUTP into DNA. The *DUT* gene encoding dUTPase in *A. thaliana* has been cloned and expressed in *E. coli*, and the recombinant protein was purified to homogeneity and crystalized (Bajaj and Moriyama 2007), however detailed kinetics of the enzyme have not been reported.

8.4.3 Conversion of dUMP to dTMP

Thymidine nucleotide synthesis is catalysed by thymidylate synthase (TS, EC 2.1.1.45), (step 5 in Figure 8.10). The reaction is dUMP + 5,10-methylenetetrahydrofolate → dTMP + dihydrofolate (Figure 8.11).

In most prokaryotes and eukaryotes dihydrofolate reductase (DHFR, EC 1.5.1.3) and TS are encoded by independent genes. In contrast plants, including *A. thaliana*, soybean and pea, have bifunctional TS-DHFR-encoding genes (Table 8.2). Structurally, *DHFR* is at the amino terminus and *TS* is at the carboxy terminus of the gene (Lazar et al. 1993; Neuburger et al. 1996; Wang et al. 1995). The bifunctional enzyme has

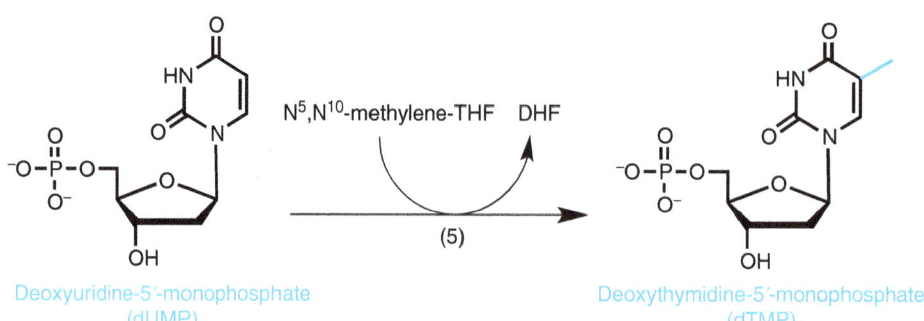

Figure 8.11 Reaction 5 in Figure 8.10. (5) Thymidylate synthase.

been purified to apparent homogeneity from pea mitochondria, which appears to be the major subcellular location of the enzyme. The TS-catalysed reaction produces dihydrofolate, which is the immediate substrate for the DHFR (Neuburger et al. 1996).

8.4.4 Thymidine Monophosphate Kinase

Thymidine monophosphate kinase (dTMP kinase, EC 2.7.4.9) catalyses the reaction: dTMP + ATP → deoxythymidine diphosphate (dTDP) + ADP (step 6 in Figure 8.10). Transcription of the *TMPK* gene of *A. thaliana* is activated during the G1/S-phase transition of the cell cycle. The single *TMPK* gene is capable of producing two proteins by alternative splicing; the longer isoform is targeted to the mitochondria, and the shorter isoform to the cytosol (Ronceret et al. 2008). This is interesting as it indicates that some genes encoding transit peptides targeted to definite organelles, may also be involved in splicing, forming isoforms without a transit peptide that appear in cytosol.

Conversion of dTDP to dTTP (step 7 in Figure 8.10) is catalysed by non-specific nucleoside diphosphate kinase which is described in Chapter 3.

8.5 Summary

The pathway of pyrimidine ribonucleotide biosynthesis *de novo* is HCO_3^- → carbamoyl-P → carbamoyl-aspartate → dihydroorotate → orotate → orotidine-5′-monophosphate → UMP. The initial reactions of the *de novo* synthesis of cytidine ribonucleotides and thymidine nucleotides is UTP → CTP and dUMP → dTMP, respectively.

References

Achar, B.S., Savithri, H.S., Vaidyanathan, C.S., and Rao, N.A. (1974). Studies on plant aspartate transcarbamylase. Purification and properties of the enzyme from mung-bean (*Phaseolus aureus*) seedlings. *Eur. J. Biochem.* 47: 15–22.

Arabidopsis-Genome-Initiative (2000). Analysis of the genome sequence of the flowering plant *Arabidopsis thaliana*. *Nature* 408: 796–815.

Ashihara, H. (1978). Orotate phosphoribosyltransferase and orotidine-5′-monophosphate decarboxylase of black gram (*Phaseolus mungo)* seedlings. *Z. Pflanzenphysiol.* 87: 225–241.

Bajaj, M. and Moriyama, H. (2007). Purification, crystallization and preliminary crystallographic analysis of deoxyuridine triphosphate nucleotidohydrolase from *Arabidopsis thaliana*. *Acta Crystallogr., Sect. F: Struct. Biol. Cryst. Commun.* 63: 409–411.

Boldt, R. and Zrenner, R. (2003). Purine and pyrimidine biosynthesis in higher plants. *Physiol. Plant.* 117: 297–304.

Christopherson, R.I. and Szabados, E. (1997). Nucleotide biosynthesis in mammals. In: *Channelling in Intermediary Metabolism* (eds. L. Agius and H.S.A. Sherratt), 315–335. London: Portland Press.

Doremus, H.D. (1986). Organization of the pathway of *de novo* pyrimidine nucleotide biosynthesis in pea (*Pisum sativum* L. cv Progress No. 9) leaves. *Arch. Biochem. Biophys.* 250: 112–119.

Doremus, H.D. and Jagendorf, A.T. (1985). Subcellular localization of the pathway of *de novo* pyrimidine nucleotide biosynthesis in pea leaves. *Plant Physiol.* 79: 856–861.

Fagan, R.L., Nelson, M.N., Pagano, P.M., and Palfey, B.A. (2006). Mechanism of flavin reduction in class 2 dihydroorotate dehydrogenases. *Biochemistry* 45: 14926–14932.

Giermann, N., Schröder, M., Ritter, T., and Zrenner, R. (2002). Molecular analysis of *de novo* pyrimidine synthesis in solanaceous species. *Plant Mol. Biol.* 50: 393–403.

Hager, S.E. and Jones, M.E. (1965). Initial steps in pyrimidine synthesis in Ehrlich ascites carcinoma *in vitro*: I. Factors affecting the incorporation of ^{14}C-bicarbonate into carbon 2 of the uracil ring of the acid-soluble nucleotides of intact cells. *J. Biol. Chem.* 240: 4556–4563.

Henderson, J.F. and Paterson, A.R.P. (1973). *Nucleotide Metabolism – An Introduction*. New York: Academic Press.

Hewitt, M.M., Carr, J.M., Williamson, C.L., and Slocum, R.D. (2005). Effects of phosphate limitation on expression of genes involved in pyrimidine synthesis and salvaging in Arabidopsis. *Plant Physiol. Biochem.* 43: 91–99.

Iwahana, H., Fujimura, M., Ii, S. et al. (1996). Molecular cloning of a human cDNA encoding a trifunctional enzyme of carbamoyl-phosphate synthetase-aspartate transcarbamoylase-dihydroorotase in *de novo* pyrimidine synthesis. *Biochem. Biophys. Res. Commun.* 219: 249–255.

Jagannatha Rao, G.S., Savithri, H.S., Seethalakshmi, S., and Appaji Rao, N. (1979). Plant aspartate transcarbamylase: an affinity chromatographic method for the purification of the enzyme from germinated seedlings. *Anal. Biochem.* 95: 401–405.

Kafer, C., Zhou, L., Santoso, D. et al. (2004). Regulation of pyrimidine metabolism in plants. *Front. Biosci.* 9: 1611–1625.

Kanamori, I., Ashihara, H., and Komamine, A. (1980). Subcellular distribution and activity of enzymes involved in uridine-5′-monophosphate synthesis in *Vinca rosea* cells. *Z. Pflanzenphysiol.* 96: 7–16.

Kanamori-Fukuda, I., Ashihara, H., and Komamine, A. (1981). Pyrimidine nucleotide biosynthesis in *Vinca rosea* cells: changes in the activity of *de novo* and salvage pathways during growth in a suspension culture. *J. Exp. Bot.* 32: 69–78.

Kapoor, M. and Waygood, E.R. (1965). Orotidine-5′-phosphate pyrophosphorylase of wheat embryos. *Can. J. Biochem.* 43: 143–151.

Khan, A.I., Chowdhry, B.Z., and Yon, R.J. (1999). Wheat-germ aspartate transcarbamoylase: revised purification, stability and re-evaluation of regulatory kinetics in terms of the Monod-Wyman-Changeux model. *Eur. J. Biochem.* 259: 71–78.

Kusama, K. and Roberts, E. (1963). Carbon dioxide incorporation into the uracil of mouse liver and Ehrlich ascites tumor cells. *Biochemistry* 2: 573–576.

Lazar, G., Zhang, H., and Goodman, H. (1993). The origin of the bifunctional dihydrofolate reductase-thymidylate synthase isogenes of *Arabidopsis thaliana*. *Plant J.* 3: 657–668.

Levitzki, A. and Koshland, D.E. Jr., (1972). Ligand-induced dimer-to-tetramer transformation in cytosine triphosphate synthetase. *Biochemistry* 11: 247–253.

Lewis, D.A. and Villafranca, J.J. (1989). Investigation of the mechanism of CTP synthetase using rapid quench and isotope partitioning methods. *Biochemistry* 28: 8454–8459.

Lieberman, I. and Kornberg, A. (1954). Enzymatic synthesis and breakdown of a pyrimidine, orotic acid: II. Dihydroorotic acid, ureidosuccinic acid, and 5-carboxymethylhydantoin. *J. Biol. Chem.* 207: 911–924.

Lin, F.-F. and Lovatt, C.J. (1986). Biosynthesis of pyrimidine nucleotides in cultured root callus of *Cucurbita pepo*. *In Vitro-Plant* 22: 1–5.

Lovatt, C.J. (1983). *De novo* purine biosynthesis in intact cells of *Cucurbita pepo*. *Plant Physiol.* 73: 766–772.

Lovatt, C.J. and Cheng, A.H. (1984). Aspartate carbamyltransferase: site of end-product inhibition of the orotate pathway in intact cells of *Cucurbita pepo*. *Plant Physiol.* 75: 511–515.

Lovatt, C.J., Albert, L.S., and Tremblay, G.C. (1979). Regulation of pyrimidine biosynthesis in intact cells of *Cucurbita pepo*. *Plant Physiol.* 64: 562–569.

Mazus, B. and Buchowicz, J. (1968). Purification and properties of dihydro-orotase from pea plants. *Acta Biochim. Pol.* 15: 317–325.

Mazuś, B. and Buchowicz, J. (1972). Activity of enzymes involved in pyrimidine metabolism in the germinating wheat grains. *Phytochemistry* 11: 77–82.

Miersch, J., Krauss, G.-J., and Metzger, U. (1986). Properties and subcellular localization of dihydroorotate dehydrogenase in cells of tomato suspension culture. *J. Plant Physiol.* 122: 55–66.

Moffatt, B.A. and Ashihara, H. (2002). Purine and pyrimidine nucleotide synthesis and metabolism. *Arabidopsis Book* 1: 1–20.

Murray, M.G. and Ross, C. (1971). Molecular weight estimations of some pyrimidine-metabolizing enzymes from pea cotyledons by gel filtration. *Phytochemistry* 10: 2645–2648.

Nara, T., Hshimoto, T., and Aoki, T. (2000). Evolutionary implications of the mosaic pyrimidine-biosynthetic pathway in eukaryotes. *Gene* 257: 209–222.

Nasr, F., Bertauche, N., Dufour, M.-E. et al. (1994). Heterospecific cloning of *Arabidopsis thaliana* cDNAs by direct complementation of pyrimidine auxotrophic mutants of *Saccharomyces cerevisiae*. I. Cloning and sequence analysis of two cDNAs catalysing the second, fifth and sixth steps of the *de novo* pyrimidine biosynthesis pathway. *Mol. Gen. Genet.* 244: 23–32.

Neuburger, M., Rébeillé, F., Jourdain, A. et al. (1996). Mitochondria are a major site for folate and thymidylate synthesis in plants. *J. Biol. Chem.* 271: 9466–9472.

Neuhard, J. and Nygaard, P. (1987). Biosynthesis and conversions of nuclotides: purine and pyrimidines. In: *Escherichia coli and Salmonella typhimurium: Cellular and Molecular Biology*, vol. 1 (ed. F.C. Neidhardt), 445–473. Washington, DC: American Society for Microbiology.

O'Neal, T.D. and Naylor, A.W. (1969). Partial purification and properties of carbamoyl phosphate synthetase of Alaska pea (*Pisum sativum* L. cultivar Alaska). Purification and general properties. *Biochem. J.* 113: 271–279.

O'Neal, T.D. and Naylor, A.W. (1976). Some regulatory properties of pea leaf carbamoyl phosphate synthetase. *Plant Physiol.* 57: 23–28.

Ong, B.L. and Jackson, J.F. (1972a). Aspartate transcarbamoylase from *Phaseolus aureus*. Partial purification and properties. *Biochem. J.* 129: 571–581.

Ong, B.L. and Jackson, J.F. (1972b). Pyrimidine nucleotide biosynthesis in *Phaseolus aureus*. Enzymic aspects of the control of carbamoyl phosphate synthesis and utilization. *Biochem. J.* 129: 583–593.

Reichard, P. and Lagerkvist, U. (1953). The biogenesis of orotic acid in liver slices. *Acta Chem. Scand.* 7: 1207–1217.

Reid, E. (1962). Factors influencing synthesis of uridine-5′-monophosphate in rat liver. *Nature* 194: 1153.

Ronceret, A., Gadea-Vacas, J., Guilleminot, J. et al. (2008). The first zygotic division in Arabidopsis requires *de novo* transcription of thymidylate kinase. *Plant J.* 53: 776–789.

Ross, C. (1965). Comparison of incorporation and metabolism of RNA pyrimidine nucleotide precursors in leaf tissues. *Plant Physiol.* 40: 65–73.

Ross, C.W. (1981). Biosynthesis of nucleotides. In: *Biochemistry of Plants*, vol. 6 (eds. P.K. Stump and E.E. Conn), 169–205. New York: Academic Press.

Ross, C. and Cole, C.V. (1968). Metabolism of cytidine and uridine in leaves. *Plant Physiol.* 43: 1227–1231.

Saha, N., Datta, S., Kharbuli, Z.Y. et al. (2007). Air-breathing catfish, *Clarias batrachus* upregulates glutamine synthetase and carbamoyl phosphate synthetase III during exposure to high external ammonia. *Comp. Biochem. Physiol. B: Biochem. Mol. Biol.* 147: 520–530.

Santoso, D. and Thornburg, R.W. (1992). Isolation and characterization of UMP synthase mutants from haploid cell suspensions of *Nicotiana tabacum*. *Plant Physiol.* 99: 1216–1225.

Sasamoto, H. and Ashihara, H. (1988). Biosynthesis of pyrimidine nucleotides and arginine in a suspension culture of *Catharanthus roseus*. *Int. J. Biochem.* 20: 87–92.

Shibata, H., Ochiai, H., Sawa, Y., and Miyoshi, S. (1986). Localization of carbamoylphosphate synthetase and aspartate carbamoyltransferase in chloroplasts. *Plant Physiol.* 80: 126–129.

Stasolla, C., Katahira, R., Thorpe, T.A., and Ashihara, H. (2003). Purine and pyrimidine nucleotide metabolism in higher plants. *J. Plant Physiol.* 160: 1271–1295.

Tatibana, M. (1978). Interrelationship of purine and pyrimidine metabolism. In: *Uric Acid* (eds. W.N. Kelley and I.M. Weiner), 125–154. Heidelberg: Springer.

Ullrich, A., Knecht, W., Piskur, J., and Loffler, M. (2002). Plant dihydroorotate dehydrogenase differs significantly in substrate specificity and inhibition from the animal enzymes. *FEBS Lett.* 529: 346–350.

Walther, R., Wald, K., Glund, K., and Tewes, A. (1984). Evidence that a single polypeptide catalyzes the two step conversion of orotate to UMP in cells from tomato suspension culture. *J. Plant Physiol.* 116: 301–311.

Wang, M., Ratnam, S., and Freisheim, J.H. (1995). Cloning, nucleotide sequence and expression of the bifunctional dihydrofolate reductase-thymidylate synthase from *Glycine max*. *Biochim. Biophys. Acta* 1261: 325–336.

Williamson, C.L. and Slocum, R.D. (1994). Molecular cloning and characterization of the *pyrB1* and *pyrB2* genes encoding aspartate transcarbamoylase in pea (*Pisum sativum* L.). *Plant Physiol.* 105: 377–384.

Wolcott, J.H. and Ross, C. (1966). Orotidine-5′-phosphate decarboxylase from higher plants. *Biochim. Biophys. Acta* 122: 532–534.

Yon, R.J. (1972). Wheat-germ aspartate transcarbamoylase. Kinetic behaviour suggesting an allosteric mechanism of regulation. *Biochem. J.* 128: 311–320.

Yon, R.J. (1973). Wheat-germ aspartate transcarbamoylase. The effect of ligands on the inactivation of the enzyme by trypsin and denaturing agents. *Biochem. J.* 131: 699–706.

Zrenner, R. and Ashihara, H. (2011). Nucleotide metabolism. In: *Plant Metabolism and Biotechnology* (eds. H. Ashihara, A. Crozier and A. Komamine), 135–162. Chichester, UK: Wiley.

Zrenner, R., Stitt, M., Sonnewald, U., and Boldt, R. (2006). Pyrimidine and purine biosynthesis and degradation in plants. *Annu. Rev. Plant Biol.* 57: 805–836.

9

Salvage Pathways of Pyrimidine Nucleotide Biosynthesis

9.1 Introduction

Although the *de novo* pyrimidine biosynthesis is very similar in prokaryotic and eukaryotic organisms (Chapter 8), pyrimidine salvage pathways exhibit more pronounced variations. In bacteria, direct conversion of uracil to uridine monophosphate (UMP) occurs, and other aspects of pyrimidine salvage are widespread in microorganisms (Munch-Petersen 1983; Neuhard and Kelln 1996).

In mammals, it is generally considered that the liver and kidneys are sites of *de novo* pyrimidine synthesis, with pyrimidine nucleosides being produced, mainly from uridine, before being transported to other organs such as the brain (Barsotti et al. 2002). Uridine kinase and uridine phosphorylase contribute to pyrimidine salvage in the brain and participate in the regulation of physiological processes most notably in the central nervous system (Balestri et al. 2007; Cansev 2006). Information on the pyrimidine salvage pathway and the importance of this process have also been obtained from studies with parasitic protozoa such as *Trypanosoma* and *Leishmania* species (Hassan and Coombs 1988).

In plants the pyrimidine base uracil, some pyrimidine ribonucleosides (uridine and cytidine) and pyrimidine deoxyribonucleosides (deoxycytidine and thymidine) can be utilized for the generation of their respective nucleoside-5′-monophosphates which are then phosphorylated producing pyrimidine nucleoside-triphosphates. uridine-triphosphate (UTP) and cytidine-triphosphate (CTP) are incorporated into RNA, and deoxythymidine triphosphate (dTTP) and deoxycytidine triphosphate (dCTP) are utilized for DNA synthesis (Ross 1981; Stasolla et al. 2003; Wasternack 1982).

9.2 Characteristics of Pyrimidine Salvage in Plants

Profiles of *in vitro* activity of enzymes involved in pyrimidine salvage have been determined using extracts from growing tubers of potato (*Solanum tuberosum*) (Katahira and Ashihara 2002) and cultured cells of white spruce (*Picea glauca*) (Ashihara et al. 2000) (Table 9.1).

Plant Nucleotide Metabolism – Biosynthesis, Degradation, and Alkaloid Formation, First Edition.
Hiroshi Ashihara, Iziar A. Ludwig and Alan Crozier.

Table 9.1 Profile of activity of enzymes involved in pyrimidine salvage in growing potato tubers (Katahira and Ashihara 2002) and cultured cells of white spruce at the logarithmic stage of growth (Ashihara et al. 2000).

Enzyme (EC number)	Reaction (step number in Figure 9.1)	Potato tubers	White spruce
1) Phosphoribosyltransferases			
Uracil phosphoribosyltransferase (2.4.2.9)	Uracil[a] + PRPP → UMP[a] + PPi (**1**)	1.3	20.3
Cytosine phosphoribosyltransferase	Cytosine[a] + PRPP → IMP[a] + PPi	nd	
2) Nucleoside kinase			
Uridine/cytidine kinase (2.7.1.48)	Uridine[a] + ATP → UMP[a] + ADP (**2**)	10.3	62.7
	Cytidine[a] + ATP → CMP[a] + ADP (**2**)	5.9	
Thymidine kinase (2.7.1.21)	Thymidine[a] + ATP → dTMP[a] + ADP (**3**)	<0.1	
3) Nucleoside phosphotransferase			
Nucleoside phosphotransferase (2.7.1.77)	Uridine[a] + AMP → UMP[a] + Adenosine (**7**)	4.5	13.7
	Cytidine[a] + AMP → CMP[a] + Adenosine (**7**)	2.8	
	Thymidine[a] + AMP → dUMP[a] + Adenosine (**7**)	17.5	

a) ^{14}C-Labelled substrates and products.
nd, not detected.
Enzyme activity is expressed as pkat mg^{-1} protein.

The salvage of uracil is carried out mainly by a one-step reaction catalysed by uracil phosphoribosyltransferase (UPRT, EC2.4.2.9) (Table 9.1), but there is also a minor, indirect uracil → uridine → UMP route in plants (Kanamori et al. 1980; Mazuś and Buchowicz 1972). Plant UPRT is distinct from orotate phosphoribosyltransferase (EC 2.4.2.10), an enzyme of *de novo* pyrimidine biosynthesis (see Chapter 8) (Ashihara 1978; Bressan et al. 1978). In contrast to uracil, cytosine is not converted to cytidine-5′-monophosphate (CMP) by UPRT (Katahira and Ashihara 2002).

Pyrimidine nucleosides, uridine, cytidine, deoxycytidine, and thymidine, are salvaged to their respective nucleotides, UMP, CMP, deoxycytidine-5′-monophosphate (dCMP), and deoxythymidine-5′-monophosphate (dTMP). A single enzyme, uridine/cytidine kinase (UK, EC 2.7.1.48) phosphorylates both uridine and cytidine. Thymidine kinase (TK, EC2.7.2.21) occurs in plants, but non-specific nucleoside phosphotransferase (NPT, EC 2.7.1.77) activity also participates in the salvage of thymidine (Stasolla et al. 2003). In growing potato tubers, uridine is a better substrate for both UK and NPT than cytidine. Activity of TK is much lower than that of NPT (Table 9.1).

Outlines of pyrimidine salvage pathways are illustrated in Figure 9.1. In contrast to purine bases (see Chapter 5), only pyrimidine bases are salvaged by a phosphoribosyltransferase, namely, UPRT (step 1 in Figure 9.1), while pyrimidine nucleosides are salvaged by UK and TK (steps 2 and 3) and/or NPT (step 7). Uridine is first hydrolysed to uracil by uridine nucleosidase (EC 3.2.2.3) (step 5) and then salvaged by UPRT (step 1). Cytidine is converted to uridine by cytidine deaminase (EC 3.5.4.5) (step 6) and then

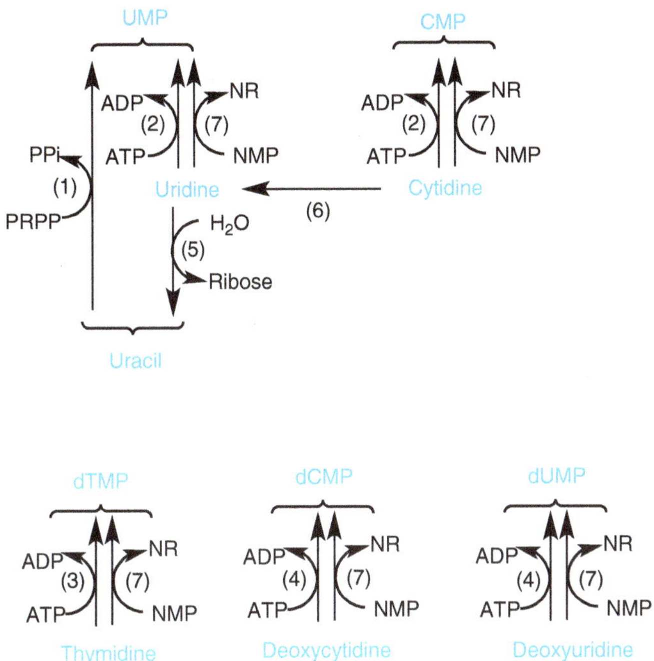

Figure 9.1 Outline of pyrimidine salvage reactions and related enzymes in plants. (1) Phosphoribosyltransferase: uracil phosphoribosyltransferase; (2) nucleoside kinases: uridine kinase (aka uridine/cytidine kinase); (3) thymidine kinase; (4) deoxyribonucleoside kinase; (5) nucleosidase: uridine nucleosidase; (6) deaminase: cytidine deaminase; (7) nucleoside phosphotransferase: non-specific nucleoside phosphotransferase.

salvaged by UK (step 2). The uridine nucleosidase and cytidine deaminase are described in Chapter 10. Cytosine and thymine are not substrates for NPT (Stasolla et al. 2003).

9.3 Enzymes of Pyrimidine Salvage

Purine base salvage by UPRT occurs in plants and bacteria, but not in mammals. A novel *UPRT* family cDNA encoding a protein with a putative UPRT domain, named human UPRT, was isolated from the human fetal brain library. However, UPRT activity is not found in organisms higher than the nematode *Caenorhabditis elegans* in which the putative UPRT protein lacks amino acids in the uracil-binding region. This suggest that the putative UPRT encoding gene is not directly related in uracil salvage.

Like purine bases, the pyrimidine base uracil can also be converted to uridine using ribose-1-phosphate by uridine phosphorylase (EC 2.4.2.3) (Balestri et al. 2007). This enzyme participates in pathways of both pyrimidine ribonucleoside degradation and salvage in animals (Liu et al. 1998), but the role of the enzyme is very limited in plants. In many plants uridine phosphorylase activity is not detectable (Katahira and Ashihara 2002; Murray and Ross 1971). Even when it has been detected in plants, the *in vitro* activity was far lower than that of UPRT in wheat embryo (Mazuś and Buchowicz 1972) and in the cell division phase of *Catharanthus roseus* (Kanamori et al. 1980).

The role of nucleoside phosphorylase in plant nucleotide metabolism thus appears to be minor.

In addition to nucleoside kinases, NPT (aka non-specific NPT, EC 2.7.1.77) also participates in purine and pyrimidine nucleoside salvage (see Chapter 5). In the case of phosphorylation of uridine and cytidine, the activity with this enzyme is lower than that of uridine/cytidine kinase (UK), but in the case of thymidine, enzyme activity is much higher in potato tuber tissues (Table 9.1). The activity of thymidine kinase (TK) is closely associated with cell division and DNA synthesis.

9.3.1 Uracil Phosphoribosyl Transferase

UPRT catalyses reaction 1 in Figure 9.1. In this reaction, the enzyme mediates UMP formation through the transfer of uracil to 5-phosphoribosyl-1-pyrophosphate (PRPP) with the release of pyrophosphate (PPi) (Figure 9.2).

A native UPRT has been partially purified from epicotyls of seedlings of pea (*Pisum sativum*). As shown in Table 9.2, the Km values of pea UPRT for uracil (0.7 μM) and for PRPP (11 μM) are relatively low and can be functional *in planta* (Bressan et al. 1978). These values are similar to those of *Escherichia coli* UPRT for uracil (4 μM) and PRPP (20 μM) (Molloy and Finch 1969). However, extremely high Km values (4–5 mM) for uracil has been found with mouse murine leukemic and the beef heart enzymes (Kessel et al. 1972). These mammalian enzymes were called pyrimidine phosphoribosyltransferase for which the best substrate is orotate (Km 50 μM) that is involved in *de novo* pyrimidine biosynthesis. The UPRT activity may be a side reaction of orotate phosphoribosyltransferase (see Chapter 8).

The genes encoding UPRT and UK have been reported to occur as a fusion gene in *Saccharomyces cerevisiae* (Kern 1990) and humans (Kashuba et al. 2002). A single non-fusion gene, has been reported in *E. coli* (Neuhard and Nygaard 1987). In *Arabidopsis thaliana*, six homologous genes are predicted to encode proteins with UPRT domains. The first gene, *UPP*, encodes a single domain UPRT protein. Five of the six genes are very similar to each other, and encode dual-domain proteins, with the *N*-terminal encoding

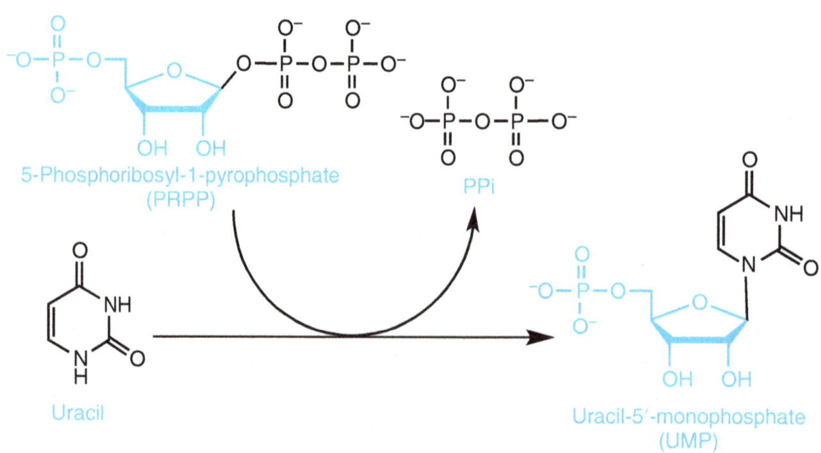

Figure 9.2 Step 1 in Figure 9.1.

Table 9.2 Properties of the native and recombinant uracil phosphoribosyltransferase (UPRT) and uridine kinase (UK) in plants.

Enzyme	Enzyme source	Isozyme	Optimum pH	Km values (μM) U	PRPP	UR	CR	ATP	Gene	TAIR Locus	Reference
UPRT	*Pisum sativum*[a]	None	8.5	0.7	11	—	—	—	—	—	Bressan et al. (1978)
	Arabidopsis thaliana[b]	None	—	[c]	—	—	—	—	*UPP*	AT3G53900	Mainguet et al. (2009)
UK	*Zea mays*[a]	None	8.0	—	—	50	125	590	—	—	Deng and Ives (1975)
	Arabidopsis thaliana[b]	UK1	—	—	—	[c]	—	[c]	*UKL1*	AT5G40870	Islam et al. (2007),
		UK2	—	—	—	[c]	—	[c]	*UKL2*	AT3G27190	Mainguet et al. (2009),
		no activity	—	—	—	—	—	—	*UKL3*	AT1G55810	Chen and Thelen (2011)]
		no activity	—	—	—	—	—	—	*UKL4*	AT4G26510	
		no activity	—	—	—	—	—	—	*UKL5*	AT3g27440	

U, Uracil; PRPP, 5-phosphoribosyl-1-pyrophosphate; UR, uridine; CR, cytidine.

a) Native enzyme.

b) Recombinant enzyme, —, not determined.

c) Activity was detected, but no information was provided on the Km value.

a uridine kinase. These genes were named *UKL1* through to *UKL5* (UK-like) (Islam et al. 2007) (Table 9.2).

Assays performed with all six *A. thaliana* genes and *E. coli* mutants, lacking the *pyrF* genes responsible for the last step of the *de novo* pyrimidine synthesis pathway, namely *UDP* (uridine phosphorylase) and *UPP* (UPRT), tested which of the genes enabled the mutant to grow with uracil as the sole pyrimidine source. Only the *UPP* gene was able to restore growth (Mainguet et al. 2009). Furthermore, characterization of T-DNA insertion mutants (i.e. the transferred DNA of the tumour-inducing [Ti] plasmid) for the *UPP*, *UKL1*, *UKL2*, and *UKL3* genes showed that the disruption of the *UPP* gene is sufficient to abolish UPRT function in *A. thaliana*. Hence, it is believed that the UPRT domains of the *UKL1-UKL5* genes are inactive, and that *UPP* is the only active UPRT in *A. thaliana*. No kinetic studies on recombinant *A. thaliana* UPRT has been reported but, arguably, its properties may be similar to pea or *E. coli* UPRT (Table 9.2).

9.3.2 Uridine/Cytidine Kinase

Phosphorylation of uridine and cytidine (step 2 in Figure 9.1) is catalysed by UK). The reactions are illustrated in Figure 9.3. UK was partially purified from *Zea mays* (maize) seeds. The enzyme uses uridine and cytidine as substrates. The Km values for uridine and cytidine are 53 and 125 μM, respectively. The reaction velocity with uridine is ~5 times faster than cytidine. ATP and GTP are phosphate donors, but CTP and UTP act

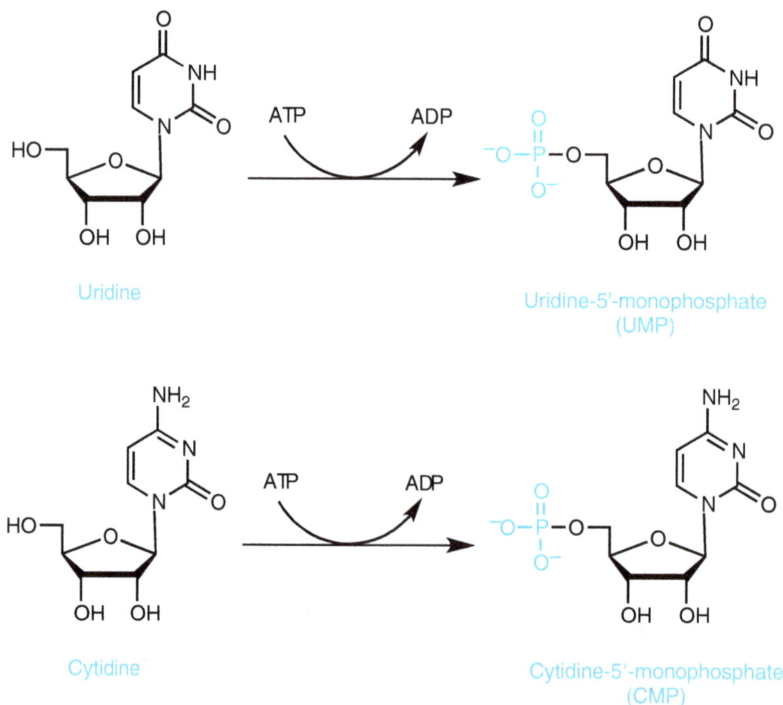

Figure 9.3 Step 2 in Figure 9.1.

as competitive inhibitors against ATP (Ki values are 60 and 240 μM, respectively) (Deng and Ives 1975).

The two *UK*-like genes, *UKL1*, and *UKL2*, which tentatively encode UK and UPRT bifunctional enzymes (Table 9.2) were studied in *A. thaliana*. T-DNA insertions in *UKL1* and *UKL2* reduced transcript expression and increased plant tolerance to the toxic analogues 5-fluorouridine and 5-fluorouracil. Enzyme activity assays using purified recombinant proteins indicated that UKL1 and UKL2 have UK but not UPRT activity. Subcellular localization using a *C*-terminal enhanced yellow fluorescent protein fusion indicated that UKL1 and UKL2 are localized in plastids (Chen and Thelen 2011). There are no reports of kinetic studies on the recombinant *A. thaliana* UK.

9.3.3 Thymidine Kinase

TK catalyses the phosphorylation of thymidine with ATP (Figure 9.4). No kinetic studies on TK from plants have been reported. Several papers on the topic were published during 1970s but they inadvertently involved non-specific NPT activity. The crude TK preparations were contaminated with strong NPT activity, which readily hydrolysed ATP, the true substrate of TK, to AMP, a substrate of NPT (Ross 1981). TK may be the principal activity in young tissues and its apparent absence is due to the high instability of the enzyme in plant extracts (Brunori et al. 1974). As a consequence, the estimation of TK activity in plants is difficult.

In the *A. thaliana* genome, TK activity was encoded by two thymidine kinase 1-like genes (*AtTK1a* and *AtTK1b*). T-DNA insertion lines of *AtTK1a* and *AtTK1b* mutant genes had normal growth, although *AtTK1a* and *AtTK1b* double mutants died at an early stage, which indicates that AtTK1a and AtTK1b catalyse redundant reactions (Table 9.3). The results suggest a crucial role for the salvage of thymidine during early plant development (Clausen et al. 2012). Xu et al. (2015) identified two TK genes named *AtTK1* and *AtTK2*. Expression of both genes was ubiquitous, but *AtTK1* was strongly expressed in high-proliferation tissues. *AtTK1* was associated with the cytosol, whereas *AtTK2* was found in mitochondria. Mutant analysis indicated that the two genes function in coordination to sustain normal plant development. The two TK proteins shared similar catalytic specificity for pyrimidine nucleosides.

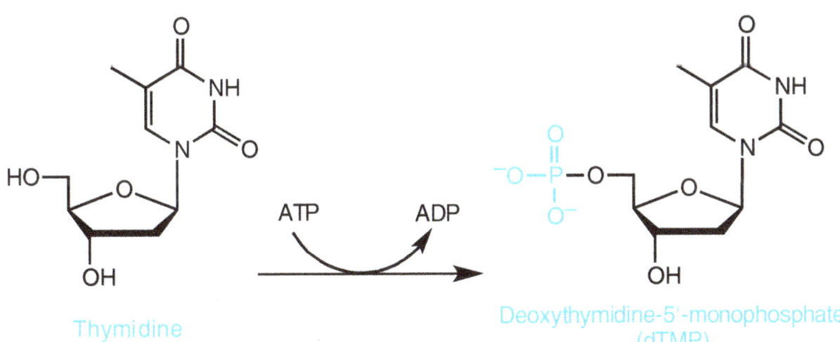

Figure 9.4 Step 3 in Figure 9.1.

Table 9.3 Properties of the recombinant thymidine kinase (TK) and deoxynucleoside kinase (NdRK) in plants.

Enzyme	Enzyme source	Isozyme	Optimum pH	Km values (μM)					Gene	TAIR Locus	Reference
				TdR	UdR	CdR	AdR	GdR			
TK	*Arabidopsis thaliana*	TK1a	—	0.6	18				*TK1a*	AT3G07800	Clausen et al. (2012)
		TK1b	—	1.4	6.6				*TK1b*	AT5G23070	Xu et al. (2015)
NdRK	*Arabidopsis thaliana*	None	—	—	—	16	14	26	*dNK*	AT1G72040	Clausen et al. (2008)

TdR, Thymidine; UdR, deoxyuridine; CdR, deoxycytidine; AdR, deoxyadenosine; GdR, deoxyguanosine.
—, Not determined.

9.3.4 Deoxyribonucleoside Kinase

Deoxyribonucleoside kinase (NdRK, aka multi-substrate deoxyribonucleoside kinase, EC 2.7.1.145) catalyses the reaction: 2′-deoxyribonucleoside + ATP → 2′-deoxyribonucleoside-5′-monophosphate + ADP (Figure 9.5). This enzyme differs from other 2′-deoxyribonucleoside kinases, namely deoxyadenosine kinase and deoxyguanosine kinase, initially reported from embryonic cells of the fruit fly, *Drosophila melanogaster*, in its broad specificity for all four common 2′-deoxyribonucleosides (Munch-Petersen et al. 1998). NdRK was identified and characterized as a unique multi-substrate deoxyribonucleoside kinase from *A. thaliana* (Table 9.3). Efficiency and specificity, based on Kcat/Km values, are deoxyuridine > deoxyadenosine > deoxycytidine > deoxyguanosine. Thymidine is not a substrate of this enzyme (Clausen et al. 2008).

9.3.5 Nucleoside Phosphotransferase

NPT uses both purine and pyrimidine nucleosides as substrates. The reactions with ribonucleoside (reaction 1) and deoxyribonucleoside (reaction 2) are shown in Figure 9.6. The reaction with purine nucleosides are described in the Section 5.5 of Chapter 5. The barley NPT is highly ordered in its nucleoside specificity. Although it phosphorylates all of the normal nucleic acid precursors, the apparent Km values

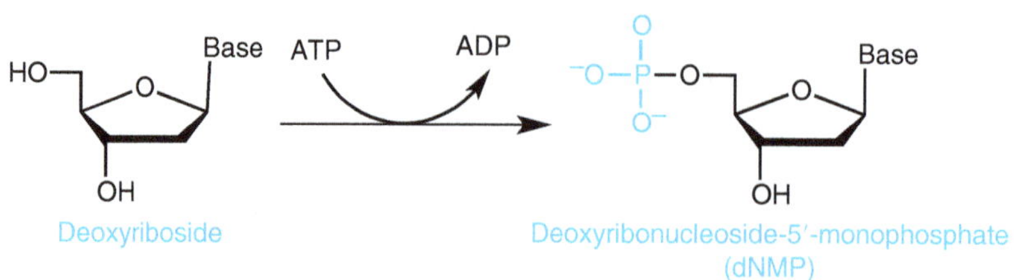

Figure 9.5 Step 4 in Figure 9.1.

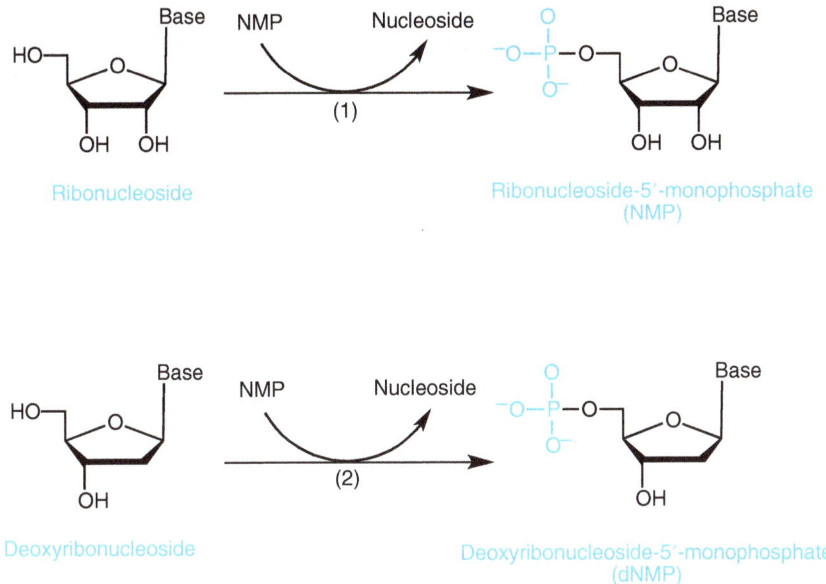

Figure 9.6 Step 7 in Figure 9.1.

vary over a range of two orders of magnitude, and the catalytic rate constants show a ~10-fold variation. Purine nucleosides are much better substrates than pyrimidine nucleosides. The Km values for pyrimidines, thymidine (0.9 mM), uridine (3.9 mM), cytidine (6.4 mM), and deoxycytidine (4.2 mM) are much higher than those of the purines, deoxyadenosine (0.09 mM) and adenosine (0.54 mM) (Prasher et al. 1982).

All purine and pyrimidine ribonucleosides and deoxyadenosine appear to be good phosphate acceptors and 5′-ribonucleotides are effective phosphate donors for NPT from the cotyledons of yellow lupin seedlings. The initial velocities were dependent on the phosphate donors, if 5′-AMP was used, activity with adenosine was higher than uridine, but with 5′-UMP the activity with adenosine and uridine is similar (Guranowski 1979).

9.4 Role of Pyrimidine Salvage in Plants

Pyrimidine salvage, like purine salvage, is energetically advantageous for nucleotides and nucleic acids. The salvage activity is closely associated with physiological events in plants. For example, the uracil salvage pathway to pyrimidine nucleotides synthesis appears only in cotyledons of legumes during the early stages of germination (Ashihara 1977), and uracil and uridine salvage activity is high in a 'turnover stage', during the lag phase of plant cell growth (Kanamori-Fukuda et al. 1981). Physiological aspects of pyrimidine salvage has been reviewed by Stasolla et al. (2003). Molecular genetic studies such as the characterization of two insertion mutants in the *UPP* gene revealed that loss of UPRT activity strongly affects early development of *A. thaliana* (Mainguet et al. 2009). In *A. thaliana*, inhibition of *de novo* synthesis at the aspartate carbamoyltransferase step leads to an increase in *UPRT* transcripts (Chen and Slocum 2008). Similarly, inhibition of the *de novo* synthesis at the UMP synthase step resulted in the stimulation of the uracil salvage pathway in potato (Geigenberger et al. 2005). These findings suggest that the uracil salvage pathway is essential when *de novo* synthesis is limiting, which may be the case during the initiation of germination. In the case of pyrimidines, one of the most important events seems to be thymidine salvage by TK, because it is closely related to DNA synthesis and repair. The product dTMP is quickly converted to dTTP. It has been reported that TK plays an important catalytic role in regulating the dNTP pool in human cells (Rampazzo et al. 2007). A similar function may also occur in plants and maintain appropriate growth and development. In a salt-resistant mangrove cells, *in vitro* activity of UK and UPRT, and the expression of one of the *UKL* homologs of the *A. thaliana* gene, were increased with NaCl treatment. Hence, pyrimidine salvage activity might be closely related to reduction of energy and the amount of substrates available for *de novo* pyrimidine biosynthesis. Salt-resistant plants use substantial amounts energy for the removal of salts from the cytoplasm of the cells, and have to supply substrates for the synthesis of compatible solutes to lower osmotic potential (Sato and Ashihara 2009).

9.5 Summary

A diversity of pyrimidine salvage occurs in different organisms. In plants, uracil is salvaged to UMP by UPRT, while uridine, cytidine, thymidine, and deoxycytidine are salvaged to their respective nucleoside-5′-mononucleotides by UK, TK, and deoxyribonucleoside kinase. Pyrimidine nucleosides are also converted to nucleoside-5′-monophosphate by NPT.

References

Ashihara, H. (1977). Changes in the activities of the *de novo* and salvage pathways of pyrimidine nucleotide biosynthesis during germination of black gram (*Phaseolus mungo*) seeds. *Z. Pflanzenphysiol.* 81: 199–211.

Ashihara, H. (1978). Orotate phosphoribosyltransferase and orotidine-5′-monophosphate decarboxylase of black gram (*Phaseolus mungo)* seedlings. *Z. Pflanzenphysiol.* 87: 225–241.

Ashihara, H., Stasolla, C., Loukanina, N., and Thorpe, T.A. (2000). Purine and pyrimidine metabolism in cultured white spruce (*Picea glauca*) cells: metabolic fate of ^{14}C-labeled precursors and activity of key enzymes. *Physiol. Plant.* 108: 25–33.

Balestri, F., Barsotti, C., Lutzemberger, L. et al. (2007). Key role of uridine kinase and uridine phosphorylase in the homeostatic regulation of purine and pyrimidine salvage in brain. *Neurochem. Int.* 51: 517–523.

Barsotti, C., Tozzi, M.G., and Ipata, P.L. (2002). Purine and pyrimidine salvage in whole rat brain: utilization of ATP-derived ribose-1-phosphate and 5-phosphoribosyl-1-pyrophosphate generated in experiments with dialyzed cell-free extracts. *J. Biol. Chem.* 277: 9865–9869.

Bressan, R.A., Murray, M.G., Gale, J.M., and Ross, C.W. (1978). Properties of pea seedling uracil phosphoribosyltransferase and its distribution in other plants. *Plant Physiol.* 61: 442–446.

Brunori, A., Hotta, Y., and Stern, H. (1974). Instability of thymidine kinase in plant extracts. *Plant Cell Physiol.* 15: 421–427.

Cansev, M. (2006). Uridine and cytidine in the brain: their transport and utilization. *Brain Res. Rev.* 52: 389–397.

Chen, C.T. and Slocum, R.D. (2008). Expression and functional analysis of aspartate transcarbamoylase and role of de novo pyrimidine synthesis in regulation of growth and development in Arabidopsis. *Plant Physiol. Biochem.* 46: 150–159.

Chen, M. and Thelen, J.J. (2011). Plastid uridine salvage activity is required for photoassimilate allocation and partitioning in Arabidopsis. *Plant Cell* 23: 2991–3006.

Clausen, A.R., Girandon, L., Knecht, W. et al. (2008). A multisubstrate deoxyribonucleoside kinase from plants. *Nucleic Acids Symp. Ser.*: 489–490.

Clausen, A.R., Girandon, L., Ali, A. et al. (2012). Two thymidine kinases and one multisubstrate deoxyribonucleoside kinase salvage DNA precursors in *Arabidopsis thaliana*. *FEBS J.* 279: 3889–3897.

Deng, Q.-I. and Ives, D.H. (1975). Non-allosteric regulation of the uridine kinase from seeds of *Zea mays*. *Biochim. Biophys. Acta* 377: 84–94.

Geigenberger, P., Regierer, B., Nunes-Nesi, A. et al. (2005). Inhibition of *de novo* pyrimidine synthesis in growing potato tubers leads to a compensatory stimulation of the pyrimidine salvage pathway and a subsequent increase in biosynthetic performance. *Plant Cell* 17: 2077–2088.

Guranowski, A. (1979). Nucleoside phosphotransferase from yellow lupin seedling cotyledons. *Biochim. Biophys. Acta* 569: 13–22.

Hassan, H.F. and Coombs, G.H. (1988). Purine and pyrimidine metabolism in parasitic protozoa. *FEMS Microbiol. Lett.* 54: 47–83.

Islam, M.R., Kim, H., Kang, S.W. et al. (2007). Functional characterization of a gene encoding a dual domain for uridine kinase and uracil phosphoribosyltransferase in *Arabidopsis thaliana*. *Plant Mol. Biol.* 63: 465–477.

Kanamori, I., Ashihara, H., and Komamine, A. (1980). Subcellular distribution and activity of enzymes involved in uridine-5′-monophosphate synthesis in *Vinca rosea* cells. *Z. Pflanzenphysiol.* 96: 7–16.

Kanamori-Fukuda, I., Ashihara, H., and Komamine, A. (1981). Pyrimidine nucleotide biosynthesis in *Vinca rosea* cells: changes in the activity of *de novo* and salvage pathways during growth in a suspension culture. *J. Exp. Bot.* 32: 69–78.

Kashuba, E., Kashuba, V., Sandalova, T. et al. (2002). Epstein-Barr virus encoded nuclear protein EBNA-3 binds a novel human uridine kinase/uracil phosphoribosyltransferase. *BMC Cell Biol.* 3: 23.

Katahira, R. and Ashihara, H. (2002). Profiles of pyrimidine biosynthesis, salvage and degradation in disks of potato (*Solanum tuberosum* L.) tubers. *Planta* 215: 821–828.

Kern, L. (1990). The URK1 gene of *Saccharomyces cerevisiae* encoding uridine kinase. *Nucleic Acids Res.* 18: 5279–5279.

Kessel, D., Deacon, J., Coffey, B., and Bakamjian, A. (1972). Some properties of a pyrimidine phosphoribosyltransferase from Murine Leukemia cells. *Mol. Pharmacol.* 8: 731–739.

Liu, M., Cao, D., Russell, R. et al. (1998). Expression, characterization, and detection of human uridine phosphorylase and identification of variant uridine phosphorolytic activity in selected human tumors. *Cancer Res.* 58: 5418–5424.

Mainguet, S.E., Gakière, B., Majira, A. et al. (2009). Uracil salvage is necessary for early Arabidopsis development. *Plant J.* 60: 280–291.

Mazuś, B. and Buchowicz, J. (1972). Activity of enzymes involved in pyrimidine metabolism in the germinating wheat grains. *Phytochemistry* 11: 77–82.
Molloy, A. and Finch, L.R. (1969). Uridine-5′-monophosphate pyrophosphorylase activity from *Escherichia coli*. *FEBS Lett.* 5: 211–213.
Munch-Petersen, A. (1983). *Metabolism of Nucleotides, Nucleosides and Nucleobases in Microorganisms*. London: Academic Press.
Munch-Petersen, B., Piskur, J., and Sondergaard, L. (1998). Four deoxynucleoside kinase activities from *Drosophila melanogaster* are contained within a single monomeric enzyme, a new multifunctional deoxynucleoside kinase. *J. Biol. Chem.* 273: 3926–3931.
Murray, M.G. and Ross, C. (1971). Molecular weight estimations of some pyrimidine-metabolizing enzymes from pea cotyledons by gel filtration. *Phytochemistry* 10: 2645–2648.
Neuhard, J. and Kelln, R.A. (1996). Biosynthesis and conversion of pyrimidines. In: *Escherichia coli and Salmonella typhimurium: Cellular and Molecular Biology*, 2e (eds. J. Neidhardt, R. Curtiss III, J.L. Ingraham, et al.), 580–599. Washington, D.C.: American Society for Microbiology.
Neuhard, J. and Nygaard, P. (1987). Biosynthesis and conversions of nuclotides: purine and pyrimidines. In: *Escherichia coli and Salmonella typhimurium: Cellular and Molecular Biology*, vol. 1 (ed. F.C. Neidhardt), 445–473. Washington, DC: American Society. Microbiology.
Prasher, D.C., Carr, M.C., Ives, D.H. et al. (1982). Nucleoside phosphotransferase from barley. Characterization and evidence for ping pong kinetics involving phosphoryl enzyme. *J. Biol. Chem.* 257: 4931–4939.
Rampazzo, C., Fabris, S., Franzolin, E. et al. (2007). Mitochondrial thymidine kinase and the enzymatic network regulating thymidine triphosphate pools in cultured human cells. *J. Biol. Chem.* 282: 34758–34769.
Ross, C.W. (1981). Biosynthesis of nucleotides. In: *Biochemistry of Plants*, vol. 6 (eds. P.K. Stump and E.E. Conn), 169–205. New York: Academic Press.
Sato, Y. and Ashihara, H. (2009). Long-term effect of NaCl on the activity of uridine and uracil salvage for nucleotide synthesis in cultured mangrove (*Bruguiera sexangula*) cells. *Plant Sci.* 176: 383–389.
Stasolla, C., Katahira, R., Thorpe, T.A., and Ashihara, H. (2003). Purine and pyrimidine nucleotide metabolism in higher plants. *J. Plant Physiol.* 160: 1271–1295.
Wasternack, C. (1982). Metabolism of pyrimidines and purines. In: *Encyclopedia of Plant Physiology*, new ser, vol. 14B (eds. A. Pirson and M.H. Zimmermann), 263–301. Berlin: Springer.
Xu, J., Zhang, L., Yang, D.-L. et al. (2015). Thymidine kinases share a conserved function for nucleotide salvage and play an essential role in *Arabidopsis thaliana* growth and development. *New Phytol.* 208: 1089–1103.

10

Interconversion of Pyrimidine Nucleotides

10.1 Introduction

Three pyrimidine skeletons, namely uracil, cytosine, and thymine, occur as bases, nucleosides and ribonucleotides, and/or deoxyribonucleotides. Like purines, these pyrimidines are also interconverted by several enzymes (Figure 10.1). Some of these conversions are carried out by non-pyrimidine specific reactions (Katahira and Ashihara 2002), which have already been covered in Chapters 8 and 9.

In this chapter, information on the enzymes involved in the pyrimidine conversions, namely, deaminases, nucleosidases, and phosphorylases are reviewed. *In situ* metabolism of ^{14}C-labelled pyrimidine metabolism and possible conversion routes are also discussed as shown in the case of purine metabolism in Chapter 6. Profiles of enzyme activities of deaminase and nucleosidase in young leaves and tubers of potato (*Solanum tuberosum*) are presented in Table 10.1 and the native and recombinant enzymes which have been characterized in plant materials are summarized in Table 10.2.

10.2 Deaminase Reactions

The conversion of cytosine to uracil rings, is carried out by deamination reactions. Three enzymes which catalyse deamination of the cytidine ring have been reported in different organisms; they are cytidine deaminase (steps 1 and 2), deoxycytidylate (dCMP) deaminase (step 3), and cytosine deaminase (step 4) in Figure 10.1.

10.2.1 Cytidine Deaminase

Cytidine deaminase (CDA, EC 3.5.4.5) catalyses the reactions: (i) cytidine to uridine (Figure 10.2) and (ii) deoxycytidine to deoxyuridine (Figure 10.3).

Partially purified native CDA from leaves of *Zea mays* catalyses deamination of both cytidine and deoxycytidine. The respective Km values for these substrates are 550 and 230 μM (Le Floc'h and Guillot 1974). The recombinant CDA of *Arabidopsis thaliana* is a homodimer, containing one zinc atom per subunit (Vincenzetti et al. 1999). This recombinant enzyme catalyses the deamination of both cytidine and deoxycytidine, and is inhibited competitively by cytosine-containing compounds (Faivre-Nitschke et al. 1999).

Plant Nucleotide Metabolism – Biosynthesis, Degradation, and Alkaloid Formation, First Edition.
Hiroshi Ashihara, Iziar A. Ludwig and Alan Crozier.

Table 10.1 Profile of activity of enzymes involved in pyrimidine interconversion in growing leaves and tubers of potato (Katahira and Ashihara 2002, 2006).

		Potato	
Enzyme (EC number)	**Reaction (step number in Figure 10.1)**	**Leaves**	**Tubers**
Cytidine deaminase (3.5.4.5)	Cytidine[a] → Uridine[a] (**1**)	20.1	9.5
	Deoxycytidine[a] → Deoxycytidine[a] (**2**)		5.2
Cytosine deaminase (3.5.4.1)	Cytosine[a] → Uracil[a] (**4**)	—	—
Uridine nucleosidase (3.2.2.3)	Uridine[a] → Uracil[a] (**5**)	17.3	4.3

a) ^{14}C-labelled substrate and product.
Enzyme activity is expressed as pkat mg^{-1} protein. —, not detected.

Table 10.2 Properties of a native and recombinant cytidine deaminases, native uridine nucleosidases and recombinant nucleosidases (NSH1) from plants.

				Km (μM)					
Enzyme	**Enzyme source**	**Isozyme**	**Expression**	**CR**	**CdR**	**UR**	**mUR**	**XR**	**Reference**
Cytidine deaminase	*Zea mays*		(native)	550	230	—	—	—	Le Floc'h and Guillot (1974)
	Arabidopsis thaliana	CDA	*E. coli*	350	110	—	—	—	Chen et al. (2016)
		CDAL4	*E. coli*	29 000	3700	—	—	—	
	Glycine max	CDA1	*E. coli*	1400	240	—	—	—	
		CDA2	*E. coli*	3000	330	—	—	—	
Uridine nucleosidase	*Phaseolus radiatus*		(native)	—	—	a)	—	—	Achar and Vaidyanathan (1967)
	Hordeum vulgare		(native)	—	—	100	—	—	Lee (1985)
Nucleosidase	*Arabidopsis thaliana*	NSH1[b)]	Yeast	—	—	2100	a)	a)	Jung et al. (2009)
			E. coli	—	—	800	440	1690	Jung et al. (2011)

CR, Cytidine; CdR, deoxycytidine; UR, uridine; mUR, methyluridine; XR, xanthosine.
a) No Km value available. —, not applicable.
b) Nucleoside hydrolase, NSH1 (former designation URH1).

The *CDA* gene family of *A. thaliana* comprises nine members, one of which encodes an active CDA. All other *CDA* family members are pseudogenes (*AtCDAL1–8*). The kinetics of recombinant proteins of CDA and CDA-like 4 are shown in Table 10.2. The Km values of CDA (350 μM for cytidine and 110 μM for deoxycytidine) are similar to the native enzyme, so this *A. thaliana* enzyme is functional *in planta*. In contrast, extremely

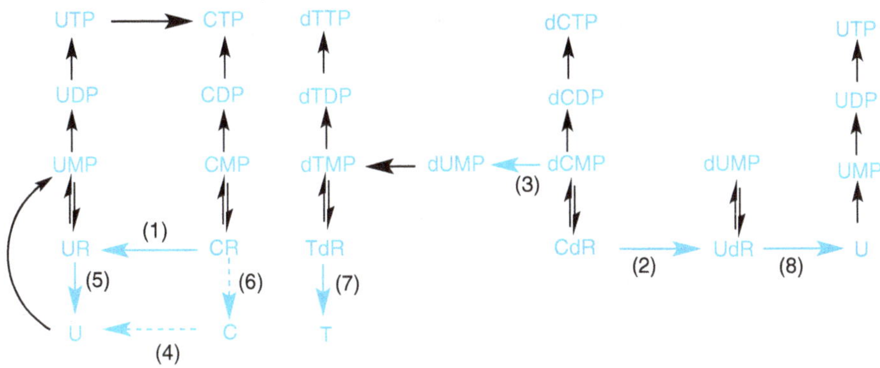

Figure 10.1 The interconversion of pyrimidines in plants. In the *de novo* pyrimidine nucleotide biosynthesis, UTP and CTP are formed (Figure 8.1). In the subsequent reaction steps, dUDP and dTMP are produced, respectively, from UDP (step 2) and dUMP (step 5) as shown in Figure 8.4. The enzymes in the interconversion are as follows: (**1** and **2**) cytidine deaminase; (**3**) dCMP deaminase; (**4**) cytosine deaminase; (**5**) uridine nucleosidase; (**6**) cytidine nuceosidase; (**7**) thymidine phosphorylase or thymidine nucleosidase; (**8**) uridine nucleosidase. C, cytosine; CR, cytidine; CdR, deoxycytidine; dCMP, deoxycytidylate; T, thymine; TdR, thymidine; U, uracil; UR, uridine; UdR, deoxyuridine.

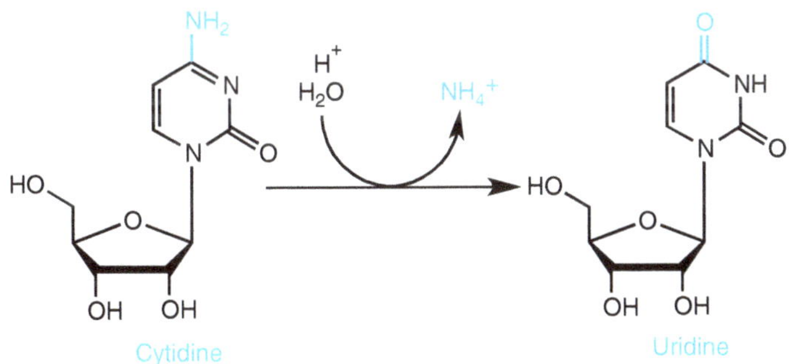

Figure 10.2 Reaction 1 in Figure 10.1.

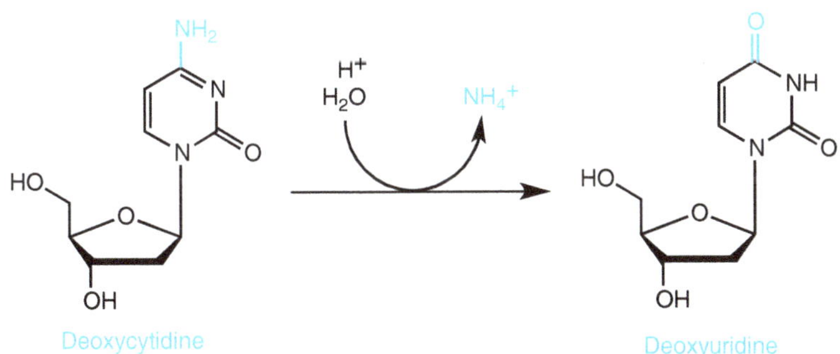

Figure 10.3 Reaction 2 in Figure 10.1.

low activity was detected with CDA-L4, and the efficiency rate (Kcat/Km) of CDA-L4 is 50–2500-fold lower than the rate of CDA. Soybean (*Glycine max*) possesses three *CDA* genes, but only two encode functional enzymes. CDAs from *A. thaliana* and soybean appear to be located in the cytosol. The functionality of the AtCDA enzyme *in vivo* was demonstrated with loss-of-function mutants accumulating high amounts of cytidine as well as CMP, cytosine, and some uridine in seeds (Chen et al. 2016).

The *A. thaliana* CDA shows higher sequence similarity to the *Escherichia coli* enzyme than to known CDAs of eukaryotes (Carter 1995). Unlike the mammalian CDAs (Laliberté and Momparler 1994), the *A. thaliana* CDA was incapable of binding RNAs nor did it show any RNA editing activities (C to U editing) *in vitro* (Faivre-Nitschke et al. 1999).

10.2.2 Cytosine Deaminase

In bacteria, such as *E. coli, Saccharomyces cerevisiae,* and *Klebsiella pneumoniae,* cytosine deaminase (EC 3.5.4.1) is an enzyme in the pyrimidine interconversion pathway. The enzyme catalyses the reaction: cytosine + $H_2O \rightarrow$ uracil + NH_3 and it allows the cell to utilize cytosine for pyrimidine nucleotide synthesis (Ahmad and Pritchard 1972). This enzyme activity has not been detected in plants, including leaves and tubers of potato (Katahira and Ashihara 2002; Katahira and Ashihara 2006) (Table 10.1).

10.2.3 Deoxycytidylate Deaminase

The dCMP deaminase (EC 3.5.4.12) catalyses the deamination of dCMP to dUMP (step 3 in Figure 10.1). Although *in vitro* activity of this enzyme has not been reported in plants, the product, UMP, is the nucleotide substrate for thymidylate synthase (EC 2.1.1.45) which catalyses the conversion of dUMP to dTMP (see Figure 8.5). The dCMP deaminase may specifically be involved in the production of dTTP from dCMP (Figure 10.1). The enzyme is activated allosterically by dCTP and inhibited by dTTP, and is, thus, regulated by the relative amounts of dCTP to dTTP in the cell (Sanchez et al. 2012). The enzyme was first reported to be present in unfertilized eggs of sea urchins (Scarano et al. 1960). Evidence has been presented that dCMP deaminase is a major contributor of dUMP for thymidylate synthase in animals and microorganisms (Chiu et al. 1977; Hou et al. 2008; Jackson 1978). In plants, a putative *dCMP deaminase* gene has been reported in rice (Xu et al. 2014), but there are no reports of biochemical enzymatic investigations being carried out.

10.3 Nucleosidase and Phosphorylase Reactions

Glycosidic bond cleavage, which releases the ribose from nucleosides, is accomplished either hydrolytically or phosphorolytically. The hydrolytic cleavage by nucleosidases is common in plants (see Chapter 6).

10.3.1 Uridine Nucleosidase

Uridine nucleosidase (uridine ribohydrolase, EC 3.2.2.3) catalyses the reaction: uridine to uracil (step 5 in Figure 10.1) (Figure 10.4).

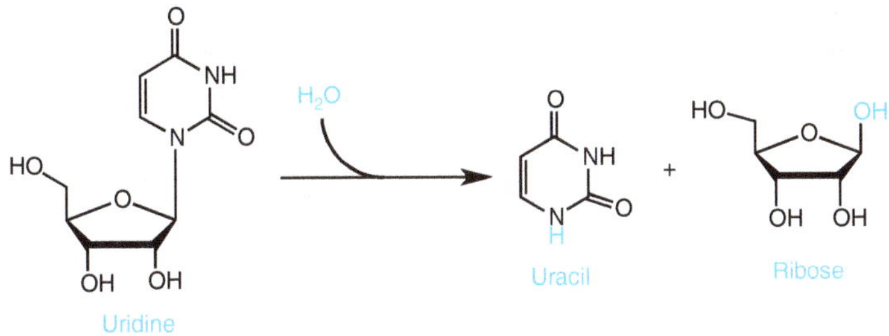

Figure 10.4 Reaction 5 in Figure 10.1.

The native enzyme purified from mung bean (*Vigna radiata)* is highly specific to uridine. It does not cleave deoxyuridine, thymidine, cytidine, deoxycytidine, adenosine, guanosine, and inosine (Achar and Vaidyanathan 1967). Uridine nucleosidase (EC 3.2.2.3) has been purified from barley (*Hordeum vulgare)* malt. The Km for uridine was 100 μM, and the activity has a very narrow specificity (Lee 1985).

Recombinant uridine ribohydrolase (URH1) has been cloned from *A. thaliana*. In contrast to native uridine nucleosidase, it has extremely broad substrate specificity with uridine and xanthosine, but not cytidine, as the preferred substrates. Although methyluridine is also hydrolysed by this recombinant enzyme, thymidine was not investigated (Jung et al. 2009; Jung et al. 2011). In contrast to plants, in *S. cerevisiae* URH1 activity is important for recycling, via the salvage pathway that converts pyrimidine deoxy- and ribonucleosides into uracil which is required for the growth of strains lacking the *de novo* pyrimidine synthesis pathway. The yeast uses uridine, cytidine, and deoxycytidine as substrates for this enzyme (Kurtz et al. 2002) (steps 5, 6, and 8 in Figure 10.1).

10.3.2 Thymidine Phosphorylase

Thymidine phosphorylase (EC 2.4.2.4) catalyses the reaction: thymidine + phosphate → thymine + deoxyribose-1-phosphate. In some mammalian tissues the enzyme also catalyses deoxyribosyltransferase reactions and possesses nucleoside deoxyribosyltransferase activity (Zimmerman and Seidenberg 1964). Potential genes encoding uridine phosphorylase (EC 2.4.2.3) and thymidine phosphorylase (EC 2.4.2.4) have not found any matches in the Arabidopsis Information Resource (TAIR) database of *A. thaliana*. Feeding studies with ^{14}C-thymidine indicated that conversion of thymidine to thymine (step 7 in Figure 10.1) occurs in plants (Katahira and Ashihara 2002). Thus, thymidine phosphorylase and/or thymidine nucleosidase probably occur in plants, although, to date, no *in vitro* activity has been reported.

10.4 *In Situ* Metabolism of ^{14}C-Labelled Pyrimidines

The possible interconversion reactions of pyrimidine nucleotides, nucleosides, and bases are estimated from the occurrence of enzyme activities noted above. However, activity of these reactions *in planta* can be determined quantitatively using radiolabelled pyrimidines. ^{14}C-Labelled orotate, uridine, and uracil have been often used to evaluate the activity of *de novo*, salvage, and degradation pathways of pyrimidine nucleotides, but there have been very few studies of other pyrimidines in plant tissues. The metabolic fates of pyrimidines have been investigated in various plant materials (Table 10.3). The ^{14}C-radiolabelling approach allows accurate determination of label uptake since the total radiolabel content is easy to detect. In plant studies, a wide variety of pyrimidine nucleosides and bases have been comprehensively investigated in growing potato tubers (Katahira and Ashihara 2002) and the findings will be discussed in the ensuing sections.

10.4.1 Metabolic Fate of Orotate

Although orotate is an intermediate of the *de novo* pyrimidine nucleotide biosynthesis and is not strictly involved in pyrimidine conversion. [2-^{14}C]Orotate is often utilized to investigate *in situ* pyrimidine metabolism (Table 10.4). The estimated metabolic fate of the radiolabelled orotate indicates the metabolic fate of the intermediate of the *de novo* pathway produced by HCO_3^-, glutamine, and aspartate (see Chapter 8). For growing potato tubers, ~40% of radioactivity was metabolized and the remainder was retained as unmetabolized orotate. Most of metabolites were distributed in UDP-glucose (20%), uracil ribonucleotides (ΣUN = UTP + UDP + UMP), mainly UTP (8%), and RNA (6%) (Katahira and Ashihara 2002). It is noteworthy that UDP-glucose is formed from UTP, which is synthesized by the *de novo* pyrimidine biosynthesis.

UTP is always present in higher amounts than UMP and UDP (Farre et al. 2001; Loef et al. 1999). The conversion of UMP to UTP by nucleoside monophosphate kinase and nucleoside diphosphate kinase (see Chapter 8) is very fast. Likewise, the synthesis of UDP-glucose from UTP by UDP-glucose pyrophosphorylase (EC2.7.7.9) is also rapid (Sowokinos et al. 1993). In many plant materials, including potato tubers, UDP-glucose is the most abundant pyrimidine (Loef et al. 1999).

10.4.2 Metabolic Fate of Uridine and Uracil

In situ metabolism of pyrimidine nucleosides and bases, especially uridine and uracil, has been investigated in many plant cells and tissues (Table 10.3). The metabolic fate of these ^{14}C-labelled compounds varies quantitatively in different organs and plant species and the stages of growth, while the metabolic profiles of individual purine nucleosides and nucleobases tend to be similar. Usually uridine is salvaged more readily than uracil while, in contrast, uracil is catabolized more rapidly than uridine (see Chapter 11). The metabolic fates of [2-^{14}C]uridine and [2-^{14}C]uracil were compared in the discs of growing potato tubers (Katahira and Ashihara 2002). The results obtained are summarized in Table 10.4.

In contrast to [2-^{14}C]orotate, an intermediate of the *de novo* pathway (Section 10.4.1), all [2-^{14}C]pyrimidine nucleosides including [2-^{14}C]uridine were readily taken up by potato discs. More than 90% of the radioactivity taken up by the discs was incorporated into salvage products, nucleotides, and nucleic acids. In the case of [2-^{14}C]uridine, ~40% of the radioactivity was recovered as uracil nucleotides (ΣU=UMP + UDP + UTP), a further 40% was incorporated into UDP-glucose, and small amounts of radioactivity were found as cytosine nucleotides (0.5%) and CDP-choline (4%). A substantial amount of radioactivity (14%) was incorporated into RNA. Almost 75% of radioactivity was

Table 10.3 Plant materials used for overall *in situ* ^{14}C-pyrimidine tracer experiments.

Plant species	Common name	Material	^{14}C-Pyrimidines	References
Picea glauca	White spruce	Cell culture	O, UR, U	Ashihara et al. (2000) Ashihara et al. (2001)
		Embryos	CR, CdR, TdR	Stasolla et al. (2003)
Pinus radiata		Shoot	O, U, UR	Stasolla et al. (2007)
Daucus carota	Carrot	Embryogenesis	O, UR, U	Ashihara et al. (1981)
Solanum tuberosum	Potato	Tubers	O, UR, U, CR, C, TdR, CdR	Katahira and Ashihara (2002)
Nicotiana tabacum	Tobacco	Leaves	U	Ashihara (1981)
Pisum sativum	Pea	Cotyledons	O, UR	Ross et al. (1971)
Vigna (= Phaseolus) mungo	Black gram	Seedlings	O, UR, U, TdR, T	Ashihara (1977) Kameyama et al. (1985)
Phaseolus vulgaris	Bean	Leaves	O, UR, U	Ross (1965)
Catharansus roseus (= Vinca rosea)	Madagascar periwinkle	Cell culture	O, UR, U, TdR, T	Kanamori-Fukuda et al., (1981), Kanamori et al. (1979), Sasamoto et al. (1987)
Rhizophora stylosa	Spotted mangrove	Leaves	UR	Sato and Ashihara (2008)
Bruguiera sexangula	Upriver orange mangrove	Cell culture	O, U, UR, CR	Suzuki-Yamamoto et al. (2006), Sato and Ashihara (2009)
		Leaves	U, UR, CR, DHU	Sato and Ashihara (2008)
Bruguiera gymnorrhiza	Black mangrove	Leaves	UR	Sato and Ashihara (2008)
Kandelia obovata (=K. candel)	Mehirugi mangrove	Leaves	UR	Sato and Ashihara (2008)
Avicennia marina	Grey mangrove	Leaves	UR	Sato and Ashihara (2008)

Tracers: orotate (O), uridine (UR), uracil (U), cytidine (CR), cytosine (C), thymidine (TdR), thymine (T), deoxycytidine (CdR), dihydrouracil (DHU).

Table 10.4 Metabolic fate of [2-^{14}C]pyrimidine bases and ribonucleosides in growing potato tubers.

^{14}C-Metabolites (%)	Orotate	Uridine	Uracil	Cytidine	Cytosine
Salvage product					
ΣUMP (UMP + UDP + UTP)	8	40	21	4	—
ΣCTP (CMP + CDP + CTP)	—	0.5	—	20	—
UDP-Glucose	20	38	12	16	—
UDP-Choline	—	4	—	2	—
RNA (U-residue)	6	11	4	46	—
RNA (C-residue)	—	4	1	5	—
DNA	—	—	—	1	—
Nucleosides and bases					
Uridine	2	3	3	3	—
Cytidine	—	—	—	3	—
Uracil	—	—	10	—	—
Cytosine	—	—	—	—	100
Catabolites					
CO_2	1	0.4	48	<0.1	—
Others	—	—	<1	—	0

Incubation time is 4 h. Values are expressed as % of radioactivity and total uptake.
Source: Adapted from Katahira and Ashihara (2002).

distributed in uracil residues of RNA and the remainder (~25%) was in cytosine residues. Release of $^{14}CO_2$ from [2-^{14}C]uridine was low.

The cytosine nucleotides noted above play a pivotal role in the synthesis and turnover of phospholipids (Wagner and Backer 1992). A major ^{14}C-labelled lipid-related metabolite, CDP-choline, is an intermediate in the generation of phosphatidylcholine from choline in plants (Suzuki et al. 2003). Formation of CDP-choline from CTP is catalysed by choline-phosphate cytidylyltransferase (EC 2.7.7.15).

In the case of [2-^{14}C]uracil-treated potato tuber discs, ~40% of the total radioactivity was incorporated into the salvage products, uracil nucleotides, UDP-glucose and RNA. The radioactivity in RNA comprised uracil residues (80%), and cytosine residues (20%). Approximately half of the total radioactivity was recovered as degradation metabolites, β-ureidopropionate (<0.1%), and CO_2 (~50% of the total radioactivity). The $^{14}CO_2$ was released from [2-^{14}C]uracil at the degradation step from β-ureidopropionate to β-alanine (see Chapter 11). In contrast to uridine salvage, uracil salvage activity is typically low in plant material (Ashihara et al. 2000; Kanamori-Fukuda et al. 1981). The postulated pathway of uridine and uracil metabolism is shown in Figure 10.5.

10.4.3 Metabolic Fate of Cytidine and Cytosine

More than half of the radioactivity taken up from [2-^{14}C]cytidine by potato tuber discs was incorporated into RNA and <1% into DNA (Katahira and Ashihara 2002). Possible

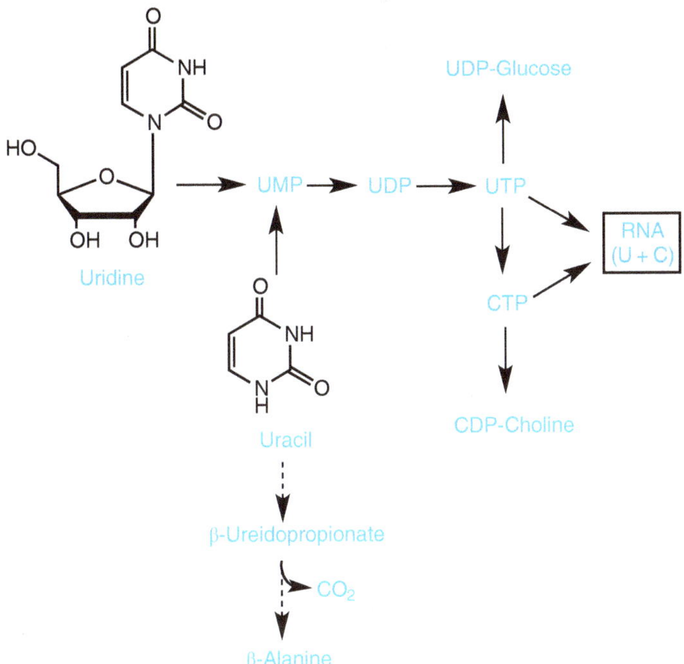

Figure 10.5 *In situ* metabolism of [2-^{14}C]uridine and [2-^{14}C]uracil in plants.

routes from cytidine to RNA and DNA are illustrated in Figure 10.6. Nearly 20% was recovered as cytosine nucleotides. Some radioactivity was also distributed in uracil nucleotides (~4%), UDP-glucose (~16%), and CDP-choline (~2%). The radioactivity in the RNA fraction was distributed in cytosine (>90%) and in uracil residues (<10%). A small amount of radioactivity (~3%) was recovered as uridine. The data indicate that [2-^{14}C]cytidine is deaminated to uridine by cytidine deaminase and then converted to UMP. A part of metabolic fate of cytidine shown in Table 10.4 may be the same as that of [2-^{14}C]uridine (Figure 10.6).

In contrast to cytidine, [2-^{14}C]cytosine taken up by the potato discs remained unmetabolized (Table 10.2). As mentioned in Section 10.2.2, cytosine deaminase, which occurs in yeast and bacteria (Ipata et al. 1971; Katsuragi et al. 1986) has not been detected in plants including potato (Table 10.1). The overall conversion routes of cytidine and cytosine are illustrated in Figure 10.6.

10.4.4 Metabolic Fate of Deoxycytidine

Uptake of [2-^{14}C]deoxycytidine by potato discs was rapid. Salvage of [2-^{14}C] deoxycytidine occurred, but substantial amounts of radioactivity were distributed, not in deoxycytidine derivatives, but in uracil ribonucleotides (15%), UDP-glucose (40%) and RNA (9%). Approximately 75% of the radioactivity in RNA was recovered as uracil residues, and the remainder (~25%) as cytosine residues. Approximately 17% of the radioactivity was recovered as free deoxyuridine (Table 10.5).

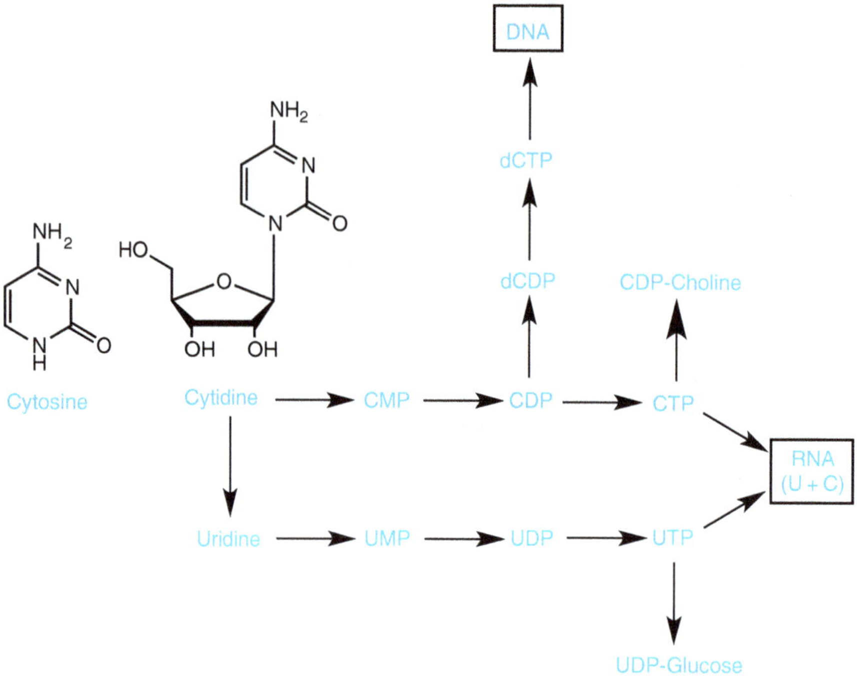

Figure 10.6 *In situ* metabolism of [2-^{14}C]cytidine and [2-^{14}C]cytosine in plants.

These results suggest that substantial amounts of deoxycytidine were converted to UDP-glucose by a novel route. The labelling pattern indicates that UDP-glucose synthesis from deoxycytidine proceeds by a deoxycytidine → deoxyuridine → uridine → UMP → UDP → UTP → UDP-glucose pathway as illustrated in Figure 10.7. Incorporation of [2-^{14}C]deoxycytidine into uracil residues of RNA also supports the operation of this pathway. This was the first report showing the involvement of this pathway in the production of uracil ribonucleotides from deoxycytidine. Conversion of deoxycytidine to deoxyuridine is catalysed by cytidine deaminase, and the next step, deoxyuridine to uridine is catalysed by pyrimidine deoxyribonucleoside 2′-hydroxylase (EC 1.14.11.3). This enzyme exists in some microorganisms, such as *Nuerospora crassa* (Bankel et al. 1972), but its occurrence has not been reported in plants or animals.

Although the metabolic rates are limited, <6% of the radioactivity from [2-^{14}C]deoxycytidine was incorporated directly into deoxyribonucleotides and DNA by potato discs. In contrast, catabolism of deoxycytidine seems not to be functional, since release of $^{14}CO_2$ from [2-^{14}C]deoxycytidine was negligible (Table 10.4; Figure 10.7).

10.4.5 Metabolic Fate of Thymidine

Uptake of [2-^{14}C]thymidine by potato discs is rapid but ~40% of thymidine remains unmetabolized. Less than 5% of [2-^{14}C]thymidine was salvaged to thymine nucleotides and more limited amounts were incorporated into DNA and RNA. Almost all of the metabolized thymidine was recovered as thymine (15%), uracil (12%), and CO_2 (30%)

Table 10.5 Metabolic fate of [2-^{14}C]deoxyribonucleosides in growing potato tubers.

^{14}C-Metabolites (%)	Deoxycytidine	Thymidine
Salvage product		
ΣUMP	15	—
ΣCMP + ΣdCMP	6	—
dTMP	—	3
UDP-Glucose	40	—
UDP-Choline	4	—
RNA (U-residue)	7	1
RNA (C-residue)	2	—
DNA	<0.1	<0.1
Nucleosides and bases		
Deoxyuridine	17	3
Uridine	1	—
Deoxycytidine	5	—
Thymidine	1	38
Uracil	—	12
Thymine	—	15
Catabolites		
CO_2	<0.1	30
Others	<0.1	<0.1

Incubation time is 4 h. Values are expressed as % of radioactivity and total uptake.
Source: Adapted from Katahira and Ashihara (2002).

(Table 10.4). Although the *in situ* metabolism of [2-^{14}C]thymine in potato tubers was not determined, the fate of this pyrimidine base has been investigated in cotyledons and embryonic axes of *Phaseolus mungo* seedlings (Figure 10.8). [2^{14}C]Thymine was degraded extensively into $^{14}CO_2$ via β-ureidoisobutyrate. This suggests thymidine is hydrolysed to thymine and enters the reductive pathway of pyrimidine degradation (Kameyama et al. 1985). The degradation of pyrimidine is discussed in Chapter 11.

10.5 Summary

A diversity of pyrimidine interconversions occur in different organisms. In plants, cytidine, deoxycytidine, and dCMP are converted to uridine, deoxyuridine, and dUMP. In mammals, conversion of dCMP to dTMP via dUMP also occurs. Uracil is hydrolysed from uridine and deoxyuridine. Thymine is produced from thymidine. Cytosine is not metabolized. In addition to nucleic acid formation, uridine and cytidine nucleotides are mainly utilized for UDP-glucose and CDP-choline formation in plants.

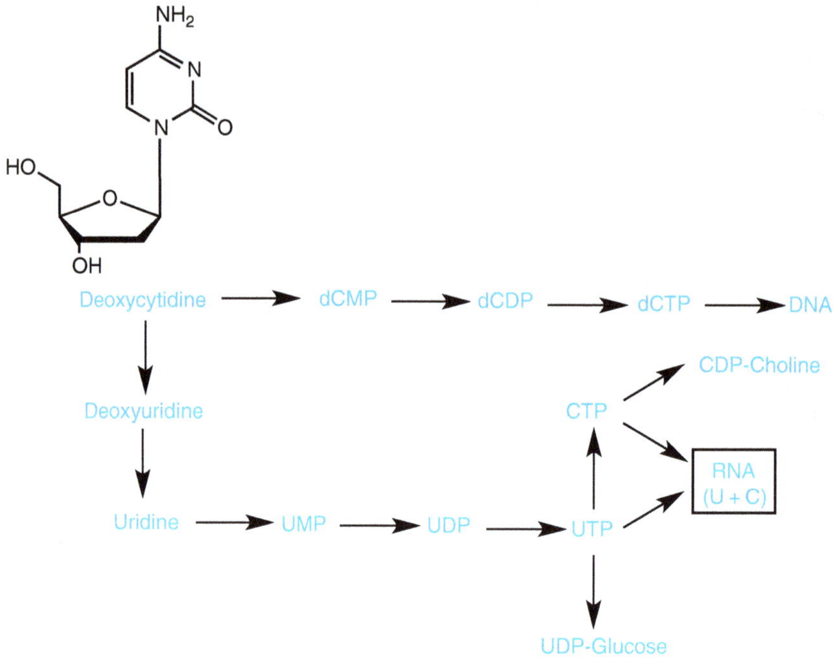

Figure 10.7 *In situ* metabolism of [2-^{14}C]deoxycytidine in plants.

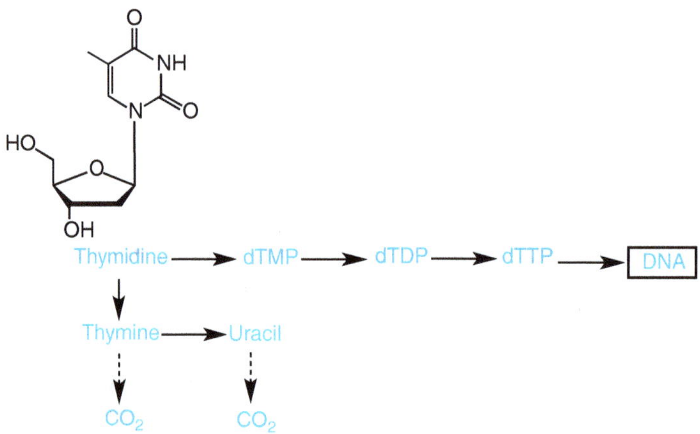

Figure 10.8 *In situ* metabolism of [2-^{14}C]thymidine in plants.

References

Achar, B.S. and Vaidyanathan, C.S. (1967). Purification and properties of uridine hydrolase from mung-bean (*Phaseolus radiatus*) seedlings. *Arch. Biochem. Biophys.* 119: 356–362.

Ahmad, S.I. and Pritchard, R.H. (1972). Location of gene specifying cytosine deaminase in *Escherichia coli*. *Mol. Gen. Genet.* 118: 323–325.

Ashihara, H. (1977). Changes in the activities of the *de novo* and salvage pathways of pyrimidine nucleotide biosynthesis during germination of black gram (*Phaseolus mungo*) seeds. *Z. Pflanzenphysiol.* 81: 199–211.

Ashihara, H. (1981). Pattern of uracil metabolism during ageing of tobacco leaves. *Acta Physiol. Plant.* 3: 77–84.

Ashihara, H., Fujimura, T., and Komamine, A. (1981). Pyrimidine nucleotide biosynthesis during somatic embryogenesis in a carrot cell suspension culture. *Z. Pflanzenphysiol.* 104: 129–137.

Ashihara, H., Stasolla, C., Loukanina, N., and Thorpe, T.A. (2000). Purine and pyrimidine metabolism in cultured white spruce (*Picea glauca*) cells: metabolic fate of ^{14}C-labeled precursors and activity of key enzymes. *Physiol. Plant.* 108: 25–33.

Ashihara, H., Loukanina, N., Stasolla, C., and Thorpe, T.A. (2001). Pyrimidine metabolism during somatic embryo development in white spruce (*Picea glauca*). *J. Plant Physiol.* 158: 613–621.

Bankel, L., Lindstedt, G., and Lindstedt, S. (1972). Thymidine 2′-hydroxylation in *Neurospora crassa*. *J. Biol. Chem.* 247: 6128–6134.

Carter, C.W. (1995). The nucleoside deaminases for cytidine and adenosine: structure, transition state stabilization, mechanism, and evolution. *Biochimie* 77: 92–98.

Chen, M., Herde, M., and Witte, C.-P. (2016). Of the nine cytidine deaminase-like genes in Arabidopsis, eight are pseudogenes and only one is required to maintain pyrimidine homeostasis *in vivo*. *Plant Physiol.* 171: 799–809.

Chiu, C.S., Ruettinger, T., Flanegan, J.B., and Greenberg, G.R. (1977). Role of deoxycytidylate deaminase in deoxyribonucleotide synthesis in bacteriophage T4 DNA replication. *J. Biol. Chem.* 252: 8603–8608.

Faivre-Nitschke, S.E., Grienenberger, J.M., and Gualberto, J.M. (1999). A prokaryotic-type cytidine deaminase from *Arabidopsis thaliana*. *Eur. J. Biochem.* 263: 896–903.

Farre, E.M., Tiessen, A., Roessner, U. et al. (2001). Analysis of the compartmentation of glycolytic intermediates, nucleotides, sugars, organic acids, amino acids, and sugar alcohols in potato tubers using a nonaqueous fractionation method. *Plant Physiol.* 127: 685–700.

Hou, H.-F., Liang, Y.-H., Li, L.-F. et al. (2008). Crystal structures of *Streptococcus mutans* 2′-deoxycytidylate deaminase and its complex with substrate analog and allosteric regulator dCTP·Mg^{2+}. *J. Mol. Biol.* 377: 220–231.

Ipata, P.L., Marmocchi, F., Magni, G. et al. (1971). Baker's yeast cytosine deaminase. Enzymic properties and allosteric inhibition by nucleosides and nucleotides. *Biochemistry* 10: 4270–4276.

Jackson, R.C. (1978). The regulation of thymidylate biosynthesis in Novikoff hepatoma cells and the effects of amethopterin, 5-fluorodeoxyuridine, and 3-deazauridine. *J. Biol. Chem.* 253: 7440–7446.

Jung, B., Flörchinger, M., Kunz, H.-H. et al. (2009). Uridine-ribohydrolase is a key regulator in the uridine degradation pathway of *Arabidopsis*. *Plant Cell* 21: 876–891.

Jung, B., Hoffmann, C., and Möhlmann, T. (2011). Arabidopsis nucleoside hydrolases involved in intracellular and extracellular degradation of purines. *Plant J.* 65: 703–711.

Kameyama, Y., Tokoro, T., and Ashihara, H. (1985). Metabolism of [2-^{14}C]thymine and [2-^{14}C]thymidine in germinating black gram (*Phaseolus mungo*) seeds. *Radioisotopes* 34: 214–221.

Kanamori, I., Ashihara, H., and Komamine, A. (1979). Changes in the activities of the pentose phosphate pathway and pyrimidine nucleotide biosynthesis during the growth of *Vinca rosea* cells in suspension culture. *Z. Pflanzenphysiol.* 93: 437–448.

Kanamori-Fukuda, I., Ashihara, H., and Komamine, A. (1981). Pyrimidine nucleotide biosynthesis in *Vinca rosea* cells: changes in the activity of *de novo* and salvage pathways during growth in a suspension culture. *J. Exp. Bot.* 32: 69–78.

Katahira, R. and Ashihara, H. (2002). Profiles of pyrimidine biosynthesis, salvage and degradation in disks of potato (*Solanum tuberosum* L.) tubers. *Planta* 215: 821–828.

Katahira, R. and Ashihara, H. (2006). Dual function of pyrimidine metabolism in potato (*Solanum tuberosum*) plants: pyrimidine salvage and supply of β-alanine to pantothenic acid synthesis. *Physiol. Plant.* 127: 38–43.

Katsuragi, T., Sakai, T., and Tonomura, K. (1986). Affinity chromatography of cytosine deaminase from *Escherichia coli* with immobilized pyrimidine compounds. *Agric. Biol. Chem.* 50: 1713–1719.

Kurtz, J.E., Exinger, F., Erbs, P., and Jund, R. (2002). The URH1 uridine ribohydrolase of *Saccharomyces cerevisiae*. *Curr. Genet.* 41: 132–141.

Laliberté, J. and Momparler, R.L. (1994). Human cytidine deaminase: purification of enzyme, cloning, and expression of its complementary DNA. *Cancer Res.* 54: 5401–5407.

Le Floc'h, F. and Guillot, A. (1974). La desoxycytidine aminohydrolase des feuilles de *Zea mays*. *Phytochemistry* 13: 2503–2509.

Lee, W.J. (1985). Uridine Nucleosidase from malt. *J. Am. Soc. Brew. Chem.* 45: 131–134.

Loef, I., Stitt, M., and Geigenberger, P. (1999). Orotate leads to a specific increase in uridine nucleotide levels and a stimulation of sucrose degradation and starch synthesis in discs from growing potato tubers. *Planta* 209: 314–323.

Ross, C. (1965). Comparison of incorporation and metabolism of RNA pyrimidine nucleotide precursors in leaf tissues. *Plant Physiol.* 40: 65–73.

Ross, C., Coddington, R.L., Murray, M.G., and Bledsoe, C. (1971). Pyrimidine metabolism in cotyledons of germinating pea. *Plant Physiol.* 47: 71–75.

Sanchez, A., Sharma, S., Rozenzhak, S. et al. (2012). Replication fork collapse and genome instability in a deoxycytidylate deaminase mutant. *Mol. Cell. Biol.* 32: 4445–4454.

Sasamoto, H., Saito, K., and Ashihara, H. (1987). Metabolism of pyrimidines in protoplasts from cultured *Catharanthus roseus* cells. *Ann. Bot.* 60: 417–420.

Sato, Y. and Ashihara, H. (2008). Pyrimidine salvage and catabolism in leaves of mangrove species. *Plant Sci.* 174: 140–148.

Sato, Y. and Ashihara, H. (2009). Long-term effect of NaCl on the activity of uridine and uracil salvage for nucleotide synthesis in cultured mangrove (*Bruguiera sexangula*) cells. *Plant Sci.* 176: 383–389.

Scarano, E., Bonaduce, L., and De Petrocellis, B. (1960). The enzymatic deamination of 6-aminopyrimidine deoxyribonucleotides. II. Purification and properties of a 6-aminopyrimidine deoxyribonucleoside 5′-phosphate deaminase from unfertilized eggs of sea urchine. *J. Biol. Chem.* 235: 3556–3561.

Sowokinos, J.R., Spychalla, J.P., and Desborough, S.L. (1993). Pyrophosphorylases in *Solanum tuberosum* (IV. Purification, tissue localization, and physicochemical properties of UDP-glucose pyrophosphorylase). *Plant Physiol.* 101: 1073–1080.

Stasolla, C., Loukanina, N., Ashihara, H. et al. (2003). Pyrimidine deoxyribonucleotide metabolism during maturation and germination of white spruce (*Picea glauca*) somatic embryos: metabolic fate of C-labelled cytidine, deoxycytidine and thymidine. *Physiol. Plant.* 118: 499–506.

Stasolla, C., Loukanina, N., Ashihara, H. et al. (2007). Comparative studies on pyrimidine metabolism in excised cotyledons of *Pinus radiata* during shoot formation *in vitro*. *J. Plant Physiol.* 164: 429–441.

Suzuki, M., Yasumoto, E., Baba, S., and Ashihara, H. (2003). Effect of salt stress on the metabolism of ethanolamine and choline in leaves of the betaine-producing mangrove species *Avicennia marina*. *Phytochemistry* 64: 941–948.

Suzuki-Yamamoto, M., Mimura, T., and Ashihara, H. (2006). Effect of short-term salt stress on the metabolic profiles of pyrimidine, purine and pyridine nucleotides in cultured cells of the mangrove tree, *Bruguiera sexangula*. *Physiol. Plant.* 128: 405–414.

Vincenzetti, S., Cambi, A., Neuhard, J. et al. (1999). Cloning, expression, and purification of cytidine deaminase from *Arabidopsis thaliana*. *Protein Expression Purif.* 15: 8–15.

Wagner, K.G. and Backer, A.I. (1992). Dynamics of nucleotides in plants studied on a cellular basis. In: *International Review of Cytology*, vol. 134 (eds. K.W. Jeon and M. Friedlander), 1–84. San Diego: Academic Press.

Xu, J., Deng, Y., Li, Q. et al. (2014). STRIPE2 encodes a putative dCMP deaminase that plays an important role in chloroplast development in rice. *J. Genet. Genom.* 41: 539–548.

Zimmerman, M. and Seidenberg, J. (1964). Deoxyribosyl transfer. I. Thymidine phosphorylase and nucleoside deoxyribosyltransferase in normal and malignant tissues. *J. Biol. Chem.* 239: 2618–2621.

11

Degradation of Pyrimidine Nucleotides

11.1 Introduction

The catabolism of pyrimidine nucleotides, like that of purine nucleotides, involves dephosphorylation, deamination, and glycosidic bond cleavage. However, in contrast to purine catabolism, the pyrimidine bases in most organisms are subjected to reduction rather than oxidation. Reductive pyrimidine base catabolism occurs in most microorganisms, plants, and animals. An oxidative pyrimidine degradation pathway is only found in a very limited number of bacteria such as *Corynebacterium* and *Mycobacterium* (Wasternack 1978).

In plants, the first feeding experiments using ^{14}C-labelled uracil and thymine were performed with germinating seeds of *Brassica napus* (oilseed rape, aka canola) (Evans and Axelrod 1960). Radioactivity from [2-^{14}C]uracil was recovered in dihydrouracil, β-ureidopropionate, and CO_2, while that from [2-^{14}C]thymine was incorporated into dihydrothymine, β-ureidoisobutyrate, and CO_2. This is in keeping with the operation of uracil → dihydrouracil → β-ureidopropionate → β-alanine + CO_2 + NH_3, and thymine → dihydrothymine → β-ureidoisobutyrate → β-aminoisobutyrate + CO_2 + NH_3 pathways (Figure 11.1). The degradation of uracil and thymine involves a hydrogenation step followed by hydrolytic conversions. It is, therefore, of interest that the exclusion of oxygen from germinating seeds represses the catabolism of these substrates.

The uracil and thymine degradation pathways are catalysed by the same enzymes. The reactions start with a coenzyme-dependent reduction, the pyrimidine ring is hydrolytically opened, and then the ureido-group is split into CO_2, NH_3, and β-alanine or β-aminoisobutyrate (Stasolla et al. 2003; Wasternack 1982; Zrenner and Ashihara 2011; Zrenner et al. 2006). The expression of genes encoding these pyrimidine degradation enzymes is coordinated with the expression of genes associated with *de novo* and salvage synthesis pathway enzymes, presumably to maintain pyrimidine homeostasis in response to changing metabolic demands (Zrenner et al. 2009). In plants, the end-product of uracil catabolism, β-alanine, is utilized for the synthesis of pantothenate (Katahira and Ashihara 2006) and β-alanine betaine (Duhazé et al. 2003). β-Alanine may further serve as an amino donor with 2-oxoglutarate and other amino acceptors (Zrenner et al. 2009). Radioactivity of [U-^{14}C]uracil is incorporated into common amino acids (Katahira and Ashihara 2006). The pyrimidine catabolic pathway also supplies NH_3 and CO_2. This pathway may ensure N and C recycling in plants, especially in developing leaves and germinating seeds.

Plant Nucleotide Metabolism – Biosynthesis, Degradation, and Alkaloid Formation, First Edition.
Hiroshi Ashihara, Iziar A. Ludwig and Alan Crozier.

11.2 Enzymes Involved in the Degradation Routes of Pyrimidines

In plants, the pyrimidine bases, uracil, and thymine, derived from uridine monophosphate (UMP) and deoxythymidine-5′-monophosphate (dTMP) are directly catabolized by a reductive degradation pathway. The catabolism of cytidine-5′-monophosphate (CMP) must take place after conversion of cytidine to uridine by cytidine deaminase, which is followed by uracil formation (Chapter 10). Uracil and thymine are degraded by the sequential reactions (steps 1–3, Figure 11.1) catalysed by dihydropyrimidine dehydrogenase (DHPDH, EC 1.3.1.2), dihydropyrimidinase (DHPase, EC 3.5.2.2) and

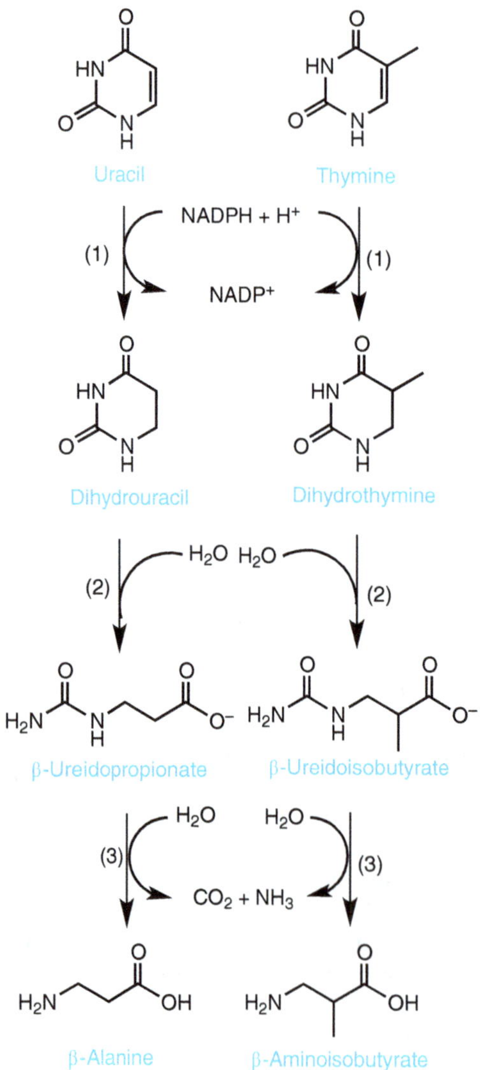

Figure 11.1 The catabolic pathway of the pyrimidines uracil and thymine in plants. The enzymes involved are: (1) dihydropyrimidine dehydrogenase, (2) dihydropyrimidinase, and (3) β-ureidopropionase.

Table 11.1 Summary of enzymes involved in the reductive pathway of pyrimidine degradation illustrated in Figure 11.1.

Step	Enzyme name	EC	Abbreviation	Gene	TAIR locus
1	Dihydropyrimidine dehydrogenase ($NADP^+$)	1.3.1.2	DHPDH	RYD1	AT3G17810
2	Dihydropyrimidinase	3.5.2.2	DHPase	PYD2	AT5G12200
3	β-Ureidopropionase	3.5.1.6	β-UPase	PYD3	AT5G64370

β-ureidopropionase (β-UPase) (Table 11.1). These enzymes are well characterized in mammals but not in plants.

11.2.1 Dihydropyrimidine Dehydrogenase

DHPDH (EC 1.3.1.2) catalyses the reaction: uracil + NADPH + H^+ → 5,6-dihydrouracil + $NADP^+$. This enzyme also acts on thymine (step 1 in Figure 11.1). *In vivo* studies have shown that DHPDH is the initial and rate-limiting enzyme in the catabolic pathway of pyrimidines in plants (Tintemann et al. 1985), however, as yet no purified native plant enzyme has been reported. In mammals, DHPDH has been purified from liver, which is the major site of pyrimidine catabolism. The enzymes from livers of human, pig, rat, and cow have very similar Km values. The Km values for uracil, thymine, and NADPH are within very limited concentration ranges 2–6 μM (uracil), 3–7 μM (thymine) and 8–30 μM (NADPH).

Although the intact DHPDH has not been characterized in plants, mutant analyses indicate the importance of this enzyme for uracil catabolism in *Arabidopsis thaliana* (Zrenner et al. 2009). The DHPDH activity *in vitro* could not found for the recombinant protein, PYD1. This may due to the fact that a compulsory protein interaction partner for PYD1 remains to be discovered in *A. thaliana* (Zrenner et al. 2009).

11.2.2 Dihydropyrimidinase

Dihydropyrimidinase (DHPase, EC 3.5.2.2) catalyses the reaction: 5,6-dihydrouracil + H_2O → β-ureidopropionate (=3-ureidopropanoate) (step 2 in Figure 11.1).

In crude wheat germ extracts, DHPase activity has been demonstrated with the conversion of [^{14}C]-dihydrouracil to [^{14}C]β-ureidopropionate (Mazuś and Buchowicz 1972). However, purification and characterization of native enzyme have not been reported with plant materials. In animals, bovine liver DHPase has been purified and characterized as a zinc metalloenzyme. The Km values of DHPase exhibit a good affinity to the liver enzyme with 8 μM for dihydrouracil and 15 μM for dihydrothymine (Brooks et al. 1983).

Some mutant studies have been undertaken to characterize plant DHPase. Although at the genetic level, the importance of this enzyme for uracil catabolism has been suggested (Gojkovic et al. 2003; Zrenner et al. 2009), there is only limited information available on plant DHPase.

11.2.3 β-Ureidopropionase

β-Ureidopropionase (β-UPase, EC 3.5.1.6) catalyses the reaction: β-ureidopropionate (aka *N*-carbamoyl-β-alanine) + H_2O → β-alanine + CO_2 + NH_3 (step 3 in Figure 11.1). β-UPase has been partially purified from *Euglena gracilis*, the single-celled eukaryote algae. The enzyme showed simple Michaelis–Menten kinetics and the Km was 38 μM (Wasternack et al. 1979). In higher plants, native β-UPase was firstly purified from the shoots of etiolated maize (*Zea mays*) seedlings. The enzyme had respective Km values for β-ureidopropionate and β-ureidoisobutyrate of 11 and 6 μM (Walsh et al. 2001). β-UPase was inactivated by dialysis against 1,10-phenanthroline and activity could be partially restored by the addition of Zn^{2+}. Some mammalian amidohydrolases, including liver β-UPase, are Zn^{2+}-dependent (Kvalnes-Krick and Traut 1993). Maize β-UPase was very sensitive to inactivation by iodoacetamide. This is prevented by addition of substrate, indicating the presence of an active Cys site. The enzyme was strongly inhibited by short chain aliphatic acids and aryl propionates, the most potent inhibitor of which was 2-(2, 6-dinitrophenoxy)-propionate (I_{50} = 0.5 μM).

The β-*UPase* (*PYD3*) gene from *A. thaliana* encoding an enzyme with β-UPase activity was isolated and expressed in *E. coli*. The recombinant protein was shown to convert β-ureidopropionate into β-alanine (Walsh et al. 2001).

11.3 The Metabolic Fate of Uracil and Thymine

The feeding experiments with ^{14}C-labelled uracil and thymine were first carried out using *B. napus* seedlings by Evans and Axelrod (1960) and Tsai and Axelrod (1965). Subsequently, the incorporation of the radioactivity from [2-^{14}C]uracil into dihydrouracil, β-ureidopropionate, and CO_2, and from [2-^{14}C]thymine into dihydrothymine, β-ureidoisobutyrate, and CO_2 has been observed in many plant materials which were listed in Table 10.3.

In plants, a dual function of pyrimidine metabolism has been proposed. In addition to the pyrimidine salvage for nucleotides and nucleic acid synthesis, a degradation product of uracil, β-alanine, is used for pantothenic acid (vitamin B5) synthesis. In fully expanded leaves of potato plants, radioactivity from [U-^{14}C]uracil is incorporated into β-ureidopropionate, CO_2, β-alanine, pantothenate, and several common amino acids (Katahira and Ashihara 2006).

In higher plants, three possible pathways of β-alanine biosynthesis have been reported; these are the spermidine (Rastogi and Davies 1989; Terano and Suzuki 1978), propionate (Hatch and Stumpf 1962; Rathinasabapathi 2002), and uracil routes (Duhazé et al. 2003; Walsh et al. 2001). As mentioned above, uracil contributes to pantothenate synthesis, at least in potato leaves. Therefore, a uracil → dihydrouracil → β-ureidopropionate → β-alanine → pantothenate pathway may operate, at least in some plants.

The synthesis of pantothenate from β-alanine (Figure 11.2) is catalysed by pantothenate synthetase (pantoate–β-alanine ligase, adenosine-5′-monophosphate [AMP]-forming, EC 6.3.2.1). The reaction is (*R*)-pantoate + β-alanine + ATP → (*R*)-pantothenate + AMP + PPi (Leonardi and Jackowski 2007). Pantothenate is the

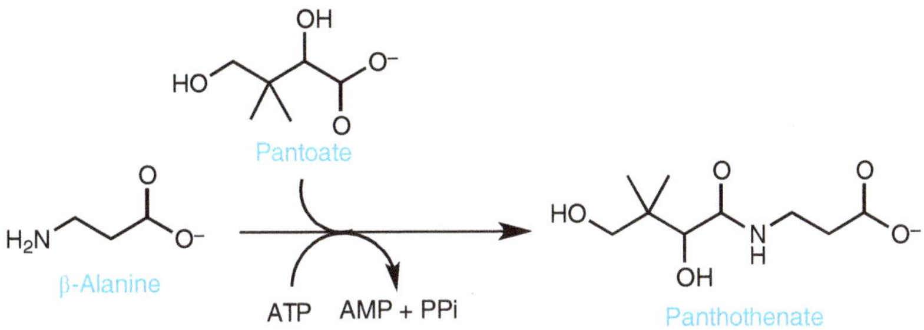

Figure 11.2 The synthesis of pantothenate (vitamin B5) from β-alanine.

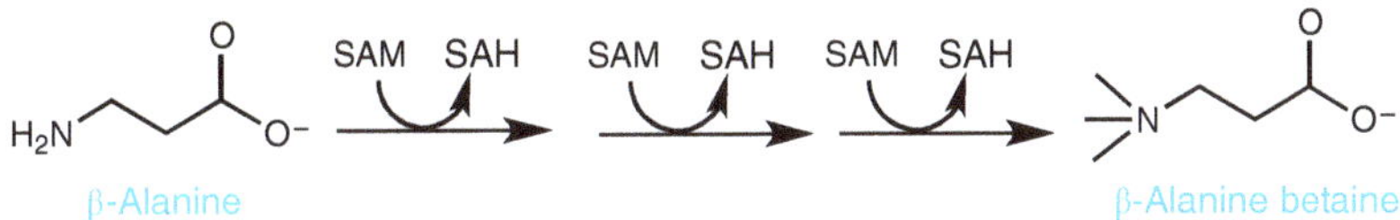

Figure 11.3 The synthesis of β-alanine betaine from β-alanine.

essential precursor to CoA and acyl carrier protein, which are cofactors required in many energy-yielding reactions and a large number of metabolic enzymes (Genschel et al. 1999).

Furthermore, it has been reported that uracil is one of the multiple sources of β-alanine in *Limonium latifolium* (sea lavender) (Duhazé et al. 2003). The β-alanine is converted to β-alanine betaine by three *S*-adenosyl-L-methionine (SAM)-dependent *N*-methyltransferase reactions (Figure 11.3). A novel type β-alanine *N*-methyltransferase has been discovered in plants, which accumulate betaine as an osmoprotectant (Raman and Rathinasabapathi 2003).

It has been proposed that β-alanine is converted to acetyl-CoA, which enters the tricarboxylic acid (TCA) cycle (Tsai and Axelrod 1965). The incorporation of radioactivity from [4,5-^{14}C]uracil into glutamine, glutamate, and aspartate has also been reported (Barnes and Naylor 1962). In potato leaves, part of the [U-^{14}C]uracil enters the TCA cycle and is utilized in the synthesis of amino acids, such as glutamate (Katahira and Ashihara 2006). Hence, a portion of pyrimidine ring catabolites is recovered as amino acid-related compounds.

11.4 Summary

In plants, like most microorganisms and animals, pyrimidines are catabolized reductively via uracil → dihydrouracil → β-ureidopropionate → β-alanine + CO_2 + NH_3, and thymine → dihydrothymine → β-ureidoisobutyrate → β-aminoisobutyrate + CO_2 + NH_3 pathways. In plants, β-alanine produced by these routes is associated with the biosynthesis of pantothenate and β-alanine betaine.

References

Barnes, R.L. and Naylor, A.W. (1962). Formation of ß-alanine by pine tissues supplied withintermediates in uracil and orotic acid metabolism. *Plant Physiol.* 37: 171–175.

Brooks, K.P., Jones, E.A., Kim, B.-D., and Sander, E.G. (1983). Bovine liver dihydropyrimidine amidohydrolase: purification, properties, and characterization as a zinc metalloenzyme. *Arch. Biochem. Biophys.* 226: 469–483.

Duhazé, C., Ganeul, D., Leport, L. et al. (2003). Uracil as one of the multiple sources of β-alanine in *Limonium latifolium*, a halotolerant β-alanine betaine accumulating Plumbaginaceae. *Plant Physiol. Biochem.* 41: 993–998.

Evans, W.R. and Axelrod, B. (1960). Pyrimidine metabolism in germinating seedlings. *Plant Physiol.* 36: 9–13.

Genschel, U., Powell, C.A., Abell, C., and Smith, A.G. (1999). The final step of pantothenate biosynthesis in higher plants: cloning and characterization of pantothenate synthetase from *Lotus japonicus* and *Oryza sativum* (rice). *Biochem. J.* 341: 669–678.

Gojkovic, Z., Rislund, L., Andersen, B. et al. (2003). Dihydropyrimidine amidohydrolases and dihydroorotases share the same origin and several enzymatic properties. *Nucleic Acids Res.* 31: 1683–1692.

Hatch, M.D. and Stumpf, P.K. (1962). Fat metabolism in higher plants. XVIII. Propionate metabolism by plant tissues. *Arch. Biochem. Biophys.* 96: 193–198.

Katahira, R. and Ashihara, H. (2006). Dual function of pyrimidine metabolism in potato (*Solanum tuberosum*) plants: pyrimidine salvage and supply of β-alanine to pantothenic acid synthesis. *Physiol. Plant.* 127: 38–43.

Kvalnes-Krick, K.L. and Traut, T.W. (1993). Cloning, sequencing, and expression of a cDNA encoding β-alanine synthase from rat liver. *J. Biol. Chem.* 268: 5686–5693.

Leonardi, R. and Jackowski, S. (2007). Biosynthesis of pantothenic acid and coenzyme A. *EcoSal Plus* 2 https://doi.org/10.1128/ecosalplus.1123.1126.1123.1124.

Mazuś, B. and Buchowicz, J. (1972). Activity of enzymes involved in pyrimidine metabolism in the germinating wheat grains. *Phytochemistry* 11: 77–82.

Raman, S.B. and Rathinasabapathi, B. (2003). β-Alanine *N*-methyltransferase of *Limonium latifolium*. cDNA cloning and functional expression of a novel *N*-methyltransferase implicated in the synthesis of the osmoprotectant β-alanine betaine. *Plant Physiol.* 132: 1642–1651.

Rastogi, R. and Davies, P.J. (1989). Polyamine metabolism in ripening tomato fruit. *Plant Physiol.* 94: 1449–1455.

Rathinasabapathi, B. (2002). Propionate, a source of ß-alanine, is an inhibitor of ß-alanine methylation in *Limonium latifolium*, Plumbaginaceae. *J. Plant Physiol.* 159: 671–674.

Stasolla, C., Katahira, R., Thorpe, T.A., and Ashihara, H. (2003). Purine and pyrimidine nucleotide metabolism in higher plants. *J. Plant Physiol.* 160: 1271–1295.

Terano, S. and Suzuki, Y. (1978). Formation of β-alanine from spermine and spermidine in maize shoots. *Phytochemistry* 17: 148–149.

Tintemann, H., Wasternack, C., Benndorf, R., and Reinbothe, H. (1985). The rate-limiting step of uracil degradation in tomato cell suspension cultures and *Euglena gracilis in vivo* studies. *Comp. Biochem. Physiol.* 82B: 787–792.

Tsai, C.S. and Axelrod, B. (1965). Catabolism of pyrimidines in rape seedlings. *Plant Physiol.* 40: 39–44.

Walsh, T.A., Green, S.B., Larrinua, I.M., and Schmitzer, P.R. (2001). Characterization of plant β-ureidopropionase and functional over expression in *Escherichia coli*. *Plant Physiol.* 125: 1001–1011.

Wasternack, C. (1978). Degradation of pyrimidines-enzymes, localization and role in metabolism. *Biochem. Physiol. Pflanz.* 173: 467–499.

Wasternack, C. (1982). Metabolism of pyrimidines and purines. In: *Encyclopedia of Plant Physiology*, New Ser, vol. 14B (eds. A. Pirson and M.H. Zimmermann), 263–301. Berlin: Springer.

Wasternack, C., Lippmann, G., and Reinbotte, H. (1979). Pyrimidine-degrading enzymes. Purification and properties of β-ureidopropionase of *Euglena gracilis*. *Biochim. Biophys. Acta* 570: 341–351.

Zrenner, R. and Ashihara, H. (2011). Nucleotide metabolism. In: *Plant Metabolism and Biotechnology* (eds. H. Ashihara, A. Crozier and A. Komamine), 135–162. Chichester, UK: Wiley.

Zrenner, R., Riegler, H., Marquard, C.R. et al. (2009). A functional analysis of the pyrimidine catabolic pathway in Arabidopsis. *New Phytol.* 183: 117–132.

Zrenner, R., Stitt, M., Sonnewald, U., and Boldt, R. (2006). Pyrimidine and purine biosynthesis and degradation in plants. *Annu. Rev. Plant Biol.* 57: 805–836.

Part IV

Physiological Aspects of Nucleotide Metabolism

12

Growth and Development

12.1 Introduction

A majority of the investigations on nucleotide metabolism during growth and development of plants have been carried out by monitoring the fate of radiolabelled purine and pyrimidine bases and nucleosides, which are readily taken up by tissues and either metabolized to nucleotides or degraded to CO_2 and NH_3. These studies have been useful in determining the activity of the *de novo*, salvage, and degradation pathways of both purine and pyrimidine metabolism during different stages of growth. Major research on this topic have been conducted during maturation and germination of somatic and zygotic embryos, shoot organogenesis *in vitro*, development of storage organs and sprouting, fruit ripening, and leaf senescence (Stasolla et al. 2003a). In nucleotide studies on growth and development, species such as *Picea glauca*, *Catharanthus roseus,* and some Fabaceae seeds, have been used. The plants used for nucleotide metabolism are list up in Table 12.1. Recently, molecular biology studies on nucleotide metabolism have been initiated principally with *Arabidopsis thaliana*, however, as yet they have yielded little information on the functional aspects of nucleotides. Hence, this chapter will focus principally on biochemical studies on the involvement of nucleotide metabolism in plant growth and development.

12.2 Embryo Maturation

Studies on embryogenesis and nucleotide metabolism have been carried out mainly on asexual, somatic embryos, because their zygotic counterparts are embedded in the maternal tissue and, hence, are difficult to isolate. *In vitro* somatic embryogenesis in *P. glauca* (white spruce) and *Daucus carota* (carrot) have been widely utilized as model systems (Stasolla et al. 2003a).

By monitoring the metabolic fate of ^{14}C-labelled adenine, adenosine, and inosine it was established that both purine salvage and degradation pathways (Figure 12.1) operate during the development of *P. glauca* somatic embryos (Ashihara et al. 2001b). It emerged that changes in salvage activity represent a potential metabolic switch required for the termination of cell proliferation and the initiation of embryo development. Extensive utilization of adenine and adenosine for nucleotide and nucleic acid synthesis, as well

Plant Nucleotide Metabolism – Biosynthesis, Degradation, and Alkaloid Formation, First Edition.
Hiroshi Ashihara, Iziar A. Ludwig and Alan Crozier.

Table 12.1 Plant materials used for physiological nucleotide metabolism studies.

Family	Scientific name	Common name	Section	Sample reference
Pinaceae	*Picea glauca*	White spruce	12.1, 12.7	Stasolla et al. (2001d)
	Pinus mugo	Pine	12.2	Stasolla et al. (2007)
	Pinus radiata	Radiata pine	12.3	Stasolla et al. (2006)
Apiaceae	*Daucus carota*	Carrot	12.1, 12.7	Stasolla et al. (2003b)
Brassicaceae	*Brassica napus*	Rapeseed	12.1	Belmonte et al. (2011)
	Arabidopsis thaliana	Thale cress	12.2, 12.7	Yin et al. (2014)
Apocynaceae	*Catharanthus roseus, Vinca rosea*	Madagascar periwinkle	12.7	Shimano and Ashihara (2006)
Sapindaceae	*Acer pseudoplatanus*	Sycamore	12.7	Shimizu et al. (1977)
Fabaceae	*Phaseolus mungo*	Black gram	12.2	Ashihara (1983)
	Glycine max	Soybean	12.2, 12.7	DeKlerk-Kiebert and Van del Plus (1984)
	Lupinus luteus	Yellow lupin	12.2	Guranowski and Barankiewicz (1979)
	Pisum sativum	Pea	12.2	Ross and Murray (1971)
Euphorbiaceae	*Ricinus communis*	Castor bean	12.2	Kombrink and Beevers (1983)
Poaceae	*Triticum aestivum*	Wheat	12.2	Schimpff et al. (1978)
	Zea mays	Maize or corn	12.2	Wanka and Walboomers (1966)
Oleaceae	*Fraxinus excelsior*	Ash or European ash	12.4	Barnola et al. (1986)
Rosaceae	*Prunus persica*	Peach	12.4	Balandier et al. (1993)
	Fragaria × ananassa	Garden strawberry	12.4	Robert et al. (1997)
Rutaceae	*Citrus sinensis*	Orange	12.5	Tomlinson and Lovatt (1987)
Asteraceae	*Helianthus tuberosus*	Jerusalem artichoke	12.6	Le Floc'h and Lafleuriel (1984)
Solanaceae	*Solanum tuberosum*	Potato	12.6	Katahira and Ashihara (2002)
	Datura innoxia	Thorn-apple	12.7	Mitsui and Ashihara (1988)
	Nicotiana tabacum	Tobacco	12.7	Meyer and Wagner (1985a)

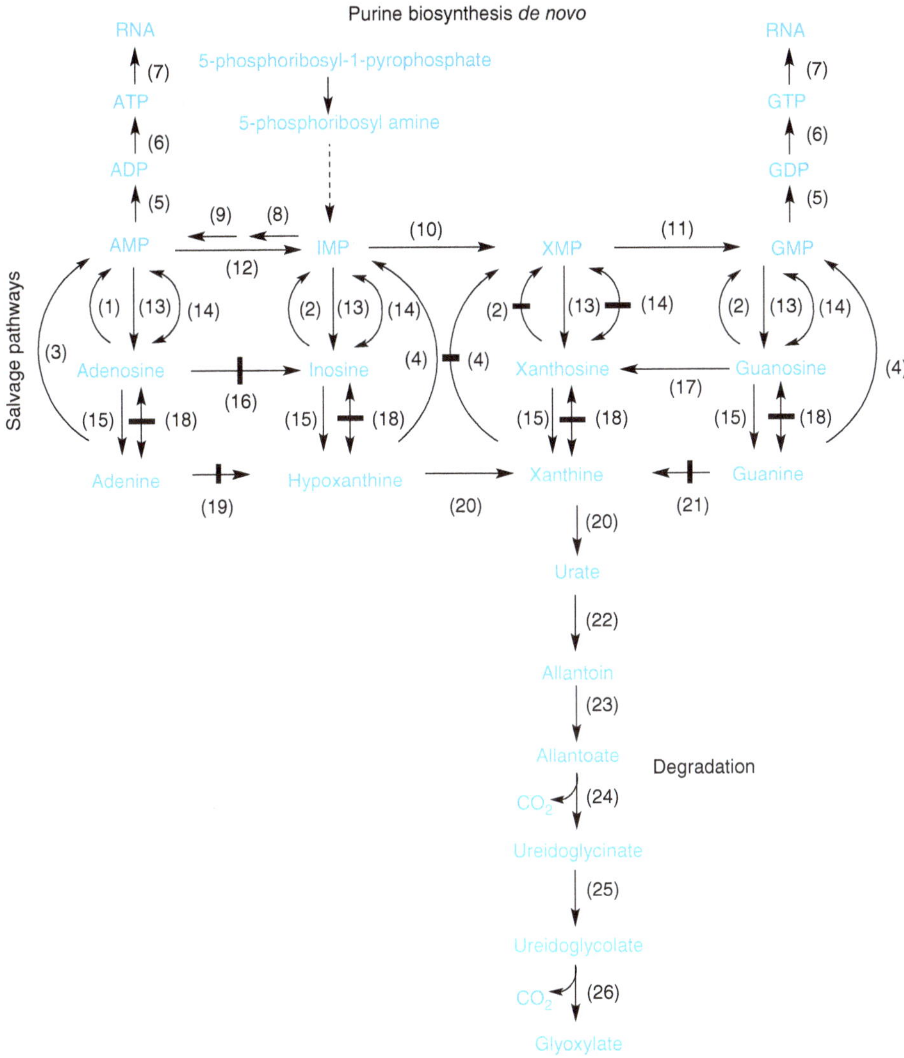

Figure 12.1 Metabolic pathways of purine compounds in plants based the fate of exogenously supplied purine-related compounds and enzymes. Reactions and enzymes involved in purine salvage and degradation are as follows. The numbers show the enzymes participating in individual reactions. Arrows broken with a vertical bar represent reactions which are not present in plants. Enzymes: (1) adenosine kinase; (2) inosine/guanosine kinase; (3) adenine phosphoribosyltransferase; (4) hypoxanthine/guanine phosphoribosyltransferase; (5) nucleoside monophosphate kinase; (6) nucleoside diphosphate kinase; (7) RNA polymerase; (8) adenylosuccinate synthetase; (9) adenylosuccinate lyase; (10) IMP dehydrogenase; (11) GMP synthetase; (12) AMP deaminase; (13) 5′-nucleotidase; (14) nucleoside phosphotransferase; (15) adenosine nucleosidase and/or inosine-guanosine nucleosidase; (16) adenosine deaminase; (17) guanosine deaminase; (18) purine nucleoside phosphorylase; (19) adenine deaminase; (20) xanthine dehydrogenase; (21) guanine deaminase; (22) allantoin synthase; (23) allantoin amidohydrolase; (24) allantoate amidohydrolase; (25) ureidoglycine aminohydrolase; (26) ureidoglycolate amidohydrolase.

as high activities of the salvage enzymes, adenosine kinase (step 1 in Figure 12.1) and adenine phosphoribosyltransferase (step 3 in Figure 12.1) were observed during the initial phases of embryo development, which is characterized by rapid cell proliferation. During the subsequent stages, which feature a reduction of cell proliferation and the initiation of somatic embryo development, decreases in salvage activity of both adenine and adenosine were apparent (Ashihara et al. 2001a).

Changes in purine metabolism also appear to be important during the imposition of desiccation on the *P. glauca* somatic embryos, required for the termination of the developmental mode and the initiation of germination (Kermode 1990). During this transition, although the incorporation of both supplied adenine and adenosine declined, the activity of adenine phosphoribosyltransferase increased (Stasolla et al. 2001d). Although operative throughout the maturation period of *P. glauca* embryos, the degradation pathway, estimated by the rate of inosine degradation (Figure 12.1), appears to not to be correlated to any particular stage of development.

Studies on pyrimidine metabolism conducted with maturing *P. glauca* somatic embryos demonstrated that the *de novo* pathway (Figure 12.2) is very active throughout the embryogenic process, as more than 80% of supplied orotate is utilized for nucleotide and nucleic acid synthesis at the different stages of embryo development (Ashihara et al. 2001a). The salvage pathway was also operative, and uridine was a better precursor for nucleotides and nucleic acids than uracil. The extensive catabolism of uracil, degraded to CO_2 via β-ureidopropionate, was due mainly to the low activity of uracil phosphoribosyltransferase relative to that of uridine kinase (Ashihara et al. 2001a). As the changing pattern of these enzymes did not correlate with the utilization of their respective precursors, it was suggested that other fine control mechanisms, including the fluctuation of levels of substrates and/or effectors might be a determinant for the control of pyrimidine metabolism during embryo development (Ashihara et al. 2001a). As was observed with a purine enzyme, adenine phosphoribosyltransferase, the activity of uridine kinase was found to increase during the subsequent imposition of desiccation (Stasolla et al. 2001d). High activity of this enzyme was also detected in *P. glauca* zygotic embryos dissected from dry seeds (Stasolla et al. 2002).

Differences in pyrimidine metabolism are observed during the early maturation stages of dried seeds of conifers and flowering plants. An active pyrimidine *de novo* pathway, estimated by the utilization of orotate for nucleotide and nucleic acid biosynthesis, operates in both somatic and zygotic *P. glauca* embryos (Stasolla et al. 2001d, 2002), but not in dried seeds of *Phaseolus mungo* (black gram) (Ashihara 1977). In *P. mungo* seeds, this situation was associated with a low content of 5-phosphoribosyl-1-pyrophosphate (PRPP) (Ashihara and Kameyama 1989). Although not measured directly, PRPP does not appear to be a limiting factor in dried conifer embryos, as the activity of PRPP synthase doubled during the imposed drying regime in *P. glauca* somatic embryos (Stasolla et al. 2001d).

Despite the paucity of information regarding the biosynthesis of deoxyribonucleotides in plants, the existence of a *de novo* and salvage pathway during the embryogenic process in both angiosperms and gymnosperms has been demonstrated (Schimpff et al. 1978; Stasolla et al. 2003b, 2003c). Evidence that the *de novo* pathway of deoxyribonucleotide biosynthesis is present in plant embryos comes from studies of Schimpff et al. (1978), who measured ribonucleotide reductase activity in embryos of wheat (*Triticum aestivum*). Incorporation of radioactivity from ^{14}C-labelled cytidine

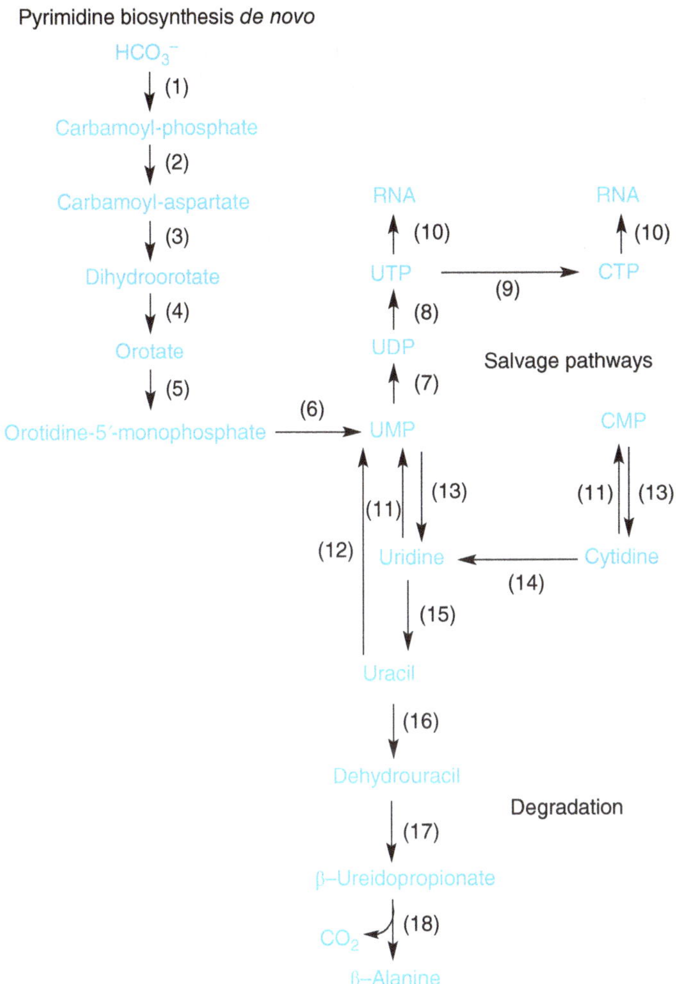

Figure 12.2 Metabolic pathways of pyrimidine compounds in plants based on the fate of exogenously supplied purine-related compounds and enzymes. Reactions and enzymes involved in pyrimidine biosynthesis *de novo*, salvage and degradation are as follows. The numbers show the enzymes participating in individual reactions. Enzymes: (1) carbamoylphosphate synthetase; (2) aspartate carbamoyltransferase; (3) dihydroorotase; (4) dihydroorotate dehydrogenase; (5) orotate phosphoribosyltransferase; (6) orotidine-5′-monophosphate decarboxylase; [(5-6) UMP synthase]; (7) UMP/CMP kinase; (8) nucleoside diphosphate kinase; (9) CTP synthase; (10) RNA polymerase; (11) uridine/cytidine kinase; (12) uracil phosphoribosyltransferase; (13) nucleoside phosphotransferase; (14) cytidine deaminase; (15) uridine nucleosidase; (16) dihydropyrimidine dehydrogenase; (17) dihydropyrimidinase; and (18) β-ureidopropionase.

into the DNA fraction of maturing *P. glauca* and carrot somatic embryos (Stasolla et al. 2003b, 2003c) also supports this concept. In carrot, the increased utilization of cytidine for DNA synthesis as maturation progresses, suggests that *de novo* synthesis of deoxyribonucleotides may be required to support active embryonic growth (Stasolla et al. 2003b).

Tracer experiments conducted with thymidine and deoxycytidine using *P. glauca* and carrot also indicate that the salvage synthesis of deoxyribonucleotides is operative during embryo development. Both precursors are rapidly metabolized by the tissue and incorporated into nucleotides and nucleic acids. Compared to deoxycytidine, thymidine was more actively incorporated into DNA rather than RNA in both systems (Stasolla et al. 2003b, 2003c). Salvage synthesis of deoxyribonucleotides may be required during the initial stages of embryo development, as low salvage activity for both deoxycytidine and thymidine was observed in habituated carrot cells that were unable to undergo embryo development upon removal of exogenous 2,4-dichlorophenoxyacetic acid (Stasolla et al. 2003b).

Two nucleotide studies were carried out using the *Brassica napus* (oilseed rape) as a model system for *in vitro* embryogenesis (Belmonte et al. 2011; Elhiti et al. 2011). Labelled ^{14}C-purine and pyrimidine metabolism and activity of enzymes were investigated in microspore-derived embryos treated with the plant hormone brassinolide to increase the embryogenic yield. Brassinolide elevated the activities of the purine (adenine, adenosine, and inosine) and pyrimidine (orotate, uridine, and uracil) pathways. The results indicate that utilization of adenine, adenosine, and uridine for nucleotides and nucleic acids increased significantly in brassinolide-treated embryos at day 15 and remained high throughout the culture period. Distinct fluctuations in nucleotide metabolism were also observed in *B. napus* microspore-derived embryos overexpressing the meristem marker gene *SHOOTMERISTEMLESS* (Elhiti et al. 2011). This suggests that alterations in nucleotide metabolism are linked to the ability of microspores to redirect their fate from a gametophytic to an embryogenesis path.

12.3 Germination

In the past, there have been many papers on nucleic acid synthesis during seed germination (Deltour 1985). In contrast, information on purine and pyrimidine metabolism during germination is somewhat limited. Availability of nucleotides during the early phases of imbibition appears to be a critical factor for successful embryo germination. Besides their involvement in the myriad of bio-energetic processes required for the mobilization of storage products, active nucleotide biosynthesis is also needed to provide the embryos with sufficient purine and pyrimidine nucleotides to support nucleic acid synthesis (Deltour 1985). Hence, it is not surprising that a tight regulation between *de novo*, salvage, and degradation pathways of both purines and pyrimidines exists at germination.

12.3.1 Purine Metabolism in Germination

The metabolic pattern of purine metabolism during germination, as proposed by Ashihara (1983), can be divided in three distinct segments: the (i) 'unmetabolized', (ii) 'salvage', and (iii) 'ureide formation' phases. Synthesis of purine nucleotides via the salvage pathways (Figure 12.1) during germination occurs mainly at the expense of adenine, adenosine, guanine, and guanosine production (Anderson 1977; Ashihara 1983; Ashihara et al. 1997; Guranowski 1979; Guranowski and Barankiewicz 1979;

Nobusawa and Ashihara 1983; Stasolla et al. 2001b). Changes in purine salvage activity during seed germination have been investigated in wheat (*T. aestivum*) and some Fabaceae species, including castor bean (*Ricinus communis*). One of the earliest detectable metabolic processes occurring in imbibed seeds is the rapid increase of ATP. In some instances this increase is accounted for by a decrease in AMP, yet in many cases there is insufficient AMP to account for all of the ATP that accumulates. It has, however, been shown that the net increase in adenine nucleotides is due mainly to the salvage of adenosine and adenine catalysed by adenosine kinase and adenine phosphoribosyltransferase (Anderson 1977).

In the Fabaceae plant, *P. mungo*, the relative rate of purine salvage and its degradation in the cotyledons and embryonic axes were determined by ^{14}C-tracer experiments during germination (Ashihara 1983). In the cotyledons, the salvage of adenine, adenosine, and guanine, assessed through the incorporation of ^{14}C into nucleotides and nucleic acids, is higher 12–48 hours after imbibition and the initiation of germination, than in the earlier 0–12 hours period and subsequent 48–96 hours. In contrast, appreciable ureide (allantoin and allantoate) formation from purines is found after 48 hours of germination. In the embryonic axes, a higher incorporation of ^{14}C-labelled purines into salvage products is observed during the early stage of germination, whereas significant incorporation into ureides is detected in the later stage. The pathways of ureide synthesis from purines are shown as 'degradation' in Figure 12.1. ^{14}C-Labelled adenine taken up by the cotyledons during imbibition is translocated to the embryonic axes where the radioactivity is incorporated into ureides.

These findings suggest that purine derivatives stored in the cotyledons of *P. mungo* seeds are utilized to produce nucleotides and nucleic acids, especially during the early stage of germination, and that they are degraded to allantoin and allantoate by the conventional purine catabolic pathway in the later stages of germination and reutilized as a nitrogen source for seedling growth. The level of PRPP is extremely low in dry seeds, but after imbibition it increases rapidly in cotyledons (Ross and Murray 1971; Ashihara and Kameyama 1989). Production of purine nucleotides via the salvage pathway during the early stages of germination of Fabaceae seeds occurs mainly at the expense of adenine, adenosine, guanine, and guanosine, which are stored in the cotyledons (Brown 1963).

During germination of *Lupinus luteus* (yellow lupin) and *P. mungo* seeds, salvage of purines is catalysed predominantly by adenine phosphoribosyltransferase and adenosine kinase (Guranowski and Barankiewicz 1979; Nobusawa and Ashihara 1983). The alternative route of adenine salvage utilizing purine nucleoside phosphorylase and adenosine kinase, is unlikely to operate in seeds because they appear to lack purine-nucleoside phosphorylase activity (Guranowski and Barankiewicz 1979). During the early phases of germination, there are indications that adenosine salvage is catalysed by adenosine kinase (step 1 in Figure 12.1). The activity of this enzyme increased in the cotyledons during germination of legume seeds (Guranowski and Barankiewicz 1979; Nobusawa and Ashihara 1983). Adenosine salvage mediated by adenosine nucleosidase and adenine phosphoribosyltransferase (steps 2 and 3 in Figure 12.1) becomes operative only during the later stages of germination, because adenosine nucleosidase activity is absent in dry seeds (Guranowski and Pawelkiewicz 1978). The contribution of nucleoside phosphotransferase appears to be important only in the cotyledons during the later period of *L. luteus* germination (Guranowski and Barankiewicz 1979). This suggests that purine salvage enzymes adenine phosphoribosyltransferase and adenosine kinase may contribute to rapid ATP production and the subsequent synthesis of nucleic acids. Alternative enzymes, adenosine nucleosidase and nucleoside phosphotransferase, may be related to the salvage of purines which is derived from catabolites of nucleic acids and nucleotides in senescing cotyledons.

The pattern of purine salvage in wheat embryos is slightly different from that occurring in Fabaceae seeds. The activities of adenine phosphoribosyltransferase and hypoxanthine/guanine phosphoribosyltransferase increase during imbibition, while adenosine kinase and PRPP synthetase activities are high dry embryos (Price and Murray 1969). Castor bean (*R. communis*) is a Euphorbiaceous plant in which the endosperm is a major storage organ, and during germination, purine nucleosides and bases are transported from the endosperm to cotyledons (Kombrink and Beevers 1983). Therefore, in growing parts of seedlings purine salvage occurs using these transported precursors. In the later stage of germination, degradation of nucleic acids and nucleotides takes place accompanied by senescence of endosperm and development of growing tissues. Ureides may be transported to newly growing parts of the young seedlings and used as a nitrogen source. The total content of nucleic acids and nucleotides declines rapidly between day 4 and day 8 of seedling growth. The function of adenosine transported into developing cotyledons from the endosperm in castor bean was estimated from the effect of adenosine applied to cotyledons of five-day-old seedlings after removal of the endosperm (Florchinger et al. 2006). Most of ^{14}C-labelled adenosine administered to detached cotyledons was incorporated into nucleotides, RNA, and DNA. Import and conversion of adenosine improved the energy content of cotyledons as revealed by a substantially increased ATP/ADP ratio. This effect was accompanied by increases in respiratory activity and a stimulation of glycolytic flux by activation of phosphofructokinase. Adenosine also stimulates the biosynthesis of starch. In caster bean, adenosine imported from the endosperm into the developing cotyledons fulfils an important function as it promotes anabolic reactions in the rapidly developing tissue.

A relationship between purine salvage activity and germination has been suggested from genetic-based studies. It has been reported that the APT-deficient mutants of *A. thaliana* germinate and grow more slowly than the wild-type (Moffatt and Somerville 1988). Furthermore, functional analysis of hypoxanthine/guanine phosphoribosyltransferase (HGPT) revealed that HGPT expression is dynamically regulated in growth during seed germination (Liu et al. 2007). The germination rates of wild-type and knockout mutant (*athgpt*1-1) seeds of *A. thaliana* are similar during the early stages (up to 48 hours after sowing) but the rate is reduced ~20% in the mutant after 60 hours. The germination rate (20–40%) of the overexpression mutants (*Ho3-8* and *Ho4-4*) is significantly higher than the rate of wild-type (~10%) 36 hours after imbibition. The contrasting germination rates exhibited by the wild-type, and the AtHGPT knockout and overexpression mutants, are probably caused by differences in AtHGPT enzyme activity levels among the three genotypes. Although no direct measurement of enzyme activity was performed, this possibility is supported indirectly by the differential sensitivities of the three genotypes to the cytotoxic compound 8-azaguanine, which can be efficiently salvaged through HGPT enzyme activity. These mutant studies support the theory that salvage reactions catalysed by adenine phosphoribosyltransferase and hypoxanthine/guanine phosphoribosyltransferase contribute to seed germination as proposed on the basis of early physiological studies.

12.3.2 Pyrimidine Metabolism in Germination

Although both the *de novo* and salvage pyrimidine pathways (Figure 12.2) are operative at germination, their contribution to the enlargement of the nucleotide pool at this juncture appears to be tightly regulated. As reported by Ashihara (1977) and Stasolla et al. (2001a), metabolism of pyrimidines can be divided in two distinct phases namely 'salvage synthesis', at the inception of germination, and '*de novo* synthesis', during later stages of germination.

Active production of pyrimidine nucleotides following imbibition occurs via the salvage pathway. In *P. mungo* cotyledons, for example, utilization of uracil and uridine for RNA and nucleotide biosynthesis increases markedly during the first 24 hours of germination, before declining (Ashihara 1977). A similar event was reported in germinating *P. glauca* somatic and zygotic embryos (Stasolla et al. 2001a, 2002). In somatic embryos, uridine incorporation into nucleic acids and nucleotides was high in dried embryos, and it slowly declined during the first six days of germination (Stasolla et al. 2001a). In this study, salvage of uridine was mainly catalysed by uridine kinase, and to a lesser extent by non-specific phosphotransferases, whose activities were also observed during germination of maize (*Zea mays*) seeds (Wanka and Walboomers 1966) and wheat grains (Mazuś and Buchowicz 1972).

As germination progresses, the contribution of the salvage pathway to the overall pyrimidine nucleotide pool declines. The mechanism limiting the activity of salvage synthesis during the later stages of germination has not been fully elucidated, although conversion of UMP to UDP may represent the limiting step. In pea (*Pisum sativum*) cotyledons incubated with labelled uracil and uridine, more radioactivity was recovered in the UMP fraction than in the UDP plus UDP-glucose fraction (Ross and Murray 1971). Alternatively, the rate of conversion of pyrimidine bases and nucleosides into nucleotides may be determined by the activity of the relative salvage enzymes. In germinating *P. glauca* somatic and zygotic embryos, for example, a decrease in uridine salvage paralleled a pronounced decline in uridine kinase activity (Stasolla et al. 2001c, 2002). A general picture emerging from germination studies is that uridine is a more rapidly utilized substrate for the synthesis of salvage products than uracil, as a large percentage of uracil is often degraded through a reductive catabolic pathway (Ashihara 1977; Evans and Axelrod 1960; Stasolla et al. 2001c).

The decrease in salvage activity as germination progresses is accompanied by an increased activity of the *de novo* pathway. Utilization of orotate for nucleotide and nucleic acid synthesis increases in cotyledons of germinating *P. mungo* (Ashihara 1977) and pea seeds (Ross et al. 1971), as well as in germinating *P. glauca* somatic embryos (Stasolla et al. 2001c). The restoration of the *de novo* pathway seems to be critical for ensuring that the embryos have sufficient nucleotides to support active nucleic acid synthesis. Increasing levels of nucleotides occur during germination of both pine pollen tubes (Nygaard 1973) and bacterial spores (Setlow and Kornberg 1970).

The restoration of the *de novo* machinery as germination progresses may be due to the increased activity of several enzymes involved in the *de novo* pathway, including orotate phosphoribosyltransferase, which, in concert with orotidine-5′-monophosphate decarboxylase, catalyses the conversion of orotate to UMP (steps 5–6 in Figure 12.2). In support of this concept, the activity of orotate phosphoribosyltransferase is generally low at the beginning of germination but increases at later stages of development (Ross et al. 1971; Ross and Murray 1971; Stasolla et al. 2001c).

As well as orotate phosphoribosyltransferase, the levels of other enzymes involved in the construction of the pyrimidine ring, including carbamoyl phosphate synthetase and aspartate carbamoyltransferase, increase as germination progresses (Mazuś and Buchowicz 1972). An alternative explanation for the restored *de novo* synthesis of pyrimidine nucleotides was proposed by Brown and Wray (1968), who observed increased activity of the pentose phosphate pathway over the first six days of germination in cotyledons of pea. This pathway is responsible for the production of ribose-5-phosphate, a substrate for the synthesis of PRPP, which is required for the conversion of orotate to UMP.

Studies on deoxyribonucleotides have also revealed that both *de novo* and salvage pathways are operative during germination (Figure 12.3). In germinating *P. glauca* somatic embryos, the activity of the *de novo* pathway is characterized by the low, but significant utilization of cytidine for DNA synthesis (steps 1–4 in Figure 12.3). Conversion of this precursor to CMP was mainly regulated by the activity of cytidine

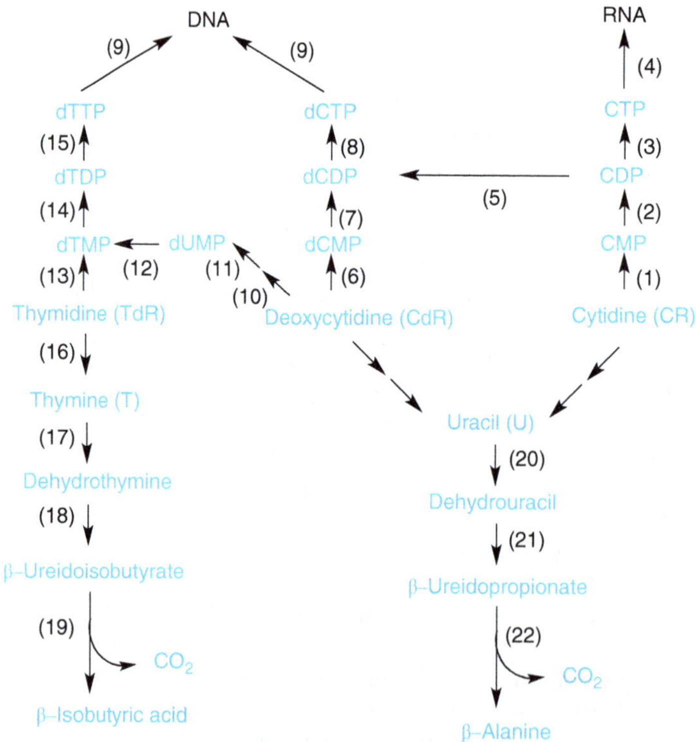

Figure 12.3 The *de novo* and salvage pathways of deoxyribonucleotides during germination of plants. Salvage pathways of cytidine, deoxycytidine, and thymidine are steps 1-3, 6-8, and 13-15 respectively. Degradation pathways of thymidine and uracil are steps 16-19 and 20-22. Detailed enzyme names are shown in Chapters 10 and 11.

kinase (step 1 in Figure 12.3), and to a lesser extent, by non-specific phosphotransferases (Stasolla et al. 2003c). The role played by cytidine kinase in cytidine metabolism during germination was also demonstrated in maize seedlings where a large fraction of cytidine was also incorporated into the RNA fraction (Wanka and Bauer 1967). In germinating *P. glauca* somatic embryos, utilization of cytidine for RNA synthesis has been demonstrated to occur either directly after conversion to CMP, or indirectly via deamination to uridine and ammonia by deaminase activity (Stasolla et al. 2003c). The presence of this latter route in plant cells was originally demonstrated by Ross and Cole (1968).

Salvage synthesis of deoxyribonucleotides at germination has also been demonstrated by monitoring the metabolic fate of deoxycytidine and thymidine (Kameyama et al. 1985; Stasolla et al. 2002, 2003c; Wanka and Walboomers 1966). In germinating *P. glauca* somatic embryos, utilization of deoxycytidine (CdR) for DNA synthesis (steps 6–9 in Figure 12.3) was higher than in fully matured embryos (Stasolla et al. 2003c). This study also demonstrated that both CdR kinase and non-specific phosphotransferases contributed equally to the salvage of CdR.

Compared to CdR, a large fraction of supplied thymidine is degraded to CO_2 during germination (steps 15–18 in Figure 12.3) (Kameyama et al. 1985; Stasolla et al. 2002). Increasing utilization of this precursor for nucleic acid synthesis was observed in germinating *P. mungo* seeds (Kameyama et al. 1985), as well as in germinating zygotic embryos of *P. glauca* (Stasolla et al. 2002). In somatic embryos of this plant, thymidine salvage appears to be critical for post-embryonic growth, as experimental manipulations of the culture conditions that decrease the germination frequency of the embryos also result in a decrease of thymidine anabolism (Stasolla et al. 2003c).

Conversion of thymidine to dTMP is mediated by thymidine kinase, as well as by non-specific phosphotransferases (step 13 in Figure 12.3). The contribution of these two enzymes to the amount of thymidine being salvaged appears to be species dependent. Wanka and Walboomers (1966) reported the activity of thymidine kinase in extracts of maize seedlings. However, it has been shown that thymidine kinase measured in crude extracts may be due to combined activities of the nucleoside phosphotransferases and ATP phosphohydrolase (Mullin and Fites 1978). Thymidine kinase activity was not detected in germinating *P. mungo* embryos (Kameyama et al. 1985), although relatively high activity has been observed in both somatic and zygotic germinating embryos of *P. glauca* (Stasolla et al. 2002, 2003c).

12.4 Organogenesis

The only two published studies on nucleotide biosynthesis during *in vitro* organogenesis are those dealing with shoot bud formation from epicotyls of *P. glauca* and excised cotyledons of *Pinus radiata* (Monterey pine) (Stasolla et al. 2003a, 2006). Investigations on both purine and pyrimidine metabolism indicate that the production of nucleotides via the salvage synthesis plays an important role during shoot formation. In *P. glauca*, epicotyls cultured under shoot-forming conditions more actively utilize adenine and adenosine for nucleotide and nucleic acid synthesis, than those cultured on a non-shoot forming medium. This higher salvage synthesis, due to increased activity of adenine phosphoribosyl transferase and adenosine kinase, occurs after 10 days in culture when shoot formation is initiated. The importance of the salvage activity during the organogenic process was also demonstrated with excised cotyledon explants of *P. radiata*, where an increased incorporation of uracil into the nucleic acid fraction, as well as a higher activity of the uracil salvage enzyme, uracil phosphoribosyltransferase, was observed during the formation of shoot primordia (Stasolla et al. 2007).

12.5 Breaking Bud Dormancy

In many perennial plants a dormant period is observed during the vegetative cycle. The resting state that generally occurs during winter is broken by low temperature and growth is restored the following spring. In trees of ash (*Fraxinus excelsior*), and peach (*Prunus persica*), and plants of strawberry (*Fragaria* × *ananassa*), breaking of dormancy is accompanied by an increased capacity of the salvage pathways of adenosine and adenine (Balandier et al. 1993; Barnola et al. 1986; Robert and Petel 2001). During the breaking of bud dormancy, an increase in the levels of nucleoside triphosphates, such as ATP and GTP, is observed after treatment with adenosine. This is referred to as the 'nucleotide test' (Lavarenne et al. 1982; Robert et al. 1997). The breaking of dormancy is accompanied by an increase in the activity of adenosine kinase, adenine phosphoribosyltransferase and/or adenosine nucleosidase (Faye and Le Floc'h 1997; Robert and Petel 2001).

12.6 Fruit Ripening

Changes in the activity of nucleotide biosynthesis during the development of orange (*Citrus sinensis*) fruits have been monitored using various ^{14}C-labelled precursors (Tomlinson and Lovatt 1987). Pyrimidine biosynthesis *de novo* in navel orange increased during the cell-division phase of fruit development. The capacity of peel tissue to synthesize pyrimidine nucleotides *de novo* decreased with fruit development up to five months of age. In contrast, synthesis of purine nucleotides was not related to any specific stage of fruit development.

12.7 Storage Organ Development and Sprouting

Studies on the enzymes and general metabolism of nucleotides in storage organs, such as roots and tubers, have been carried out mainly with tubers of Jerusalem artichoke (*Helianthus tuberosus*) and potato (*Solanum tuberosum*). The key enzymes of purine nucleotide metabolism, AMP deaminase (Le Floc'h and Lafleuriel 1983) and guanylate kinase (Le Floc'h and Lafleuriel 1990) were purified and their regulatory properties and subcellular localization investigated. Gendraud (1975) reported that dormant Jerusalem artichoke buds had adenosine salvage activity. However, conversion of salvaged adenine nucleotides to other nucleotides appeared only in non-dormant buds. When non-dormant Jerusalem artichoke tubers were transferred to a temperature high enough to allow growth, adenosine nucleosidase activity increased (Le Floc'h and Lafleuriel 1984).

In potato tubers nucleotide-sugars, such as ADP-glucose and uridine diphosphate glucose, play an important role in starch formation and sugar-starch conversion. Loef et al. (1999, 2001) argued that increased levels of pyrimidine and purine nucleotides, formed from exogenously supplied orotate and adenine, stimulate the conversion of sucrose to starch. Detailed studies on purine and pyrimidine metabolism in potato tubers have been reported by Katahira and Ashihara (2002, 2006).

12.8 Suspension-Cultured Cells

Suspension cultured plant cells represent a valuable tool for studying growth and differentiation. Transfer of starved cells into fresh medium results in synchronous growth, as well as synchronous uptake and utilization of metabolites (King and Street 1977; Wylegalla et al. 1985). Cells cultured under these conditions display very similar growth behaviour, which can be divided into (i) lag phase; (ii) cell division phase, characterized by an increase in cell mass; (iii) cell elongation phase caused by increased vacuolation; and (iv) stationary/starvation phase, delineated by a decreased availability of nutrients in the medium.

12.8.1 Nucleotide Pools

Biochemical studies conducted with synchronously growing cultured plant cells have revealed that the concentration of nucleotides in a particular phase may represent a regulatory signal affecting the transition of the cells to the next phase of development (Wagner and Backer 1992). Following inoculation of starved cells into fresh medium a pronounced increase in nucleotide levels has been observed in a number of species, including *Acer pseudoplatanus* (sycamore) (Shimizu et al. 1977), *C. roseus* (Sasamoto and Ashihara 1988; Shimano and Ashihara 2006; Yin et al. 2007), *Glycine max* (soybean) (DeKlerk-Kiebert and Van del Plus 1984), and *Datura innoxia* (thorn-apple) (Meyer and Wagner 1985b). A similar, but less pronounced increase, was also observed with guanine nucleotides (Meyer and Wagner 1985b). Increased nucleotide biosynthesis at the lag phase seems to be a prerequisite for the initiation of cell proliferation, where marked DNA and RNA synthesis occurs (Wylegalla et al. 1985). In cultured cells of *D. innoxia*, for example, the intracellular increase of nucleoside triphosphates is similar, but slightly precedes that of the RNA content, which reaches its maximum in the cell division phase (Wylegalla et al. 1985).

During cell division, the nucleotide pool decreases, as well as the energy charge, possibly because of ATP consumption. The adenylate energy charge ([ATP]+1/2 [ADP]/[ATP]+[ADP]+[AMP]), in agreement with respiratory activity data (Givan and Collin 1967), usually increases in the first hours of the lag phase, and then decreases during the cell division phase, before attaining constant levels in the following cell elongation and starvation phases (Meyer and Wagner 1985b; Shimazaki et al. 1982). Based on these findings, Shimazaki et al. (1982) divided adenylate metabolism of *C. roseus* cells into an 'energy generating stage' during the lag phase, and an 'energy utilizing stage' which is initiated by the onset of the cell division.

Qualitative differences within the overall nucleotide pool also appear to exist during the growth of cultured cells. A comparison of the different components of the nucleotide pools of different cell culture systems, including sycamore, *Datura innoxia* (Mitsui and Ashihara 1988), tobacco (Meyer and Wagner 1985a), and *C. roseus* (Shimano and Ashihara 2006), revealed strikingly similar results. In general, the lower fraction was composed of cytosine nucleotides (<3%), and the highest was uracil nucleotides (>50%) with the UDP-sugars as the main component. It has been speculated that metabolic adjustments of the uracil and adenine nucleotide pools denote the presence of regulatory mechanisms controlling the endogenous levels of purines and pyrimidine nucleotides (Stasolla et al. 2003a).

12.8.2 Nucleotide Biosynthesis

A major synthesis of nucleotides occurs during the lag phase of cell growth, in preparation for cell proliferation. By measuring nucleotide levels in *Datura* cells at six-hour intervals after inoculation into fresh medium, Meyer and Wagner (1985b) demonstrated a biphasic behaviour in the increase of the overall nucleotide pool. In the case of the adenine and uracil nucleotide pools, the first increase was observed at 6 and 15 hours, respectively, whereas a second increase occurred a few days later. A similar pattern was observed during the lag phase with the uptake of inorganic phosphate (Wylegalla et al. 1985). These results, as interpreted by Kanamori-Fukuda et al. (1981), suggest that nucleotide synthesis during cell growth in suspension culture can be divided in two stages: (i) a 'turn-over or salvage phase' in the lag phase, and (ii) a 'true biosynthetic stage', characterized by the reactivation of the *de novo* pathway, at the inception of cell division.

This concept was substantiated in experiments with radiolabelled substrates. Pyrimidine salvage, as estimated by the incorporation of uracil and uridine into RNA, as well as salvage of purine precursors, including adenine, adenosine, guanine, and hypoxanthine, was high during the lag phase of cultured cells and decreased in the subsequent stages. This extensive utilization of bases and nucleosides for nucleotide synthesis was ascribed mainly to the high activities of the major purine and pyrimidine salvage enzymes (Hirose and Ashihara 1984; Kanamori-Fukuda et al. 1981). Activation of the *de novo* pathway appears to be a later event. In cultured *C. roseus* cells, orotate utilization for RNA synthesis increased markedly during the cell division phase (Kanamori et al. 1979). This delay in the restoration of the *de novo* machinery is probably due to a lag in the reactivation of enzymes participating in the *de novo* pathway of both purines and pyrimidines. The activities of purine enzymes involved in the *de novo* pathway, including glycinamide ribonucleotide synthetase and amidophosphoribosyltransferase (see Figure 4.1 in Chapter 4) showed a similar pattern in cultured carrot cells (Ashihara and Nygaard 1989). Although much research on nucleic acid metabolism during cell growth has been conducted with angiosperms, the activities of *de novo* and salvage pathways have been demonstrated in cultured cells of the conifer, *P. glauca* (Ashihara et al. 2001a).

12.8.3 Nucleotide Availability

Studies conducted on uptake and utilization of phosphate have demonstrated that cellular phosphate may act as the signal for the initiation and termination of cell division, possibly by regulating the availability of nucleoside triphosphates within the system. In *C. roseus* and *D. innoxia* cells, the uptake of phosphate from the medium is completed within the lag phase. At the beginning of cell proliferation, the phosphate pool is consumed and this parallels a decrease of the cellular nucleoside-triphosphate pool. Depletion of intracellular phosphate and nucleoside triphosphates below a threshold may represent a metabolic switch, which terminates cell division and initiates cell elongation (Wagner and Backer 1992).

Studies on nucleotide metabolism during the stationary phase have been conducted mainly with *D. innoxia* cells, where the length of this phase can be extended by up to two weeks, without affecting cell viability. In general, the stationary phase shows a reduction of both DNA and RNA (Ashihara and Nygaard 1989; Ukaji and Ashihara 1986; Wylegalla et al. 1985), which is probably the result of a decreased nucleotide pool. It is interesting to notice that the overall energy charge of the nucleotide pool is generally high and this high ratio may be a metabolic condition for survival (Stasolla et al. 2003a). Purine and pyrimidine nucleotide metabolism studies have been carried out using ^{14}C-precursors in suspension cultured cells of *A. thaliana* (Yin et al. 2014) and the mangrove plant, *Bruguiera sexangula* (Sato and Ashihara 2009).

12.9 Molecular Studies

Most of the extensive nucleotide metabolism information on plant growth and development deals with physiological and biochemical processes. Recently, the molecular events regulating these nucleotide pathways have been investigated in *A. thaliana* and other plants, such as maize, using Arabidopsis genes. Some papers on nucleotide metabolism in *A. thaliana* have been published including information on: a salvage enzyme, hypoxanthine-guanine phosphoribosyltransferase (Liu et al. 2007), and a purine nucleotide catabolism enzyme, guanosine deaminase (Dahncke and Witte 2013). For pyrimidines, a pyrimidine *de novo* synthesis enzyme, aspartate transcarbamoylase (Chen and Slocum 2008), *de novo* and salvage pyrimidine nucleotide synthesis (Witz et al. 2012; Pedroza-García et al. 2019), and the pyrimidine salvage enzymes, uracil phosphoribosyltransferases (Mainguet et al. 2009) and thymidine kinase (Le Ret et al. 2018), as well as the pyrimidine degradation enzyme, uridine ribohydrolase (Jung et al. 2009) have been the principle focus of studies.

12.10 Summary

This chapter has examined physiological aspects of nucleotide metabolism in various areas of growth and organization of development in plants, including embryo maturation and germination, *in vitro* organogenesis, storage organ development and sprouting, leaf senescence, cultured plant cells, and notes recent molecular studies.

References

Anderson, J.D. (1977). Responses of adenine nucleotides in germinating soybean embryonic axes to exogenously applied adenine and adenosine. *Plant Physiol.* 60: 689–692.

Ashihara, H. (1977). Changes in activities of the *de novo* and salvage pathways of pyrimidine nucleotide biosynthesis during germination of black gram (*Phaseolus mungo*) seeds. *Z. Pflanzenphysiol.* 81: 199–211.

Ashihara, H. (1983). Changes in activities of purine salvage and ureide synthesis during germination of black gram (*Phaseolus mungo*) seeds. *Z. Pflanzenphysiol.* 113: 47–60.

Ashihara, H. and Kameyama, Y. (1989). Biosynthesis of pyrimidines during germination of black gram seeds: changes in the level of participant enzymes. *Curr. Sci.* 58: 889–891.

Ashihara, H. and Nygaard, P. (1989). Purine nucleotide synthesis and RNA synthesis in suspension cultured cells of carrots. *Physiol. Plant.* 75: 31–36.

Ashihara, H., Takasawa, Y., and Suzuki, T. (1997). Metabolic fate of guanosine in higher plants. *Physiol. Plant.* 100: 909–916.

Ashihara, H., Loukanina, N., Stasolla, C., and Thorpe, T.A. (2001a). Pyrimidine metabolism during somatic embryo development in white spruce (*Picea glauca*). *J. Plant Physiol.* 158: 613–621.

Ashihara, H., Stasolla, C., Loukanina, N., and Thorpe, T.A. (2001b). Purine metabolism during white spruce somatic embryo development: salvage of adenine, adenosine, and inosine. *Plant Sci.* 160: 647–657.

Balandier, P., Gendraud, M., Rageau, R. et al. (1993). Bud break delay on single node cuttings and bud capacity for nucleotide accumulation as parameters for endo- and paradormancy in peach trees in a tropical climate. *Sci. Horticul.* 55: 249–261.

Barnola, P., Lavarenne, S., Gendraud, M., and Jallut, N. (1986). Dormance des bourgeons apicaux du frêne (*Fraxinus excelsior* L.) : évaluation du pool des nucléosides triphosphates et éventail des températures actives sur le débourrement des bourgeons en période de dormance. *Ann. For. Sci.* 43: 339–350.

Belmonte, M., Elhiti, M., Ashihara, H., and Stasolla, C. (2011). Brassinolide-improved development of *Brassica napus* microspore-derived embryos is associated with increased activities of purine and pyrimidine salvage pathways. *Planta* 233: 95–107.

Brown, E. (1963). Purine and pyrimidine derivatives in mature pea seeds. *Biochem. J.* 88: 498–504.

Brown, E.G. and Wray, J.L. (1968). Correlated changes of some enzyme activities and co-factor and substrate contents of pea cotyledon tissue during germination. *Biochem. J.* 108: 437–444.

Chen, C.T. and Slocum, R.D. (2008). Expression and functional analysis of aspartate transcarbamoylase and role of de novo pyrimidine synthesis in regulation of growth and development in Arabidopsis. *Plant Physiol. Biochem.* 46: 150–159.

Dahncke, K. and Witte, C.P. (2013). Plant purine nucleoside catabolism employs a guanosine deaminase required for the generation of xanthosine in Arabidopsis. *Plant Cell* 25: 4101–4109.

DeKlerk-Kiebert, Y.M. and Van del Plus, L.H.W. (1984). The course of adenine nucleotide contents in white and green soybean cells during growth in batch suspension culture. *Plant Sci. Lett.* 33: 155–162.

Deltour, R. (1985). Nuclear activation during early germination of the higher plant embryo. *J. Cell Sci.* 75: 43–83.

Elhiti, M., Ashihara, H., and Stasolla, C. (2011). Distinct fluctuations in nucleotide metabolism accompany the enhanced in vitro embryogenic capacity of *Brassica cells* over-expressing SHOOTMERISTEMLESS. *Planta* 234: 1251–1265.

Evans, W.R. and Axelrod, B. (1960). Pyrimidine metabolism in germinating seedlings. *Plant Physiol.* 36: 9–13.

Faye, F. and Le Floc'h, F. (1997). Adenosine kinase of peach tree flower buds: purification and properties. *Plant Physiol. Biochem.* 35: 15–22.

Florchinger, M., Zimmermann, M., Traub, M. et al. (2006). Adenosine stimulates anabolic metabolism in developing castor bean (*Ricinus communis* L.) cotyledons. *Planta* 223: 340–348.

Gendraud, M. (1975). Contribution à l'étude du métabolisme des nucléosides diet triphosphates de tubercules de topinambour cultivés in vitro en rapport avec leurs potentialités morphogénétiques. *Plant Sci. Lett.* 4: 53–59.

Givan, C.V. and Collin, H.A. (1967). Studies on the growth in culture of plant cells: II. Changes in respiration rate and nitrogen content associated with the growth of *Acer pseudoplatanus* L. cells in suspension culture. *J. Exp. Bot.* 18: 321–331.

Guranowski, A. (1979). Plant adenosine kinase: purification and some properties of the enzyme from *Lupinus luteus* seeds. *Arch. Biochem. Biophys.* 196: 220–226.

Guranowski, A. and Barankiewicz, J. (1979). Purine salvage in cotyledons of germinating lupin seeds. *FEBS Lett.* 104: 95–98.

Guranowski, A. and Pawelkiewicz, J. (1978). Adenosylhomocysteinase and adenosine nucleosidase activities in *Lupinus luteus* cotyledons during seed formation and germination. *Planta* 139: 245–247.

Hirose, F. and Ashihara, H. (1984). Changes in activity of enzymes involved in purine "salvage" and nucleic acid degradation during growth of *Catharanthus roseus* cells in suspension culture. *Physiol. Plant.* 60: 532–538.

Jung, B., Flörchinger, M., Kunz, H.-H. et al. (2009). Uridine-ribohydrolase is a key regulator in the uridine degradation pathway of *Arabidopsis*. *Plant Cell* 21: 876–891.

Kameyama, Y., Tokoro, T., and Ashihara, H. (1985). Metabolism of [2-^{14}C]thymine and [2-^{14}C]thymidine in germinating black gram (*Phaseolus mungo*) seeds. *Radioisotopes* 34: 214–221.

Kanamori, I., Ashihara, H., and Komamine, A. (1979). Changes in the activities of the pentose phosphate pathway and pyrimidine nucleotide biosynthesis during the growth of *Vinca rosea* cells in suspension culture. *Z. Pflanzenphysiol.* 93: 437–448.

Kanamori-Fukuda, I., Ashihara, H., and Komamine, A. (1981). Pyrimidine nucleotide biosynthesis in *Vinca rosea* cells: changes in the activity of *de novo* and salvage pathways during growth in a suspension culture. *J. Exp. Bot.* 32: 69–78.

Katahira, R. and Ashihara, H. (2002). Profiles of pyrimidine biosynthesis, salvage and degradation in disks of potato (*Solanum tuberosum* L.) tubers. *Planta* 215: 821–828.

Katahira, R. and Ashihara, H. (2006). Profiles of purine biosynthesis, salvage and degradation in disks of potato (*Solanum tuberosum* L.) tubers. *Planta* 225: 115–126.

Kermode, A.R. (1990). Regulatory mechanisms involved in the transition from seed development to germination. *Crit. Rev. Plant Sci.* 9: 155–195.

King, P.J. and Street, H.E. (1977). Growth patterns in cell cultures. In: *Plant Tissue and Cell Culture*, vol. 2 (ed. H.E. Street), 307–387. Oxford: Blackwell.

Kombrink, E. and Beevers, H. (1983). Transport of purine and pyrimidine bases and nucleosides from endosperm to cotyledons in germinating castor bean seedlings. *Plant Physiol.* 73: 370–376.

Lavarenne, S., Champciaux, M., Barnola, P., and Gendraud, M. (1982). Métabolisme des nucléotides et dormance des bourgeons chez le Frêne. *Physiol. Vég.* 20: 371–376.

Le Floc'h, F. and Lafleuriel, J. (1983). The particulate AMP aminohydrolase of Jerusalem artichoke tubers: partial purification and properties. *Physiol. Veg.* 21: 15–27.

Le Floc'h, F. and Lafleuriel, J. (1984). Évolution des activités enzymatiques des voies de recyclage des bases et nucléosides puriques dans les tubercules de Topinambour, au cours de la levéede la dormance. *C. R. Acad. Sc. Paris* 298: 69–72.

Le Floc'h, F. and Lafleuriel, J. (1990). Purification and properties of guanylate kinase of mitochondria from tubers of Jersalem artichoke. *Plant Physiol. Biochem.* 28: 191–201.

Le Ret, M., Belcher, S., Graindorge, S. et al. (2018). Efficient replication of the plastid genome requires an organellar thymidine kinase. *Plant Physiol.* 178: 1643–1656.

Liu, X., Qian, W., Liu, X. et al. (2007). Molecular and functional analysis of hypoxanthine-guanine phosphoribosyltransferase from *Arabidopsis thaliana*. *New Phytol.* 175: 448–461.

Loef, I., Stitt, M., and Geigenberger, P. (1999). Orotate leads to a specific increase in uridine nucleotide levels and a stimulation of sucrose degradation and starch synthesis in discs from growing potato tubers. *Planta* 209: 314–323.

Loef, I., Stitt, M., and Geigenberger, P. (2001). Increased levels of adenine nucleotides modify the interaction between starch synthesis and respiration when adenine is supplied to discs from growing potato tubers. *Planta* 212: 782–791.

Mainguet, S.E., Gakière, B., Majira, A. et al. (2009). Uracil salvage is necessary for early Arabidopsis development. *Plant J.* 60: 280–291.

Mazuś, B. and Buchowicz, J. (1972). Activity of enzymes involved in pyrimidine metabolism in the germinating wheat grains. *Phytochemistry* 11: 77–82.

Meyer, R. and Wagner, K.G. (1985a). Analysis of the nucleotide pool during growth of suspension cultured cells of *Nicotiana tabacum* by high performance liquid chromatography. *Physiol. Plant.* 65: 439–445.

Meyer, R. and Wagner, K.G. (1985b). Nucleotide pools in suspension-cultured cells of *Dature innoxia* I. Changes during growth of the batch culture. *Planta* 166: 439–445.

Mitsui, K. and Ashihara, H. (1988). Nucleotide pools and purine metabolism in tissue cultures of wild-type *Dature innoxia* and adenine-requiring auxotroph. *Plant Cell Physiol.* 29: 1171–1183.

Moffatt, B. and Somerville, C. (1988). Positive selection for male-sterile mutants of Arabidopsis lacking adenine phosphoribosyl transferase activity. *Plant Physiol.* 86: 1150–1154.

Mullin, B.C. and Fites, R.C. (1978). Thymidine phosphorylation in *Arachis hypogaea* by combined activities of adenosine triphosphate phosphohydrolase and nucleoside phosphotransferase. *Physiol. Plant.* 42: 395–400.

Nobusawa, E. and Ashihara, H. (1983). Purine metabolism in cotyledons and embryonic axes of black gram (*Phaseolus mungo* L.) seedlings. *Int. J. Biochem.* 15: 1059–1065.

Nygaard, P. (1973). Nucleotide metabolism during pine pollen germination. *Physiol. Plant.* 28: 361–371.

Pedroza-García, J.A., Nájera-Martínez, M., Mazubert, C. et al. (2019). Role of pyrimidine salvage pathway in the maintenance of organellar and nuclear genome integrity. *Plant J.* 97: 430–446.

Price, C.E. and Murray, A.W. (1969). Purine metabolism in germinating wheat embryos. *Biochem. J* 115: 129–133.

Robert, F. and Petel, G. (2001). Nucleotide synthesis capability as a marker of the petiole elongation of strawberry plants (*Fragaria* X *ananassa* Duch.). *J. Horticul. Sci. Biotechnol.* 75: 690–696.

Robert, F., Petel, G., Risser, G., and Gendraud, M. (1997). Determination of the growth potential of strawberry plants (*Fragaria* × *ananassa* Duch.) by morphological and nucleotide measurements in relation to chilling. *Can. J. Plant Sci.* 77: 127–132.

Ross, C. and Cole, C.V. (1968). Metabolism of cytidine and uridine in leaves. *Plant Physiol.* 43: 1227–1231.

Ross, C. and Murray, M.G. (1971). Development of pyrimidine-metabolizing enzymes in cotyledons of germinating peas. *Plant Physiol.* 48: 626–630.

Ross, C., Coddington, R.L., Murray, M.G., and Bledsoe, C. (1971). Pyrimidine metabolism in cotyledons of germinating pea. *Plant Physiol.* 47: 71–75.

Sasamoto, H. and Ashihara, H. (1988). Biosynthesis of pyrimidine nucleotides and arginine in a suspension culture of *Catharanthus roseus*. *Int. J. Biochem.* 20: 87–92.

Sato, Y. and Ashihara, H. (2009). Long-term effect of NaCl on the activity of uridine and uracil salvage for nucleotide synthesis in cultured mangrove (*Bruguiera sexangula*) cells. *Plant Sci.* 176: 383–389.

Schimpff, G., Muller, H., and Follmann, H. (1978). Age-dependent DNA labeling and deoxyribonucleotide synthesis in wheat seeds. *Biochim. Biophys. Acta* 520: 70–81.

Setlow, P. and Kornberg, A. (1970). Biochemical studies of bacterial sporulation and germination XXII. Energy metasbolism in early stages of germination of *Bacillus megaterium*. *J. Biol. Chem.* 245: 3637–3644.

Shimano, F. and Ashihara, H. (2006). Effect of long-term phosphate starvation on the levels and metabolism of purine nucleotides in suspension-cultured *Catharanthus roseus* cells. *Phytochemistry* 67: 132–141.

Shimazaki, A., Hirose, F., and Ashihara, H. (1982). Changes in adenine nucleotide levels and adenine salvage during the growth of *Vinca rosea* cells in suspension culture. *Z. Pflanzenphysiol.* 106: 191–198.

Shimizu, T., Clifton, A., Komamine, A., and Fowler, M.W. (1977). Changes in metabolite levels during growth of *Acer pseudoplatanus* (sycamore) cells in batch. *Physiol. Plant.* 40: 125–129.

Stasolla, C., Loukanina, N., Ashihara, H. et al. (2001a). Changs in pyrimidine nucleotide biosynthesis during germination of white spruce (*Picea glauca*) somatic embryos. *In Vitro Cell. Dev. Biol. Plant* 37: 285–292.

Stasolla, C., Loukanina, N., Ashihara, H. et al. (2001b). Ascorbic acid changes the pattern of purine metabolism during germination of white spruce somatic embryos. *Tree Physiol.* 21: 359–367.

Stasolla, C., Loukanina, N., Ashihara, H. et al. (2001c). Changes in pyrimidine nucleotide biosynthesis during germination of white spruce (*Picea glauca*) somatic embryos. *In Vitro Cell. Dev. Biol. Plant* 37: 285–292.

Stasolla, C., Loukanina, N., Ashihara, H. et al. (2001d). Purine and pyrimidine metabolism during the partial drying treatment of white spruce (*Picea glauca*) somatic embryos. *Physiol. Plant.* 111: 93–101.

Stasolla, C., Loukanina, N., Ashihara, H. et al. (2002). Pyrimidine nucleotide biosynthesis and nucleic acid metabolism in embryos and megagametophytos of white spruce (*Picea glauca*) during germination. *Physiol. Plant.* 115: 155–165.

Stasolla, C., Katahira, R., Thorpe, T.A., and Ashihara, H. (2003a). Purine and pyrimidine nucleotide metabolism in higher plants. *J. Plant Physiol.* 160: 1271–1295.

Stasolla, C., Loukanina, N., Ashihara, H. et al. (2003b). Changes in deoxyribonucleotide biosynthesis during carrot somatic embryogenesis. *Plant Physiol. Biochem.* 41: 779–785.

Stasolla, C., Loukanina, N., Ashihara, H. et al. (2003c). Pyrimidine deoxyribonucleotide metabolism during maturation and germination of white spruce (*Picea glauca*) somatic embryos: metabolic fate of C-labelled cytidine, deoxycytidine and thymidine. *Physiol. Plant.* 118: 499–506.

Stasolla, C., Loukanina, N., Ashihara, H. et al. (2006). Changes of purine and pyrimidine nucleotide biosynthesis during shoot initiation from epicotyl explants of white spruce (*Picea glauca*). *Plant Sci.* 171: 345–354.

Stasolla, C., Loukanina, N., Ashihara, H. et al. (2007). Comparative studies on pyrimidine metabolism in excised cotyledons of *Pinus radiata* during shoot formation *in vitro*. *J. Plant Physiol.* 164: 429–441.

Tomlinson, P.T. and Lovatt, C.J. (1987). Nucleotide metabolism in 'Washington' navel orange fruit: II. Pathway capacities during development. *J. Amer. Soc. Hort. Sci.* 112: 535–539.

Ukaji, T. and Ashihara, H. (1986). Changes in levels of cellular constituents in suspension culture of *Catharanthus roseus* associated with inorganic phosphate depletion. *Z. Naturforsch.* 41c: 1045–1051.

Wagner, K.G. and Backer, A.I. (1992). Dynamics of nucleotides in plants studied on a cellular basis. In: *International Review of Cytology*, vol. 134 (eds. K.W. Jeon and M. Friedlander), 1–84. San Diego: Academic Press.

Wanka, F. and Bauer, F.W. (1967). On phosphorylation and deamination of pyrimidine and deoxy-pyrimidine nucleosides by enzymes from germinating corn seedlings. *Z. Pflanzenphysiol.* 58: 175–186.

Wanka, F. and Walboomers, J.M. (1966). Thymidine kinase and uridine kinase in corn seedlings. *Z. Pflanzenphysiol.* 55: 458–465.

Witz, S., Jung, B., Fürst, S., and Möhlmann, T. (2012). *De Novo* pyrimidine nucleotide synthesis mainly occurs outside of plastids, but a previously undiscovered nucleobase importer provides substrates for the essential salvage pathway in *Arabidopsis*. *Plant Cell* 24: 1549–1559.

Wylegalla, C., Meyer, R., and Wagner, K.G. (1985). Nucleotide pools in suspension-cultured cells of *Datura innoxia* II. Correlation with nutrient uptake and macromolecular synthesis. *Planta* 166: 446–451.

Yin, Y., Shimano, F., and Ashihara, H. (2007). Involvement of rapid nucleotide synthesis in recovery from phosphate starvation of *Catharanthus roseus* cells. *J. Exp. Bot.* 58: 1025–1033.

Yin, Y., Katahira, R., and Ashihara, H. (2014). Metabolism of purine nucleosides and bases in suspension-cultured *Arabidopsis thaliana* cells. *Eur. Chem. Bull.* 3: 925–934.

13

Environmental Factors and Nucleotide Metabolism

13.1 Introduction

Environmental factors that affect plant growth include light, temperature, water, humidity, and nutrition. When they are in limited supply or in excess, plants are subjected to various stresses. In this chapter, the involvement of purine and pyrimidine nucleotides in response to different types of stress is discussed. In nucleotide metabolism studies, the effects of phosphate, salt, water, wounding, iron, and light have been investigated. The plants used for these investigations are listed in Table 13.1.

13.2 Effect of Phosphate on Nucleotide Metabolism

Cultured plant cells have in the main been utilized for detailed investigations on the metabolic role of nutrients. Inorganic phosphate (Pi) has been shown to participate in many metabolic processes that affect plant cell growth and development. Complete uptake of Pi by cultured cells is generally achieved during the first day after inoculation of fresh medium (Wylegalla et al. 1985), although immediate uptake was reported in *Catharanthus roseus* cells grown in the absence of Pi for a 67 hour period (Ashihara and Ukaji 1986).

Overall changes in cellular metabolism have been observed as a result of Pi starvation. These include reduced rates of respiration (Li and Ashihara 1990), decreased protein and nucleic acid synthesis (Li and Ashihara 1989; Ukaji and Ashihara 1987b), development of bypasses of respiration (Duff et al. 1989; Nagano and Ashihara 1993; Nagano et al. 1994), as well as changes in nucleotide pool size and composition (Dancer et al. 1990; Li and Ashihara 1990; Shimano and Ashihara 2006; Ukaji and Ashihara 1987b). In *C. roseus* cells the endogenous ATP level of Pi-starved cells was <35 nmol g^{-1} f.w. Inoculation of cultures with 10–20 mmol l^{-1} Pi, resulted in an increase to 180 nmol g^{-1} f.w. (Ashihara and Ukaji 1986). A similar result was obtained by Dancer et al. (1990), who observed a reduction in the overall adenine and uracil nucleotide pools, as well as a decrease in the ATP/ADP and uridine triphosphate (UTP)/uridine diphosphate (UDP) ratios in *Chenopodium rubrum* (red goosefoot) cell suspension cultures grown in the absence of Pi. These results, together with the inability of cells inoculated into a Pi-free medium to

Plant Nucleotide Metabolism – Biosynthesis, Degradation, and Alkaloid Formation, First Edition.
Hiroshi Ashihara, Iziar A. Ludwig and Alan Crozier.

Table 13.1 Plant materials used for environmental factors of nucleotide metabolism.

Factors	Plant name	Sample references
Phosphate stress	*Catharanthus roseus*	Ashihara et al. (1988), Shimano and Ashihara (2006)
	Chenopodium rubrum (red goosefoot)	Dancer et al. (1990)
	Arabidopsis thaliana	Muller et al. (2004)
Salt stress	*Zea mays* (maize)	Peterson et al. (1987, 1988)
	Capsicum annuum (pepper)	Nieman et al. (1988)
	Carthamus tinctorius (safflower)	Nieman et al. (1988)
	Sonneratia alba (mangrove)	Akatsu et al. (1996)
	Avicennia marina (mangrove)[a]	Suzuki et al. (2003)
	Bruguiera sexangula (mangrove)[a]	Sato and Ashihara (2009)
	Rhizophora stylosa (mangrove)[a]	Ashihara et al. (2003)
	Catharanthus roseus	Kubota and Ashihara (1993)
	Spinacia oleracea (spinach)	Weretilnyk et al. (2001)
	Beta vulgaris (sugar beet)	Weretilnyk et al. (2001)
Water stress	*Eleusine coracana* (ragi)	Kandpal and Rao (1984)
	Pisum sativum (pea)	Raveneau et al. (2017)
Wound stress	*Solanum tuberosum* (potato)	Katahira and Ashihara (2006)
Iron stress	*Hordeum vulgare* (barley)	Itai et al. (2000)
Light stress	*Cucumis sativus* (cucumber)	Šebesta et al. (1964)
	Pharbitis nil (morning glory)	Nguyen (1979)
	Zea mays (maize)	Pilet and Gabathuler (1982)

a) Mangroves are shrubs that grow in coastal saline water.

increase their endogenous level of ATP (Ukaji and Ashihara 1986), clearly suggest that the rise in nucleotide content observed during the lag phase, that is necessary for the subsequent cell division phase, is dependent upon Pi availability in the medium (Wagner and Backer 1992).

The effects of Pi availability on the *de novo* and salvage synthesis of nucleotides have been investigated (Ashihara et al. 1988). Comparative studies on nucleotide synthesis with *C. roseus* cells grown with and without Pi revealed increased activities of *de novo* and salvage pathways of both purines and pyrimidine nucleotides in the presence of Pi. In the same study, it was also shown that the rate of uridine degradation was reduced significantly in cells grown with Pi in the incubation medium. Although the role played by Pi during the synthesis of nucleotides needs clarification, several reports indicate that the activities of both *de novo* and salvage pathways are regulated by the 5-phosphoribosyl-1-pyrophosphate (PRPP) concentration, which was increased by Pi. The level of PRPP is very low in the stationary phase of cultured *C. roseus* cells, but increases rapidly after reinoculation into fresh medium (Hirose and Ashihara 1983). However, this increase is not observed if the cells are transferred into a medium devoid of Pi (Ukaji and Ashihara 1987a).

The effect of long-term Pi starvation of up to three weeks on the levels of purine nucleotides and related compounds has been examined using suspension-cultured *C. roseus* cells. Levels of adenine and guanine nucleotides, especially ATP and GTP, were markedly reduced during Pi-starvation. There was an increase in the activity of RNase, DNase, 5′- and 3′-nucleotidases and acid phosphatase, which may participate in the hydrolysis of nucleic acids and nucleotides. Accumulation of adenosine, adenine, guanosine, and guanine was observed during the long-term Pi starvation. Long-term Pi starvation markedly depressed the flux of transport of exogenously supplied ^{14}C-labelled adenosine and adenine. However, radioactivity that was taken up by the Pi-starved cells was readily converted to adenine nucleotides although RNA synthesis from the labelled substrates was reduced substantially. The activities of adenosine kinase, adenine phosphoribosyltransferase, and adenosine nucleosidase are maintained at a high level in long-term Pi starved cells (Shimano and Ashihara 2006). Involvement of rapid nucleotide synthesis in recovery from Pi starvation of *C. roseus* cells has been demonstrated (Yin et al. 2007). Growth of suspension-cultured *C. roseus* cells ceased during Pi starvation, but recommenced upon addition of Pi even after long-term starvation. The feeding experiments with ^{33}Pi and Pi-starved cells showed the most heavily labelled organic compounds were nucleotides, followed by sugar phosphates. The RNA, protein, and free nucleotide content all decreased gradually during Pi starvation, however, these compounds, especially nucleotides, increase markedly in the 24 hours after addition of Pi. These responses are found in short- and long-term Pi-starved cells, although the total amounts of these compounds are lower in the long-term Pi-deficient cells. With the nucleotides, a large increase is observed in nucleoside triphosphates rather than nucleoside mono- and diphosphates. The transcript level of phosphate transporter and the activities of acid phosphatase, 5′- and 3′-nucleotidase, and adenosine nucleosidase are all reduced by the addition of Pi. In contrast, the activity of adenine phosphoribosyltransferase is increased substantially in the Pi-fed cells. Little or no increase is observed in adenosine kinase.

Several studies with plants have suggested that expression of many genes depends on the Pi level. For example, the phosphate transporter gene is strongly expressed in Pi-deficient plants (Dong et al. 1999; Kai et al. 2002). This is probably an adaptation to obtain more Pi from the surrounding cells. The synthesis of many mRNAs still occurs in Pi-deficient cells. However, levels of nucleotide triphosphates, which are substrates for RNA synthesis, are decreased markedly in the cells, and net synthesis of RNA must, therefore, be greatly reduced by Pi deficiency. This implies that studies on the gene expression and investigation at the metabolite level are both important in determining the Pi-dependent mechanism of control of cell growth. Based on the ^{33}P feeding study, it appears that an initial event in recovery from Pi depletion involves the synthesis of nucleotides, especially nucleoside triphosphates.

In Pi-starved cells, adenosine-5′-monophosphate (AMP) and ADP dominate the adenine nucleotides, and the adenylate energy charge is much lower than in growing cells (Shimano and Ashihara 2006). When Pi is supplied to Pi deficient cultures ATP is formed from ADP mainly by respiration (stage 1 – see Figure 13.1). Turnover, i.e. utilization and regeneration of ATP is accelerated, and phosphate groups at the γ-position of ATP are transferred to various nucleoside mono- and diphosphates and free sugars, so that many nucleoside triphosphates and sugar phosphates are formed (stage 2). The net increase in nucleotides is initiated by salvage reactions utilizing

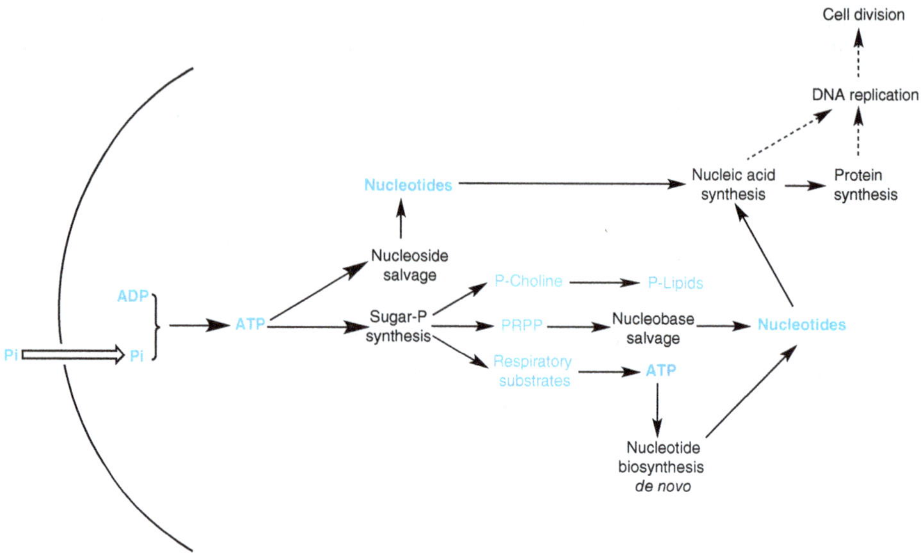

Figure 13.1 Possible metabolic events during recovery from Pi starvation in suspension-cultured *Catharanthus roseus* cells. When Pi is supplied to the Pi-deficient culture, ATP is formed from ADP and Pi mainly by respiration (stage 1). Turnover of ATP is accelerated, and phosphate groups at the g-position of ATP molecules are transferred to various nucleoside mono- and diphosphates and free sugars, so that many nucleoside triphosphates and sugar phosphates are formed (stage 2). Once the total nucleotide level has increased, it triggers the net synthesis of RNA and protein (stage 3). DNA duplication then takes place and cell division begins (stage 4). Source: adapted from Yin et al. (2007).

the nucleosides and nucleobases, which accumulate in the Pi-starved cells. ATP and PRPP are used for nucleoside and nucleobase salvage reactions by kinases and phosphoribosyltransferases, respectively. Once the total nucleotide level has increased, it triggers the net synthesis of RNA and protein (stage 3). It has been reported that free amino acids accumulate in Pi-starved cells where protein synthesis is restricted, so that amino acid synthesis may not be required for protein synthesis during the early phase of recovery. DNA duplication then takes place and cell division begins (stage 4).

A substantial increase in nucleotide level and a net increase in RNA and protein content and in cell numbers are detected in 24-hour Pi-fed cells. A possible sequence of metabolic activation beginning with cessation of Pi starvation is shown in Figure 13.1. The activities of hydrolysing enzymes, acid phosphatase, and nucleotidases are reduced markedly when the availability of free Pi increases. These findings are supported by the fact that expression of the genes encoding acid phosphatase and some hydrolysing enzymes are rapidly reduced in Pi-fed *Arabidopsis thaliana* seedlings (Muller et al. 2004). The activities of adenosine kinase, adenosine nucleosidase, and adenine phosphoribosyltransferase are maintained at a high level even in long-term Pi starved cells, and would appear to be involved in the limited turnover of nucleotides (Shimano and Ashihara 2006). These enzymes may also have an important function in the resumption of cell growth via the increase in net synthesis of nucleotides during the early stage of recovery from Pi starvation. Of the three enzymes measured, adenine phosphoribosyltransferase and adenosine kinase activities increased or are maintained by the addition of Pi, whereas adenosine nucleosidase is reduced. This differing response suggests that adenosine nucleosidase is not involved in purine salvage but takes part in the degradation process. These observations indicate that rapid reformation of nucleotides by salvage reactions are key events in recovery from Pi starvation.

13.3 Effect of Salts on Nucleotide Metabolism

Plants require some salts for growth, but concentrations above the optimum often result in reduced growth and yield. Expenditure of energy derived from oxidation of carbohydrates and salt pumping is often observed in salt-stressed plants (Nieman et al. 1988). Changes in the synthesis and utilization of nucleotides, especially ATP, in response to salt stress have been reported in several systems, including maize (*Zea mays*) (Peterson et al. 1987, 1988), pepper (*Capsicum annuum*) and safflower (*Carthamus tinctorius*) (Nieman et al. 1988), and the mangrove species *Sonneratia alba* (Akatsu et al. 1996). From these studies, it emerges that salt stress decreases the total nucleotide pool, but increases the adenylate energy charge. For examples, in *S. alba* the energy charge of cells cultured in the presence of 100 mmol l^{-1} NaCl was 0.82, compared to 0.72 with control cells (Akatsu et al. 1996). This result, similar to that described by Kubota and Ashihara (1993) with *C. roseus* cells, is indicative of the presence of an active adenylate kinase, which is responsible for the enlargement of the endogenous ATP pool. High levels of ATP may be necessary for removal of NaCl from the cytosol of cells to the vacuoles and/or the external environment.

Some plant species accumulate glycine betaine in leaves as a compatible osmolyte. Choline, a direct precursor of glycine betaine, is produced from ethanolamine via three *S*-adenosyl-L-methionine (SAM)-dependent *N*-methyltransferase reactions. *S*-adenosyl-L-homocysteine (SAH) produced in these reactions is a potent inhibitor of SAM-dependent methyltransferases (Figure 13.2). SAH is hydrolysed to homocysteine and adenosine by SAH hydrolase. In glycine betaine-forming spinach (*Spinacia oleracea*) and sugar beet (*Beta vulgaris*), SAH hydrolase and adenosine kinase activities increase in leaves subjected to 100–300 mM of NaCl and these changes reflect increased levels of SAH hydrolase and adenosine kinase protein, and transcripts. In contrast, tobacco (*Nicotiana tabacum*) and canola (*Brassica* sp.), which do not accumulate glycine betaine, do not show comparable changes in these enzymes (Weretilnyk et al. 2001). Since SAH hydrolase activity is inhibited by its products, adenosine kinase is not a stress responsive enzyme *per se*, but plays a pivotal role in sustaining *trans*-methylation reactions in general by serving as a coarse metabolic control to reduce the cellular concentration of free adenosine and SAH (Moffatt and Weretilnyk 2001).

Similar effects have been observed with *Avicennia marina*, a halophytic mangrove shrub that grows in tidal swamps (Suzuki et al. 2003). It accumulates considerable quantities of glycine betaine in its leaves and the levels are increased by salt stress (Ashihara et al. 1997). The feeding experiments revealed that salt stress stimulates the incorporation of radioactivity from [1,2-^{14}C]ethanolamine into glycine betaine, and that of [methyl-^{14}C]SAM into choline, phosphocholine, and glycine betaine. *In vitro* activity of adenosine kinase and adenosine nucleosidase increases after leaf disks are incubated with 250 and 500 mM NaCl. Adenosine nucleosidase participates in the removal of adenosine from the SAM cycle (Suzuki et al. 2003).

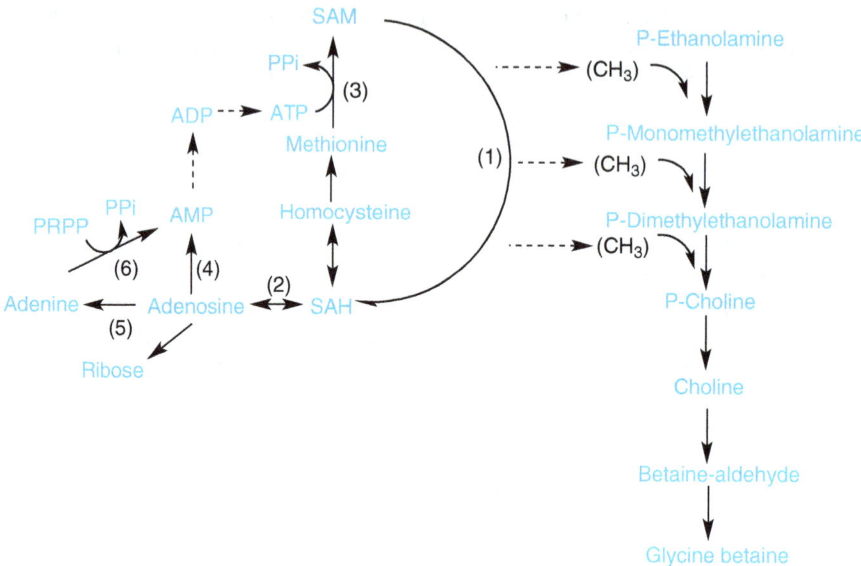

Figure 13.2 Pathways for the generation and utilization of *S*-adenosyl-L-methionine (SAM) in *Avicennia marina* leaves. Three moles of SAM are used to convert P-ethanolamine (phosphoethanolamine) to P-choline (phosphocholine) by *N*-methyltransferase reactions (step 1). SAH (*S*-adenosylhomocysteine is hydrolysed by SAH hydrolase (step 2). Homocysteine, one of the products of SAH hydrolase, is utilized for the regeneration of SAM (step 3); adenosine is directly salvaged by adenosine kinase (step 4) or hydrolysed by adenosine nucleosidase (step 5). Adenine is salvaged by adenine phosphoribosyltransferase (step 6). Ribose may be converted to PRPP (phosphoribosylpyrophosphate). Source: adapted from Suzuki et al. (2003).

In addition to the ability to remove adenosine from the SAM cycle (Figure 13.2), salt-tolerant halophytes, such as mangrove species, usually possess an active purine salvage mechanism, which may require an adequate supply of ATP for removal of salts from the cytoplasm of their cells. The salvage activity of adenosine has been compared in seven different mangrove species and in poplar (Ashihara et al. 2003). In four mangrove species, *Rhizophora stylosa, Bruguiera gymnorrhiza, Kandelia candel,* and *S. alba,* the adenosine salvage activity was higher than in poplar and ATP synthesis from adenosine was stimulated by 250 mM NaCl. An adequate supply of ATP appears to be essential for the normal growth of these mangrove trees. In contrast, the adenosine salvage activity of leaves of the mangrove shrubs, *A. marina* and *Lumnitzera racemosa,* is low. The difference in adenosine salvage ability observed between large mangrove trees and small mangrove shrubs is not easily explained, but it might be due to the different requirements of energy for growth.

Effect of short-term (three hours) salt stress on the metabolic profiles of nucleotides in cultured cells of the mangrove tree, *Bruguiera sexangula* was studied by Ashihara et al. (2003) by determining the uptake and overall metabolism of radiolabelled intermediates involved in the *de novo* pathways and substrates of salvage pathways for nucleotide biosynthesis in the presence and absence of 100 mM NaCl. The activity of the *de novo* pyrimidine biosynthesis, estimated by [2-^{14}C]orotate metabolism, was not affected significantly by the salt. About 20–30% of [2-^{14}C]uridine, [2-^{14}C]uracil, and more than 50% of [2-^{14}C]cytidine were salvaged for pyrimidine nucleotide biosynthesis. Substantial quantities of these compounds were degraded to $^{14}CO_2$ via β-ureidopropionate, and degradation of β-ureidopropionate was enhanced by salt treatment.

The long-term (up to 21 days) effect of 100 mM NaCl on the salvage activity of uridine and uracil was investigated using salt-resistant cultured cells of the mangrove species, *B. sexangula* (Sato and Ashihara 2009). These *in situ* studies indicated that [2-^{14}C]uridine and [2-^{14}C]uracil were efficiently converted to uridine nucleotides, UDP-sugars, and RNA (Figure 13.3). Conversion was higher in the NaCl-treated cells than in control cells after culturing for 14 and 21 days. The activity of uridine kinase and uracil phosphoribosyltransferase, and the expression of one of the *UK/uracil phosphoribosyltransferase* (*UPRT*) homologs of the *A. thaliana* gene (Islam et al. 2007) encoding a dual functional UPRT (steps 1 and 2 in Figure 13.3), were all elevated by NaCl. In contrast, catabolism of [2-^{14}C]uridine and [2-^{14}C]uracil and uridine nucleosidase activity was lower in the salt-treated cells. The effect of long-term salt stress on

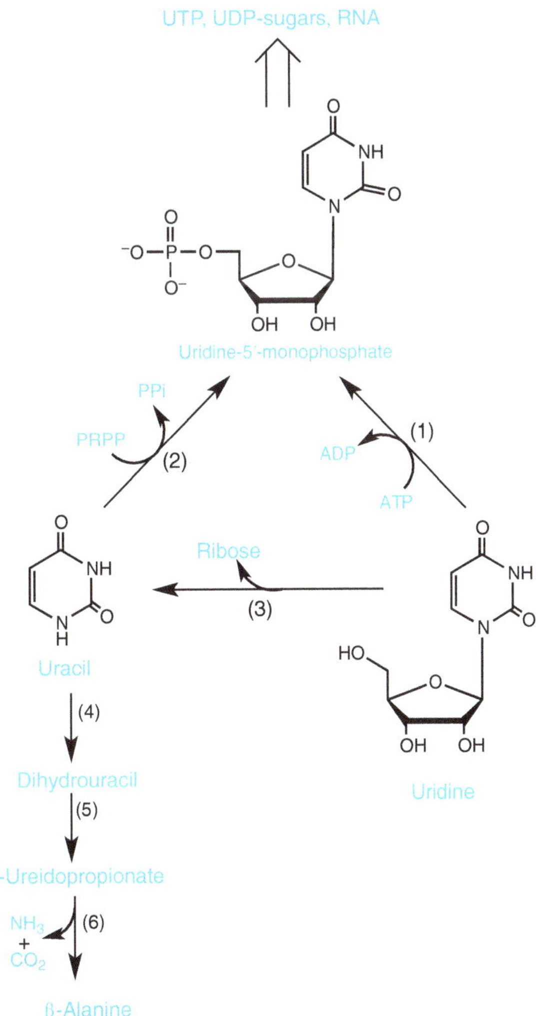

Figure 13.3 Possible metabolic fate of [2-^{14}C]uridine and [2-^{14}C]uracil in cultured *Bruguiera sexangula* cells. Enzymes measured in this study are (1) uridine kinase; (2) uracil phosphoribosyltransferase; (3) uridine nucleosidase. Source: adapted from Sato and Ashihara (2009).

pyrimidine salvage activity may indicate that continuously supplied NaCl stimulates the activity of pyrimidine salvage enzymes. The expression of *AtUK/UPRT1*, was also stimulated. Thus, the increased activity of uridine kinase and uracil phosphoribosyltransferase may be due to the amount of the newly synthesized enzyme as a result of the increased gene expression. These processes appeared to be closely related to the increase in uridine and uracil salvage in *B. sexangula* cells induced by long-term salt stress. Conversion of uridine to uracil by uridine nucleosidase, therefore, appears to be a key reaction to shift from pyrimidine salvage to catabolism. Interestingly, uridine nucleosidase activity was reduced in the NaCl-treated cells.

13.4 Effect of Water Stress

Alteration of the regulation of *de novo* pyrimidine biosynthesis by water stress has been suggested in experiments with ragi (*Eleusine coracana*) seedlings. Feedback inhibition of aspartate transcarbamoylase (see step 2, Figure 8.1) by uridine-5′-monophosphate (UMP) found in control plants disappeared in water-stressed seedlings (Kandpal and Rao 1984). In some proteins, water stress caused a change in the ratio of intramolecular SH, S-S exchange reaction (SH/SS bonds). Therefore, a similar change might lead to the desensitization of the regulatory properties of aspartate transcarbamoylase. The physiological significance of this change has yet to be elucidated. Water and life are linked, but some organisms lose almost all cellular water and enter a non-metabolic state of anhydrobiosis. Adenylate metabolism was investigated during seed imbibition and drying using intact or fragmented seeds of pea (*Pisum sativum*) (Raveneau et al. 2017). AMP was confirmed as the major adenylate stored in dry seeds, and the normal adenylate balance was rapidly restored upon rehydration. Conversely, redrying of fully imbibed seeds reversed the balance towards AMP accumulation. During early imbibition, adenylate kinase would first operate in the AMP + ATP → 2ADP direction, which fits with apparent thermodynamic data measured during seed fragment imbibition, thus providing ADP for oxidative phosphorylation and leading to rapid re-equilibration of the adenylate pools. Subsequently, during late imbibition of whole seeds, there is an increase in the total amount of adenylates, which probably reflects a general increase in metabolic activities prior to germination. Such an increase could result from *de novo* biosynthesis, or from the purine salvage pathway which recycles adenosine released by degradation of nucleotides and reactions related to methionine and SAM metabolism (Stasolla et al. 2003).

13.5 Effect of Wound Stress

Activation of plant storage tissue after excision and incubation is known as the 'wounding' or 'ageing' phenomenon (Kahl 1974). 'Wound respiration' of potato tubers has been studied since the late nineteenth century (Muller-Thungau 1882). Many investigations have examined the effect of wounding on cyanide-sensitive or cyanide-resistant respiration and mitochondrial membrane proteins (Hiser and McIntosh 1990), but there are only a few studies on the synthesis of ATP during the wounding process. One study reported by Katahira and Ashihara (2006) determined the effect of wound stress in potato tubers on the purine salvage pathway. It was shown that levels of ATP and other nucleotides increased two- to seven-fold in 'wounded' potato tuber slices maintained on moist paper for 24 hours (Table 13.2). Furthermore, the relative expression intensity of genes encoding adenosine kinase and adenine phosphoribosyltransferase in the tuber slices was higher than that of genes of the *de novo* pathway, encoding glycineamide ribonucleotide formyltransferase (GART) (step 1 in Figure 13.4) and 5-aminoimidazole ribonucleotide synthetase (AIRS) (step 2 in Figure 13.4) (line 1 in Table 13.3). *In vitro* activities of adenosine kinase (step 3 in Figure 13.4) and adenine phosphoribosyltransferase (step 4 in Figure 13.4) doubled following wounding. Adenosine nucleosidase activity was negligible in freshly prepared slices, but its activity increased substantially as the slices aged (part 2 in Table 13.3). *In situ* adenosine salvage activity, estimated from the incorporation of radioactivity from [8-^{14}C]adenosine into nucleotides and RNA, increased more than fivefold in the wounded slices (part 3 in Table 13.3) (Katahira and Ashihara 2006). These results strongly suggest that increased expression of the genes encoding enzymes of adenosine salvage during wounding is closely related to the increased supply of adenine nucleotides in the wounded slices. From an energetics viewpoint, the wounding process of potato tuber slices is similar to the dormancy break of intact potato tubers.

Accelerated energy generation is required for these processes. In order to maintain high levels of ATP in potato tuber slices after wounding, the salvage pathways of adenine and adenosine appear to be very effective. It appears that the enhanced adenine and adenosine salvage activity in the wounded tuber slices is directly attributable to the increased expression of the genes encoding adenosine kinase and adenine phosphoribosyltransferase activities. The substantial increase in adenosine and adenine salvage activities in wounded tubers may participate in the salvage of adenine and/or adenosine,

Table 13.2 Comparison of nucleotide content in fresh and wounded slices of potato tubers.

Nucleotides	Fresh (nmol g^{-1})	Wounded (nmol g^{-1})
ATP	26.7	71.0
ADP	4.5	13.8
AMP	9.4	13.3
GTP	3.6	12.5
GDP	1.1	2.2
GMP	1.0	5.7
UTP	11.5	17.4
UDP	3.0	4.9
CTP	0.7	5.0
CDP	0.3	0.8

ATP, Adenosine triphosphate; ADP, adenosine diphosphate; AMP, adenosine-5′-monophosphate; GTP, guanosine triphosphate; GDP, guanosine diphosphate; GMP, guanosine-5′-monophosphate; UTP, uridine triphosphate; UDP, uridine diphosphate; CTP, cytidine triphosphate; CDP, cytidine diphosphate.
Source: Adapted from Katahira and Ashihara (2006).

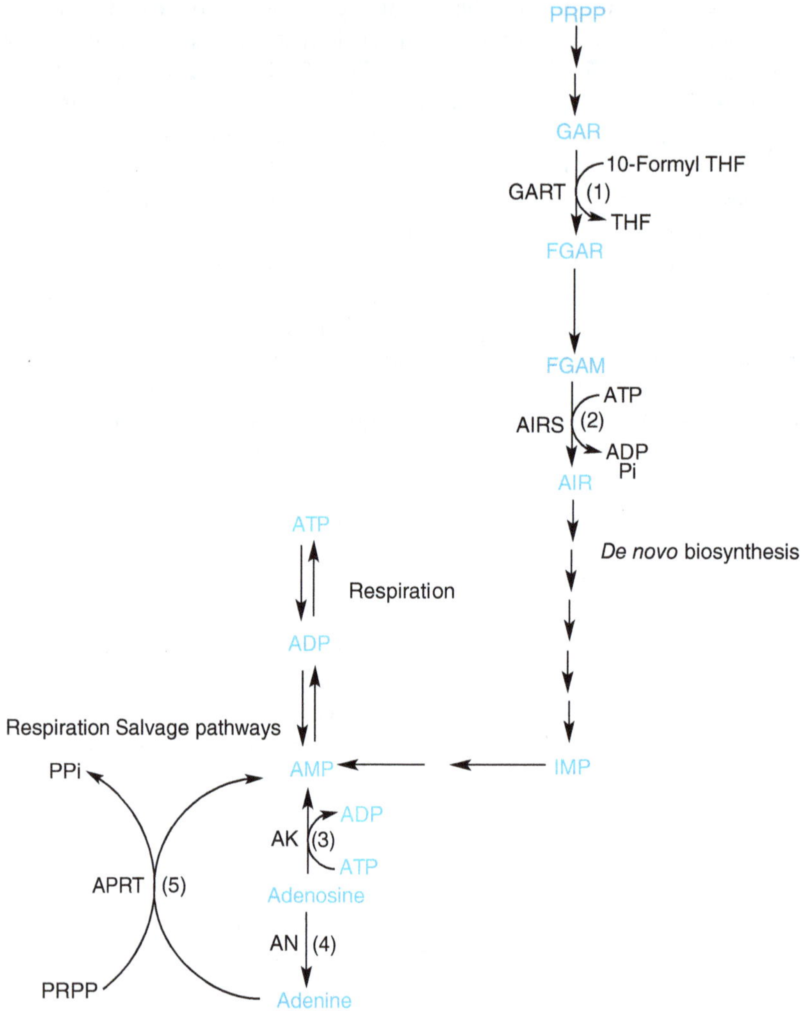

Figure 13.4 Metabolic pathways for ATP biosynthesis. The *de novo* and salvage pathways of adenine nucleotide synthesis are shown. Expression of genes shown in italics and activity of enzymes in numbers were studied in this study. GART (1) and AIRS (2), which respectively, encode glycineamide ribonucleotide formyl transferase (10-formyltetrahydrofolate: 5′-phosphoribosylglycinamide *N*-formyltransferase) and 5-aminoimidazole ribonucleotide synthetase (2-[formamido]-*N*1-[5-phosphoribosyl] acetamidine cycloligase [ADP-forming]), AK (3) adenosine kinase, AN (4) adenosine nucleosidase, and APRT (5) adenine phosphoribosyltransferase. AIR, 5-aminoimidazole ribonucleotide; FGAM, formylglycine amidine ribonucleotide; FGAR, formylglycineamide ribonucleotide; GAR, glycineamide ribonucleotide; PRPP, 5-phosphoribosyl-1-pyrophosphate; THF, tetrahydrofolate. Source: adapted from Katahira and Ashihara (2006).

Table 13.3 Relative transcript levels, activities of enzymes, and *in situ* biosynthesis of adenine nucleotides and nucleic acids in fresh and wounded slices of potato tubers.

	Fresh	Wounded
1) Relative transcript levels (% of the values observed in fresh slices)		
GART (Glycineamide ribonucleotide formyl transferase)	100	270
AIRS (5-Aminoimidazole ribonucleotide synthetase)	100	247
AK (Adenosine kinase)	100	4980
APRT (Adenine phosphoribosyltransferase)	100	2645
2) Activities of enzymes (activities are pkat mg^{-1} protein)		
AK (Adenosine kinase)	97	176
APRT (Adenine phosphoribosyltransferase)	117	206
AN (Adenosine nucleosidase)	1	93
3) In situ *metabolism of adenosine* (incorporations of ^{14}C are nmol g^{-1} fresh wt)		
ATP + ADP	1.2	7.4
AMP	0.21	0.60
Nucleic acid	0.18	0.72
Ureides	0.07	0.30
CO_2	0.00	0.01

Source: Adapted from Katahira and Ashihara (2006).

which are produced by rapid turnover of ATP (Katahira and Ashihara 2006). These salvage pathways may also contribute to the recycling of nucleosides derived from RNA. A fourfold increase in activity of RNases has been observed 24 hours after wounding induced by the slicing of potato tubers (Sacher et al. 1972).

13.6 Effect of Iron Deficiency

In Graminaceous species, production of the mugineric acid family phytosiderophores, which solubilize inorganic Fe(III) compounds by chelation, increases following Fe deficiency. Itai et al. (2000) isolated a protein induced by Fe deficiency from roots of barley (*Hordeum vulgare*) and also showed that both, adenine phosphoribosyltransferase activity and expression of a gene encoding adenine phosphoribosyltransferase, were enhanced by Fe deficiency. As phytosiderophores are produced from SAM, and adenine is released from the methionine cycle for regeneration of SAM, adenine recycling by adenine phosphoribosyltransferase for ATP synthesis would appear to be required for the maintenance of an adequate supply of ATP for the production of SAM.

13.7 Effect of Light

Early studies described the effects of light on RNA synthesis through the *de novo* and salvage pathways of pyrimidine synthesis. In cucumber (*Cucumis sativus*) seedlings, incorporation of radioactivity from [6-^{14}C]orotate, but not [6-^{14}C]uracil, into RNA pyrimidine bases was stimulated by the light (Šebesta et al. 1964). The effect of light on deamination and oxidation of adenine compounds in cotyledons of *Pharbitis nil* was reported by Nguyen (1979). Since labelling of ureides from [8-^{14}C]adenine was higher in darkness than in light, the influence of light on the deamination and the oxidation of AMP by cotyledon disks was investigated. Among the three possible precursors for the deaminative step, adenine was found to be the best compound as adenosine was easily hydrolysed into adenine, and AMP underwent complete hydrolysis before entering the cells. By analysis of [8-^{14}C]adenine metabolism it was found that the rate of deamination of adenine compounds was faster in light than in darkness. In contrast, the activity of xanthine dehydrogenase was much higher in darkness. The level of the activity of uricase was unaffected by light.

Two millimetre apical segments of maize primary roots were analysed in relation to the effects of light and gravity on adenine nucleotide content. ATP content is very sensitive to these stimuli. ATP levels are lower in roots exposed to light than in those maintained in darkness. The energy charge decreases markedly after exposure to light and gravity. With vertical roots the energy charge was stable. Light and gravity may act on cell metabolism, modifying the energy requirements (Gabathuler and Pilet 1981). The effect of brief light illumination by microbeams on the adenine nucleotide levels in maize primary roots has been investigated. Total adenylate pool was increased up to 150% of the initial value within two minutes (Pilet and Gabathuler 1982). Further studies are needed to elucidate the mechanism of these light effects on nucleotide metabolism.

13.8 Summary

Effect of phosphate, salts, water, wound, iron, and light stress on function and control of nucleotide metabolism in plants is provided from the research on plant biology.

References

Akatsu, M., Hosoi, Y., Sasamoto, H., and Ashihara, H. (1996). Purine metabolism in cells of a mangrove plant, *Sonneratia alba*, in tissue culture. *J. Plant Physiol.* 149: 133–137.

Ashihara, H. and Ukaji, T. (1986). Inorganic phosphate absorption and its effect on adenosine 5′-triphosphate level in suspension cultured cells of *Catharanthus roseus*. *J. Plant Physiol.* 124: 77–85.

Ashihara, H., Li, X.N., and Ukaji, T. (1988). Effect of inorganic phosphate on the biosynthesis of purine and pyrimidine nucleotides in suspension-cultured cells of *Catharanthus roseus*. *Ann. Bot.* 61: 225–232.

Ashihara, H., Adachi, K., Otawa, M. et al. (1997). Compatible solutes and inorganic ions in the mangove plant *Avicennia marina* and their effects on the activities of enzymes. *Z. Naturforsch.* 52c: 433–440.

Ashihara, H., Wakahara, S., Suzuki, M. et al. (2003). Comparison of adenosine metabolism in leaves of several mangrove plants and a poplar species. *Plant Physiol. Biochem.* 41: 133–139.

Dancer, J., Veith, R., Feil, R. et al. (1990). Independent changes of inorganic pyrophosphate and ATP/ADP or UTP/UDP ratios in plant cell suspension cultures. *Plant Sci.* 66: 59–63.

Dong, B., Ryan, P.R., Rengel, Z., and Delhaize, E. (1999). Phosphate uptake in *Arabidopsis thaliana*: dependence of uptake on the expression of transporter genes and internal phosphate concentrations. *Plant Cell Environ.* 22: 1455–1461.

Duff, S.M.G., Moorhead, G.B.G., Lefebvre, D.D., and Plaxton, W.C. (1989). Phosphate starvation inducible "bypasses" of adenylate- and phosphate-dependent glycolytic enzymes in *Brassica nigra* suspension cells. *Plant Physiol.* 90: 1275–1278.

Gabathuler, R. and Pilet, P.-E. (1981). Light and gravity effects on adenine nucleotide content and energy charge in maize roots. *Plant Cell Physiol.* 22: 1499–1506.

Hirose, F. and Ashihara, H. (1983). Content and availability of 5-phosphoribosyl-1-pyrophosphate in cultured cells of *Catharanthus roseus*. *Z. Pflanzenphysiol.* 110: 183–190.

Hiser, C. and McIntosh, L. (1990). Alternative oxidase of potato is an integral membrane protein synthesized de novo during aging of tuber slices. *Plant Physiol.* 93: 312–318.

Islam, M.R., Kim, H., Kang, S.W. et al. (2007). Functional characterization of a gene encoding a dual domain for uridine kinase and uracil phosphoribosyltransferase in *Arabidopsis thaliana*. *Plant Mol. Biol.* 63: 465–477.

Itai, R., Nakanishi, H., Nishizawa, N.K. et al. (2000). Induced activity of adenine phosphoribosyltransferase (APRT) in iron-deficient barley roots: a possible role for phytosiderophore production. *J. Exp. Bot.* 51: 1179–1188.

Kahl, G. (1974). Metabolism of plant storage tissue slices. *Bot. Rev.* 40: 263–314.

Kai, M., Takazumi, K., Adachi, H. et al. (2002). Cloning and characterization of four phosphate transporter cDNAs in tobacco. *Plant Sci.* 163: 837–846.

Kandpal, R.P. and Rao, A. (1984). Water stress-induced desensitization of aspartate transcarbamylase from ragi (*Eleusine coracana*) seedlings. *Biochem. Int.* 9: 307–314.

Katahira, R. and Ashihara, H. (2006). Role of adenosine salvage in wound-induced adenylate biosynthesis in potato tuber slices. *Plant Physiol. Biochem.* 44: 551–555.

Kubota, K. and Ashihara, H. (1993). Levels of metabolites related to glycolysis in *Catharanthus roseus* cells during culture. *Phytochemistry* 34: 1509–1513.

Li, X.N. and Ashihara, H. (1989). Effect of inorganic phosphate on the utilization of sucrose by suspension-cultured *Catharanthus roseus* cells. *Ann. Bot.* 64: 33–36.

Li, X.N. and Ashihara, H. (1990). Effect of inorganic phosphate on sugar catabolism by suspension-cultured *Catharanthus roseus*. *Phytochemistry* 29: 497–500.

Moffatt, B.A. and Weretilnyk, E.A. (2001). Sustaining *S*-adenosyl-L-methionine-dependent methyltransferase activity in plant cells. *Physiol. Plant.* 113: 435–442.

Muller, R., Nilsson, L., Krintel, C., and Hamborg Nielsen, T. (2004). Gene expression during recovery from phosphate starvation in roots and shoots of *Arabidopsis thaliana*. *Physiol. Plant.* 122: 233–243.

Muller-Thungau, H. (1882). Uber Zuckranhaufung in Pflanzentheilen in Forge niederer Temperatur. *Landw. Jahrb.* 5: 751–828.

Nagano, M. and Ashihara, H. (1993). Long-term phosphate starvation and respiratory metabolism in suspension-cultured *Catharanthus roseus* cells. *Plant Cell Physiol.* 34: 1219–1228.

Nagano, M., Hachiya, A., and Ashihara, H. (1994). Phosphoenolpyruvate carboxylase from heterotrophically cultured *Catharanthus roseus* cells. *Ann. Bot.* 74: 417–422.

Nguyen, J. (1979). Effect of light on deamination and oxidation of adenylic compounds in cotyledons of *Pharbitis nil*. *Physiol. Plant.* 46: 255–259.

Nieman, R.H., Clark, R.A., Pap, D. et al. (1988). Effects of salt stress on adenine and uridine nucleotide pools, sugar and acid-soluble phosphate in shoot of pepper and safflower. *J. Exp. Bot.* 39: 301–309.

Peterson, T.A., Nieman, R.H., and Clark, R.A. (1987). Nucleotide metabolism in salt-stressed *Zea mays* L. root tips I. Adenine and uridine nucleotides. *Plant Physiol.* 85: 984–989.

Peterson, T.D., Lovatt, C.J., and Nieman, R.H. (1988). Salt stress causes acceleration of purine catabolism and inhibition of pyrimidine salvage in *Zea mays* root tips. *J. Exp. Bot.* 207: 1389–1395.

Pilet, P.E. and Gabathuler, R. (1982). Localized white light effect on root adenine nucleotides. *Plant Sci. Lett.* 26: 121–125.

Raveneau, M.-P., Benamar, A., and Macherel, D. (2017). Water content, adenylate kinase, and mitochondria drive adenylate balance in dehydrating and imbibing seeds. *J. Exp. Bot.* 68: 3501–3512.

Sacher, J.A., Towers, G.H.N., and Davies, D.D. (1972). Effect of light and ageing on enzymes, particularly phenylalanine ammonia lyase, in discs of storage tissue. *Phytochemistry* 11: 2383–2391.

Sato, Y. and Ashihara, H. (2009). Long-term effect of NaCl on the activity of uridine and uracil salvage for nucleotide synthesis in cultured mangrove (*Bruguiera sexangula*) cells. *Plant Sci.* 176: 383–389.

Šebesta, K., Bauerová, B., Martinovic, B., and Šormová, Z. (1964). The effect of light on the utilization of uracil and orotic acid as the precursors of pyrimidine bases of nucleic acids in plants. *Collect. Czech. Chem. Commun.* 29: 801–806.

Shimano, F. and Ashihara, H. (2006). Effect of long-term phosphate starvation on the levels and metabolism of purine nucleotides in suspension-cultured *Catharanthus roseus* cells. *Phytochemistry* 67: 132–141.

Stasolla, C., Katahira, R., Thorpe, T.A., and Ashihara, H. (2003). Purine and pyrimidine nucleotide metabolism in higher plants. *J. Plant Physiol.* 160: 1271–1295.

Suzuki, M., Yasumoto, E., Baba, S., and Ashihara, H. (2003). Effect of salt stress on the metabolism of ethanolamine and choline in leaves of the betaine-producing mangrove species *Avicennia marina*. *Phytochemistry* 64: 941–948.

Ukaji, T. and Ashihara, H. (1986). Changes in levels of cellular constituents in suspension culture of *Catharanthus roseus* associated with inorganic phosphate depletion. *Z. Naturforsch.* 41c: 1045–1051.

Ukaji, T. and Ashihara, H. (1987a). Effect of inorganic phosphate on synthesis of 5-phosphoribosyl-1-pyrophosphate in cultured plant cells. *Int. J. Biochem.* 19: 1127–1131.

Ukaji, T. and Ashihara, H. (1987b). Effect of inorganic phosphate on the level of amino acids in suspension-cultured cells of *Catharanthus roseus*. *Ann. Bot.* 60: 109–114.

Wagner, K.G. and Backer, A.I. (1992). Dynamics of nucleotides in plants studied on a cellular basis. *Int. Rev. Cytol.* 134: 1–84.

Weretilnyk, E.A., Alexander, K.J., Drebenstedt, M. et al. (2001). Maintaining methylation activities during salt stress. The involvement of adenosine kinase. *Plant Physiol.* 125: 856–865.

Wylegalla, C., Meyer, R., and Wagner, K.G. (1985). Nucleotide pools in suspension-cultured cells of *Datura innoxia* II. Correlation with nutrient uptake and macromolecular synthesis. *Planta* 166: 446–451.

Yin, Y., Shimano, F., and Ashihara, H. (2007). Involvement of rapid nucleotide synthesis in recovery from phosphate starvation of *Catharanthus roseus* cells. *J. Exp. Bot.* 58: 1025–1033.

Part V

Purine Alkaloids

14

Occurrence of Purine Alkaloids

14.1 Introduction

The most well-known secondary metabolites synthesized from nucleotides are purine alkaloids (aka, xanthine alkaloids). They are methylxanthines and methylurates whose structures are based on the skeletons of xanthine and urate. The major purine alkaloid typically is caffeine, which is found in popular non-alcoholic beverages such as coffee, tea, and maté. Theobromine accumulates in the seeds of *Theobroma cacao* (Zheng et al. 2004) and is, therefore, a constituent of cocoa drinks and chocolate products. Theacrine, a methylurate, is a less well-known purine alkaloid. However, it accumulates in leaves of *Camellia assamica* var. *kucha* that are used to produce a tea called kucha (Lu et al. 2009; Zheng et al. 2002), which is used as a Chinese herbal medicine that is claimed to have anti-inflammatory and analgesic effects (Ashihara et al. 2017).

Historically, the isolation of caffeine from coffee seeds was first reported independently by the German researchers, Runge (1820) and Von Giese (1820). Caffeine was identified as 'thein' in tea leaves by Oudry (1927). Since then, the presence of caffeine had been reported in at least 80 species in plants (see Ashihara and Crozier 1999). This number was based mainly on a review by Kihlman (1977), which quoted Willaman and Schubert (1961) and O'Connell (1969), and subsequent information published by Stewart (1985) and Kretschmar and Baumann (1999). However, caffeine and related purine alkaloids occur in many species of the same or similar genus, such as *Coffea, Camellia, Theobroma,* and *Ilex* (Anthony et al. 1993; Deng et al. 2017; Mazzafera and Carvalho 1992). Phylogenetic relationships indicate that purine alkaloid-accumulating plants include at least 13 orders in the plant kingdom (Ashihara et al. 2017).

The biosynthetic pathway of caffeine from purine nucleotides was established by extensive purification of tea caffeine synthase (Kato et al. 1999) and the cloning of the gene encoding this enzyme (Kato et al. 2000). The purine alkaloid biosynthetic pathways are described in Chapter 15.

14.2 Chemical Structure of Purine Alkaloids

The structures of methylxanthines and methylurates which occur in plants are illustrated in Figures 14.1 and 14.2. Xanthine, xanthosine, and xanthosine-5′-monophosphate are, respectively, the nucleobase, nucleoside, and nucleotide involved in nucleotide

Plant Nucleotide Metabolism – Biosynthesis, Degradation, and Alkaloid Formation, First Edition.
Hiroshi Ashihara, Iziar A. Ludwig and Alan Crozier.

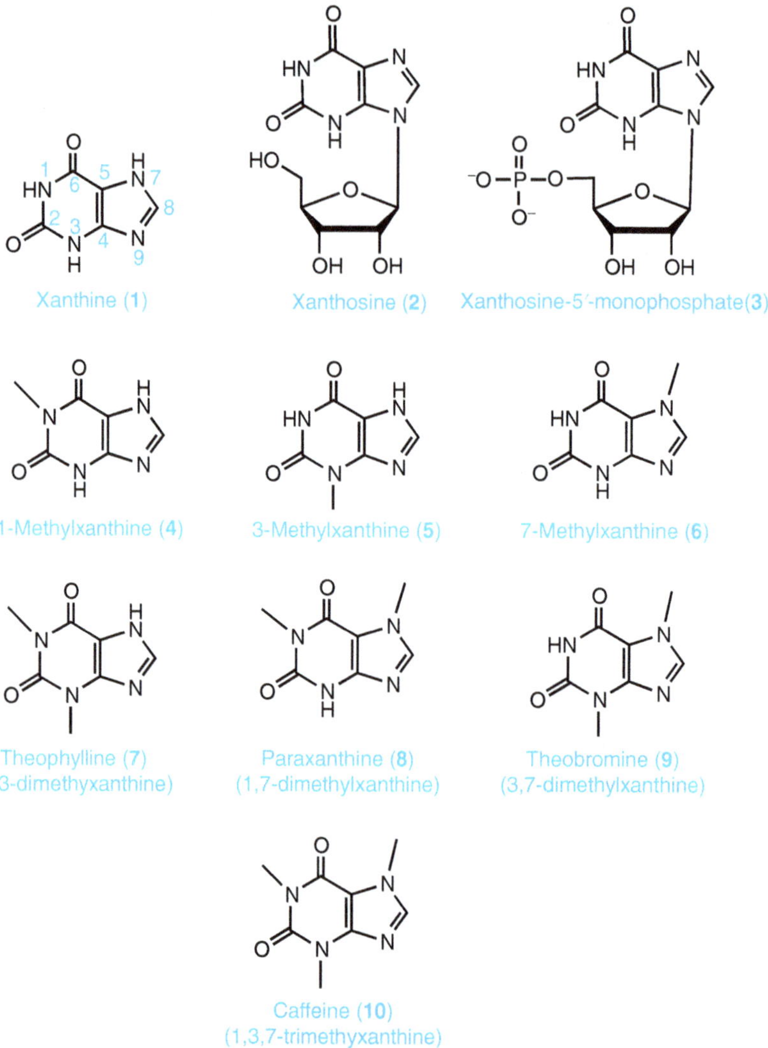

Figure 14.1 Structure of methylxanthines and related metabolites of plants. Xanthine, xanthonosine, xanthosine-5′-monophosphate, and mono-, di- and trimethylxanthines are illustrated. In others, 7-methylxanthine (7mXR) and 7-methylxanthosine-5′-monophosphate (7mXMP) also occur in plants.

metabolism. The monomethylxanthines, 1-methylxanthine, 3-methylxanthine, and 7-methylxanthine, and the dimethylxanthines, theophylline, paraxanthine, and theobromine are metabolites involved in the biosynthesis and/or catabolism of caffeine (Figure 14.1). The methylurates, 1,3,7-trimethylurate (aka in medical literature as 8-oxy-caffeine), theacrine, methylliberine, and libertine (Figure 14.2) are found in a limited number of species.

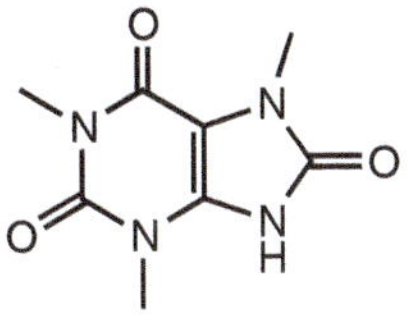

1,3,7-Trimethylurate (**11**)

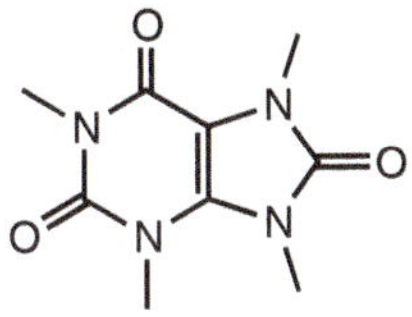

Theacrine (**12**)
[1,3,7,9-Tetramethylurate]

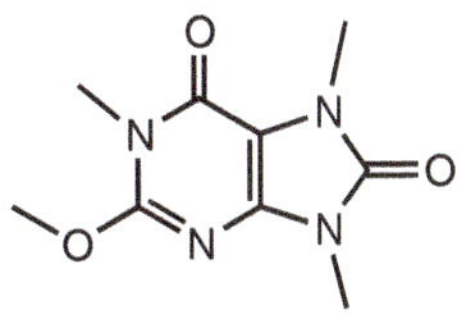

Methylliberine(**13**)
[*O*(2),1,7,9-Tetramethylurate]

Libertine (**14**)
[*O*(2),1,9-Trimethylurate]

Figure 14.2 Structure of urates and methylurates observed in plants.

14.3 Occurrence of Purine Alkaloids in Plants

Phylogenetic relationships of orders which include purine alkaloid-accumulating plants, are shown in Figure 14.3, where the order of the plant species is according to Smith et al. (2006) for ferns and APG III for Angiospermae (The Angiosperm Phylogeny 2009). Typical species and their common names and the major purine alkaloids present are summarized in Table 14.1.

14.3.1 Purine Alkaloids in Tea and Related Species

The purine alkaloid contents of *Camellia* species are presented in Table 14.2. Only four species of *Camellia* contain caffeine in young leaves (Nagata and Sakai 1984). Sizable amounts of caffeine (2–3% d.w.) occur in cultivated tea species, *Camellia sinensis* var. *sinensis*, *C. sinensis* var. *assamica* (aka *C. assamica*) and the wild Yunnan large leaf tea, *Camellia taliensis* (Table 14.2). Small amounts of caffeine (<0.02%) occur in *Camellia kissi*. Theobromine, accumulates in *Camellia ptilophylla* and *Camellia irrawadiensis*. A small amount of theobromine is found in young leaves of *C. sinensis*, where it is an intermediate in caffeine biosynthesis. The tetramethylurate, theacrine accumulates in leaves of *C. assamica* var. *kucha* where it comprises 75% of the total purine alkaloid pool. In addition to these *Camellia* species, detectable amounts of caffeine have been found in 17 samples in the Theaceae and Pentaphyllaceae families grown at the Botanical Gardens of the University of Tokyo (Ashihara et al. 2017; Deng et al. 2013).

It has been reported that more than 99% of the total caffeine of tea seedlings is found in the leaves, and only a small amount occurs in other parts of the plant. The concentration of caffeine is highest in mature leaves, while theobromine has been detected only in

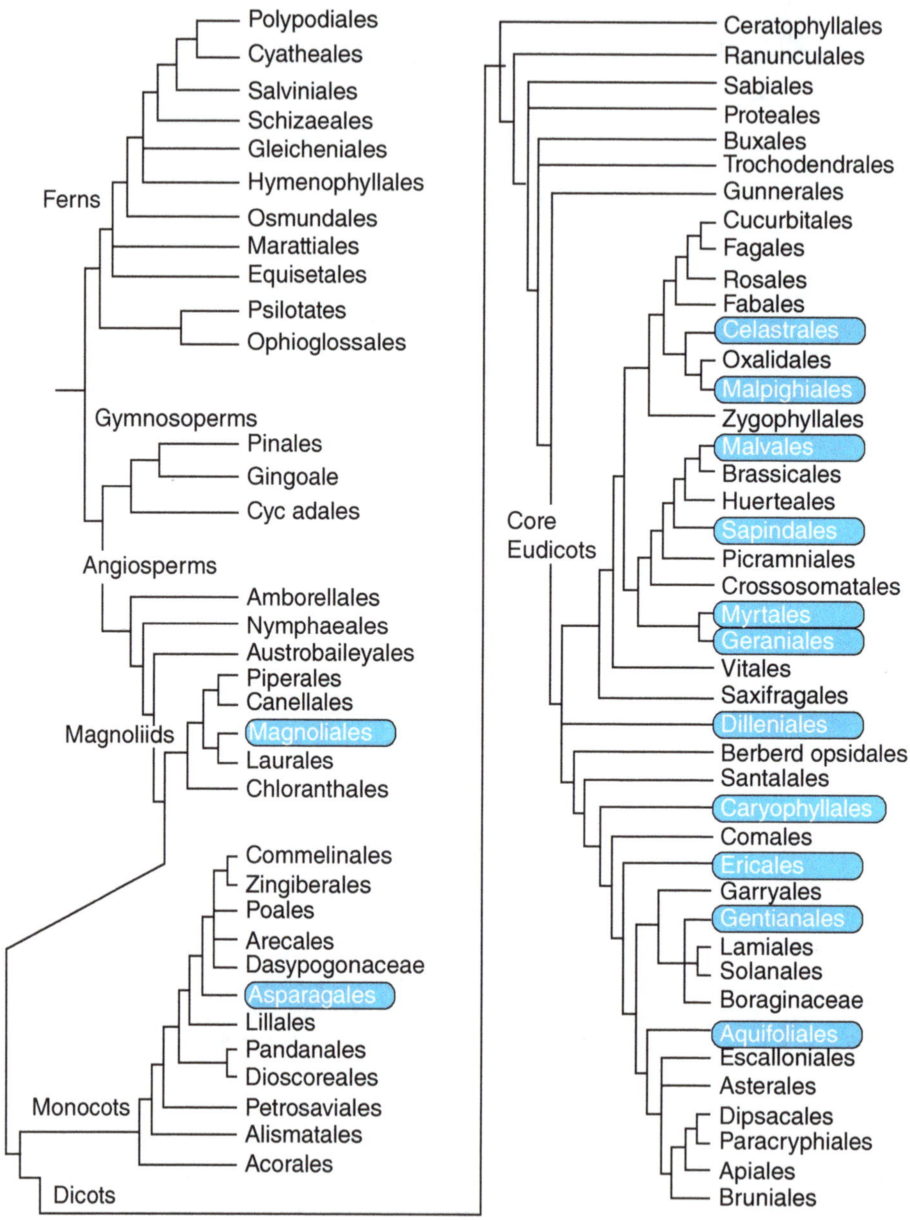

Figure 14.3 Phylogenetic relationship of xanthine alkaloid-accumulating plant orders. The blue boxes show the orders that include species accumulating caffeine and related xanthine alkaloids. Source: Adapted from Ashihara et al. (2017).

Table 14.1 Occurrence of purine alkaloids in the plant kingdom.

Order	Typical species	Common name	Products
Magnoliales	*Annona cherimola*	Cherimoya	Cf
Asparagales	*Scilla maritima*	Red squill	Cf
Celastrales	*Maytenus* sp.	Maytenus	Cf
Malpighiales	*Banisteriopsis caapi*	Caapi	Cf
Malvales	*Cola acuminata*	Kola nut	Cf > Tb
	Herrania purpurea	Monkey cocoa	Tc
	Theobroma cacao	Cacao (cocoa)	Tb > Cf
Sapindales	*Citrus paradisi*	Grapefruit	Cf > Tp > Tb > Px
	Paullinaia cupana	Guaraná	Cf > Tb > Tp
Myrtales	*Combretum jacquinii*	Red bushwillow	Cf
Geraniales	*Erodium cicutarium*	Pinweed	Cf
Dilleniales	*Davilla rugosa*	Fire vine	Cf
Caryophallales	*Cereus jamacaru*	Cactus	Cf
Ericales	*Camellia sinensis*	Tea	Cf
	Camellia assamica[a)]	Kucha	Tc > Cf
Gentinales	*Coffea arabica*	Coffee (arabica)	Cf
	Coffea canephora	Coffee (robusta)	Cf
Aquifoliales	*Ilex paraguariensis*	Maté	Cf
	Villaresia mucronata	Chilean citronella tree	Cf

Tb – theobromine, Tp – theophylline, Px – paraxanthine, Cf – caffeine, Tc – theacrine.
a) To simplify the description, *Camellia sinensis* var. *sinensis* and *C. sinensis* var. *assamica* are often referred to as *C. sinensis* and *C. assamica*.
Source: Adapted from Ashihara et al. (2017).

Table 14.2 Purine alkaloid content of the leaves of *Camellia* species. The values are shown as % d.w.

Species or cultivar	Cf	Tb	Tc
C. sinensis var. *sinensis*	2.8		
C. sinensis var. *assamica*[a)]	2.4		
C. taliensis	2.5		
C. kissi	<0.02		
C. irrawadiensis		0.8	
C. ptilophylla (Cocoa Tea)		6.8	
C. assamica var. *kucha* (kucha)	0.6	0.4	2.8
C. sasanqua			
C. japonica			

Cf, caffeine; Tb, theobromine; Tc, theacrine.
a) *C. sinensis* var. *assamica* is referred to as *C. assamica*.
Source: Based on the data of Ashihara et al. (2017).

Table 14.3 Endogenous levels of metabolites of purine alkaloids in *Camellia sinensis* leaves.

Metabolites	Young[a)]	Mature[b)]	Aged[c)]
XMP	29	43	38
Xanthosine	22	15	13
Xanthine	6	4	4
7-Methylxanthosine	179	23	22
7-Methylxanthine	67	16	16
3-Methylxanthine	5	38	2
1-Methylxanthine	2	14	8
Theobromine	1 170	228	174
Caffeine	36 400	20 600	13 700

Values are expressed as nmol g^{-1} f.w.
a) Young leaves, newly emerged small expanding leaves.
b) Mature leaves, young fully expanded leaves.
c) Aged leaves, 1-year-old leaves.
Source: Based on the data of Ashihara and Crozier (1999).

young leaves. In tea seeds almost all caffeine is located in the seed coat at a concentration of ~5 μmol g^{-1} f.w. (Ashihara and Kubota 1986).

In addition to theobromine, other intermediates involved in the biosynthesis and catabolism of caffeine, have been detected in *C. sinensis* leaves (Table 14.3). Young tea leaves contain 7-methylxanthosine, 7-methylxanthine, 3-methylxanthine and 1-methylxanthine. In mature leaves, 7-methylxanthosine and 7-methylxanthine levels decline, while 3-methylxanthine and 1-methylxanthine increase (Ashihara and Crozier 1999).

The anatomical localization of caffeine within young leaves of *C. sinensis* has been investigated using immunohistochemical methods and confocal scanning laser microscopy. The results demonstrated that caffeine is localized within vascular bundles, mainly in the precursor phloem (van Breda et al. 2013).

14.3.2 Purine Alkaloids in Coffee and Related Species

The genus *Coffea* has around 100 native species distributed in tropical and subtropical areas in Africa, and the most important economic species are *Coffea arabica* and *Coffea canephora*. *C. arabica* is unusual within the genus since it is the only species so far analysed that is self-compatible, and a natural polyploid (tetraploid) with $2n = 4x = 44$ chromosomes; it is considered to be a segmental allopolyploid because it presents a disomic inheritance and a regular meiotic behaviour. All other species in the genus are self-incompatible diploids with $2n = 2x = 22$ (Pinto-Maglio 2006). Commercial arabica coffee (*C. arabica*) is the major coffee produced in Brazil, Colombia, Mexico, Ethiopia, and Guatemala, while robusta coffee (*C. canephora*) is cultivated mainly in Vietnam, Indonesia, and the Ivory Coast (Petracco 2005).

Table 14.4 Purine alkaloid content of seeds of *Coffea* species.

Species or cultivar	Caffeine
C. arabica	0.8–1.8
C. canephora	2.0–3.0
C. liberica	1.4[a)]
C. dewevrei	1.2[b)]
C. racemosa	1.2
C. kapakata	0.72
C. eugenioides	0.55
C. sessiliflora	0.46
C. salvatrix	0.28
C. farafanganensis	0.09
C. homollei	0.06
C. pseudozanguebariae	0.02
C. perrieri	0.00

Values are expressed as % d.w.

a) *C. liberica* leaves, in addition to caffeine, contain <2-4 mg of theacrine and libertine

b) In *C. dewevrei*, small amounts of theacrine and libertine were also detected (Petermann and Baumann 1983).

Source: Based on Ashihara et al. (2017).

The caffeine content of the seeds of most cultivars of *C. arabica* and *C. canephora* range, respectively, from 0.8–1.8% and 2.0–3.0% d.w. (Table 14.4). Some *Coffea* species, *Coffea dewevrei* and *Coffea liberica*, have a similar caffeine content as *C. canephora*. In contrast, the seeds of other species contain less caffeine than *C. arabica*. In addition to the methylxanthines, mature leaves of *C. liberica*, *C. dewevrei*, and *Coffea abeokutae* contain the methyluric acids, theacrine and methylliberine (see structures in Figure 14.2) (Baumann et al. 1976; Petermann and Baumann 1983; Petermann et al. 1977).

Caffeine is distributed principally in leaves (41%) and cotyledons (58%) of coffee seedlings although trace amounts (<1%) are also found in the stems and roots (Zheng and Ashihara 2004). Young expanding leaves of *C. arabica* also contain theobromine. The content is slightly higher than that of caffeine in the young small-sized developing leaves, but the caffeine content becomes higher than that of theobromine during leaf development (Frischknecht et al. 1986; Fujimori and Ashihara 1994).

In fruits consisting of pericarp and seeds, Keller et al. (1972) reported that the caffeine content of the pericarp fell from 2.0 to 0.2% d.w., while it remained constant at 1.3% during the development of seeds in *C. arabica* fruits. Koshiro et al. (2006) monitored changes in caffeine levels in the pericarp and seeds of *C. arabica* and *C. canephora* during fruit ripening. In ripened fruits, 80% and 89% of total caffeine of *C. arabica* and *C. canephora*, respectively, was found in the seeds and the remainder was in the pericarp.

Table 14.5 Changes in caffeine contents in pericarp (or whole fruits) and seeds during development of fruits (stage A to G) of *Coffea arabica* and *Coffea canephora*.

	C. arabica (cv. Mokka)		*C. arabica* (cv. Catimor)		C. canephora	
Stage	Pericarp	Seeds	Pericarp	Seeds	Pericarp	Seeds
A	0.45		0.62		1.8	
B	1.4		2.8		4.0	
C	2.6		5.4		8.1	
D	7.6		11.7		11.6	
E	1.8	11.7	5.7	19.4	4.2	22.6
F	4.4	15.2	8.7	21.8	8.3	21.3
G	4.6	18.0	10.2	24.2	3.4	28.3

Caffeine contents are expressed as μmol per part. In stages A to D, caffeine contents in total fruit are shown as pericarp. The caffeine concentration if the mature stage (G) of *C. arabica* cv. Mokka, and *C. arabica* cv. Catimor and *C. canephora* are 1.0, 1.0, and 1.9% d.w., respectively (Koshiro et al. 2006).

Caffeine accumulates gradually in the seeds, but a similar build up does not occur in the pericarp (Table 14.5).

14.3.3 Purine Alkaloids in Maté

Young leaves of maté, aka, Yerba mate, (*Ilex paraguariensis*) are used to produce a beverage. It is traditionally consumed in central and southern regions of South America, primarily in Paraguay, as well as in Argentina, Uruguay, southern and central-western Brazil, and the Chaco region of Bolivia and southern Chile (IUCN 1998), although it is slowly increasing in popularity in North America and Europe.

Clifford and Ramirez-Martinez (1990) reported that the respective concentrations of caffeine and theobromine in dried maté leaves were 0.9–1.7% and 0.5–0.9% d.w. The presence of theobromine in various *Ilex* species was reported by Bohinc et al. (1975). They detected theobromine in the following *Ilex* species: *Ilex aquifolium*, *Ilex crenatu*, *Ilex caroliunu*, *Ilex cassine*, *Ilex perudo*, and *Ilex umbigua*. Filip et al. (1983) reported that theobromine, but not caffeine, was present in *Ilex argentina*. Edwards and Bennett (2005) examined the diversity of methylxanthine levels in *I. cassine* and *Ilex vomitoria*. The caffeine and theobromine contents were 0.12% and 0.22% in *I. cassine* and 0.56% and 0.11% d.w. in *I. vomitoria*.

The distribution of purine alkaloids within maté plants has been reported (Mazzafera 1994). Caffeine was found in developing and old leaves (0.2–0.9% d.w.), bark (0.2%), wood (0.06%), and in immature and mature fruits (0.01–0.04%). Smaller amounts of theobromine (<0.4%) and theophylline (<0.02%) were present in leaves, but theobromine was not detected in mature fruits, and theophylline did not accumulate in the fruits, bark, and wood. The concentrations of purine alkaloids in young leaves and immature fruits were higher than that occurring in old leaves and mature fruits. Young and old leaves collected from branches containing fruits showed significantly lower amounts of theobromine and caffeine than those from the branches not bearing fruits.

14.3.4 Purine Alkaloids in Cacao and Related Species

Seeds of cacao (*T. cacao*) are used in the manufacture of chocolate products (Crozier et al. 2012). Despite the similarities in their terms, cacao and cocoa are different. Cacao is a plant name while cocoa is the name of its derived commercial products. The three major cultivated cacao varieties are forastero, criollo, and trinitario (Pereira-Caro et al. 2013). Theobromine is the dominant purine alkaloid in seeds of cacao. Timbie et al. (1978) analysed seeds of different *T. cacao* varieties and the total purine alkaloid content ranged from 2.4 to 5.0% of defatted cocoa with an average of 3.7%. Theobromine comprised 52–99% of total purine alkaloids averaging 87%. Hammerstone Jr et al. (1994) examined the purine alkaloid content of several genotypes representing the three horticultural races of *T. cacao*. Theobromine and caffeine concentrations were varied as follows: criollo type (1.5–3.3% theobromine, 0.3–1.3% of defatted seeds), forastero type (0.9–1.2% theobromine, 0.08–1.2% of defatted seeds), and trinitario type (0.9–1.2% theobromine, 0.3–0.6% of defatted seeds).

Theacrine is a purine alkaloid found in seeds of many *Theobroma* species except for *Theobroma cacao*: *Theobroma speciosum* (0.12% of defatted seed), *Theobroma microcarpum* (0.02%), *Theobroma angustifolium* (0.06%), *Theobroma grandiforum* (cupuassu, 0.26%), *Theobroma subincanaum* (0.28%), and *Theobroma mammosum* (0.06%). Immature seeds of *Theobroma bicolor* (jaguar tree) accumulate theobromine (0.26%), but this is replaced by theacrine in mature seeds (0.35%) (Hammerstone Jr et al. 1994). *T. bicolor* has been one of the most useful species, aside from *T. cacao*, as it has been used as a food since prehistoric times (Kufer and McNeil 2009).

Herrania species, belonging to the Sterculioideae, a subfamily of the Malvaceae, are morphologically similar to *Theobroma* plants. Hammerstone et al. (1994) reported that nine species of *Herrania* contained theacrine (0.23–2.0%). The species with a relatively high concentration of theacrine were *Herrania purpurea* (2.0%), *Herrania cuatrecasana* (1.9%), *Herrania balaensis* (1.8%), and *Herrania columbia* (1.7%). Trace amounts of caffeine were found in some species, but no theobromine was detected in any of the *Herrania* species.

Changes in the levels of theobromine and caffeine during the development of leaves of cacao have been reported (Hammerstone et al. 1994; Koyama et al. 2003). The theobromine content is highest in the young leaves and decreases gradually with age. In contrast, the level of caffeine remained almost constant, but both of the purine alkaloids disappeared in the oldest leaves.

Koyama et al. (2003) surveyed several methylxanthines, in addition to theobromine and caffeine, in leaves at four different stages of developmental: young small red leaves (stage I), developing pale green intermediate sized leaves (stage II), fully developed green leaves (stage III) from flush shoots, and aged leaves (stage IV) from 1-year-old shoots (Table 14.6). A high concentration of purine alkaloids was found in young leaves (stage I), where the major alkaloid was theobromine followed by caffeine. In addition, small amounts of 1-methylxanthine, 3-methylxanthine, 7-methylxanthine,

Table 14.6 Endogenous levels of metabolites of purine alkaloids in cacao leaves.

Metabolites	Stage I[a)]	Stage II[b)]	Stage III[c)]	Stage IV[d)]
1-Methylxanthine		25	8	3
3-Methylxanthine	26	23	2	2
7-Methylxanthine	114	15	9	5
7-Methylxanthosine	9			
Theophylline	2	1	6	12
Theobromine	4530	112	33	77
Paraxanthine				
Caffeine	754	112	36	10

The contents are expressed as nmol g^{-1} f.w.
a) Young small red leaves.
b) Developing pale green leaves.
c) Fully developed green leaves.
d) Aged leaves.
Source: Koyama et al. (2003).

7-methylxanthosine, and theophylline were also present although paraxanthine did not accumulate in detectable quantities. The concentrations and the absolute amounts of these purine alkaloids decreased markedly with leaf development. More than 95% of purine alkaloids present in the stage I leaves were absent in aged leaves.

The respective theobromine and caffeine concentrations in young leaves (stage I) were 0.037 and 0.007% d.w. Consistent with this observation, Gurney et al. (1991) reported that small amounts of theobromine (0.051% d.w.) and trace amounts of caffeine were present only in very young, red flush leaves of cacao plants, and that levels declined by ~90% when the leaves became green. Thus, small amounts of theobromine and caffeine appear in young cacao leaves, but their levels fall substantially with leaf development. This trend is, therefore, different from those occurring in tea (Table 14.3) and coffee leaves where the decline in purine alkaloid concentrations with age is much less marked

In marked contrast to the leaves, the seeds of cacao accumulate purine alkaloids (Bucheli et al. 2001; Gurney et al. 1991; Hammerstone Jr et al. 1994; Pereira-Caro et al. 2013; Senanayake and Wijesekera 1971; Sotelo and Alvarez 1991; Zheng et al. 2004). Zheng et al. (2004) compared the purine alkaloid contents in cacao fruits at different developmental stages: stage A (young small size fruits, ~2 g f.w.), stage B (medium size fruits, ~100 g f.w.), and stage C (~500 g f.w.). For example, whole fruits (stage A), pericarp and seed tissue (mainly endosperm) (stage B), and pericarp, placenta, embryo (mainly cotyledons), and seed coat (stage C) were used (Table 14.7).

In stage A fruits, theobromine was the major purine alkaloid, followed by caffeine. In stage B fruits, theobromine was the predominant purine alkaloid in the pericarp but caffeine comprised 74% of the purine alkaloids in the seeds. In stage C fruits, the highest concentration of purine alkaloids was found in the seeds (cotyledons and an embryonic axis), in which the major component was theobromine (22 µmol g^{-1} f.w.) and the next was caffeine (4.9 µmol g^{-1} f.w.). Compared with the seeds, the total alkaloid content of the pericarp was very low (0.05 µmol g^{-1} f.w.); the major components were 3-methylxanthine (44%), caffeine (38%), and theobromine (15%). On a dry weight

Table 14.7 Endogenous levels of metabolites of purine alkaloids in cacao fruits.

Stage	Stage A[a] Fruit	Stage B[b] Pericarp	Seed	Stage C[c] Pericarp	Seed coat	Cotyledon	Placenta
1-Methylxanthine					4	21	
3-Methylxanthine	1			23	5	25	
7-Methylxanthine	1	2		1	6	23	
7-Methylxanthosine						12	
Theophylline	2	2	18		2		11
Theobromine	671	206	10	8	779	21 900	238
Paraxanthine	1				3	2	
Caffeine	89	80	82	20	161	4860	87

The contents are expressed as nmol g^{-1} f.w.
a) Young small size fruits (~2 g f.w.).
b) Developing medium size fruits (~100 g f.w.).
c) Fully developed fruits (~500 g f.w.).
Source: Zheng et al. (2004).

basis, the theobromine content in seeds of stage C fruits was equivalent to 0.62% d.w. This value is comparable to the 0.43–0.89% content of theobromine in maturing cacao seeds 125 days after pollination (Bucheli et al. 2001). As well as these major alkaloids, small amounts of 1-, 3-, and 7-methylxanthines, 7-methylxanthosine, theophylline, and paraxanthine were also detected (Table 14.7).

14.3.5 Purine Alkaloids in Cola Species

Like cacao, cola is a member of the Sterculiaceae in which they belong to a tropical African genus. The genus comprises about 140 species and the most commonly consumed are *Cola acuminata*, *Cola nitida*, and *Cola anomala*. Belliardo et al. (1985) reported that caffeine occurred in seeds of *Cola* spp. at 1.4–2.2% d.w. Subsequently, a more detailed investigation was performed by Niemenak et al. (2008). The respective mean contents of caffeine and theobromine were 1.1% and 0.04% in *C. acuminata*, 1.4% and 0.05% in *C. nitida*, and 0.7% and 0.06% in *C. anomala*. Species in this genus are sometimes referred to as Kola tree or Kola nut. The caffeine-containing fruit produced by the trees is often used as a flavouring ingredient in beverages (Duke 2000).

14.3.6 Purine Alkaloids in Guaraná and Related Species

Sapindales, *Paullinia* (Sapindaceae) and *Citrus* (Rutaceae) species also contain purine alkaloids. Seeds of guaraná (*Paullinia cupana* var. *sorbilis*), which are utilized commercially in South America as a beverage and flavouring ingredient, contain high concentrations (2.5–6.5% d.w.) of caffeine with lower amounts of theobromine and theophylline (Schimpl et al. 2014). Weckerle et al. (2003) screened 34 species of *Paullinia* and related genera and found only three species, *P. cupana*, *Paullinia yoco*, and *Paullinia pachycarpa* contained purine alkaloids. In *P. cupana* seeds (cotyledons), caffeine (4.3% d.w.) was the major alkaloid along with very small amounts

Table 14.8 Occurrence of purine alkaloids in different organs of guaraná.

	Theobromine	Theophylline	Caffeine
Young leaves	1.26		<0.01
Mature leaves	0.03		
Axis cortex	<0.01		
Stem cortex	0.09		0.05
Floral bud	0.14		0.02
Seed (cotyledon)	0.02	<0.01	4.28
Testa	0.03	<0.01	1.64
Aril			
Pericarp	0.20	<0.01	0.02
Septa	0.22	0.08	0.67

The contents are shown as % d.w.
Source: Adapted from data of Weckerle et al. (2003).

of theobromine (0.015%) and theophylline (0.007%). Caffeine also occurs in the testa and septa while, in contrast, theobromine was detected mainly in the leaves, stem, and pericarp (Table 14.8) (Baumann et al. 1995; Weckerle et al. 2003).

In *Paullinia yoco*, caffeine is found principally in the stem cortex (0.45% d.w.), stem wood (0.28%), and floral buds (0.19%). Smaller amounts of theobromine (0.03–0.05%) accumulate in the mature leaves, axis cortex, and floral buds, while theophylline has not been detected in any part of this plant. *P. pachycarpa* accumulated only small amounts of theobromine (<0.03%) in the leaves and stems, and no purine alkaloids were detected in the fruits (Weckerle et al. 2003).

14.3.7 Purine Alkaloids in Citrus Species

Stewart (1985) reported the presence of a low concentration of caffeine in the flower buds of eight cultivars of citrus and in the leaves of Valencia orange (*Citrus sinensis*). The caffeine content of flower buds varied from 19 to 50 $\mu g\, g^{-1}$ d.w. and the highest amount was found in Lisbon lemon flowers. The concentration in leaves was 6 $\mu g\, g^{-1}$ d.w.

Detailed studies of the distribution of purine alkaloids in citrus flowers were reported by Kretschmar and Baumann (1999). They detected substantial amounts of caffeine and theophylline and trace amounts of theobromine and paraxanthine in the flowers of *Citrus paradisi* (grapefruit), *Citrus maxima* (shaddock), *Citrus limon* (lemon), and *Citrus trifoliata* (syn. *Poncirus trifoliata*, trifoliate orange). Caffeine was undetectable in the small buds but it appeared at anthesis. Purine alkaloids, mainly caffeine, were located in the stamens. For example, the highest amount of purine alkaloids was detected in the stamens (96.5%) of *C. limon* flowers and the remainder occurred in the petals (2.5%) and pistils (1.0%). Concentrations of caffeine in different floral organs of *Citrus* species are presented in Table 14.9. In two instances, stamens were separated into the filament and anthers and this revealed that caffeine and theophylline were located mainly in the anthers at concentrations in *C. paradisi* and *C. maxima* which corresponded to 0.7–0.8% d.w. Purine alkaloids in pollen were analysed in *Citrus*

Table 14.9 Occurrence of purine alkaloids in different organs of *Citrus* sp.

	Grapefruit (*C. paradise*)		Shaddock (*C. maxima*)		Trifoliate orange (*C. trifoliate*)		Lemon (*C. limon*)	
	Cf	Tp	Cf	Tp	Cf	Tp	Cf	Tp
Stamens	3.23	0.17	1.10	0.31	0.081	0.06	1.42	0.12
Filament	1.92	0.14	0.85	0.01	nd	nd	nd	nd
Anther	8.55	1.49	7.90	2.75	nd	nd	nd	nd
Pollen	nd	nd	6.86	1.92	nd	nd	nd	nd
Nectar	0.49	0.06	0.09	<0.01	nd	nd	0.06	nd

Content is $\mu mol\ g^{-1}$ f.w. Cf, caffeine; Tp, theophylline; nd, data not obtained.
Source: Adapted from data of Kretschmar and Baumann (1999).

medica (citron). Caffeine (6.9 $\mu mol\ g^{-1}$ f.w.) and theophylline (1.9 $\mu mol\ g^{-1}$ f. w.) were found in the microspores and small amounts (<0.6 $\mu mol\ ml^{-1}$) of purine alkaloids, mainly caffeine, were also detected in the nectar of citrus flowers (Kretschmar and Baumann 1999).

14.3.8 Purine Alkaloids in Other Plants

In addition to the species in the five orders of plants mentioned above, older literature has reported the occurrence of caffeine in plants involved in eight other orders (Kihlman 1977). The species and family names are as follows:

- Magnoliales: *Annona cherimola* (Annonaceae)
- Asparagales: *Scilla maritima* (syn. *Drimia maritima*) (Asparagaceae)
- Celastrales: *Maytenus* sp. (Celastraceae)
- Malpighiales: *Banisteriopsis caapi* (Malpighiaceae), *Turnera ulmifolia* (Passifloraceae), *Populus alba* (Salicaceae)
- Myrtales: *Combretum jacquinii*, *C. loeflingii* (Combretaceae)
- Geraniales: *Erodium cicutarium* (Geraniaceae)
- Dilleniales: *Davilla rugosa* (Dilleniaceae)
- Caryophallales: *Cereus jamacaru*, *Harrisia adscendens*, *Leocereus bahiensis*, *Pilocereus gounellei*, *Trichocereus* sp. (Cactaceae)

The occurrence of purine alkaloids has been investigated mainly in plants of economic significance (Ashihara et al. 2013) and medicinal plants grown on the American continent (Kihlman 1977). Their actual occurrence in the plant kingdom may well be more diverse. On the other hand, most of the identifications and quantifications reported in the older literature were based on paper chromatography and colorimetric-based techniques, and have not been confirmed with more modern analytical methodology.

Recently, two phytochemical studies made use of HPLC coupled with tandem mass spectrometry to investigate the occurrence of caffeine. Pierattini et al. (2016) reported the presence of caffeine (<0.4 $\mu g\ g^{-1}$ f.w.), theobromine (<1.5 $\mu g\ g^{-1}$ f.w.), and theophylline (~60 $\mu g\ g^{-1}$ f.w.) in the roots of *P. alba* plantlets (Order Malpighiales). These purine alkaloids were also detected in the leaves and stems at lower concentrations. Small amounts of caffeine (86 $ng\ g^{-1}$ d.w.) and theobromine (70 $ng\ g^{-1}$ d.w.) have been detected in a cactus, *Epiphyllum oxypetalum* (order Caryophallales) (Pereira-Caro and Ashihara 2016, purine alkaloids in cactus, personal notebook).

Strictly speaking, methylxanthine-synthesizing and accumulating plants are not the same. Therefore, a different approach, such as the determination of the occurrence of gene expression and enzyme activity may be required to clarify whether or not the capacity to synthesize purine alkaloids is present in a particular species. This type of confirmation is necessary for chemotaxonomic analysis of caffeine and related alkaloids. Caffeine, used in beverages, pharmaceutical, and personal care products, is a recognized contaminant of aquatic ecosystems (Bruton et al. 2010; Pierattini et al. 2016). Thus, small amounts might be found in plants grown in the wild that do not synthesize purine alkaloids. To determine the occurrence of biosynthesis is therefore necessary to confirm the origin of the xanthine alkaloids.

Biosynthesis of purine alkaloids is well-documented in *Camellia* and *Coffea* species, although different routes may be operative in other species. Convergent evolution may have occurred in the case of caffeine biosynthesis. The biosynthesis pathways are described in Chapter 15.

14.4 Summary

Purine alkaloids, namely methylxanthines and methylurates, occur in a variety of plant species. High concentrations do accumulate; caffeine in coffee (*C. arabica* and *C. canephora*), tea (*C. sinensis*), maté (*I. paraguariensis*), guaraná (*P. cupana*) and *Citrus* sp.; and theobromine in cacao (*T. cacao*) and cocoa tea (*Camellia ptilophylla*). The methylurate, theacrine occurs in kucha (*C. assamica*, cv. *kucha*). Most purine alkaloids are located in leaves, seeds, and flowers of the ~100 species that have been investigated.

References

Anthony, F., Clifford, M.N., and Noirot, M. (1993). Biochemical diversity in the genus *Coffea* L.: chlorogenic acids, caffeine and mozambioside contents. *Genetic Resour. Crop Evol.* 40: 61–70.

Ashihara, H. and Crozier, A. (1999). Biosynthesis and metabolism of caffeine and related purine alkaloids in plants. *Adv. Bot. Res.* 30: 117–205.

Ashihara, H. and Kubota, H. (1986). Patterns of adenine metabolism and caffeine biosynthesis in different parts of tea seedlings. *Physiol. Plant.* 68: 275–281.

Ashihara, H., Yokota, T., and Crozier, A. (2013). Biosynthesis and catabolism of purine alkaloids. *Adv. Bot. Res.* 68: 111–138.

Ashihara, H., Mizuno, K., Yokota, T., and Crozier, A. (2017). Xanthine alkaloids: occurrence, biosynthesis, and function in plants. In: *Progress in the Chemistry of Organic Natural Products 105* (eds. A.D. Kinghorn, H. Falk, S. Gibbons and J.i. Kobayashi), 1–88. Cham: Springer.

Baumann, T.W., Oechslin, M., and Wanner, H. (1976). Caffeine and methylated uric acids: chemical patterns during vegatative development of *Coffea liberica*. *Biochem. Physiol. Pflanz.* 170: 217–225.

Baumann, T.W., Schulthess, B.H., and Hanni, K. (1995). Guarana (*paullinia cupana*) rewards seed dispersers without intoxicating them by caffeine. *Phytochemistry* 39: 1063–1070.

Belliardo, F., Martelli, A., and Valle, M.G. (1985). HPLC determination of caffeine and theophylline in *Paullinia cupana* Kunth (guarana) and *Cola spp.* samples. *Z. Leben. Unt. Forsch.* 180: 398–401.

Bohinc, P., Korbar–Šmid, J., and Bizilj–Kremžar, L. (1975). Theobromin in *Ilex perado blättern. Planta Med.* 28: 374–378.

van Breda, S.V., Merwe, C.F., Robbertse, H., and Apostolides, Z. (2013). Immunohistochemical localization of caffeine in young *Camellia sinensis* (L.) O. Kuntze (tea) leaves. *Planta* 237: 849–858.

Bruton, T., Alboloushi, A., Garza, B.d.l. et al. (2010). Fate of caffeine in the environment and ecotoxicological considerations. In: *Contaminants of Emerging Concern in the Environment: Ecological and Human Health Considerations*, vol. 1048 (ed. R.U. Halden), 257–273. American Chemical Society.

Bucheli, P., Rousseau, G., Alvarez, M. et al. (2001). Developmental variation of sugars, carboxylic acids, purine alkaloids, fatty acids, and endoproteinase activity during maturation of *Theobroma cacao* L. seeds. *J. Agric. Food Chem.* 49: 5046–5051.

Clifford, M.N. and Ramirez-Martinez, J.R. (1990). Chlorogenic acids and purine alkaloids contents of Maté (*Ilex paraguariensis*) leaf and beverage. *Food Chem.* 35: 13–21.

Crozier, A., Ashihara, H., and Tomas-Barberan, F. (2012). *Teas, Cocoa and Coffee: Plant Secondary Metabolites and Health.* Oxford: Wiley-Blackwell.

Deng, W.-W., Jin, Y., Yuan, Y., and Zhang, Z.-Z. (2013). Profile of purine metabolism and purine alkaloid biosynthesis in *Schima* and *Eurya* plants. *Bull. Bot. Res.* 33: 410–415.

Deng, W.-W., Rakotomalala, J.J., Nagai, C., and Ashihara, H. (2017). Caffeine biosynthesis and purine metabolism in leaves of *Mascarocoffea* species. *Eur. Chem. Bull.* 6: 223–228.

Duke, J.A. (2000). *Handbook of Nuts: Herbal Reference Library.* Boca Raton FL: CRC Press.

Edwards, A.L. and Bennett, B.C. (2005). Diversity of methylxanthine content in *Ilex cassine* L. and *Ilex vomitoria* ait.: assessing sources of the north American stimulant cassina. *Econom. Bot.* 59: 275–285.

Filip, R., de Iglesias, D.I.A., Rondina, R.V.D., and Coussio, J.D. (1983). Análisis de las Hojas y Tallos de *Ilex argentina* Lillo. *I. Xantinas. Acta Farm. Bonaer.* 2: 87–90.

Frischknecht, P.M., Ulmer-Dufek, J., and Baumann, T.W. (1986). Purine alkaloid formation in buds and developing leaflets of *Coffea arabica*: expression of an optimal defence strategy? *Phytochemistry* 25: 613–616.

Fujimori, N. and Ashihara, H. (1994). Biosynthesis of theobromine and caffeine in developing leaves of *Coffea arabica*. *Phytochemistry* 36: 1359–1361.

Gurney, K.A., Evans, L.V., and Robinson, D.S. (1991). Extraction of purine alkaloids from cocoa tissues and determination by high-performance liquid chromatography. *Phytochem. Anal.* 2: 15–19.

Hammerstone, J.F. Jr., Romanczyk, L.J. Jr., and Aitkent, W.M. (1994). Purine alkaloid distribution within *Herrania* and *Theobroma*. *Phytochemistry* 35: 1237–1240.

IUCN, 1998. *Ilex paraguariensis.* The IUCN Red List of Threatened Species 1998: e.T32982A9740718. Doi: 10.2305/IUCN.UK.1998.RLTS.T32982A9740718.en.

Kato, M., Mizuno, K., Fujimura, T. et al. (1999). Purification and characterization of caffeine synthase from tea leaves. *Plant Physiol.* 120: 579–586.

Kato, M., Mizuno, K., Crozier, A. et al. (2000). Caffeine synthase gene from tea leaves. *Nature* 406: 956–957.

Keller, H., Wanner, H., and Baumann, T.W. (1972). Kaffeinsynthese in Fruchten und Gewebekulturen von *Coffea arabica*. *Planta* 108: 339–350.

Kihlman, B.A. (1977). Occurence and biosynthesis of methylated oxypurines in plants. In: *Caffeine and Chromosome*, 11–18. Amsterdam: Elsevier.

Koshiro, Y., Zheng, X.Q., Wang, M. et al. (2006). Changes in content and biosynthetic activity of caffeine and trigonelline during growth and ripening of *Coffea arabica* and *Coffea canephora* fruits. *Plant Sci.* 171: 242–250.

Koyama, Y., Tomoda, Y., Kato, M., and Ashihara, H. (2003). Metabolism of purine bases, nucleosides and alkaloids in theobromine-forming *Theobroma cacao* leaves. *Plant Physiol. Biochem.* 41: 977–984.

Kretschmar, J.A. and Baumann, T.W. (1999). Caffeine in citrus flowers. *Phytochemistry* 52: 19–23.

Kufer, J. and McNeil, C.L. (2009). The jaguar tree (*Theobroma bicolor* Bonpl.). In: *Chocolate in Mesoamerica: A Cultural History of Cacao* (ed. C.L. McNeil), 1–542. Oxford: Oxford University Press.

Lu, J.-L., Wang, D.-M., Shi, X.-G. et al. (2009). Determination of purine alkaloids and catechins in different parts of *Camellia assamica* var. *kucha* by HPLC-DAD/ESI-MS/MS. *J. Sci. Food Agric.* 89: 2024–2029.

Mazzafera, P. (1994). Caffeine, theobromine and theophylline distribution in *Ilex paraguariensis*. *Rev. Bras. Fis. Veg.* 6: 149–151.

Mazzafera, P. and Carvalho, A. (1992). Breeding for low seed caffeine content of coffee (*Coffea* L.) by interspecific hybridization. *Euphytica* 59: 55–60.

Nagata, T. and Sakai, S. (1984). Differences in caffeine, flavonols and amino acids contents in leaves of cultivated species of *Camellia*. *Jpn. J. Breed.* 34: 459–467.

Niemenak, N., Onomo, P.E., Fotso et al. (2008). Purine alkaloids and phenolic compounds in three *Cola* species and *Garcinia kola* grown in Cameroon. *S. Afr. J. Bot.* 74: 629–638.

O'Connell, F.D. (1969). Isolation of caffeine from Banisteriopsis inebrians (Malpighiaceae). *Naturwissenschaften* 56: 139.

Oudry, V. (1927). Thein, eine organische Salzbase im thee (*Thea chinesis*). *Mag. Pharm.* 19: 49–50.

Pereira-Caro, G., Borges, G., Nagai, C. et al. (2013). Profiles of phenolic compounds and purine alkaloids during the development of seeds of *Theobroma cacao* cv. Trinitario. *J. Agric. Food Chem.* 61: 427–434.

Petermann, J. and Baumann, T.W. (1983). Metabolic relations between methylxanthines and methyluric acids in *Coffea*. *Plant Physiol.* 73: 961–964.

Petermann, J., Baumann, T.W., and Wanner, H. (1977). A new tetramethyluric acid from *Coffea liberica* and *C. Dewevre*. *Phytochemistry* 16: 620–621.

Petracco, M. (2005). Our everyday cup of coffee: the chemistry behind its magic. *J. Chem. Educ.* 82: 1161.

Pierattini, E.C., Francini, A., Raffaelli, A., and Sebastiani, L. (2016). Degradation of exogenous caffeine by *Populus alba* and its effects on endogenous caffeine metabolism. *Environ. Sci. Pollut. Res.* 23: 7298–7307.

Pinto-Maglio, C.A.F. (2006). Cytogenetics of coffee. *Braz. J. Plant Physiol.* 18: 37–44.

Runge, F.F. (1820). *Neueste phytochemische Entdeckungen zur Begründung einer wissenschaftlichen Phytochemie Phytochemische Entdeckungen*, 204. Berlin: Reimer.

Schimpl, F.C., Kiyota, E., Mayer, J.L.S. et al. (2014). Molecular and biochemical characterization of caffeine synthase and purine alkaloid concentration in guarana fruit. *Phytochemistry* 105: 25–36.

Senanayake, U.M. and Wijesekera, R.O.B. (1971). Theobromine and caffeine content of the cocoa bean during its growth. *J. Sci. Food Agr.* 22: 262–263.

Smith, A.R., Pryer, K.M., Schuettpelz, E. et al. (2006). A classification for extant ferns. *Taxon* 55: 705–731.

Sotelo, A. and Alvarez, R.G. (1991). Chemical composition of wild Theobroma species and their comparison to the cacao bean. *J. Agric. Food Chem.* 39: 1940–1943.

Stewart, I. (1985). Identification of caffeine in citrus flowers and leaves. *J. Agric. Food Chem.* 33: 1163–1165.

The Angiosperm Phylogeny Group (2009). An update of the angiosperm phylogeny group classification for the orders and families of flowering plants: APG III. *Bot. J. Linn. Soc.* 161: 105–121.

Timbie, D.J., Sechrist, L., and Keeney, P.G. (1978). Application of high-pressure liquid chromatography to the study of variables affecting theobromine and caffeine concentrations in cocoa beans. *J. Food Sci.* 43: 560–565.

Von Giese, F. (1820). Vermischte Notizen. 1. Kaffeestoff und Salzgehalt des Quassia Extrakts. *Annu. Chem.* 4: 240–241.

Weckerle, C.S., Stutz, M.A., and Baumann, T.W. (2003). Purine alkaloids in Paullinia. *Phytochemistry* 64: 735–742.

Willaman, J. J., Schubert, B. G., 1961. Alkaloid-bearing plants and their contained alkaloids. Technical Bulletin No. 1234. Agricultural Reserch Service, US Department of Agriculture.

Zheng, X.Q. and Ashihara, H. (2004). Distribution, biosynthesis and function of purine and pyridine alkaloids in *Coffea arabica* seedlings. *Plant Sci.* 166: 807–813.

Zheng, X.Q., Ye, C.X., Kato, M. et al. (2002). Theacrine (1,3,7,9-tetramethyluric acid) synthesis in leaves of a Chinese tea, kucha (*Camellia assamica* var. kucha). *Phytochemistry* 60: 129–134.

Zheng, X.Q., Koyama, Y., Nagai, C., and Ashihara, H. (2004). Biosynthesis, accumulation and degradation of theobromine in developing *Theobroma cacao* fruits. *J. Plant Physiol.* 161: 363–369.

15

Biosynthesis of Purine Alkaloids

15.1 Introduction

Although caffeine was isolated from tea and coffee in the early 1820s, the biosynthetic pathway by which it is produced from purine nucleotides has been the topic of much speculation and was not fully established until a highly purified *N*-methyltransferase, caffeine synthase, was isolated from tea leaves (Kato et al. 1999) after which a gene encoding the enzyme was cloned (Kato et al. 2000). The pathway of caffeine biosynthesis from xanthosine, a catabolite of purine nucleotides, has now been elucidated in both coffee and tea.

15.2 A Brief History of Caffeine Biosynthesis Research

Among the many alkaloids produced by plants, arguably caffeine is the one which has attracted most interest over the years. James Bonner speculated in his famous book, '*Plant Biochemistry*' that caffeine was synthesized by a xanthine → heteroxanthine (7-methylxanthine) → theobromine → caffeine pathway (Bonner 1950). The proposed pathway was based on the following: (i) the occurrence of heteroxanthine in *Beta vulgaris* (Bresler 1904), and (ii) the conversion of theobromine to caffeine by *Theobroma cacao* seedlings that had been reported by Weevers (1930).

Anderson and Gibbs (1962) and Inoue and co-workers (Inoue 1972; Inoue and Kawamura 1961; Inoue et al. 1960) demonstrated the incorporation of various precursors into caffeine by leaves of coffee and tea. Up to 1960 the carbon precursors utilized in the biosynthesis of the purine ring had been identified in birds, mammals, bacteria, and yeast, but there was no definitive information on the biosynthesis of purines in plants. In addition to general purine compounds, such as ATP and GTP, it was known that some higher plants synthesized methylated purines, including caffeine, but there was no information on the relationship between formation of the purine ring and its constituent methyl groups.

Anderson and Gibbs (1962), using ^{14}C-labelled CO_2, glycine, serine, HCOONa (sodium formate), HCHO (methanal), CH_3OH (methanol), and methionine, demonstrated that caffeine was synthesized from the same carbon precursors utilized for purines while the methyl groups are derived from a different source. Inoue et al. (1960) reported incorporation of root-absorbed $(^{15}NH_4)_2SO_4$ (ammonium sulfate) into caffeine in leaves of tea seedlings. They also demonstrated the biosynthesis of caffeine

Plant Nucleotide Metabolism – Biosynthesis, Degradation, and Alkaloid Formation, First Edition.
Hiroshi Ashihara, Iziar A. Ludwig and Alan Crozier.

from $(^{15}NH_4)_2SO_4$ and ^{14}C-glycine by excised shoots (Inoue 1972; Inoue and Kawamura 1961). In addition, it was established that the methyl groups are derived from [methyl-^{14}C]methionine (Inoue and Adachi 1962). Caffeine biosynthesis from several purine precursors was investigated by Russian scientists using tea leaves, and similar conclusions were drawn (Proiser and Serenkov 1963; Serenkov and Proiser 1961).

In the 1970s research on caffeine biosynthesis was carried out using intact plant tissues and tissue cultures. These studies confirmed that the purine ring of caffeine was produced from the same precursors that were used for general purine metabolism, and several caffeine biosynthesis pathways from purine compounds were postulated. Ogutuga and Northcote (1970) at the University of Cambridge proposed two routes: one derived from nucleic acids, namely methylated nucleic acids → 7-methyl-GMP (guanosine monophosphate) → 7-methylguanosine → 7-methylxanthosine → 7-methylxanthine → theobromine → caffeine; with another leading from xanthine in the purine pool: xanthine → 3-methylxanthine → theophylline → caffeine. In contrast, Suzuki and Takahashi (1976c) at Kyoto University proposed other pathways: (i) nucleic acids (tRNA) → 1-methyl-AMP (adenosine monophosphate) → 1-methylxanthine → theophylline → caffeine, and (ii) purine nucleotides → 7-methyl-AMP or 1-methyl-GMP → 7-methylxanthine → theobromine → caffeine. However, several pieces of evidence seriously questioned the validity of these proposals (Looser et al. 1974; Suzuki and Takahashi 1976b).

In contrast to these studies with plant tissues, cell-free systems proved more amenable in obtaining direct proof for the involvement of *S*-adenosyl-L-methionine (SAM) and the sequence of methylations of the purine ring. At that time, one of the difficulties in extracting active enzymes from tea and coffee leaves was that they contained high concentrations of contaminating phenolic compounds. This was overcome with the use of insoluble polyvinylpyrrolidone (PVP) to remove the phenolics (Loomis 1969). Suzuki and Takahashi (1975) prepared cell-free extracts from tea leaves using insoluble PVP and detected two methyltransferase activities that catalysed the transfer of methyl groups from SAM to 7-methylxanthine and theobromine, resulting, respectively, in the production of theobromine and caffeine. These findings suggested that caffeine was synthesized from 7-methylxanthine via theobromine but not from xanthine, xanthosine, xanthosine monophosphate (XMP), or hypoxanthine. In contrast to tea extracts, the extraction of active enzyme activity from coffee was less straight forward and Suzuki and Takahashi (1975) reported that their attempts to obtain *in vitro* enzyme activity from coffee leaves that was capable of synthesizing theobromine were unsuccessful.

Active cell-free extracts from coffee tissues were first obtained by Roberts and Waller (1979). This was achieved with the use of insoluble PVP, and rigorous exclusion of oxygen via the presence of solid CO_2 and liquid N_2, and rapid gel filtration to remove residual phenolic materials. The resultant enzyme preparation was not only able to convert 7-methylxanthosine to theobromine but also theobromine to caffeine.

In the 1980s, *N*-methyltransferase activity for caffeine biosynthesis was demonstrated in tissue and cell cultures of coffee by Baumann et al. (1983) at the University of Zurich and Waller et al. (1983) at the Oklahoma State University. Their enzyme systems catalysed the conversion of 7-methylxanthosine to caffeine. The 7-methylxanthosine synthase activity that catalysed the first methylation of the xanthine ring of caffeine was initially demonstrated in tea extracts by Negishi et al. (1985b) at the University of Tsukuba in Japan. Later, *N*-methylnucleosidase activity, which converted

7-methylxanthosine to 7-methylxanthine, was isolated from extracts of young tea leaves (Negishi et al. 1988). Thus, a xanthosine → 7-methylxanthosine → 7-methylxanthine → theobromine → caffeine pathway was proposed as the major route of caffeine biosynthesis.

The initial methyl acceptor for the main caffeine biosynthesis pathway in tea and coffee is xanthosine, and there are several pathways leading to xanthosine. Based on studies on general purine metabolism in plants (see Chapters 4 and 5), xanthosine can be produced (i) from xanthosine-5′-monophosphate, (ii) by *de novo* purine biosynthesis, and (iii) from cellular pools of AMP and GMP, which are derived from both *de novo* and some purine salvage pathways.

Suzuki and Takahashi (1976a) reported that adenine is the most efficient precursor for caffeine synthesis in tea shoots. Ashihara and Kubota (1986) demonstrated that [8-^{14}C]adenine was converted to AMP by the salvage enzyme, adenine phosphoribosyltransferase (EC 2.4.2.7), and then to xanthosine by an AMP → IMP (inosine monophosphate) → XMP → xanthosine pathway. Subsequently, there were several reports that purine precursors, such as adenosine, hypoxanthine, and inosine, were also utilized as precursors for caffeine biosynthesis. It was proposed that these compounds entered the caffeine biosynthesis pathway after their conversion to AMP and/or GMP (Ashihara 1993; Ashihara and Kubota 1987; Negishi et al. 1992). By the early 1990s, three routes: the *de novo* route, the AMP route, and the GMP route, had been proposed as the source of xanthosine (Suzuki et al. 1992). Later, a new pathway that utilized adenosine released from the SAM cycle was proposed by Koshiishi et al. (2001). Four routes had, therefore, been suggested for the formation of the purine skeleton of caffeine (Ashihara and Crozier 2001).

Other alternative routes of caffeine biosynthesis were put forward in the 1990s. Nazario and Lovatt (1993) argued that theobromine was not the immediate precursor of caffeine. They hypothesised that theobromine was synthesized from adenine nucleotides without a xanthine species serving as an intermediate, whereas caffeine was likely to be derived from a methylated xanthine. However, they could not identify the steps by which xanthine was methylated. Ashihara et al. (1996b) provided strong evidence against the Nazario and Lovatt proposal and confirmed the operation of an AMP → IMP → XMP (or GMP → guanosine) → xanthosine → 7-methylxanthosine → 7-methylxanthine → theobromine → caffeine pathway in young coffee leaves, using pulse-chase experiments with ^{14}C-precursors.

Baumann's group in Zurich reported the results of their studies on the initial methylation step of caffeine biosynthesis (Schulthess and Baumann 1995; Schulthess et al. 1996). They proposed that caffeine biosynthesis starts with XMP, not xanthosine, with the metabolically channelled conversion of XMP to 7-methyl-XMP which in turn was metabolized to 7-methylxanthosine. However, this pathway was not supported by subsequent Japanese genetic research (Mizuno et al. 2003a).

Attempts to purify the native *N*-methyltransferases involved in caffeine biosynthesis were reported by several research groups. However, it was very difficult to purify these enzymes because they are extremely labile. Partially purified enzymes were reported from laboratories in Brazil (Mazzafera et al. 1994), Japan (Ashihara et al. 1996a; Kato et al. 1996), Switzerland (Mosli Waldhauser et al. 1997), and the UK (Gillies et al. 1995). Eventually, Kato et al. (1999) successfully isolated a highly purified *N*-methyltransferase that catalysed the last two steps of the caffeine biosynthesis pathway. The enzyme

Table 15.1 Summary of enzymes involved in caffeine biosynthesis in plants.

Step[a)]	Enzyme name	EC	Reaction
1	7-Methylxanthosine synthase (= xanthosine *N*-methyltransferase)	2.1.1.158	Xanthosine + SAM → 7-methylxanthine + SAH
2	*N*-Methyl nucleosidase	3.2.2.25	7-Methylxanthosine + H_2O → 7-methylxanthine + ribose
3	Theobromine synthase (= 7-methylxanthine *N*-methyltransferase)	2.1.1.159	7-Methylxanthine + SAM → theobromine + SAH
4	Caffeine synthase (= theobromine *N*-methyltransferase)	2.1.1.160	Theobromine + SAM → caffeine + SAH Paraxanthine + SAM → caffeine + SAH 7-Methylxanthine + SAM → theobromine + SAH

SAM, *S*-adenosyl-L-methionine; SAH, *S*-adenosyl-L-homocysteine.
a) Steps in Figure 15.1.

referred to as caffeine synthase was purified to apparent homogeneity and detailed properties, including the *N*-amino-terminal sequence were reported.

In 2000, using the *N*-terminal sequence of the highly purified caffeine synthase (Kato et al. 1999), a gene encoding caffeine synthase was cloned from tea leaves (Kato et al. 2000). The substrate specificity of the recombinant enzyme was very similar to that of the native enzyme purified from young tea leaves. Genes encoding *N*-methyltransferases involved in caffeine biosynthesis were then cloned from coffee plants by two independent Japanese research groups (Mizuno et al. 2001, 2003a,b; Ogawa et al. 2001; Uefuji et al. 2003). Currently, three different coffee genes encoding *N*-methyltransferases associated with caffeine biosynthesis have been cloned, namely, 7-methylxanthine synthase, theobromine synthase, and caffeine synthase (Ashihara et al. 2008). These findings support the operation of the four-step caffeine biosynthesis pathway leading from xanthosine as proposed by Suzuki and Takahashi (1975), Negishi et al. (1985a), and Ashihara et al. (1996b). A summary of the four enzymes involved in caffeine biogenesis is shown in Table 15.1. There is now widespread agreement that a xanthosine → 7-methylxanthosine → 7-methylxanthine → theobromine → caffeine pathway is the main route of caffeine biosynthesis in tea and coffee plants.

15.3 Caffeine Biosynthesis Pathway

Purine alkaloids are formed from purine nucleotides in plants. As noted above, current data from a series of studies on *in situ* metabolism of labelled precursors and characterization of enzymes and genes have established that the main caffeine biosynthesis pathway in tea and coffee plants is a four-step sequence consisting of three methylations and one nucleosidase reaction with xanthosine serving as the initial substrate. The conversion of xanthosine to caffeine involves three *N*-methylation steps (steps 1, 3, and 4 in Figure 15.1) and the removal of a ribose moiety (step 2 in Figure 15.1). The main pathway is via theobromine. However, a route with paraxanthine (1,7-dimethylxanthine)

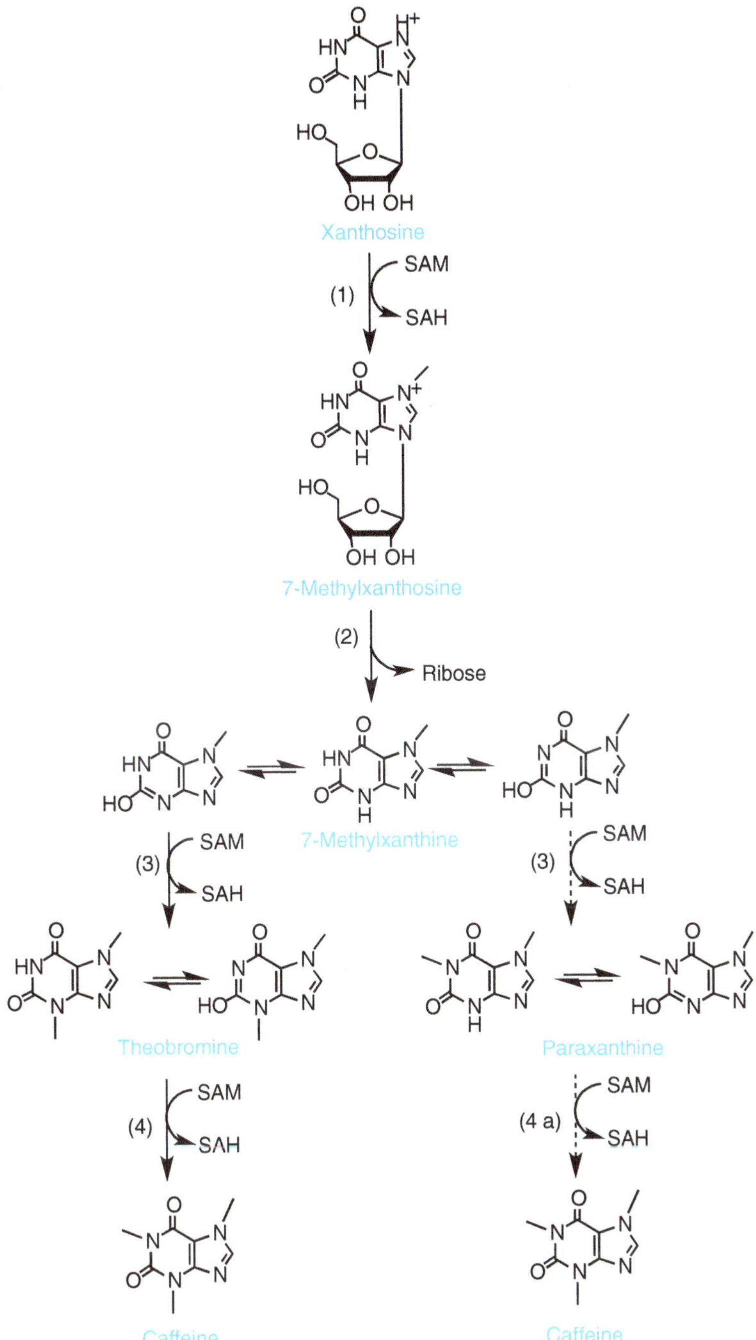

Figure 15.1 The biosynthesis pathways of caffeine from xanthosine. The major pathway consisting of four steps are shown (steps 1–4). Three types of *N*-methyltransferases have been found in coffee plants. Catalysed steps are as follows: step 1, 7-methylxanthosine synthase (EC 2.1.1.158); step 3, theobromine synthase (EC 2.1.1.159); and steps 3 and/or 4, caffeine synthase (EC 2.1.1.160). Conversion of 7-methylxanthosine to 7-methylxanthine (step 2) is catalysed by *N*-methylnucleosidase (EC 3.2.2.25). Minor pathways (steps 3a and 4a) shown with dotted arrows, may occur because of the broad substrate specificities of caffeine synthase.

(steps 3a and 4a in Figure 15.1) as an intermediate is also feasible. 7-Methylxanthine, theobromine, and paraxanthine may exist as different tautomers. Three tautomers for 7-methylxanthine are shown in Figure 15.1, but the tautomer with the hydroxyl group at C-6 is not considered here, because there seems to be no potential general base/acid catalyst for the methyl transfer in the X-ray structure (Yue and Guo 2014). Different tautomers can undergo *N*-3 and *N*-1 methylation. The available evidence indicates that the pathway is essentially the same in tea and coffee, but some differences may occur in other plant species. Therefore, this chapter will consider separately data from tea (*Camellia sinensis*), coffee (*Coffea arabica* and *Coffea canephora*), and other species.

15.3.1 *N*-Methyltransferase Nomenclature

There is no consistency in the literature on the nomenclature of *N*-methyltransferases involved in caffeine biosynthesis. Kato et al. (1999) used product names of each enzyme: 7-methylxanthosine synthase, theobromine synthase, and caffeine synthase. Since the enzyme names were registered with the IUBMB enzyme nomenclature, EC numbers were created in 2007; 7-methylxanthosine synthase (EC 2.1.1.158), theobromine synthase (EC 2.1.1.159), and caffeine synthase (EC 2.1.1.160). However, when comparing the various *N*-methyltransferases, substrate names, *S*-adenosyl-L-methionine (SAM): xanthosine *N*-methyltransferase, SAM: 7-methylxanthine *N*-methyltransferase and SAM: theobromine *N*-methyltransferase are also used, in keeping with the nomenclature employed for methyltransferases (Table 15.1). Ogawa et al. (2001) also use substrate names: xanthosine methyltransferase, monomethylxanthine methyltransferase, and dimethylxanthine methyltransferase.

15.3.2 Formation of 7-Methylxanthine from Xanthosine

Feeding experiments with tea suggest that the initial substrate in the main caffeine biosynthesis pathway is xanthosine derived from purine nucleotides. This reaction was first demonstrated using *in situ* ^{14}C-feeding experiments with excised tea shoots (Negishi et al. 1985a). Approximately 50% of [2-^{14}C]xanthosine taken up by shoots was incorporated into caffeine within 24 hours, with radioactivity appearing transiently in theobromine within 1 hour. Radioactivity was also incorporated in small amounts into 7-methylxanthosine and 7-methylxanthine at early points in the incubation. Negishi et al. (1985b) also detected 7-methylxanthosine synthase activity that catalysed the conversion of xanthosine to 7-methylxanthosine in cell-free extracts from tea shoots. When XMP was added to the incubation medium, instead of xanthosine, small amounts of 7-methylxanthosine (12% of the amount obtained with xanthosine) were detected but there was an absence of 7-methyl-XMP. This suggests that XMP was not methylated directly. The apparent activity may be due to xanthosine, being produced from XMP by the cell-free extracts as a result of contamination with nucleotidase activity. The same results have been obtained with purified *N*-methyltransferase from tea leaves (Ashihara and Crozier 2001). Xanthosine *N*-methyltransferase did not methylate XMP. Therefore, this nucleotide is unlikely to be the direct methyl acceptor in tea as initially suggested by Negishi et al. (1985b). In contrast to the results obtained with tea leaves, there is one report of an XMP *N*-methyltransferase in coffee leaves. With this enzyme, XMP rather than xanthosine is the methyl acceptor (Schulthess et al. 1996). With the tea enzymes, in contrast to xanthosine, xanthine, is only a weak methyl acceptor yielding trace quantities of 3-methylxanthine (Kato et al. 1996; Negishi et al. 1985b).

15.3.3 7-Methylxanthosine Synthase

7-Methylxanthosine synthase (EC 2.1.1.158) catalyses the conversion of xanthosine to 7-methylxanthosine (Figure 15.2). The systematic name is SAM: xanthosine 7*N*-methyltransferase. This enzyme is specific for xanthosine, and neither XMP nor xanthine act as substrates as it does not have either 1*N*- or 3*N*-methylation activity.

Native 7*N*-methyltransferase enzyme activity has been detected *in vitro* in crude and partially purified extracts from tea (Kato et al. 1996; Negishi et al. 1985b) and coffee (Mosli Waldhauser et al. 1997; Schulthess et al. 1996). In the earlier paper by Suzuki and Takahashi (1975), which demonstrated the existence of *N*-methyltransferases involved in caffeine biosynthesis in cell-free tea leaf extracts, it was stated that it was difficult to detect the first enzyme which catalyses the methylation of xanthosine. However, this is probably due to the fact that 7-methylxanthine synthase activity is low compared with

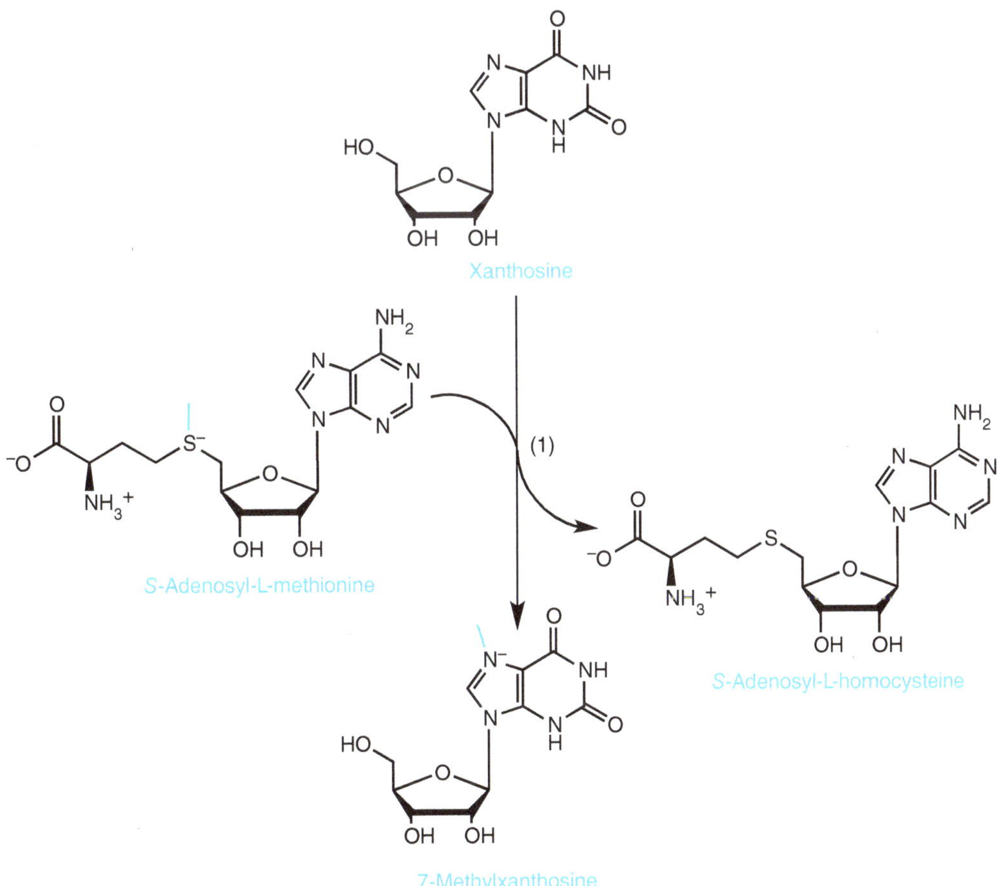

Figure 15.2 Step 1 in Figure 15.1.

Table 15.2 Seasonal changes in the levels of three *N*-methyltransferases of caffeine biosynthesis in tea leaves.

	N-methyltransferases			
Sampling time	**(XR)**	**(7mX)**	**(Tb)**	**PRPP synthase[a)]**
April	18	288	30	1115
May	13	195	20	763
June	0	15	<1	67
July	0	<1	0	73
August	0	0	0	42

Enzyme activity is expressed as pkat $g^{-a)}$f.w. The substrates of enzymes are xanthosine (XR); 7-methylxanthine (7mX), and theobromine (Tb).

a) The activity of PRPP synthetase, which participates in the synthesis of purine, pyrimidine, and other nucleotides, was still be maintained at detectable levels in July and August.

Source: Adapted from Fujimori et al. (1991).

other *N*-methyltransferases and, furthermore, it is extremely unstable *in vitro*. The levels of activity of the enzymes involved in the three *N*-methyltransferases steps change seasonally. As shown in Table 15.2, enzyme activities appear only in young tea leaves and low 7-methylxanthosine synthase activity is detected only in April and May, and its activity is less than that of the two other *N*-methyltransferase (Fujimori et al. 1991).

Kato et al. (1996) partially purified tea leaf caffeine biosynthesis *N*-methyltransferases by anion-exchange and gel-filtration column chromatography. When xanthosine was used as a substrate only the 7N position was methylated. In this study, 7-methylxanthosine synthase could not be separated from the other *N*-methyltransferases. However, subsequent extensive purification revealed that 7-methylxanthosine synthase, which is involved in the first methylation step, is distinct from the enzymes that catalyse second and third methylation steps of biosynthesis of caffeine (Kato et al. 1999).

Schulthess et al. (1996) reported that in addition to xanthosine, XMP could act as a substrate for 7*N*-methyltransferase. As shown in Table 15.3, XMP was converted to 7-methyl-XMP, 7-methylxanthosine and 7-methylxanthine in crude enzyme extracts of young coffee leaves, with SAM acting as the methyl donor. Conversion to 7-methyl-XMP was enhanced at higher concentrations of XMP in the presence of sodium molybdate, a nucleotidase inhibitor. On the basis of these data and *in situ* feeding experiments, which failed to detect ^{14}C-labelled 7-methyxanthosine in coffee leaves, a XMP → 7-methyl-XMP → 7-methylxanthosine → 7-methyxanthine pathway was proposed.

The genes encoding 7-methylxanthosine synthase were initially isolated from *C. arabica* and the properties of the recombinant enzymes reported by Mizuno et al. (2003a) and Uefuji et al. (2003). A single gene encodes a polypeptide consisting of 372 amino acids with an apparent molecular mass of 41.8 kDa. It is expressed uniformly in most parts of coffee plants, including the leaves, floral buds, and immature but not mature beans.

A recombinant 7-methylxanthosine synthase protein has been prepared using the coffee gene sequences, and its biochemical properties characterized. The recombinant

Table 15.3 Native and recombinant 7-methylxanthosine synthase from tea and coffee source.

Enzyme source	Gene	Substrate	Km value (μM) XR	SAM	Product	(%)	Reference
Camellia sinensis	Native	XR			7mXR	100	Kato et al. (1996)
Coffea arabica	Native	XR			7mXR	29	Schulthess et al. (1996)
					7mX	71	
		XMP			7mXMP	18	
					7mXR	57	
					7mX	23	
C. arabica	*CmXRS1*	XR	74		7mXR	100	Mizuno et al. (2003a,b)
C. arabica	*CaXMT1*	XR	78	13	7mXR	100	Uefuji et al. (2003)
Coffea canephora	*XMT*	XR			7mXR 7mX	100[a]	McCarthy and McCarthy (2007)

a) The results were corrected by the authors (see text).

enzyme uses xanthosine as a substrate, but neither XMP nor xanthine (Mizuno et al. 2003a). The Km for xanthosine (74–78 μM) and SAM (13 μM) seem to be appropriate values (Table 15.3).

These findings do not support the hypothesis of Schulthess et al. (1996) that caffeine biosynthesis in coffee starts with the metabolically channelled formation of 7-methyl-XMP. In the Schulthess et al. study the 7*N* position of xanthosine and XMP were methylated. XMP *N*-methyltransferase activity, which was measured in the presence of sodium molybdate, an inhibitor of nucleotidase activity, which strongly reduced hydrolysis of both XMP and 7-methyl-XMP. The results obtained with this assay are shown in Table 15.3. When XMP was used as a substrate, 7-methyl-XMP (18%), 7-methylxanthosine (57%), and 7-methylxanthine (23%) were produced and the results were used as the basis of the novel hypothesis that caffeine biosynthesis begins with the methylation of XMP rather than xanthosine. This hypothesis has been the subject of more recent discussion by Baumann (2015).

There has also been debate about the product of 7-methylxanthosine synthase. Mizuno et al. (2003a) and Uefuji et al. (2003) reported that 7-methylxanthosine is the product of recombinant 7-methylxanthosine synthase from *C. arabica*, in keeping with the native enzyme also yielding 7-methylxanthosine (Schulthess et al. 1996). In contrast, McCarthy and McCarthy (2007), who used a recombinant 7-methylxanthosine synthase from *C. canephora*, proposed that the methyl transfer and nucleoside cleavage were coupled and catalysed by a single enzyme, 7-methylxanthosine synthase/nucleosidase (Table 15.3). Uefuji et al. (2003) noted that crude recombinant enzyme, but not the purified enzyme, catalysed the conversion of xanthosine to 7-methylxanthine, possibly as the result of the presence *Escherichia coli*-derived nucleoside phosphorylase(s) contamination of the bacterial extracts.

Recently, the conclusions of McCarthy and McCarthy (2007) were revised by the group after they obtained data showing that the main product of the 7-methylxanthosine synthase is 7-methylxanthosine, and not 7-methylxanthine as originally reported (A.A. McCarthy, personal communication, 2017). Thus, *N*-methylnucleosidase appears to be involved in caffeine biosynthesis in coffee as well as in tea (see below).

15.3.4 *N*-Methylnucleosidase

7-Methyxanthosine is converted to 7-methylxanthine by *N*-methylnucleosidase (7-methylxanthosine nucleosidase, EC 3.2.2.25). The reaction is 7-methylxanthosine + H_2O → 7-methylxanthine + D-ribose (Figure 15.3). Negishi et al. (1988) partially purified the native enzyme from tea leaves. The substrate specificity of this enzyme was broad, and high activity was found with several methyl purine nucleosides. The enzyme preferentially hydrolysed 3- and 7-methylpurine nucleosides, such as 3-methylxanthosine, 3-methyladenosine and 7-methylguanosine (Table 15.4). The estimated molecular weight was 55 000 Da and the pH optimum was 8.0–8.5. This enzyme is distinct from adenosine nucleosidase (EC 3.2.2.7), which has a pH optimum of 4.5. There were several ribonucleosidase activities in the crude extract from tea leaves. Among them, adenosine nucleosidase activity, measured at pH 4.5, was ~60 times higher than 7-methylxanthosine nucleosidase activity at pH 8.5. Ribonucleosidase activity with guanosine and inosine was negligible, while xanthosine nucleosidase activity at pH 4.5 was approximately 30% of that of 7-methylxanthosine nucleosidase activity at pH 8.5.

The results obtained with the native tea leaf enzymes suggest that the conversion of xanthosine to 7-methylxanthine, via 7-methylxanthosine, is catalysed by two

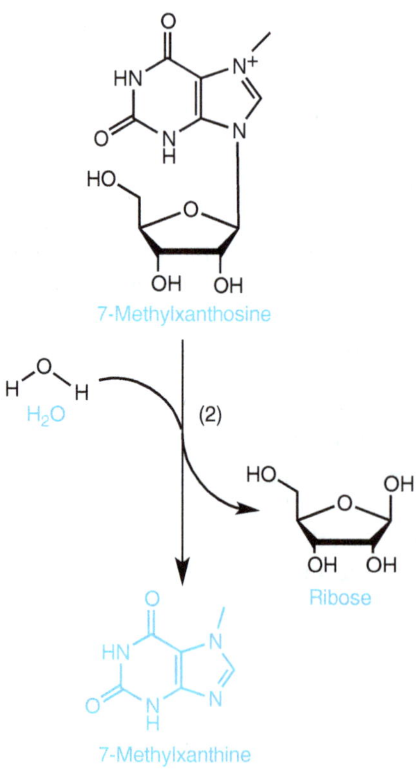

Figure 15.3 Step 2 in Figure 15.1.

Table 15.4 Substrate specificity of *N*-methylnucleosidase and adenosine nucleosidase from tea leaves.

Substrate	*N*-Methyl nucleosidase	Adenosine nucleosidase
7-Methylxanthosine	100[a]	0
3-Methylxanthosine	198	0
1-Methylxanthosine	12	0
Xanthosine	8	0
7-Methyladenosine	13	515
3-Methyladenosine	178	7
1-Methyladenosine	22	1
Adenosine	6	100[a]
7-Methylguanosine	189	0
3-Methylguanosine	32	0
1-Methylguanosine	0	0
Guanosine	0	0
7-Methylinosine	121	0
3-Methylinosine	128	0
1-Methylinosine	9	0
Inosine	5	0

N-Methyl nucleosidase and adenosine nucleosidase activities were measured at pH 7.5 and 5.5, respectively.

a) Data presented as a per cent of the value obtained with 7-methylxanthosine for *N*-methyl nucleosidase and adenosine for adenosine nucleosidase.

Source: Adapted from Negishi et al. (1988).

enzymes, 7-methylxanthosine synthase and *N*-methylnucleosidase. However, the tea gene encoding the *N*-methylnucleosidase has not, as yet, been cloned. A putative *N*-methylnucleosidase is also likely to be functional for the second step of caffeine biosynthesis in coffee but no *in vitro* enzyme activity has been isolated.

15.3.5 Formation of Caffeine from 7-Methylxanthine

Feeding experiments using tea leaf segments established that caffeine is synthesized from 7-methylxanthine via theobromine. Negishi et al. (1985a) reported that [2-^{14}C]7-methylxanthine was metabolized to theobromine and then further converted to caffeine in young tea shoots. Ashihara et al. (1997a) found that [2-^{14}C]theobromine was efficiently incorporated into caffeine in young and mature tea leaf segments. These feeding experiments support the operation of a 7-methylxanthine → theobromine → caffeine pathway in tea (steps 2–4 in Figure 15.1).

15.3.6 Caffeine Synthase

The second and third methylation steps in the caffeine biosynthesis pathway (steps 3 and 4 in Figure 15.1) are catalysed by SAM-dependent *N*-methyltransferase(s), namely, theobromine synthase, and/or caffeine synthase. Caffeine synthase is a dual-functional enzyme that catalyses the two-step reaction, 7-methylxanthine → theobromine → caffeine, while theobromine synthase catalyses the single-step conversion of 7-methylxanthine to theobromine. In tea only, a caffeine synthase has been found, while both theobromine and caffeine synthases occur in coffee.

Caffeine synthase (EC 2.1.1.160, aka, dimethylxanthine methyltransferase) catalyses three reactions (Figure 15.4), namely theobromine → caffeine (step 4); paraxanthine → caffeine (step 4a); and 7-methylxanthine to theobromine (step 3). Highly purified, native caffeine synthase was obtained from young tea leaves by Kato et al. (1999). The enzyme was purified 520-fold to apparent homogeneity by fractionation with ammonium sulfate and hydroxyapatite, anion-exchange chromatography, affinity chromatography with adenosine agarose, and gel-filtration chromatography. The enzyme displayed a sharp pH optimum at 8.5. The final preparation exhibited 3*N*-

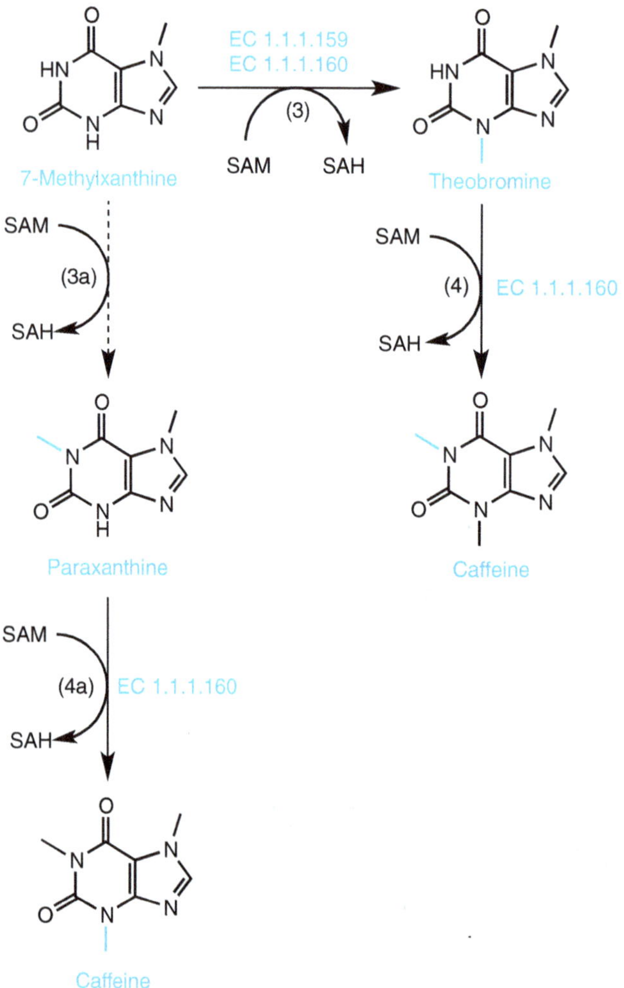

Figure 15.4 Steps in Figure 15.1 in which caffeine synthase (EC 1.1.1.160) and theobromine synthase (EC 1.1.1.159) participate are illustrated.

and 1*N*-methyltransferase activity and had a broad substrate specificity, showing high activity towards paraxanthine, 7-methylxanthine, and theobromine, and low activity with 3-methylxanthine and 1-methylxanthine (Table 15.5). Neither xanthosine nor XMP were methylated by the enzyme. The 20-amino acid *N*-terminal sequence was determined. To date highly purified native enzymes have not been obtained from other plants including coffee.

Other caffeine and theobromine synthases have been obtained as recombinant enzymes. Rapid amplification of complementary DNA ends (RACEs) was used with degenerate gene-specific primers based on the amino-terminal sequence of caffeine synthase to obtain a 1.31-kb sequence of cDNA. The 5′ untranslated sequence of the cDNA fragment was isolated by 5′-RACE. The isolated cDNA termed *TCS1* (Kato et al. 2000) was expressed in *E. coli* and the lysates of the bacterial cells incubated with a variety of xanthine substrates in the presence of SAM. The substrate specificity of the recombinant enzyme (Kato et al. 2000) was similar to that of the native enzyme purified from young tea leaves (Kato et al. 1999). Both the recombinant and native tea enzymes principally catalyse 3*N*- and 1*N*-methylation of the purine ring of mono- and dimethylxanthines (Table 15.5).

In coffee, plural genes encoding methylxanthine *N*-methyltransferases have been cloned. The recombinant enzymes obtained from these genes show different substrate specificities. The recombinant coffee caffeine synthase (EC 2.1.1.160), namely, CCS1 (Mizuno et al. 2003b) and CaDXMT1 (Uefuji et al. 2003), utilize paraxanthine, theobromine, and 7-methylxanthine as substrates (Table 15.5). Paraxanthine is the most active substrate of tea and coffee caffeine synthase, but *in vivo* only limited amounts of paraxanthine appear to be synthesized. The synthesis of paraxanthine from 7-methylxanthine by caffeine synthase (step 3a in Figure 15.4) has not been reported. Therefore, *in planta*, caffeine synthase is involved principally

Table 15.5 Comparison of substrate specificity of native and recombinant caffeine synthase from tea (*Camellia sinensis*), coffee (*Coffea arabica*), and guaraná (*Paullinia cupana*).

	Tea		Coffee		Guaraná
Substrate	**Native (1)**	**TCS1 (2)**	**CCS1 (3)**	**CaDXMT1 (4)**	**PcCS (5)**
7mX	100[a)]	100	100	100	100
3mX	17.6	1.0	3.4	0	0
1mX	4.2	12.3	2.2	0	0
Tb (=3,7mX)	26.8	18.5	104	380	189
Tp (=1,3mX)	<0.1	<0.1	0	0	0
Px (=1,7mX)	210	230	417	10 000	0
X	<0.1	—	—	—	—
XR	0	0	—	—	—
XMP	0	—	—	—	—

—, not determined.

a) Enzyme activity is expressed as a percentage of the activity on 7-methylxanthine.

Source: References are as follows: (1) Kato et al. (1999); (2) Kato et al. (2000); (3) Mizuno et al. (2003a,b); (4) Uefuji et al. (2003); (5) Schimpl et al. (2014).

in the conversion of 7-methylxanthine to caffeine via theobromine (steps 3 and 4 in Figure 15.4).

Schimpl et al. (2014) obtained a recombinant caffeine synthase (PcCS) from guaraná (*Paullinia cupana* var. *sorbilis*). Like the tea and coffee enzymes, it is a dual-functional enzyme able to catalyse the conversion of 7-methylxanthine to theobromine and also that of theobromine to caffeine (steps 3 and 4 in Figure 15.4). However, in contrast to tea and coffee caffeine synthases, this enzyme does not use paraxanthine as a substrate (Table 15.5). This unique substrate affinity of guaraná caffeine synthase is probably explained by the distinct amino acid residues found in the active site of the predicted protein.

15.3.7 Theobromine Synthase

Theobromine synthase (EC 2.1.1.159, aka, monomethylxanthine methyltransferase) catalyses the reaction, 7-methylxanthine → theobromine (step 3 in Figure 15.4). Comparisons of the substrate specificity of this enzyme are shown in Table 15.6. In coffee, in addition to the dual-functional caffeine synthase, genes encoding theobromine synthase (EC 2.1.1.159) have been cloned; the recombinant enzyme is specific for the conversion of 7-methylxanthine to theobromine (step 3 in Figure 15.4) (Mizuno et al. 2001; Ogawa et al. 2001). The recombinant enzyme obtained from coffee genes catalyses *3N*- but not *1N*-methylation. Furthermore, the activity of *3N*-methylation of 7-methylxanthine (step 3 in Figure 15.4) was much greater than the activity associated with *3N*-methylation of paraxanthine (step 4a in Figure 15.4).

Some purine alkaloid-producing species including *Camellia irrawadiensis*, *Camellia ptilophylla*, and *T. cacao* accumulate theobromine rather than caffeine. Yoneyama et al. (2006) cloned genes encoding theobromine synthase (*ICS1*, *PCS1*, and *BTS1*) from these three species. The recombinant enzymes had only *3N*-methyltransferase activity, mainly converting 7-methylxanthine to theobromine (step 3 in Figure 15.4). Limited activity catalysing *3N*-methylation of paraxanthine to form caffeine was detected only in *C. ptilophylla* (Table 15.6). The theobromine synthase appears to participate principally in the synthesis of theobromine in theobromine-accumulating plants. However, the genes encoding theobromine synthase have been isolated from some *Camellia* plants that do not accumulate theobromine. *In vitro*, the recombinant enzymes principally catalyse

Table 15.6 Comparison of substrate specificity of recombinant theobromine synthase from coffee (*Coffea arabica*), cacao (*Theobroma cacao*), *Camellia irrawadiensis*, and *C. ptilophylla*.

	Coffee		Cacao	Camellia	
Substrate	**CTS1 (1)**	**CaMXT1 (2)**	**BTS1 (3)**	**ICS1 (3)**	**PCS1 (3)**
7mX	100[a)]	100	100	100	100
Px (=1,7mX)	1.4	5.0	—	10.9	11.0
Tb (=3,7 mX)	—	—	—	0.7	—

—, not determined.

a) Enzyme activity is expressed as a percentage of the activity on 7-methylxanthine.

Source: References are as follows: (1) Mizuno et al. (2003a,b); (2) Uefuji et al. (2003); (3) Yoneyama et al. (2006).

the conversion of 7-methylxanthine to theobromine (Ishida et al. 2009). The role of theobromine synthase in these species is unknown, although it would appear not to be involved in the biosynthesis of xanthine alkaloids.

15.4 Genes and Proteins of Caffeine Synthase Family

To investigate the evolutionary relationship, a comparison of the amino acid sequences of the caffeine synthase family of proteins was undertaken. Figure 15.5 shows the amino acid sequences of caffeine synthases from coffee, cacao, maté, tea, and guaraná. There are four highly conserved regions, namely, motif A, motif B′, motif C, and the YFFF-region (Kato and Mizuno 2004; Kato and Nakayama 2018). Three conserved motifs, A, B, and C, of the binding site of the methyl donor of SAM, are also present in most plant SAM-dependent *O*-methyltransferases (Joshi and Chiang 1998).

The motif B′ and the YFFF region are specific to the motif B′ methyltransferase family. The consensus sequence of the motif B′ region, which is located between motif A and motif C, contains a large number of hydrophilic amino acids such as asparagine and aspartic acid, while the consensus sequence of the YFFF region, located downstream to motif C, contains aromatic amino acid residues (Mizuno et al. 2003b). These four regions may have important roles in both the *N*-methylation reaction and in discriminating the position of *N*-methylation on the purine ring. In addition to the caffeine synthase family, members of the motif B′ methyltransferase include some *C*-methyltransferases, such as salicylic acid carboxyl methyltransferase (Ross et al. 1999), benzoic acid carboxyl methyltransferase (Dudareva et al. 2000), jasmonic acid carboxyl methyltransferase (Seo et al. 2001), farnesoic acid carboxyl methyltransferase (Yang et al. 2006), indole-3-acetic acid methyltransferase (Zhao et al. 2008), gibberellic acid methyltransferase (Varbanova et al. 2007), and loganic acid carboxyl methyltransferase (Murata et al. 2008) (Table 15.7). The motif B′ methyltransferase family is sometimes referred to as the SABATH family, based on the initial letters of the names of the substrates of methyltransferases initially named by D'Auria et al. (2003).

Crystallographic data on salicylic acid carboxyl methyltransferase from *Clarkia breweri* suggest that members of this family exist as dimers in solution (Zubieta et al. 2003). Structural analysis of 7-methylxanthosine synthase and caffeine synthase from *C. canephora* has also revealed a dimeric structure (McCarthy and McCarthy 2007). Despite the marked similarity in amino acid sequences of *N*-methyltransferases, each enzyme catalyses the methylation of specific substrate(s). Some reports suggest that a single amino acid residue of the *N*-methyltransferases decides the substrate specificity (Ogawa et al. 2001; Yoneyama et al. 2006).

Amino acid sequences of the caffeine synthase family derived from coffee are more than 80% homologous but share only 40% homology with caffeine synthase from tea. Figure 15.6 shows the phylogenetic tree analysis of the motif B′ methyltransferase family. The figure shows that the amino acid sequences of several methyltransferases share a high degree of sequence identity within the same genus. This implies that the xanthine alkaloid biosynthesis pathways in coffee, tea, cacao, and guaraná might have evolved in parallel with one another. However, Denoeud et al. (2014) proposed that convergent evolution in caffeine biosynthesis occurs in different plant species. Caffeine-forming plants belong to several unrelated families, but they accumulate caffeine produced by a similar,

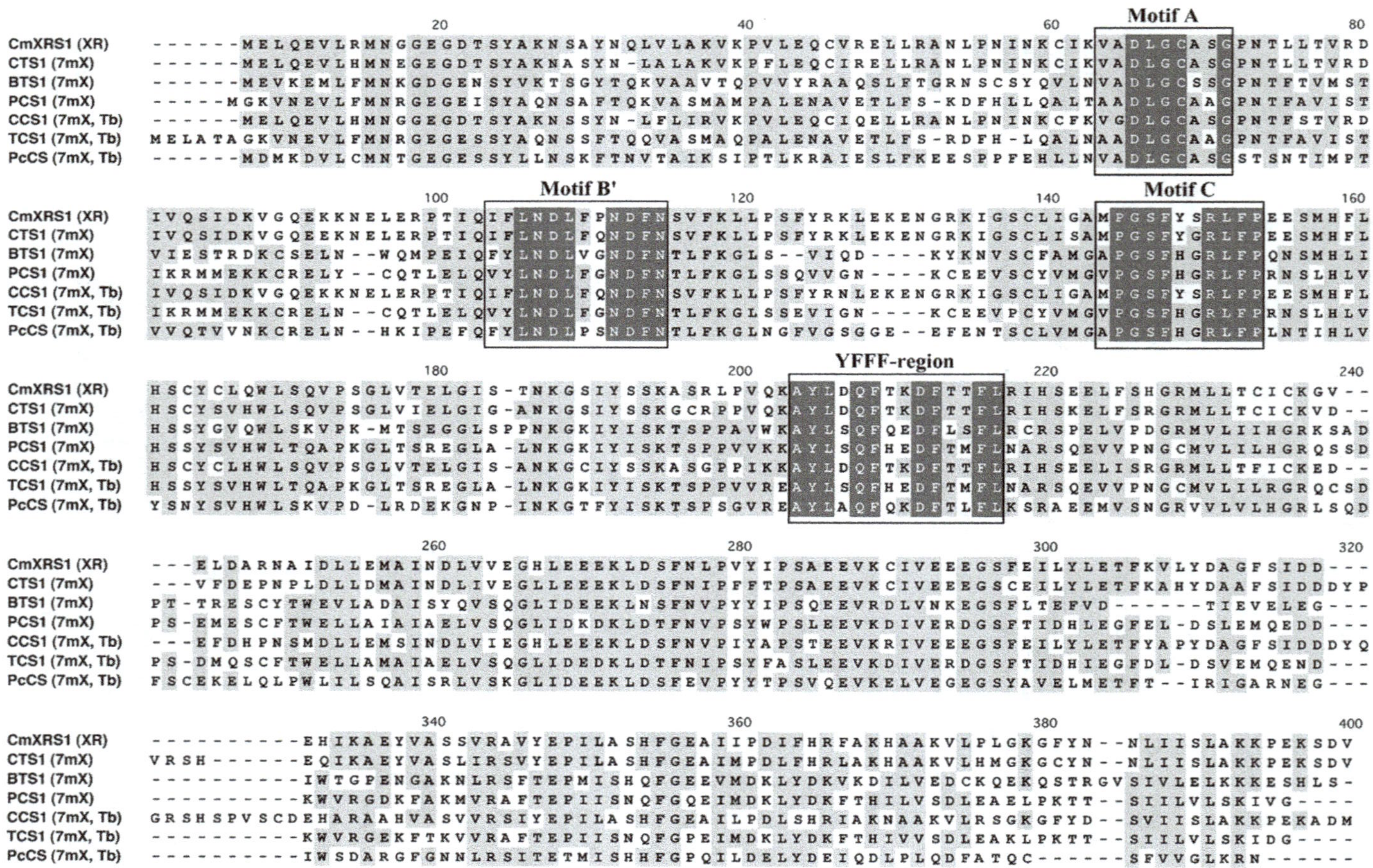

Figure 15.5 Comparison of the amino acid sequences of caffeine synthases from several caffeine containing plants. Alignment of the amino acid sequences for CmXRS1, CTS1, and CCS1 (accession numbers AB034699, AB034700, and AB086414, respectively) from coffee, BTS1 (accession number AB096699) from cacao, PCS1 (accession number AB207817) from cocoa tea (*Camellia ptilophylla*), TCS1 (accession number AB031280) from tea, and PcCS (accession number DAA64605) from guaraná. The identified substrates are shown in parentheses. The core SAM binding domain is indicated in blue . The SAM binding motifs (A, B′, and C) and YFFF conserved region are shown by black boxes. Source: Adapted from Ashihara et al. (2017).

Table 15.7 Example of the motif B′-methyltransferase family enzymes.

Species	Enzyme name	Gene name	Accession number
Coffee (*Coffea arabica*)	7-Methylxanthine synthase	*CmXRS1*	AB034699
		CaXMT1	AB048793
		CaXMT2	JX978522
	Theobromine synthase	*CTS1*	AB034700
		CaMXMT1	AB048794
		CaMXMT2	AB084126
	Caffeine synthase	*CCS1*	AB086414
		CaDXMT1	AB084125
	Tentative caffeine synthase	*CtCS7*	AB086415
	Methyltransferase-like protein	*CaMTL1*	AB039725
		CaMTL2	AB048792
	Trigonelline synthase	*CTgS*	AB054842
Coffee (*Coffea canephora*)	7-Methylxanthine synthase	*CcXMT1*	JX978518
	Theobromine synthase	*CcMXMT1*	JX978517
	Caffeine synthase	*CcDXMT*	JX978516
	Methyltransferase-like protein	*CcMTL*	JX978528
Cacao (*Theobroma cacao*)	Theobromine synthase	*BTS1*	AB096699
Tea (*Camellia sinensis*)	Caffeine synthase	*TCS1*	AB031280
		TCS2	AB031281
Cocoa tea (*Camellia ptilophylla*)	Theobromine synthase	*PCS1*	AB207817
		PCS2	AB207818
Camellia irrawadiensis	Theobromine synthase	*ICS1*	AB056108
		ICS2	AB207816
Camellia (*Camellia japonica*)	Theobromine synthase	*CjCS1*	AB297451
Guaraná (*Paullinia cupana*)	Caffeine synthase	*PcCS*	DDA64605
Fairy fans (*Clarkia breweri*)	Salicylic acid carboxyl methyltransferase	*CbSAMT*	AF133053
Snapdragon (*Antirrhinum majus*)	Benzoic acid carboxyl methyltransferase	*AmBAMT*	AF198492
Rice (*Oryza sativa*)	Indole-3-acetic acid methyltransferase	*OsIAMT*	EU375746
Thale cress (*Arabidopsis thaliana*)	Benzoic acid/salicylic acid carboxyl methyltransferase	*AtBSMT1*	NM111981
	Jasmonic acid carboxyl methyltransferase	*AtJAMT*	AY008434
	IAA carboxylmethyltransferase	*AtIAMT1*	NM124907
	Gibberellin methyl transferase	*AtGAMT1*	NM118775

Source: Adapted from Nakayama et al. (2015).

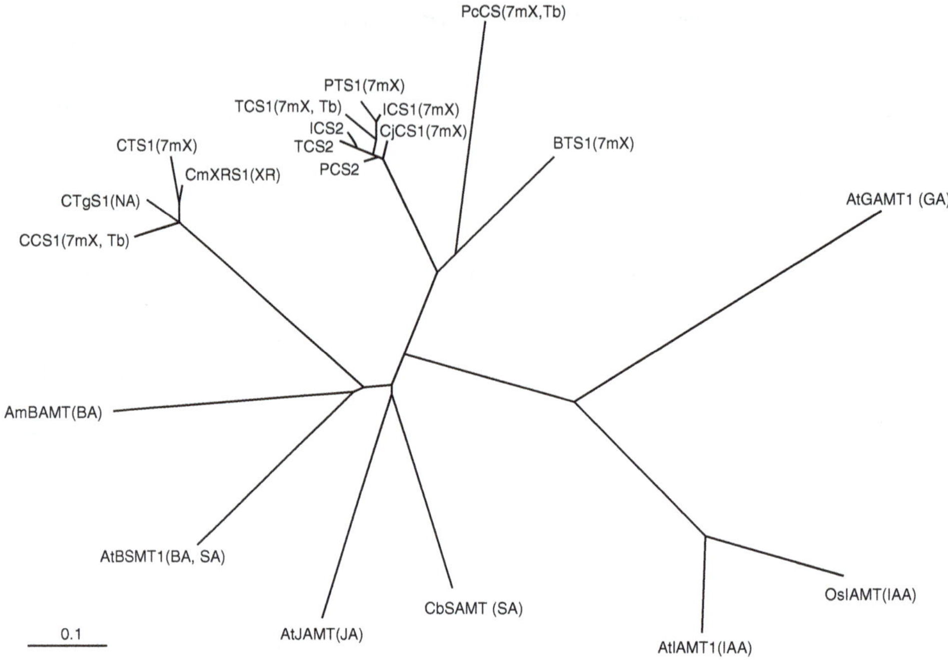

Figure 15.6 Molecular phylogenetic tree of motif B′ methyltransferase family. The identified substrates are shown in parentheses. Abbreviation of substrates are as follows: BA, benzoic acid; GA, gibberellic acid; IAA, indole-3-acetic acid; JA, jasmonic acid; 7mX, 7-methylxanthine; NA, nicotinate; SA, salicylic acid; Tb, theobromine; XR, xanthosine. The unrooted tree is created by a neighbour-joining method using Clustal W software. The sources of sequences given by the accession numbers are as follows: TCS1, AB031280; TCS2, AB031281; ICS1, AB056108; ICS2, AB207816; PCS1, AB207817; PCS2, AB207818; CjCS1, AB297451; CmXRS1, AB034699; CTS1, AB034700; CCS1, AB086414; CTgS1, AB054842; BTS1, AB096699; CbSAMT, AF133053; AmBAMT, AF198492; OsIAMT, EU375746; AtBSMT1, NM111981; AtJAMT, AY008434; AtIAMT1, NM124907; AtGAMT1, NM118775; PcCS, DDA64605. Source: Adapted from Ashihara et al. (2017).

if not identical, biosynthesis pathway. The genome sequence of caffeine biosynthesis indicates that the methyltransferase genes in some lineages have evolved independently from other branches of the SABATH (motif B′) methyltransferase gene family (Pichersky and Lewinsohn 2011).

Huang et al. (2016) have proposed that convergent caffeine production has evolved by two previously unknown pathways in cacao (*T. cacao*), guaraná (*P. cupana*), and citrus (*Citrus sinensis*) using either caffeine synthase or xanthine methyltransferase-like enzymes. In tea and coffee plants, the major precursor for the caffeine is xanthosine, not xanthine, which ^{14}C-tracer experiments indicate is the main initial substrate for conventional purine catabolism in plants. However, both ^{14}C-xanthosine and ^{14}C-xanthine are incorporated into theobromine in cacao leaves (Koyama et al. 2003) indicating, arguably, that the TcCS1 enzyme participates in the conversion of xanthine into theobromine.

Ishida et al. (2009) reported the occurrence of theobromine synthase genes in species of *Camellia* that do not accumulate xanthine alkaloids. Arguably, this represents additional evidence that monophyletic genes occur in such *Camellia* plants.

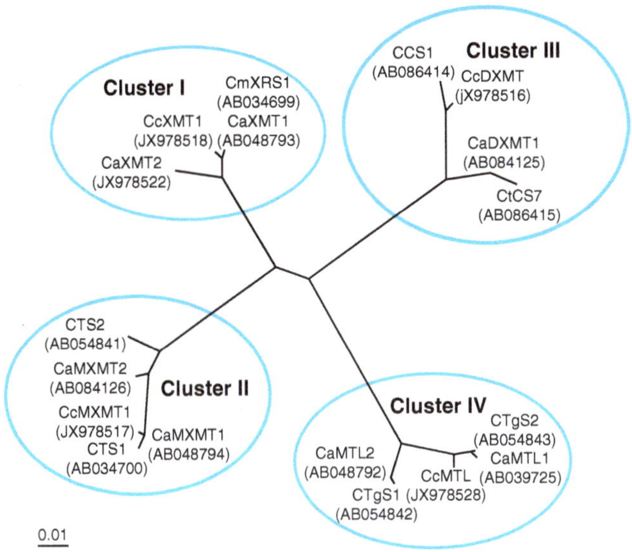

Figure 15.7 Unrooted maximum likelihood tree based on the alignment of different 18 *N*-methyltransferases involved in coffee caffeine synthase family. Clusters I, II, III, and IV correspond to 7-methylxanthosine synthases, theobromine synthases, caffeine synthases, and trigonelline synthases, respectively. The accession numbers are as follows: CmXRS1(AB034699); CaXMT1 (AB048793); CcXMT1 (JX978518); CaXMT2 (JX978522); CcMXMT1 (JX978517); CaMXMT1 (AB048794); CaMXMT2 (AB084126); CTS1 (AB034700); CTS2 (AB054841); CcDXMT (JX978516); CaDXMT1 (AB084125); CtCS7 (AB086415); CCS1 (AB086414); CTgS1 (AB054842); CaMTL1 (AB039725); CcMTL (JX978528); CTgS2 (AB054843); CaMTL2 (AB048792). Source: Adapted from Ashihara et al. (2017).

Perrois et al. (2015) proposed that the different *N*-methyltransferases involved in caffeine biosynthesis in coffee belong to three different clusters that align with the function of each enzyme. Clusters I, II, and III correspond, respectively, to 7-methylxanthosine synthase, theobromine synthase, and caffeine synthase (Figure 15.7).

Mizuno et al. (2014) reported that the conversion of nicotinate to trigonelline is also catalysed by *N*-methyltransferases belonging to the motif B′ methyltransferase family in *C. arabica*. Genes previously cloned from *C. arabica*, and tentatively named as *CtCS3* and *CtCS4*, were shown to encode trigonelline synthase 1 (*CTgS1*) and 2 (*CTgS2*). The sequence of these isozymes exhibited >95% homology while homology between the trigonelline synthase (CTgS) and caffeine synthase (CCS1) was 82%. CTgS can be classified as Cluster IV as shown in Figure 15.7. Trigonelline accumulates in coffee seeds in similar concentrations to caffeine (Ashihara 2015; Ashihara et al. 2015). Trigonelline is a pyridine alkaloid derived from nicotinate (see Chapter 21). It is interesting that both biosynthetic enzymes possess high homology.

15.5 Xanthosine Biosynthesis from Purine Nucleotides

Caffeine biosynthesis in tea and coffee is initiated by the methylation of xanthosine. Xanthosine, is supplied by several different pathways: (i) *de novo* purine biosynthesis (*de novo* route); (ii) degradation of adenine nucleotides (AMP route); (iii) the SAM cycle (SAM route); (iv) degradation of nicotinamide adenine diphosphate (NAD) or

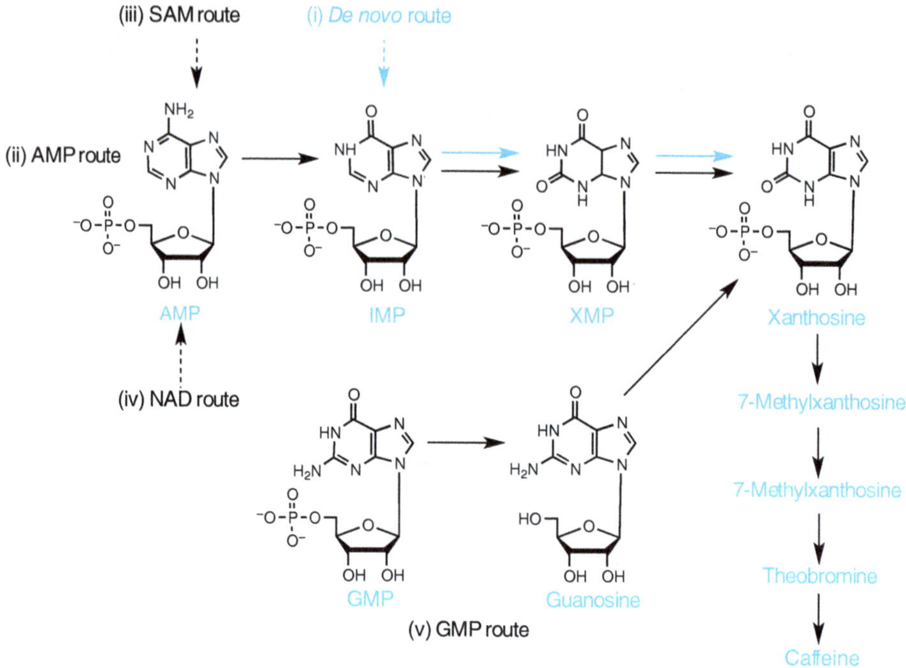

Figure 15.8 Outline of five possible pathways of xanthosine biosynthesis, a key substrate for caffeine formation in tea and coffee. (i) *De novo* route, (ii) AMP route, (iii) SAM route, (iv) NAD route, and (v) GMP route.

nicotinamide adenine dinucleotide phosphate (NADP) (NAD route), and (v) guanine nucleotides (GMP route) (Figure 15.8).

15.5.1 *De Novo* Purine Route

As described in Chapter 4, plants, as well as most other organisms, produce inosine-5′-monophosphate by *de novo* purine nucleotide biosynthesis. In the *de novo* pathway, IMP is synthesized from 5-phosphoribosyl-1-pyrophosphate (PRPP), glycine, glutamine, aspartate, and 10-formyl tetrahydrofolate (Figure 4.1).

The contribution of *de novo* purine biosynthesis to caffeine biosynthesis was demonstrated by the incorporation of several small-molecule precursors of purine biosynthesis into caffeine in coffee (Anderson and Gibbs 1962) and tea leaves (Inoue 1972; Inoue and Kawamura 1961; Inoue et al. 1960). Ito and Ashihara (1999) examined whether the *de novo* pathway of purine nucleotide biosynthesis contributed directly to the biosynthesis of caffeine in tea leaves. The ^{15}N label from [^{15}N]glycine was incorporated into theobromine and caffeine, and this was reduced markedly by azaserine and aminopterin, known inhibitors of purine biosynthesis *de novo*. Radioactivity from [2-^{14}C]5-aminoimidazole-4-carboxamide ribonucleoside (AICA), a precursor of aminoimidazolecarboxamide ribonucleotide, was also incorporated into theobromine and caffeine. Pulse-chase experiments with [^{15}N]glycine and [2-^{14}C]AICA suggested that theobromine is the immediate precursor of caffeine.

Incorporation of radioactivity from [2-^{14}C]AICA into purine alkaloids was not influenced by treatment with coformycin, an inhibitor of plant AMP deaminase. The rate of the incorporation of radioactivity into theobromine from [8-^{14}C]inosine, an immediate precursor of IMP, was more rapid than from [8-^{14}C]adenine, a precursor of AMP. Based on these results, it was concluded that most of the purine alkaloids originate from newly produced IMP formed by the *de novo* pathway for purine nucleotide biosynthesis ('route i' in Figure 15.8). Although it is difficult to completely exclude the participation of other routes, the contribution of the *de novo* pathway to caffeine biosynthesis appears to be important, especially in young tea leaves in which a very rapid accumulation of purine alkaloids takes place (Ito and Ashihara 1999).

15.5.2 Adenosine Monophosphate Route

A portion of the xanthosine used for caffeine biosynthesis is derived from adenine nucleotide pools. The cellular concentrations of adenine and guanine nucleotides are, respectively, 225 and 26 nmol g^{-1} f.w. in young tea leaves (Ashihara and Crozier 1999) and 303 and 60 nmol g^{-1} f.w. in young maté leaves (Ashihara 1993). Hence, caffeine biosynthesis from adenine nucleotides would appear to be quantitatively more important than that from guanine nucleotides. There are several potential pathways for xanthosine synthesis from AMP, but the AMP → IMP → XMP → xanthosine route is likely to predominate ('route ii' in Figure 15.8). Coformycin which as noted above is an inhibitor of AMP deaminase in plants (Yabuki and Ashihara 1991), inhibits the caffeine biosynthesis in tea flower buds (Fujimori and Ashihara 1993) and maté leaves (Ashihara 1993). This suggests an involvement of AMP deaminase in the biosynthesis of the purine alkaloids. All three enzymes participating in the conversion, namely AMP deaminase, IMP dehydrogenase, and 5′-nucleotidase, have been detected in tea leaves (Koshiishi et al. 2001).

15.5.3 S-Adenosyl-L-methionine Cycle Route

The SAM cycle route ('route iii' in Figure 15.8) is a variation in the route from adenine nucleotides discussed above. The four-step caffeine biosynthesis pathway includes three methylation steps that utilize SAM as the methyl donor. In this process, SAM is converted to *S*-adenosine-L-homocysteine (SAH), which in turn is hydrolysed to L-homocysteine and adenosine (Figure 15.9). Koshiishi et al. (2001) demonstrated that significant amounts of radioactivity from [methyl-^{14}C]methionine and [methyl-^{14}C]SAM were incorporated into theobromine and caffeine in young tea leaf segments. In addition, very high SAH hydrolase activity was found in cell-free extracts from young tea leaves. Substantial amounts of radioactivity from [adenosyl-^{14}C]SAH were also recovered as theobromine and caffeine in tea leaf segments, demonstrating that adenosine derived from SAH is utilized for the synthesis of the purine ring of caffeine. From the profiles of activity of related enzymes in tea leaf extracts, it was proposed that the major route from SAM to caffeine is a SAM → SAH → adenosine → adenine → AMP → IMP → XMP → xanthosine → 7-methylxanthosine → 7-methylxanthine → theobromine → caffeine pathway. In addition, direct adenosine kinase-catalysed formation of AMP from adenosine may participate as an alternative minor route. On the basis of these observations, Koshiishi et al. (2001) proposed that in young tea leaves, not only the methyl groups, but also the purine ring of caffeine could be derived from SAM. The capacity of the pathway is such that in theory it is possible that the purine ring of caffeine is produced exclusively by this route.

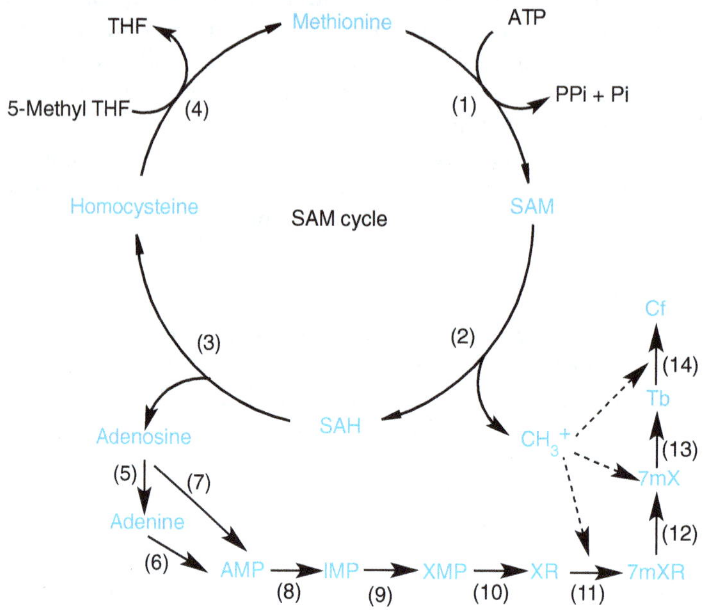

Figure 15.9 A possible caffeine biosynthesis pathway from adenosine released by the *S*-adenosyl-L-methionine (SAM) cycle. Enzymes (EC numbers) shown are: (1) SAM synthetase (2.5.1.6), (2) SAM-dependent *N*-methyltransferases for purine alkaloids (2.1.1. 158–160), (3) *S*-adenosyl-L-homocysteine (SAH) hydrolase (3.3.1.1), (4) methionine synthase (2.1.1.13), (5) adenosine nucleosidase (3.2.2.7), (6) adenine phosphoribosyltransferase (2.4.2.7), (7) adenosine kinase (2.7.1.20), (8) AMP deaminase (3.5.4.6), (9) inosine 5′-monophosphate (IMP) dehydrogenase (1.1.1.205), (10) 5′-nucleotidase (3.1.3.5), (11) 7-methylxanthosine synthase (2.1.1.158), (12) *N*-methylnucleosidase (3.2.2.25), (13) theobromine synthase (2.1.1.159) or caffeine synthase (2.1.1.160), and (14) caffeine synthase (2.1.1.160). 5-Methyl THF, 5-methyl tetrahydrofolate; THF, Tetrahydrofolate. Source: Adapted from Ashihara and Crozier (2001).

15.5.4 Nicotinamide Adenine Diphosphate Catabolism Route

The NAD catabolism route ('route iv' in Figure 15.8) is also a variation in the AMP route. If NAD is degraded to nicotinamide mononucleotide and AMP by NAD pyrophosphatase, the product, AMP, enters the AMP pathway of caffeine biosynthesis. The NAD route appears to be important in coffee because another alkaloid, trigonelline, is formed from NAD-derived nicotinate (see Figure 21.1) (Ashihara 2015; Ashihara et al. 2015). Baumann (2015) considered a hypothesis originally proposed by Kremers (1954), and speculated that NAD is a precursor for both caffeine and trigonelline. In this pathway, NAD would be transformed to nicotinate-hypoxanthine dinucleotide by the simultaneous action of nicotinamidase and a deaminase. In a second step, the 'hypoxanthine moiety' is oxidized by an IMP dehydrogenase to yield nicotinate xanthine dinucleotide. Finally, by the dual action of the same *N*-methyltransferase, trigonelline, and 7-methylxanthine are liberated. Although this is a most interesting hypothesis, there is an absence of supporting experimental evidence.

15.5.5 Guanosine Monophosphate Route

Xanthosine utilized for caffeine biosynthesis is also produced from guanine nucleotides by a GMP → guanosine → xanthosine pathway ('route v' in Figure 15.8). Guanosine deaminase (EC 3.5.4.15) activity has been found in enzyme extracts from young tea leaves (Negishi et al. 1994). The synthesis of caffeine from [8-^{14}C]guanosine in tea leaves has also confirmed the operation of this pathway (Ashihara et al. 1997b). Although GMP is converted to IMP by the action of GMP reductase in bacteria and mammals (Mager and Magasanik 1960; Renart and Sillero 1974; Stephens and Whittaker 1973), there is no evidence for the presence of GMP reductase in higher plants (Ashihara et al. 1991; Ashihara et al. 1997b; Le Floc'h et al. 1982). Therefore, the GMP → IMP → XMP → xanthosine pathway, would appear not to be functional in plants.

15.6 Summary

The main biosynthesis of caffeine is from xanthosine with a xanthosine → 7-methylxanthosine → 7-methylxanthine → theobromine → caffeine pathway. The enzymes involve two or three types of *S*-adenosyl-L-methionine-dependent *N*-methyltransferases and *N*-methylnucleosidase. The substrate, xanthosine is the catabolite of purine nucleotides produced from the *de novo* and some purine salvage pathways.

References

Anderson, L. and Gibbs, M. (1962). The biosynthesis of caffeine in the coffee plant. *J. Biol. Chem.* 237: 1941–1944.

Ashihara, H. (1993). Purine metabolism and the biosynthesis of caffeine in maté leaves. *Phytochemistry* 33: 1427–1430.

Ashihara, H. (2015). Plant biochemistry: Trigonelline biosynthesis in *Coffea arabica* and *Coffea canephora*. In: *Coffee in Health and Disease Prevention* (ed. V.R. Preedy), 19–28. San Diego: Academic Press.

Ashihara, H. and Crozier, A. (1999). Biosynthesis and metabolism of caffeine and related purine alkaloids in plants. *Adv. Bot. Res.* 30: 117–205.

Ashihara, H. and Crozier, A. (2001). Caffeine: a well known but little mentioned compound in plant science. *Trends Plant Sci.* 6: 407–413.

Ashihara, H. and Kubota, H. (1986). Patterns of adenine metabolism and caffeine biosynthesis in different parts of tea seedlings. *Physiol. Plant.* 68: 275–281.

Ashihara, H. and Kubota, H. (1987). Biosynthesis of purine alkaloids in *Camellia* plants. *Plant Cell Physiol.* 28: 535–539.

Ashihara, H., Mitsui, K., Yabuki, N., and Nygaard, P. (1991). Adenosine metabolism and growth of adenosine-requiring mutant cells of *Datura innoxia*. *Int. J. Environ. Res. Public Health* 2: 129–135.

Ashihara, H., Kato, M., and Crozier, A. (1996a). Caffeine biosynthesis in leaves of *Camellia sinensis*: Substrate specificity of *N*-methyltransferase. In: *Principles Regulating*

Biosynthesis and Storage of Secondary Products, 40–44. Halle-Wittenberg: Phytochemical Society of Europe.
Ashihara, H., Monteiro, A.M., Gillies, F.M., and Crozier, A. (1996b). Biosynthesis of caffeine in leaves of coffee. *Plant Physiol.* 111: 747–753.
Ashihara, H., Gillies, F.M., and Crozier, A. (1997a). Metabolism of caffeine and related purine alkaloids in leaves of tea (*Camellia sinensis* L.). *Plant Cell Physiol.* 38: 413–419.
Ashihara, H., Takasawa, Y., and Suzuki, T. (1997b). Metabolic fate of guanosine in higher plants. *Physiol. Plant.* 100: 909–916.
Ashihara, H., Sano, H., and Crozier, A. (2008). Caffeine and related purine alkaloids: biosynthesis, catabolism, function and genetic engineering. *Phytochemistry* 69: 841–856.
Ashihara, H., Ludwig, I., Katahira, R. et al. (2015). Trigonelline and related nicotinic acid metabolites: occurrence, biosynthesis, taxonomic considerations, and their roles in planta and in human health. *Phytochem. Rev.* 14: 765–798.
Ashihara, H., Mizuno, K., Yokota, T., and Crozier, A. (2017). Xanthine alkaloids: Occurrence, biosynthesis, and function in plants. In: *Progress in the Chemistry of Organic Natural Products*, vol. 105 (eds. A.D. Kinghorn, H. Falk, S. Gibbons and J.I. Kobayashi), 1–88. Cham, Switzerland: Springer International Publishing.
Baumann, T.W. (2015). Revisiting caffeine biosynthesis - speculations about the proximate source of its purine ring. *Nat. Prod. Commun.* 10: 793–797.
Baumann, T.W., Koetz, R., and Morath, P. (1983). *N*-Methyltransferase activities in suspension cultures of *Coffea arabica* L. *Plant Cell Rep.* 2: 33–35.
Bonner, J. (1950). *Plant Biochemistry*. New York: Academic Press.
Bresler, H.W. (1904). Uber die Bestimmung der Nucleinbasen im Safte von *Beta vulgaris*. *Hoppe Seylers Z. Physiol. Chem.* 4: 535–541.
D'Auria, J.C., Chen, F., and Pichersky, E. (2003). The SABATH family of MTS in *Arabidopsis thaliana* and other plant species. In: *Recent Advances in Phytochemistry*, vol. 37 (ed. T.R. John), 253–283. Elsevier.
Denoeud, F., Carretero-Paulet, L., Dereeper, A. et al. (2014). The coffee genome provides insight into the convergent evolution of caffeine biosynthesis. *Science* 345: 1181–1184.
Dudareva, N., Murfitt, L.M., Mann, C.J. et al. (2000). Developmental regulation of methyl benzoate biosynthesis and emission in snapdragon flowers. *Plant Cell* 12: 949–961.
Fujimori, N. and Ashihara, H. (1993). Biosynthesis of caffeine in flower buds of *Camellia sinensis*. *Ann. Bot.* 71: 279–284.
Fujimori, N., Suzuki, T., and Ashihara, H. (1991). Seasonal variations in biosynthetic capacity for the synthesis of caffeine in tea leaves. *Phytochemistry* 30: 2245–2248.
Gillies, F.M., Jenkins, G.I., Ashihara, H., and Crozier, A. (1995). *In vitro* biosynthesis of caffeine: the stability of *N*-methyltransferase activity in cell-free preparations from liquid endosperm of *Coffea arabica*. In: *Proc. 16th Int. Conf. Coffee Sci*, vol. 1, 599–605. Kyoto.
Huang, R., O'Donnell, A.J., Barboline, J.J., and Barkman, T.J. (2016). Convergent evolution of caffeine in plants by co-option of exapted ancestral enzymes. *Proc. Natl. Acad. Sci.* 113: 10613–10618.
Inoue, T. (1972). Studies on biogenesis of tea components. IV. Incorporation of glycine-2-^{14}C to caffeine. *Proc. Hoshi Coll. Pharm.* 13: 60–62.
Inoue, T. and Adachi, F. (1962). Studies on biogenesis of tea components. III. The origin of the methyl groups in caffeine. *Chem. Pharm. Bull. (Tokyo)* 10: 1212–1214.
Inoue, T. and Kawamura, Y. (1961). Studies on biogenesis of tea components II. Formation of caffeine in excised tea shoots. *Chem. Pharm. Bull. (Tokyo)* 9: 236–238.

Inoue, T., Yamashita, S., Kawamura, Y., and Sasaki, G. (1960). Studies on biogenesis of tea components I. absorption of ^{15}N to new leaves and old leaves. *Yakugaku Zasshi* 80: 548–551.

Ishida, M., Kitao, N., Mizuno, K. et al. (2009). Occurrence of theobromine synthase genes in purine alkaloid-free species of *Camellia* plants. *Planta* 229: 559–568.

Ito, E. and Ashihara, H. (1999). Contribution of purine nucleotide biosynthesis *de novo* to the formation of caffeine in young tea (*Camellia sinencis*) leaves. *J. Plant Physiol.* 254: 145–151.

Joshi, C.P. and Chiang, V.L. (1998). Conserved sequence motifs in plant *S*-adenosyl-L-methionine-dependent methyltransferases. *Plant Mol. Biol.* 37: 663–674.

Kato, M. and Mizuno, K. (2004). Caffeine synthase and related methyltransferases in plants. *Front. Biosci.* 9: 1833–1842.

Kato, M. and Nakayama, F. (2018). Caffeine. In: *Biotechnology of Natural Products* (eds. W. Schwab, B.M. Lange and M. Wüst), 131–143. Cham: Springer.

Kato, M., Kanehara, T., Shimizu, H. et al. (1996). Caffeine biosynthesis in young leaves of *Camellia sinensis*: in vitro studies on *N*-methyltransferase activity involved in the conversion of xanthosine to caffeine. *Physiol. Plant.* 98: 629–636.

Kato, M., Mizuno, K., Fujimura, T. et al. (1999). Purification and characterization of caffeine synthase from tea leaves. *Plant Physiol.* 120: 579–586.

Kato, M., Mizuno, K., Crozier, A. et al. (2000). Caffeine synthase gene from tea leaves. *Nature* 406: 956–957.

Koshiishi, C., Kato, A., Yama, S. et al. (2001). A new caffeine biosynthetic pathway in tea leaves: utilisation of adenosine released from the *S*-adenosyl-L-methionine cycle. *FEBS Lett.* 499: 50–54.

Koyama, Y., Tomoda, Y., Kato, M., and Ashihara, H. (2003). Metabolism of purine bases, nucleosides and alkaloids in theobromine-forming *Theobroma cacao* leaves. *Plant Physiol. Biochem.* 41: 977–984.

Kremers, R.E. (1954). Speculation on DPN as a biochemical precursor of caffeine and trigonelline in coffee. *J. Am. Pharm. Assoc.* 43: 423–424.

Le Floc'h, F., Lafleuriel, J., and Guillot, A. (1982). Interconversion of purine nucleotides in Jerusalem artichoke shoots. *Plant Sci. Lett.* 27: 309–316.

Loomis, W.D. (1969). Removal of phenolic compounds during the isolation of plant enzymes. In: *Methods in Enzymology*, vol. 13 (ed. J.M. Lowenstein), 555–563. Academic Press.

Looser, E., Baumann, T.W., and Wanner, H. (1974). The biosynthesis of caffeine in the coffee plant. *Phytochemistry* 13: 2515–2518.

Mager, J. and Magasanik, B. (1960). Guanosine 5′-phosphate reductase and its role in the interconversion of purine nucleotides. *J. Biol. Chem.* 235: 1474–1478.

Mazzafera, P., Wingsle, G., Olsson, O., and Sandberg, G. (1994). *S*-Adenosyl-L-methionine:theobromine 1-*N*-methyltransferase, an enzyme catalysing the synthesis of caffeine in coffee. *Phytochemistry* 17: 1577–1584.

McCarthy, A.A. and McCarthy, J.G. (2007). The structure of two *N*-methyltransferases from the caffeine biosynthetic pathway. *Plant Physiol.* 144: 879–889.

Mizuno, K., Tanaka, H., Kato, M. et al. (2001). cDNA cloning of caffeine (theobromine) synthase from coffee (*Coffea arabica* L.). In: *Proceedings of the 19th International Scientific Colloquium on Coffee*, 815–818. Association for Science and Information on Coffee.

Mizuno, K., Kato, M., Irino, F. et al. (2003a). The first committed step reaction of caffeine biosynthesis: 7-methylxanthosine synthase is closely homologous to caffeine synthases in coffee (*Coffea arabica* L.). *FEBS Lett.* 547: 56–60.

Mizuno, K., Okuda, A., Kato, M. et al. (2003b). Isolation of a new dual-functional caffeine synthase gene encoding an enzyme for the conversion of 7-methylxanthine to caffeine from coffee (*Coffea arabica* L.). *FEBS Lett.* 534: 75–81.

Mizuno, K., Matsuzaki, M., Kanazawa, S. et al. (2014). Conversion of nicotinic acid to trigonelline is catalyzed by *N*-methyltransferase belonged to motif B′ methyltransferase family in *Coffea arabica*. *Biochem. Biophys. Res. Commun.* 452: 1060–1066.

Mosli Waldhauser, S.S., Gillies, F.M., Crozier, A., and Baumann, T.W. (1997). Separation of the *N*-7 methyltransferase, the key enzyme in caffeine biosynthesis. *Phytochemistry* 45: 1407–1414.

Murata, J., Roepke, J., Gordon, H., and De Luca, V. (2008). The leaf epidermome of *Catharanthus roseus* reveals its biochemical specialization. *Plant Cell* 20: 524–542.

Nakayama, F., Mizuno, K., and Kato, M. (2015). Biosynthesis of caffeine underlying the diversity of motif B' methyltransferase. *Natural Product Communications* 10: 799–801.

Nazario, G.M. and Lovatt, C.J. (1993). Separate *de novo* and salvage purine pools are involved in the biosynthesis of theobromine but not caffeine in leaves of *Coffea arabica* L. *Plant Physiol.* 103: 1203–1210.

Negishi, O., Ozawa, T., and Imagawa, H. (1985a). Conversion of xanthosine into caffeine in tea plants. *Agric. Biol. Chem.* 49: 251–253.

Negishi, O., Ozawa, T., and Imagawa, H. (1985b). Methylation of xanthosine by tea-leaf extracts and caffeine biosynthesis. *Agric. Biol. Chem.* 49: 887–890.

Negishi, O., Ozawa, T., and Imagawa, H. (1988). *N*-methylnucleosidase from tea leaves. *Agric. Biol. Chem.* 52: 169–175.

Negishi, O., Ozawa, T., and Imagawa, H. (1992). Biosynthesis of caffeine from purine nucleotides in tea plant. *Biosci. Biotechnol. Biochem.* 56: 499–503.

Negishi, O., Ozawa, T., and Imagawa, H. (1994). Guanosine deaminase and guanine deaminase from tea leaves. *Biosci. Biotechnol. Biochem.* 58: 1277–1281.

Ogawa, M., Herai, Y., Koizumi, N. et al. (2001). 7-Methylxanthine methyltransferase of coffee plants. Gene isolation and enzymatic properties. *J. Biol. Chem.* 276: 8213–8218.

Ogutuga, D.B.A. and Northcote, D.H. (1970). Biosynthesis of caffeine in tea callus tissue. *Biochem. J.* 117: 715–720.

Perrois, C., Strickler, S., Mathieu, G. et al. (2015). Differential regulation of caffeine metabolism in *Coffea arabica* (Arabica) and *Coffea canephora* (Robusta). *Planta* 241: 179–191.

Pichersky, E. and Lewinsohn, E. (2011). Convergent evolution in plant specialized metabolism. *Annu. Rev. Plant Biol.* 62: 549–566.

Proiser, E. and Serenkov, G.P. (1963). Caffeine biosynthesis in tea leaves. *Biokhimiya (Moscow)* 28: 857–861.

Renart, M.F. and Sillero, A. (1974). GMP reductase in *Artemia salina*. *Biochim. Biophys. Acta* 341: 178–186.

Roberts, M.F. and Waller, G.R. (1979). *N*-Methyltransferases and 7-methyl-*N*9-nucleoside hydrolase activity in *Coffea arabica* and the biosynthesis of caffeine. *Phytochemistry* 18: 451–455.

Ross, J.R., Nam, K.H., D'Auria, J.C., and Pichersky, E. (1999). *S*-Adenosyl-methionine: salicylic acid carboxyl methyltransferase, an enzyme involved in floral scent production and plant defense, represents a new class of plant methyltransferases. *Arch. Biochem. Biophys.* 367: 9–16.

Schimpl, F.C., Kiyota, E., Mayer, J.L.S. et al. (2014). Molecular and biochemical characterization of caffeine synthase and purine alkaloid concentration in guarana fruit. *Phytochemistry* 105: 25–36.

Schulthess, B.H. and Baumann, T.W. (1995). Are xanthosine and 7-methylxanthosine caffeine precursors? *Phytochemistry* 39: 1363–1370.

Schulthess, B.H., Morath, P., and Baumann, T.W. (1996). Caffeine biosynthesis starts with the metabolically channelled formation of 7-methyl-XMP- a new hypothesis. *Phytochemistry* 41: 169–175.

Seo, H.S., Song, J.T., Cheong, J.J. et al. (2001). Jasmonic acid carboxyl methyltransferase: a key enzyme for jasmonate-regulated plant responses. *Proc. Natl. Acad. Sci.* 98: 4788–4793.

Serenkov, G.P. and Proiser, E. (1961). Biosynthesis of caffeine in tea leaves. *Doklady Akademica Nauka USSR* 140: 716–719.

Stephens, R.W. and Whittaker, V.K. (1973). Calf thymus GMP reductase: control by XMP. *Biochem. Biophys. Res. Commun.* 53: 975–981.

Suzuki, T. and Takahashi, E. (1975). Biosynthesis of caffeine by tea-leaf extracts: Enzymic formation of theobromine from 7-methylxanthine and of caffeine from theobromine. *Biochem. J.* 146: 87–96.

Suzuki, T. and Takahashi, E. (1976a). Caffeine biosynthesis in *Camellia sinensis*. *Phytochemistry* 15: 1235–1239.

Suzuki, T. and Takahashi, E. (1976b). Further investigation of the biosynthesis of caffeine in tea plants (*Camellia sinensis* L.): methylation of transfer ribonucleic acid by tea leaf extracts. *Biochem. J.* 160: 181–184.

Suzuki, T. and Takahashi, E. (1976c). Metabolism of methionine and biosynthesis of caffeine in the tea plant (*Cammellia sinensis* L.). *Biochem. J.* 160: 171–179.

Suzuki, T., Ashihara, H., and Waller, G.R. (1992). Purine and purine alkaloid metabolism in *Camellia* and *Coffea* plants. *Phytochemistry* 31: 2575–2584.

Uefuji, H., Ogita, S., Yamaguchi, Y. et al. (2003). Molecular cloning and functional characterization of three distinct *N*-methyltransferases involved in the caffeine biosynthetic pathway in coffee plants. *Plant Physiol.* 132: 372–380.

Varbanova, M., Yamaguchi, S., Yang, Y. et al. (2007). Methylation of gibberellins by Arabidopsis GAMT1 and GAMT2. *Plant Cell* 19: 32–45.

Waller, G.R., MacVean, C.D., and Suzuki, T. (1983). High production of caffeine and related enzyme activities in callus cultures of *Coffea arabica* L. *Plant Cell Rep.* 2: 109–112.

Weevers, T. (1930). Die Funkion der Xanthinderivate im Pflanzenstoffwechsel. *Arch. Neerl. Sci. 3B* 5: 111–195.

Yabuki, N. and Ashihara, H. (1991). Catabolism of adenine nucleotides in suspension-cultured plant cells. *Biochim. Biophys. Acta* 1073: 474–480.

Yang, Y., Yuan, J.S., Ross, J. et al. (2006). An *Arabidopsis thaliana* methyltransferase capable of methylating farnesoic acid. *Arch. Biochem. Biophys.* 448: 123–132.

Yoneyama, N., Morimoto, H., Ye, C.X. et al. (2006). Substrate specificity of *N*-methyltransferase involved in purine alkaloids synthesis is dependent upon one amino acid residue of the enzyme. *Mol. Genet. Genomics* 275: 125–135.

Yue, Y. and Guo, H. (2014). Quantum mechanical/molecular mechanical study of catalytic mechanism and role of key residues in methylation reactions catalyzed by dimethylxanthine methyltransferase in caffeine biosynthesis. *J. Chem. Inf. Model.* 54: 593–600.

Zhao, N., Ferrer, J.-L., Ross, J. et al. (2008). Structural, biochemical, and phylogenetic analyses suggest that indole-3-acetic acid methyltransferase is an evolutionarily ancient member of the SABATH family. *Plant Physiol.* 146: 455–467.

Zubieta, C., Ross, J.R., Koscheski, P. et al. (2003). Structural basis for substrate recognition in the salicylic acid carboxyl methyltransferase family. *Plant Cell* 15: 1704–1716.

16

Physiological and Ecological Aspects of Purine Alkaloid Biosynthesis

16.1 Introduction

In Chapter 15 caffeine biosynthesis was reviewed from the viewpoint of biochemistry and molecular biology. In this chapter, physiological and ecological aspects of caffeine production are summarized. There are many physiological-based publications related to organs and tissue age, cellular and subcellular location, and control mechanisms of caffeine biosynthesis. In addition, there are hypotheses concerning the ecological roles of caffeine in plants. The 'allelopathic or autotoxic function theory' proposes that caffeine in coffee berries is released into the soil where it inhibits germination of seeds around the parent plants. The 'chemical defence theory' proposes that caffeine in young leaves, fruits, and flower buds of species such as tea and coffee acts as a defence to protect young soft tissues from pathogens and herbivores.

16.2 Physiology of Caffeine Biosynthesis

There are investigations involving the caffeine content of each organ, *in situ* tracer experiments using tissues, measurement of enzyme activity, and determination of gene expression. In general, caffeine biosynthesis occurs in young tissues, in particular leaves of flush shoots, young fruits, and flower buds. In tea and coffee plants, purine alkaloid biosynthesis activity appears to be seasonal and related to the age of tissues, such as young leaves, although the xanthine degradation pathway of purines is often operating as conventional plant purine catabolism (Figure 16.1). Details of the purine catabolism pathway related to the oxidation of xanthine are described in Chapter 6.

Theobromine synthesis from purine nucleotides via xanthosine appears to be restricted to young tissues, although conversion of theobromine to caffeine (Figure 16.1) occurs in the later stages of tissue development. Transient accumulation of theobromine is, therefore, often observed in young tissues of tea and coffee plants. Caffeine appears to be produced from purine precursors in the tissues in which it accumulates. In general, the highest concentrations of caffeine are in young leaves and fruits, but total content per organ increases gradually during development and maturation. After that the amounts in leaves and pericarp decrease with age, especially in leaves. An absence of purine alkaloids in senescent leaves is observed following abscission from the parent plant.

In general, secondary metabolites like alkaloids accumulate in the vacuoles of cells of vegetative tissues and seeds and seed coats of fruits. However, biosynthesis is often

Plant Nucleotide Metabolism – Biosynthesis, Degradation, and Alkaloid Formation, First Edition.
Hiroshi Ashihara, Iziar A. Ludwig and Alan Crozier.

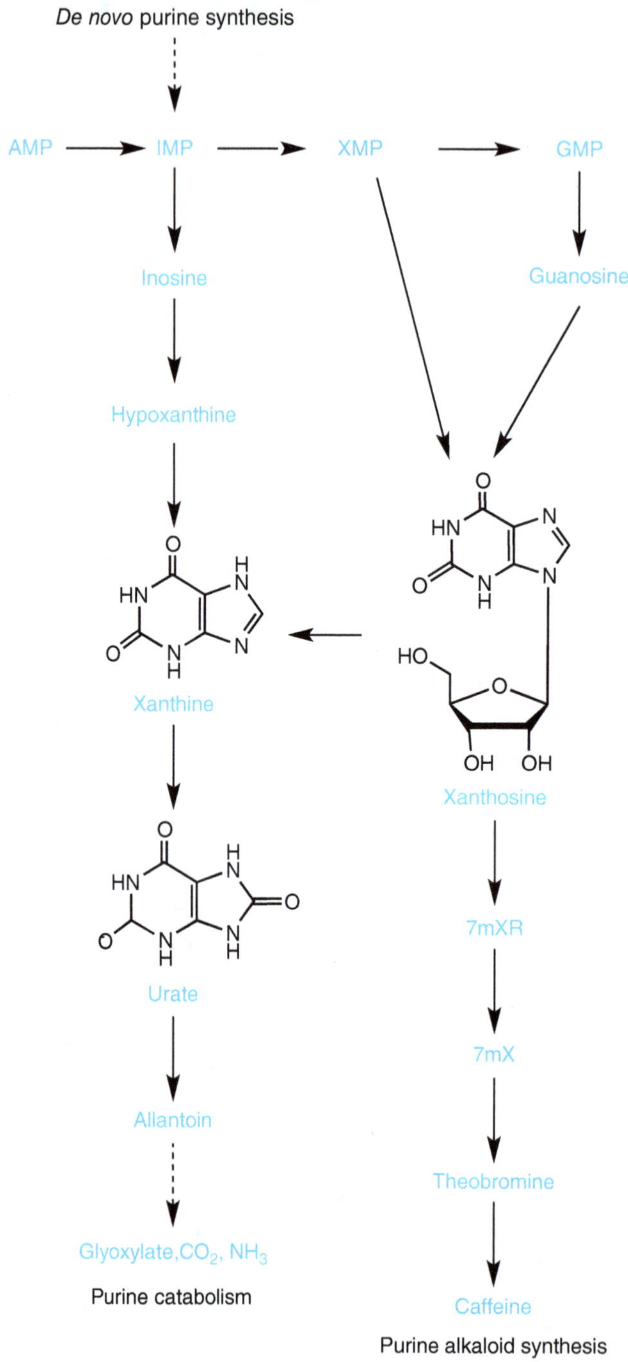

Figure 16.1 The major routes of caffeine biosynthesis and purine nucleotide catabolism in tea and coffee plants.

carried out at other sites and translocated from source cells to sink tissues. For example, the pyridine alkaloid, nicotine, is synthesized exclusively in roots, loaded into the xylem, and transported upwards in the transpiration stream to the leaves (Shoji and Hashimoto 2011). For such a translocation, at least three membrane transport events would have to be involved, namely, export from the plasma membrane in root cells, import at the plasma membrane in leaf cells, and additional transport into the vacuoles in leaf cells (Yazaki et al. 2008). In contrast to nicotine, long-distance transport of caffeine has not been observed, although intracellular transport to vacuoles may occur. Organ and subcellular localization of caffeine in biosynthesis are discussed below.

16.2.1 Purine Alkaloid Biosynthesis in Different Species

Purine alkaloid biosynthetic activity in different organs and tissues was investigated using ^{14}C-feeding experiments, determination of enzyme activity and/or expression of caffeine synthase genes. Activity has been detected in leaves, flowers, and fruits. In general, high biosynthetic activity is found in young tissues and disappears with maturation, while the activity in roots is low. The main biosynthetic pathway of purine alkaloids in tea and coffee plants has been well established, although the pathways operating in other species has not been studied in similar detail. It has been reported that the recombinant *N*-methyltransferases involved in purine alkaloid synthesis, obtained from cacao (*Theobroma cacao*), guaraná (*Paullinia cupana*), and orange (*Citrus sinensis*), show different substrate specificities (Huang et al. 2016). The potential pathways are shown in Figure 16.2, although experiments with recombinant enzymes do not necessarily accurately reflect the *in vivo* purine alkaloid synthesis pathways. Related, more physiological studies are discussed below.

16.2.2 *Camellia*

With *Camellia* plants, ^{14}C-feeding precursor experiments have often used leaf tissues and/or fruits of different ages. In tea, caffeine biosynthesis from [8-^{14}C]adenine was found to occur exclusively in the green chlorophyll-containing buds, leaves, and shoots. The biosynthesis of theobromine and caffeine was extremely high in young leaves, and decreased as four-month-old seedlings aged. In contrast, very limited incorporation of [8-^{14}C]adenine into these alkaloids was detected in the roots and cotyledons (Ashihara and Kubota 1986). Seasonal variations in purine alkaloid biosynthesis from [8-^{14}C]adenine were studied at monthly intervals over a 12-month period using leaf discs from the shoots of tea plants growing under natural field conditions. Biosynthesis of theobromine and caffeine was found only in the young leaves harvested from April to May (Fujimori et al. 1991) (Table 16.1).

Similar feeding experiments with *Camellia irrawadiensis* detected theobromine biosynthesis in stamens and petals of flowers. The biosynthetic activity was higher in the flower buds than in the flowers (Fujimori and Ashihara 1990). A detailed investigation of caffeine biosynthesis in the stamens isolated from flower buds of tea has also been reported (Fujimori and Ashihara 1993).

Caffeine biosynthesis is also found in fruits of *Camellia sinensis* plants. The activity has been detected in the pericarp, cotyledons, and seed coats (Table 16.1). The biosynthetic activity was high in young tissues of all parts of the tea fruits, especially in the pericarp

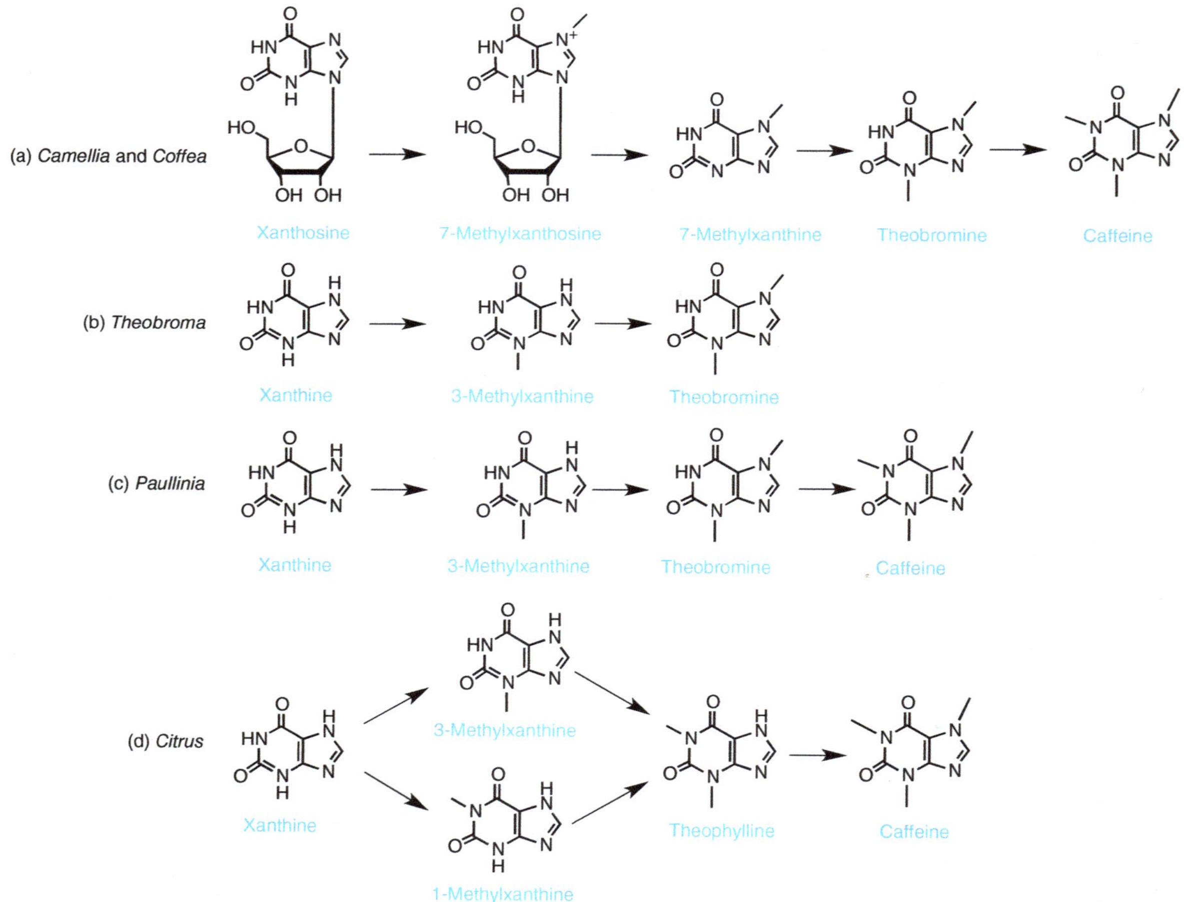

Figure 16.2 The putative reactions of caffeine biosynthesis in different plants species: (a) native and recombinant enzymes from tea and coffee: (b) major recombinant enzymes in cacao, reactions involving in xanthosine may also be involved; (c) recombinant enzymes from guaraná; and (d) recombinant enzymes from citrus.

Table 16.1 Purine alkaloid biosynthesis in different organs of tea (*Camelia sinensis*).

Samples	Organs	Stage	Synthesis (%)
Tea seedlings[a)]	Leaves	Young	100[b)]
		Middle	14
		Aged	11
	Stem	Upper	12
		Lower	4
	Cotyledons		<1
	Root		<1
Tea leaves[c)]	Leaves	April	100[b)]
		May	81
		June	<1
		After July	0
Tea fruits[d)]	Pericarp	April	68
		May	100[b)]
		June	73
		July	55
		September	20
		October	20
	Seed coat	June	23
		July	41
		September	27
		October	<1
	Cotyledon	July	5
		September	9
		October	<1

a) The biosynthetic activity of purine alkaloids (theobromine plus caffeine) from [8-^{14}C]adenine is expressed as % of the highest rate.
Source: Based on the data from [a] Ashihara and Kubota (1986); [c] Fujimori et al. (1991); and [d] Terrasaki et al. (1994).

(Terrasaki et al. 1994). The studies on the expression of genes involved in caffeine biosynthesis in the leaves, stems, and roots of four-month-old tea seedlings indicated that expression occurred in all organs, but the amounts of transcripts of caffeine synthase (TCS1) were much higher in young leaves than in other tissues (Li et al. 2008). The activity of *N*-methyltransferases involved in caffeine biosynthesis was highest in young leaves, but disappeared with subsequent development and growth (Fujimori et al. 1991). Using an RNA *in situ* hybridization technique, Li et al. (2007) reported that the caffeine synthase gene is expressed mainly in the palisade parenchyma and the epicuticle of tea leaves and less so in the spongy parenchyma and hypoderm.

16.2.3 *Coffea*

In coffee, caffeine biosynthetic activity has been detected in leaves and fruits. Transcript accumulation for the identified *N*-methyltransferase genes for caffeine biosynthesis was analysed by reverse transcription-polymerase chain reaction (RT-PCR) using gene-specific primer sets. The results indicated plural genes were expressed in different organs and tissues of coffee plants. The transcripts of *CmXRS1, CTS2,* and *CCS1* were detected in the developing endosperm, young leaves, and flower buds, and the amount of transcripts was highest in the developing endosperm. In the flower buds, marked expression of *CTS2* was detected but the expression of *CmXRS1* and *CCS1* was weak. In young leaves, the amounts of *CTS2* transcript were much lower than those of *CmXRS1* and *CCS1* (Mizuno et al. 2001, 2003). Transcripts for *CaXMT1* were found in all tissues with the exception of mature fruits. Transcripts for *CaMXMT1* and *CaMXMT2* were identified at high levels in young leaves, floral buds, and immature fruits. In contrast, those of *CaDXMT1* were detected predominantly in immature fruits (Uefuji et al. 2003).

Feeding experiments have revealed that incorporation of [8-^{14}C]adenine into theobromine and caffeine occurs in small, young leaves of *Coffea arabica* but disappears in fully developed leaves (Fujimori and Ashihara 1994). In six-month-old *C. arabica* seedlings, theobromine, and caffeine are synthesized from [8-^{14}C]adenine only in young leaves and young shoots including the buds, but not roots or aged cotyledons (Zheng and Ashihara 2004).

Caffeine biosynthesis activity was found in the pericarp and seeds of *C. arabica* and *Coffea canephora* fruits. A high biosynthetic activity, estimated from *in situ* [8-^{14}C]adenine metabolism experiments, was found in small, immature fruits, which consist of perisperm and pericarp, and in developing seeds (endosperm). The biosynthetic activities of caffeine were reduced in both the pericarp and seeds with the growth and ripening of fruits (Table 16.2) (Koshiro et al. 2006). Gene expression studies revealed that the transcripts of three *N*-methyltransferase genes for caffeine biosynthesis *CmXRS1, CTS2,* and *CCS1* were detected in every stage of growth, although the amounts of the transcripts were significantly lower in later stages of fruit ripening. The pattern of expression of genes for caffeine synthesis during growth is broadly related to the *in situ* synthesis of caffeine from adenine nucleotides. The amounts of the transcripts of the three *N*-methyltransferase genes were higher in the seeds than in the pericarp (Koshiro et al. 2006). These observations strongly suggest that caffeine accumulated in coffee seeds is synthesized predominantly in the seeds themselves. This contradicts an early report in which Baumann and Wanner (1972) proposed that in *C. arabica* fruits caffeine in the seeds was derived from the pericarp.

16.2.4 *Theobroma*

The biosynthetic activity of theobromine has been investigated in the fruits, using three different growth stages, and the leaves of cacao (*T. cacao*) (Zheng et al. 2004). Feeding experiments with [8-^{14}C]adenine showed that theobromine biosynthesis occurred in small, immature fruits, the pericarp of developing fruits, and cotyledons of large size fruits (Table 16.3). No theobromine biosynthesis was detected in the pericarp, placenta, and seed coats. Therefore, very limited amounts of purine alkaloids may be transported

Table 16.2 Biosynthesis of theobromine (Tb) and caffeine (Cf) from [8-^{14}C]adenine in fruits of different coffee species during 18 h incubation.

Stage		*Coffea arabica* (cv. Mokka)		*C. arabica* (cv. Catimor)		*C. canephora*	
		Biosynthesis		Biosynthesis		Biosynthesis	
		Tb	Cf	Tb	Cf	Tb	Cf
A	Whole fruit	—	—	1.90	1.10	0.97	—
B	Whole fruit	1.42	0.67	2.40	2.60	1.93	0.51
C	Whole fruit	2.20	1.65	1.20	1.90	2.51	1.09
D	Pericarp	0.33	0.08	0.10	—	0.54	0.13
	Seed	0.44	1.25	0.20	0.50	0.03	0.07
E	Pericarp	0.30	0.10	0.20	0.10	0.66	0.23
	Seed	0.25	0.56	0.10	0.20	0.02	0.09
F	Pericarp	0.02	—	0.10	—	—	—
	Seed	0.01	—	—	—	—	—
G	Pericarp	0.09	—	—	—	—	—
	Seed	0.01	—	—	—	—	—

—, No radioactivity was incorporated.
The values are expressed as kBq 100 mg^{-1} f.w.
Source: Based on the data of Koshiro et al. (2006).

Table 16.3 Biosynthesis of purine alkaloids from [8-^{14}C]adenine in different organs of cacao (*Theobroma cacao*) fruits.

Stage	Organs	3mX	7mX	Theobromine	Caffeine
A	Whole fruit	1.1	0.3	4.4	—
B	Pericarp	0.3	0.2	0.7	—
	Seed	—	—	—	—
C	Pericarp	—	—	—	—
	Cotyledon	—	—	26.7	—
	Placenta	—	—	—	—
	Seed coat	—	—	—	—

3mX, 3-Methylxanthine; 7mX, 7-methylxanthine. —, No radioactivity was incorporated.
The values are expressed as kBq 100 mg^{-1} f.w. during 18 h incubation.
Source: Based on the data of Zheng et al. (2004).

from the pericarp to seed tissue in young fruits, but most alkaloids appeared to be synthesized in the cotyledons of cacao seeds.

The purine metabolism in young small leaves, developing intermediate-sized leaves, and fully expanded leaves from flush shoots of cacao trees, and aged leaves from one-year-old shoots have been investigated (Koyama et al. 2003). Biosynthesis of theobromine from various purine precursors was found only in young leaves (Table 16.4) with the data indicating the following biosynthetic pathways for theobromine:

Table 16.4 Biosynthesis of purine alkaloids from [8-^{14}C]purine bases and nucleosides in young leaves of cocoa.

	3mX	7mX + 7mXR	Theobromine	Caffeine
Adenine	0.3	1.9	6.7	—
Adenosine	0.3	1.6	7.8	—
Guanine	0.6	1.3	7.6	—
Guanosine	1.7	0.8	6.9	—
Hypoxanthine	1.0	0.6	12.3	—
Inosine	1.0	0.6	8.8	—
Xanthine	0.9	—	8.7	—
Xanthosine	2.5	—	16.2	—

—, No radioactivity was incorporated. 3mX, 3-methylxanthine; 7mX + 7mXR, 7-methylxanthine and 7-methylxanthosine.
The values are expressed as % of total radioactivity taken up by the samples.
Source: The results are from Koyama et al. (2003).

(a) AMP → IMP → XMP → xanthosine → 7-methylxanthosine → 7-methylxanthine → theobromine
(b) GMP → guanosine → xanthosine → 7-methylxanthosine → 7-methylxanthine → theobromine
(c) xanthine → 3-methylxanthine → theobromine.

Although no caffeine biosynthesis from ^{14}C-labelled purine bases and nucleosides was observed during 18 hours incubations, exogenously supplied ^{14}C-theobromine was converted to caffeine by young leaves. The accumulation of small amounts of caffeine observed in cacao leaves can be explained by the fact that its conversion from theobromine is very slow as theobromine is rapidly catabolized in the leaves. The theobromine biosynthesis enzymes related to route (c) have not yet been isolated, but this activity has been found with the recombinant enzyme from cacao (Huang et al. 2016). In cacao it is, therefore, possible that theobromine is produced from xanthine via 3-methylxantine. TcCS1 can utilize xanthine, 1-methylxanthine, 7-methylxanthine, and xanthosine as substrates, and is able to convert xanthine to 3-methylxanthine, while TcCS2 catalyses the formation of theobromine from 3-methylxanthine as illustrated in Figure 16.2b.

16.2.5 Maté

Biosynthesis of purine alkaloids in maté was investigated using ^{14}C-labelled adenine, hypoxanthine, guanosine, theobromine, theophylline, and xanthine (Ashihara 1993; Yin et al. 2015). In spring, adenine, hypoxanthine, and guanosine were incorporated into caffeine in the young, pale-green leaves but not in the mature, dark-green leaves (Table 16.5). In winter, maté leaves accumulated caffeine but not theobromine, indicating that caffeine is the end product of purine alkaloid synthesis. In young maté leaves, significant amounts of [8-^{14}C]theobromine and [8-^{14}C]theophylline were utilized for caffeine biosynthesis, but the conversion was not observed in mature leaves.

Table 16.5 Biosynthesis of purine alkaloids from ^{14}C-labelled precursors in maté (*Ilex paraguariensis*) leaves during 24 h incubation.

^{14}C-Precursor	Products	Young leaves	Mature leaves	References
Adenine	Tb	3.5	—	Ashihara (1993)
	Cf	8.1	—	
Hypoxanthine	Tb	9.4	—	
	Cf	5.8	—	
Guanosine	Tb	5.8	—	
	Cf	6.0	—	
Theobromine	Cf	24.7	0.6	Yin et al. (2015)
Theophylline	Cf	32.0	—	
Xanthine	3mX	1.5	1.1	
	Tb	9.0	—	
	Tp	1.0	—	
	Cf	4.0	—	

—, No radioactivity was incorporated. Cf, caffeine; 3mX, 3-methylxanthine; Tb, theobromine; Tp, theophylline.
The values are expressed as % of radioactivity taken by the leaf segments.

Xanthine was catabolized extensively via ureides in both young and mature leaves, with only limited amounts being utilized for the synthesis of theobromine, theophylline, and caffeine (Table 16.5). Feeding experiments suggest that in addition to the main theobromine → caffeine route, an alternative minor xanthine → 3-methylxanthine → theophylline → caffeine pathway also occurs (Yin et al. 2015). The putative sequence and predicted structure of the caffeine synthase of maté has been reported (Debat et al. 2014), but the enzymatic properties have not yet been investigated.

16.2.6 Guaraná

In guaraná (*Paullinia cupana*), the caffeine synthase gene, *PcCS*, is expressed in all tissues. Transcript levels in young leaves and the apical stem are much higher than in mature leaves and roots. In fruits, expression was observed in both the seeds and pericarp. The highest expression occurred in immature seeds. Transcript levels decreased markedly as the fruits matured. These results obtained using RT-PCR were confirmed by *in situ* hybridization analysis; *PcCS* was expressed in the cotyledon of the seeds of immature and intermediate stages of fruit maturation but not in mature fruits (Schimpl et al. 2014).

A recombinant PcCS allowed functional characterization of the enzyme as a bifunctional CS, able to catalyse the methylation of 7-methylxanthine to theobromine, and theobromine to caffeine. It has been reported that two different recombinants, PcCS1 and PcCS2, are *N*-methyltransferases which, respectively, catalysed the reactions, xanthine to 3-methylxanthine and 3-methylxanthine to theobromine (Huang et al. 2016). If three recombinants PcCS1, PcCS2, and PcCS are actually operative in guaraná fruits, the pathway starts with xanthine and the route may be xanthine → 3-methylxanthine → theobromine → caffeine (Figure 16.2).

16.2.7 *Citrus*

The occurrence of caffeine and theophylline in *Citrus* flowers (Kretschmar and Baumann 1999) has already been discussed in Section 14.3.7. Phylogenetic studies found that *C. sinensis* expresses two duplicate methylxanthine synthases in the caffeine-containing flowers (Huang et al. 2016). The CisXMT1 catalyses both the formation of 1- and 3-methylxanthine from xanthine, and theophylline from 1- and 3-methylxanthine. A second enzyme, CisXMT2 catalyses caffeine synthesis from theophylline (Figures 16.1 and 16.2). In *Citrus*, theophylline appears transiently in developing flower buds, which accumulate caffeine (Kretschmar and Baumann 1999). Therefore, theophylline appears to be the dimethylxanthine intermediate in caffeine biosynthesis. In line with these findings, metabolic studies also indicate that xanthine is the principal substrate for purine alkaloid biosynthesis in citrus species.

16.3 Subcellular Localization of Caffeine Biosynthesis

Plant cells are subdivided compartments, with chloroplasts, cytoplasts, and vacuoles being the major sites in the biosynthesis and accumulation of secondary metabolites. There is no conclusive evidence on the subcellular localization of methylxanthines, although it has been speculated that caffeine accumulates in vacuoles of leaf cells. It has also been proposed that in coffee, caffeine occurs in the vacuoles as a complex with chlorogenic acids (caffeoylquinic acids) (Mosli Waldhauser and Baumann 1996). The subcellular distribution of enzymes involved in caffeine biosynthesis is summarized below.

16.3.1 Caffeine Synthase

The subcellular localization of caffeine synthase has been the subject of controversy. The results of traditional biochemical fractionation of intracellular organelles indicated that caffeine synthase was associated with chloroplasts (Kato et al. 1998; Koshiishi et al. 2001). In contrast, subsequent molecular investigations on the localization of caffeine synthase suggest that it is probably located in the cytosol (Ogawa et al. 2001) or the external surface of the vacuole (Kumar et al. 2007).

Kato et al. (1998) examined the subcellular distribution of caffeine synthase from young tea leaves using a Percoll density gradient. The vast majority of the caffeine synthase activity appeared as a sharp peak corresponding with the distribution of chlorophyll. The distribution of triose phosphate isomerase activity, a marker for both chloroplasts and cytosol, indicated that little or no caffeine synthase activity was present in the cytosol. A crude chloroplast preparation was obtained that was washed then purified using a Percoll density gradient. The purified chloroplasts were frozen with liquid N_2 and stored at −80 °C overnight. The thawed samples were centrifuged at 10 000 *g* after which caffeine synthase activity was found in the supernatant. It was concluded that the enzyme is located in the stroma of the chloroplasts (Kato et al. 1998). This is also supported by the fact that the optimum pH for tea caffeine synthase activity was 8.5, which is the typical pH optimum of various enzymes located in the stroma of chloroplasts (Foyer 1984).

A molecular approach to investigate the localization of caffeine synthase in coffee has been carried out by two groups. Ogawa et al. (2001) used a cDNA fragment covering the entire coding region of the caffeine synthase gene (*CaMXMT1*) fused to pGFP2. When the resulting plasmid was introduced into the epidermal layer of onions by particle bombardment, green fluorescence was detected in the cytoplasm. Kodama et al. (2008) confirmed this result and demonstrated that all three *N*-methyltransferases involved in caffeine biosynthesis were localized in the cytosol. Kumar et al. (2007) investigated the localization of caffeine synthase in coffee endosperm cells using the promoter for one of the *N*-methyltransferase gene families involved in caffeine biosynthesis. These constructs and *pCAMBIA 1301* bearing the intron *uidA* gene driven by the cauliflower mosaic virus *35S* promoter, were electroporated into coffee endosperm, and the activity of *β*-glucuronidase (GUS) localized. In tissues transformed with the construct-containing promoter and first exon, enzymatic activity was localized on the outer surface of the vacuole. Antibodies to coffee caffeine synthase were also specifically localized in the same region. In tissues bearing either the caffeine synthase-GUS construct without the first exon or pCAMBIA 1301 with intron GUS, GUS activity was spread throughout the cytoplasm. These observations indicate that the *N*-methyltransferase is located on the external surface of the vacuole.

16.3.2 The *De Novo* Route Enzymes

There are several reports on the subcellular localization of enzymes involved in xanthosine formation via the *de novo* route. Although no data are available on the cellular localization of *de novo* purine nucleotide synthesis in xanthine alkaloid-forming plants, it has been shown in *Arabidopsis thaliana* that all genes encoding enzymes required for the 10-step synthesis of inosine-5′-monophosphate (IMP) from 5-phosphoribosyl-1-pyrophosphate (PRPP) (steps 1–10 in Figure 4.1), contain sequences that are predicted to encode *N*-terminal plastid-transit peptides (Zrenner and Ashihara 2011; Zrenner et al. 2006). This observation suggests chloroplast localization of *de novo* IMP synthesis. Experimental evidence for such a localization of these enzymes has been obtained but is fragmentary, showing only localization of the PRPP amidotransferase in *A. thaliana* (Hung et al. 2004). In *Vigna unguiculata* (cowpea), 5-aminoimidazole ribonucleotide (AIR) synthase was immunolocalized in chloroplasts and mitochondria (Smith et al. 1998). The genes encoding inosine-5′-monophosphate dehydrogenase, which is involved in the conversion of IMP to xanthosine-5′-monophosphate (XMP), do not contain *N*-terminal transit peptides. This suggests that the synthesis of XMP from IMP may occur in the cytosol, although experimental evidence supporting this possibility in plants is, as yet, lacking.

The situation may be somewhat different in non-symbiotic plants. Atkins et al. (1997) investigated intracellular localization of *de novo* purine synthesis in ureide-forming cow-pea (*V. unguiculata*) nodules using sucrose and Percoll density gradient centrifugation. Enzyme activity analysis showed that both plastids and mitochondria had a full complement of enzymes for *de novo* purine synthesis from ribose-5-phosphate to IMP. The less soluble enzymes of purine oxidation including IMP dehydrogenase, were not associated with either of these organelles.

16.3.3 The AMP Route Enzymes

Adenosine monophosphate (AMP) is synthesized from IMP via adenylosuccinate by the *de novo* purine biosynthesis (steps 11 and 12 in Figure 4.1). The cDNAs encoding the enzymes for this conversion, adenylosuccinate lyase and adenylosuccinate synthase, have been cloned from several plant species, including *A. thaliana*, and shown to contain putative plastid transit sequences (van der Graaff et al. 2004). Therefore, all steps of AMP biosynthesis from PRPP, namely, the *de novo* purine biosynthetic pathway and the AMP branch point pathway from IMP are located in the chloroplasts. Although AMP is produced mainly in the chloroplasts, it is exported to the cytosol and other organelles. As discussed above, the AMP route of xanthosine synthesis is catalysed by AMP deaminase, IMP dehydrogenase, and 5′-nucleotidase. Since amino acid sequences of these enzymes do not contain an *N*-terminal transit peptide, the enzymes are probably located in the cytosol.

16.3.4 The SAM Route Enzymes

Using traditional biochemical isolation techniques, Koshiishi et al. (2001) showed that caffeine synthase activity and part of the activities of *S*-adenosyl-L-homocysteine (SAH) hydrolase, adenosine nucleosidase, adenine phosphoribosyltransferase, and adenosine kinase are located in tea chloroplasts. In contrast, no *S*-adenosyl-L-methionine (SAM) synthetase activity was detected in the purified chloroplast fraction. Although localization of caffeine synthase was not part of the study, chloroplastic localization of adenosine kinase, SAH hydrolase, and adenine phosphoribosyltransferase has been reported. Schoor et al. (2011) used immunogold labelling to detect adenosine kinase and SAH hydrolase isoforms in the cytosol, chloroplast, and nucleus of *A. thaliana*. Chloroplastic adenine phosphoribosyltransferase1 (P31166, AT1G27450) is registered.

16.3.5 Subcellular Localization and Transport of Intermediates

The precursors of purine alkaloid biosynthesis, namely, purine nucleotides, may be synthesized in chloroplasts with the final products, typically caffeine or theobromine, accumulating in vacuoles. In some natural product biosynthesis pathways, metabolic channelling with a multi-enzyme complex has been postulated (Jorgensen et al. 2005). Using bimolecular fluorescence complementation analyses, Kodama et al. (2008) revealed that each of the three *N*-methyltransferases involved in caffeine biosynthesis in coffee form not only homo-dimers but also hetero-dimers. They speculated that if dimerization among three closely related proteins resulted in minimal functional interference, there may be a rapid conversion of the initial substrate to the final product, effectively as a single step. Further studies are necessary to confirm the possible metabolic channelling of caffeine biosynthesis.

16.4 Regulation of Caffeine Biosynthesis

Similar to the biosynthesis of other secondary metabolites, caffeine and theobromine biosynthesis is regulated primarily by gene expression and the subsequent synthesis of enzymes. Changes in the expression rates of genes (transcript level), quantity (activity) of key enzymes of caffeine biosynthesis and accumulation have often been associated with various physiological changes.

The expression of *N*-methyltransferases is, for instance, closely associated with the accumulation of caffeine in tea leaves (Deng et al. 2015a; Kato et al. 2010; Li et al. 2008; Mohanpuria et al. 2011; Mohanpuria et al. 2009), coffee fruits (Koshiro et al. 2006; Perrois et al. 2015), stressed leaves of cacao (Bailey et al. 2005a, b), and guaraná plants (Schimpl et al. 2014). *In vitro* activity of *N*-methyltransferases is correlated with the expression of enzymes in tea leaves (Fujimori et al. 1991). There are several examples of regulation of caffeine biosynthetic genes in response to salicylic acid and methyl jasmonate (Kumar and Giridhar 2015), light (Kumar et al. 2015b), drought, and salinity (Kumar et al. 2015a). The data suggest that transcriptional regulation of caffeine biosynthesis genes has a major role in the accumulation of caffeine.

Minor modification of the rate of caffeine biosynthesis may also be regulated by 'fine control' comprising the supply of substrates and/or methyl donor and feedback control of enzyme activity. The results obtained with tea tissue culture have shown that adding precursors of purine nucleotides, adenosine, guanosine, or hypoxanthine, to the culture medium, influences the rate of growth but not the accumulation of caffeine. This suggests that caffeine biosynthesis is not influenced by the availability of purine precursors (Deng et al. 2008). In contrast, addition of paraxanthine, doubled the level of caffeine compared to controls. This implies that availability of SAM is not a principal factor in the control of caffeine biosynthesis and points to 7-methylxanthosine synthase activity being the most plausible rate-limiting factor of caffeine biosynthesis (Deng et al. 2008).

16.5 Ecological Roles of Caffeine

There are hypotheses concerning the ecological roles of caffeine in plants. The 'allelopathic or autotoxic function theory' proposes that caffeine in coffee seeds, which fall to the ground, is released into the soil where it inhibits germination of seeds around the parent bushes. The 'chemical defence theory' proposes that caffeine in young leaves, fruits, and flower buds acts as a defence to protect young soft tissues from pathogens and herbivores. There are currently several examples indicating that caffeine can have a toxic effect. The effective caffeine doses for inhibiting the development of various organisms are summarized in Table 16.6.

16.5.1 Allelopathic Function Theory

The term 'allelopathy' was first used by Molisch (1937). It is utilized for chemical interactions whereby a plant inhibits the growth of neighbouring plants. Recently, the definition was enlarged to include interaction of plants with microorganisms. This topic is covered in Section 16.5.4 on chemical defence. Allelochemicals include phenolic acids, glucosinolates, terpenes, and flavonoids. Compared to these secondary metabolites, there is little information on the allelopathic activity of alkaloids (Anaya et al. 2006; Lovett and Hoult 1998). A high allelopathic effect of alkaloids has been observed in leachates of coffee leaves and roots and extracts from the soil of coffee plantations (Anaya et al. 1987). Waller et al. (1986) reported on the presence of caffeine in soils of long-established coffee

Table 16.6 Effective caffeine dose for biological activity.

Organism	Species	Dose (%)
Bacterium	*Pseudomonas syringae*	0.04
Fungus	*Crinipellis perniciosa*	0.05
	Monacrosporium ambrosium	0.05–0.5
	Monacrosporium ambrosium	0.5–1.0
Insect (larvae)	*Pieris rapae* (small white butterfly)	0.05
	Manduca sexta (tobacco horn moth)	0.03–10
	Vanessa cardui (painted lady butterfly)	0.1–0.3
	Tenabrio ssp. (mealworm)	0.1–0.3
	Oncopeltus fasciatus (milkweed bug nymph)	0.3
	Culex ssp. (mosquito)	0.0007
	Tribolium confusum (confused flour beetle)	0.2
	T. castaneum (red flour beetle)	0.2
Mollusc	Snail	0.1
	Slug	0.01–0.1
Plant	*Amaranthus spinosum*	0.06–0.12
	Lactuca sativa (lettuce)	0.03
	Lactuca sativa (lettuce) protoplasts	0.01–0.02
	Avena fatua	0.2–1
	Vicia sativa	0.5–1

Source: Adapted from Kim et al. (2010).

plantations. As young roots of coffee plants are highly susceptible to caffeine, this may cause autotoxicity.

16.5.2 Effect of Caffeine on Plant Growth

Around 20% of the caffeine in the seeds of germinating *C. arabica* is excreted when radicles emerge (Baumann and Gabriel 1984; Friedman and Waller 1983). Friedman and Waller (1983) reported that the growth of radicles was almost completely inhibited by >10 mM caffeine. The tips of radicles were darkened within five days and subsequently died. In addition, 5–20 mM caffeine reduced hypocotyl extension by ~50%. A similar but milder inhibition was observed with theophylline. Purine alkaloids excreted by coffee seeds, thus, appear to have autotoxic properties.

The effect of xanthine alkaloids on the growth of plant species that do not produce the purine alkaloids has been investigated. Chou and Waller (1980) reported inhibitory effects on the growth of radicles of lettuce seedlings at concentrations of 0.5–2.0 μM. Inhibition by caffeine, theophylline, paraxanthine, and theobromine ranged from 54–90%, 58–81%, 62–85%, and 32–64%, respectively. Rizvi et al. (1981) reported the selective phytotoxicity of caffeine between leguminous plants and some weeds. Germination of *Amaranthus spinosus* and *Echinochloa colonum* seed was inhibited by less than ~6 mM caffeine, while in contrast, it had no significant effect on the germination of *Phaseolus mungo*. Smyth (1992) found that 2.5 mM caffeine caused, respectively, 90% and 50% inhibition in the elongation of roots and shoots of rice seedlings.

Little is known as to whether or not purine alkaloids function as allelochemicals and directly affect cell proliferation of non-purine alkaloid producing plants. Sasamoto et al. (2015) reported the effect of four purine alkaloids on the proliferation of lettuce cells using a method specifically designed to investigate allelopathy. Cell division of protoplasts was inhibited significantly by 250 μM caffeine which had a marked inhibitory effect on colony formation of cells derived from the protoplasts. These observations suggest that a relatively low concentration of caffeine can be toxic for the proliferation of plant cells.

Traditionally, caffeine has been utilized in cytological studies to induce the formation of binucleate cells (Kihlman 1977) and, as a consequence, it has frequently been used for a variety of purposes in plant cell research (Amino and Nagata 1996; Kihlman 1949; Kihlman and Leven 1949; Manandhar et al. 1996; Mineyuki et al. 1989). Caffeine is a nucleotide derivative, and exogenous caffeine may, therefore, directly influence nucleotide metabolism in plants. The short-term effect of 1 mM caffeine on nucleotide metabolism was examined in seedlings of rice (*Oryza sativa*) by Deng et al. (2015b). The exogenous caffeine increased the rate of biosynthesis of nucleotides. Increased nucleoside and salvage activity has often been observed in the lag phase of proliferation of cultured plant cells (Kanamori-Fukuda et al. 1981; Shimazaki and Ashihara 1982; Shimazaki et al. 1982; Yin et al. 2007, 2014). Therefore, it can be speculated that caffeine increased the nucleotide level for rapid proliferation of cells.

The effect of caffeine on the expression profiles of soluble proteins was carried out in shoot-root axes of rice seedlings by Deng et al. (2015c). Two-dimensional difference gel electrophoresis combined with matrix-assisted laser desorption/ionization-time of flight mass spectrometry was employed for the separation and identification of proteins. The results indicated that the size of 65 protein spots was changed by treatment with 1 mM caffeine. Among them, caffeine reduced 51 proteins and, arguably, growth inhibition might be related to the reduced expression of these proteins.

16.5.3 Allelopathy in Natural Ecosystems

Experimental evidence, as discussed above, suggests purine and related purine alkaloids have allelopathic activity. However, it is unclear to what extent caffeine is involved in allelopathy in natural ecosystems because soil bacteria, such as *Pseudomonas putida*, degrade purine alkaloids (see Chapter 19). The caffeine concentration in soil around mature, old coffee trees has been reported; levels in Mexican canopy soil indicated that substantial amounts of the alkaloid are released from the tree litter, which comprises fallen berries and young leaves, and over the years accumulate in the vicinity of the roots. The annual amount of litter from mature coffee trees is estimated to be ~150–200 g dry matter m^{-2} $year^{-1}$, and this may release ~1–2 g caffeine m^{-2} $year^{-1}$ (Epifanio 1981). Part of this may be degraded by soil bacteria, although Waller (1989) argued that the antimicrobial activity of caffeine may reduce catabolism of alkaloids in the soil and so prolong its retention and increase its accumulation.

16.5.4 Chemical Defence Theory

Chemical defence of plants against microorganisms and animals has been investigated. Nathanson (1984) was the first to report that natural and synthetic purine alkaloids inhibit insect feeding and are pesticidal at concentrations known to occur in plants. Such effects are due primarily to an inhibition of phosphodiesterase activity and an increase in intracellular cyclic AMP. At lower concentrations, methylxanthines are potent synergists of other pesticides known to activate adenylate cyclase in insects. These findings suggest that purine alkaloids may function as natural insecticides and that phosphodiesterase inhibitors, alone or in combination with other compounds, may be useful in insect control. Similar results were obtained by Hollingsworth et al. (2002, 2003) who reported that spraying tomato leaves with a 1% solution of caffeine deterred feeding by tobacco hornworms, while treatment of cabbage leaves and orchids with 0.01–0.1% solutions of caffeine acted as a neurotoxin and killed or repelled slugs and snails. Therefore, the high caffeine content of young leaves, fruits, and flower buds of species such as *C. arabica* and *C. sinensis* may act as a defence to protect young soft tissues from pathogens and herbivores (Frischknecht et al. 1986; Fujimori et al. 1991). It has been suggested that caffeine synthesized in response to infection of cacao leaves might inhibit the growth of fungal and bacterial pathogens and restrict their proliferation beyond the original site of infection (Aneja and Gianfagna 2001).

The chemical defence theory was verified experimentally by Sano and his coworkers using transgenic caffeine-forming plants. They reported that transgenic tobacco and chrysanthemum plants, which produced caffeine, have, respectively, repellent effects against tobacco cutworm (*Spodoptera litura*) and fungal resistance (Kim et al. 2006, 2011a, b; Uefuji et al. 2005). Kim and Sano (2008) found that caffeine-producing transgenic plants constitutively expressed defence-related genes encoding pathogenesis related proteins 1a (PR1a) and proteinase inhibitor II (PI-II) under non-stressed conditions. The transgenic tobacco plants were highly resistant against tobacco mosaic virus and *Pseudomonas syringae*. Expression of PR1a and PI-II was higher in transgenic plants than in wild-type plants following infection. Exogenous caffeine applied to wild-type tobacco leaves conferred similar resistance properties. These findings indicate that caffeine acts as a signal molecule activating the defence system of the host plants by directly or indirectly initiating gene expression. Recently, it was shown that caffeine induces the production of a mildly toxic secondary metabolite *in planta* that stimulates endogenous self-defence systems, thereby conferring tolerance or resistance against biotic stresses (Kim et al. 2010; Sano et al. 2013). These results, of what are essentially greenhouse-based studies, strongly suggest that caffeine can have a key role in the chemical defence of plants. However, its ecological function in plants that produce caffeine in a natural ecosystem remains to be determined. Biotechnology of transgenic caffeine synthesis for the chemical defence will be described in the next chapter.

16.6 Summary

The enzymes and genes participating in purine alkaloid biosynthesis in different plant species are discussed. Investigations of caffeine biosynthesis suggest that the potential biological effects of purine alkaloids are found in younger tissues where caffeine can inhibit the growth of fungal and bacterial pathogens as well as repelling slugs, snails, larvae, and insect feeders. Caffeine in the soil, leached from fallen leaves and seeds of caffeine-producing species, can have allopathic effects inhibiting the germination of seeds and the growth of neighbouring plants.

References

Amino, S. and Nagata, T. (1996). Caffeine-induced uncoupling of mitosis from DNA replication in tobacco BY-2 cells. *J. Plant Res.* 109: 219–222.

Anaya, A.L., Ramos, L., Hernandez, J.G., and Cruz-Ortega, R. (1987). Allelopathy in Mexico. In: *Allelochemicals: Role in Agriculture and Forestry (ACS Series 330)* (ed. G.R. Waller), 89–101. Washington, D.C.: American Chemical Society.

Anaya, A.L., Cruz-Ortega, R., and Waller, G.R. (2006). Metabolism and ecology of purine alkaloids. *Front. Biosci.* 11: 2354–2370.

Aneja, M. and Gianfagna, T. (2001). Induction and accumulation of caffeine in young, actively growing leaves of cocoa (*Theobroma cacao* L.) by wounding or infection with *Crinipellis perniciosa*. *Physiol. Mol. Plant Pathol.* 59: 13–16.

Ashihara, H. (1993). Purine metabolism and the biosynthesis of caffeine in maté leaves. *Phytochemistry* 33: 1427–1430.

Ashihara, H. and Kubota, H. (1986). Patterns of adenine metabolism and caffeine biosynthesis in different parts of tea seedlings. *Physiol. Plant.* 68: 275–281.

Atkins, C.A., Smith, P., and Storer, P.J. (1997). Reexamination of the intracellular localization of *de novo* purine synthesis in cowpea nodules. *Plant Physiol.* 113: 127–135.

Bailey, B.A., Bae, H., Strem, M.D. et al. (2005a). Developmental expression of stress response genes in *Theobroma cacao* leaves and their response to Nep1 treatment and a compatible infection by *Phytophthora megakarya*. *Plant Physiol. Biochem.* 43: 611–622.

Bailey, B.A., Strem, M.D., Bae, H. et al. (2005b). Gene expression in leaves of *Theobroma cacao* in response to mechanical wounding, ethylene, and/or methyl jasmonate. *Plant Sci.* 168: 1247–1258.

Baumann, T.W. and Gabriel, H. (1984). Metabolism and excretion of caffeine during germination of *Coffea arabica* L. *Plant Cell Physiol.* 25: 1431–1436.

Baumann, T.W. and Wanner, H. (1972). Untersuchungen Äber den transport von Kaffein in der Kaffeepflanze (*Coffea arabica*). *Planta* 108: 11–19.

Chou, C.-H. and Waller, G. (1980). Possible allelopathic constituents of *Coffea arabica*. *J. Chem. Ecol.* 6: 643–654.

Debat, H.J., Grabiele, M., Aguilera, P.M. et al. (2014). Exploring the genes of yerba mate (*Ilex paraguariensis* A. St.-Hil.) by NGS and *de novo* transcriptome assembly. *PLoS One* 9: e109835.

Deng, W.W., Li, Y., Ogita, S., and Ashihara, H. (2008). Fine control of caffeine biosynthesis in tissue cultures of *Camellia sinensis*. *Phytochem. Lett.* 1: 195–198.

Deng, W.-W., Li, M., Gu, C.-C. et al. (2015a). Low caffeine content in novel grafted tea with *camellia sinensis* as scions and camellia oleifera as stocks. *Nat. Prod. Commun.* 10: 789–792.

Deng, W.W., Katahira, R., and Ashihara, H. (2015b). Short-term effect of caffeine on purine, pyrimidine and pyridine metabolism in rice (*Oryza sativa*) seedlings. *Nat. Prod. Commun.* 10: 737–741.

Deng, W.W., Sasamoto, H., and Ashihara, H. (2015c). Effect of caffeine on the expression pattern of water-soluble proteins in rice (*Oryza sativa*) seedlings. *Nat. Prod. Commun.* 10: 733–736.

Epifanio, J. A., 1981. Ecology of the cafetalero agroecosystem (in Spanish). PhD thesis. Universidad Nacional Autonoma de Mexico. P. 166.

Foyer, C.H. (1984). *Photosynthesis*. New York: Wiley.

Friedman, J. and Waller, G.R. (1983). Seeds as allelopathic agents. *J. Chem. Ecol.* 9: 1107–1117.

Frischknecht, P.M., Ulmer-Dufek, J., and Baumann, T.W. (1986). Purine alkaloid formation in buds and developing leaflets of *Coffea arabica*: expression of an optimal defence strategy? *Phytochemistry* 25: 613–616.

Fujimori, N. and Ashihara, H. (1990). Adenine metabolism and the synthesis of purine alkaloids in flowers of *Camellia* plants. *Phytochemistry* 29: 3513–3516.

Fujimori, N. and Ashihara, H. (1993). Biosynthesis of caffeine in flower buds of *Camellia sinensis*. *Ann. Bot.* 71: 279–284.

Fujimori, N. and Ashihara, H. (1994). Biosynthesis of theobromine and caffeine in developing leaves of *Coffea arabica*. *Phytochemistry* 36: 1359–1361.

Fujimori, N., Suzuki, T., and Ashihara, H. (1991). Seasonal variations in biosynthetic capacity for the synthesis of caffeine in tea leaves. *Phytochemistry* 30: 2245–2248.

van der Graaff, E., Hooykaas, P., Lein, W. et al. (2004). Molecular analysis of "*de novo*" purine biosynthesis in solanaceous species and in *Arabidopsis thaliana*. *Front. Biosci.* 9: 1803–1816.

Hollingsworth, R.G., Armstrong, J.W., and Campbell, E. (2002). Pest control caffeine as a repellent for slugs and snails. *Nature* 417: 915–916.

Hollingsworth, R.G., Armstrong, J.W., and Campbell, E. (2003). Caffeine as a novel toxicant for slugs and snails. *Ann. Appl. Biol.* 142: 91–97.

Huang, R., O'Donnell, A.J., Barboline, J.J., and Barkman, T.J. (2016). Convergent evolution of caffeine in plants by co-option of exapted ancestral enzymes. *Proc. Nat. Acad. Sci.* 113: 10613–10618.

Hung, W.-F., Chen, L.-J., Boldt, R. et al. (2004). Characterization of Arabidopsis glutamine phosphoribosyl pyrophosphate amidotransferase-deficient mutants. *Plant Physiol.* 135: 1314–1323.

Jorgensen, K., Rasmussen, A.V., Morant, M. et al. (2005). Metabolon formation and metabolic channeling in the biosynthesis of plant natural products. *Curr. Opin. Plant Biol.* 8: 280–291.

Kanamori-Fukuda, I., Ashihara, H., and Komamine, A. (1981). Pyrimidine nucleotide biosynthesis in *Vinca rosea* cells: changes in the activity of *de novo* and salvage pathways during growth in a suspension culture. *J. Exp. Bot.* 32: 69–78.

Kato, A., Crozier, A., and Ashihara, H. (1998). Subcellular localization of the *N*-3 methyltransferase involved in caffeine biosynthesis in tea. *Phytochemistry* 48: 777–779.

Kato, M., Kitao, N., Ishida, M. et al. (2010). Expression for caffeine biosynthesis and related enzymes in *Camellia sinensis*. *Z. Naturforsch.* 65C: 245–256.

Kihlman, B. (1949). The effect of purine derivatives on chromosomes. *Hereditas* 35: 393–396.

Kihlman, B.A. (1977). *Caffeine and Chromosomes*. Amsterdam: Elsevier.

Kihlman, B. and Leven, A. (1949). The cytological effect of caffeine. *Hereditas* 35: 109–111.
Kim, Y.S. and Sano, H. (2008). Pathogen resistance of transgenic tobacco plants producing caffeine. *Phytochemistry* 69: 882–888.
Kim, Y.S., Uefuji, H., Ogita, S., and Sano, H. (2006). Transgenic tobacco plants producing caffeine: a potential new strategy for insect pest control. *Transg. Res.* 15: 667–672.
Kim, Y.S., Choi, Y.E., and Sano, H. (2010). Plant vaccination: stimulation of defense system by caffeine production in planta. *Plant Signal. Behav.* 5: 489–493.
Kim, Y.-S., Lim, S., Kang, K.-K. et al. (2011a). Resistance against beet armyworms and cotton aphids in caffeine-producing transgenic chrysanthemum. *Plant Biotechnol.* 28: 393–395.
Kim, Y.S., Lim, S., Yoda, H. et al. (2011b). Simultaneous activation of salicylate production and fungal resistance in transgenic chrysanthemum producing caffeine. *Plant Signal. Behav.* 6: 409–412.
Kodama, Y., Shinya, T., and Sano, H. (2008). Dimerization of *N*-methyltransferases involved in caffeine biosynthesis. *Biochimie* 90: 547–551.
Koshiishi, C., Kato, A., Yama, S. et al. (2001). A new caffeine biosynthetic pathway in tea leaves: utilisation of adenosine released from the *S*-adenosyl-L-methionine cycle. *FEBS Lett.* 499: 50–54.
Koshiro, Y., Zheng, X.Q., Wang, M. et al. (2006). Changes in content and biosynthetic activity of caffeine and trigonelline during growth and ripening of *Coffea arabica* and *Coffea canephora* fruits. *Plant Sci.* 171: 242–250.
Koyama, Y., Tomoda, Y., Kato, M., and Ashihara, H. (2003). Metabolism of purine bases, nucleosides and alkaloids in theobromine-forming *Theobroma cacao* leaves. *Plant Physiol. Biochem.* 41: 977–984.
Kretschmar, J.A. and Baumann, T.W. (1999). Caffeine in *Citrus* flowers. *Phytochemistry* 52: 19–23.
Kumar, A. and Giridhar, P. (2015). Salicylic acid and methyl jasmonate restore the transcription of caffeine biosynthetic *N*-methyltransferases from a transcription inhibition noticed during late endosperm maturation in coffee. *Plant Gene* 4: 38–44.
Kumar, V., Satyanarayana, K.V., Ramakrishna, A. et al. (2007). Evidence for localization of *N*-methyltransferase (MMT) of caffeine biosynthetic pathway in vacuolar surface of *Coffea canephora* endosperm elucidated through localization of GUS reporter gene driven by NMT promoter. *Curr. Sci.* 93: 383–389.
Kumar, A., Naik, G.K., Simmi, P.S., and Giridhar, P. (2015a). Salinity and drought response alleviate caffeine content of young leaves of *Coffea canephora* var. Robusta cv. S274. *J. Appl. Biol. Biotechnol.* 3: 50–60.
Kumar, A., Simmi, P.S., Naik, G.K., and Giridhar, P. (2015b). RP-HPLC and transcript profile indicate increased leaf caffeine in *Coffea canephora* plants by light. *J. Biol. Earth Sci.* 5: 1–9.
Li, Y.H., Gu, W., and Ye, S. (2007). Expression and location of caffeine synthase in tea plants. *Russ. J. Plant Physiol.* 54: 698–701.
Li, Y., Ogita, S., Keya, C.A., and Ashihara, H. (2008). Expression of caffeine biosynthesis genes in tea (*Camellia sinensis*). *Z. Naturforsch.* 63c: 267–270.
Lovett, J.V. and Hoult, A.H.C. (1998). Allelopathy in plants. In: *Alkaloids: Biochemistry, Ecology, and Medicinal Applications* (eds. M.F. Roberts and M.P. Wink), 337–347. London: Plenum Press.

Manandhar, G., Apostolakos, P., and Galatis, B. (1996). Cell division of binuclear cells induced by caffeine: spindle organization and determination of division plane. *J. Plant Res.* 109: 265–275.

Mineyuki, Y., Letham, D.S., and Hocart, C.H. (1989). New 3-substituted xanthines: potent inhibitors of cell plate formation. *Cell Biol. Int. Rep.* 13: 129–136.

Mizuno, K., Tanaka, H., Kato, M. et al. (2001). cDNA cloning of caffeine (theobromine) synthase from coffee (*Coffea arabica* L.). In: *Proceedings of the 19th International Scientific Colloquium on Coffee*, 815–818. Association for Science and Information on Coffee.

Mizuno, K., Kato, M., Irino, F. et al. (2003). The first committed step reaction of caffeine biosynthesis: 7-methylxanthosine synthase is closely homologous to caffeine synthases in coffee (*Coffea arabica* L.). *FEBS Lett.* 547: 56–60.

Mohanpuria, P., Kumar, V., Joshi, R. et al. (2009). Caffeine biosynthesis and degradation in tea [*Camellia sinensis* (L.) O. Kuntze] is under developmental and seasonal regulation. *Mol. Biotechnol.* 43: 104–111.

Mohanpuria, P., Kumar, V., Ahuja, P., and Yadav, S. (2011). Producing low-caffeine tea through post-transcriptional silencing of caffeine synthase mRNA. *Plant Mol. Biol.* 76: 523–534.

Molisch, H. (1937). *Der Einfluss Einer Pflanze Auf Die Andere-Allelopathie*. Jena: Gustav Fischer.

Mosli Waldhauser, S.S. and Baumann, T.W. (1996). Compartmentation of caffeine and related purine alkaloids depends exclusively on the physical chemistry of their vacuolar complex formation with chlorogenic acids. *Phytochemistry* 42: 985–996.

Nathanson, J.A. (1984). Caffeine and related methylxanthines: possible naturally occurring pesticides. *Science* 226: 184–187.

Ogawa, M., Herai, Y., Koizumi, N. et al. (2001). 7-Methylxanthine methyltransferase of coffee plants. Gene isolation and enzymatic properties. *J. Biol. Chem.* 276: 8213–8218.

Perrois, C., Strickler, S., Mathieu, G. et al. (2015). Differential regulation of caffeine metabolism in *Coffea arabica* (Arabica) and *Coffea canephora* (Robusta). *Planta* 241: 179–191.

Rizvi, S.J.H., Mukerji, D., and Mathur, S.N. (1981). Selective phyto-toxicity of 1,3,7-trimethylxanthine between *Phaseouls mungo* and some weeds. *Agric. Biol. Chem.* 45: 1255–1256.

Sano, H., Kim, Y.-S., and Choi, Y.-E. (2013). Like cures like: caffeine immunizes plants against biotic stresses. *Adv. Bot. Res.* 68: 273–300.

Sasamoto, H., Fujii, Y., and Ashihara, H. (2015). Effect of purine alkaloids on the proliferation of lettuce cells derived from protoplasts. *Nat. Prod. Commun.* 10: 751–754.

Schimpl, F.C., Kiyota, E., Mayer, J.L.S. et al. (2014). Molecular and biochemical characterization of caffeine synthase and purine alkaloid concentration in guarana fruit. *Phytochemistry* 105: 25–36.

Schoor, S., Farrow, S., Blaschke, H. et al. (2011). Adenosine kinase contributes to cytokinin interconversion in Arabidopsis. *Plant Physiol.* 157: 659–672.

Shimazaki, A. and Ashihara, H. (1982). Adenine and guanine salvagein cultured cells of *Catharanthus roseus*. *Ann. Bot.* 50: 531–534.

Shimazaki, A., Hirose, F., and Ashihara, H. (1982). Changes in adenine nucleotide levels and adenine salvage during the growth of *Vinca rosea* cells in suspension culture. *Z. Pflanzenphysiol.* 106: 191–198.

Shoji, T. and Hashimoto, T. (2011). *Nicotine Biosynthesis. Plant Metabolism and Biotechnology*, 191–216. Wiley.

Smith, P.C., Mann, A., Goggin, D., and Atkins, C. (1998). AIR synthetase in cowpea nodules: a single gene product targeted to two organelles? *Plant Mol. Biol.* 36: 811–820.

Smyth, D. (1992). Effect of methylxanthine treatment on rice seedling growth. *J. Plant Growth Regul.* 11: 125–128.

Terrasaki, Y., Suzuki, T., and Ashihara, H. (1994). Purine metabolism and the biosynthesis of purine alkaloids in tea fruits during development. *Plant Physiol. (Life Sci. Adv.)* 13: 135–142.

Uefuji, H., Ogita, S., Yamaguchi, Y. et al. (2003). Molecular cloning and functional characterization of three distinct *N*-methyltransferases involved in the caffeine biosynthetic pathway in coffee plants. *Plant Physiol.* 132: 372–380.

Uefuji, H., Tatsumi, Y., Morimoto, M. et al. (2005). Caffeine production in tobacco plants by simultaneous expression of three coffee *N*-methyltrasferases and its potential as a pest repellant. *Plant Mol. Biol.* 59: 221–227.

Waller, G.R. (1989). Biochemical frontiers of allelopathy. *Biol. Plant.* 31: 418–447.

Waller, G.R., Kumari, D., Friedman, J. et al. (1986). Caffeine autotoxicity in *Coffea arabica* L. In: *The Science of Allelopathy* (eds. A.R. Putnam and C.-S. Tang), 243–269. New York: Wiley.

Yazaki, K., Sugiyama, A., Morita, M., and Shitan, N. (2008). Secondary transport as an efficient membrane transport mechanism for plant secondary metabolites. *Phytochem. Rev.* 7: 513–524.

Yin, Y., Shimano, F., and Ashihara, H. (2007). Involvement of rapid nucleotide synthesis in recovery from phosphate starvation of *Catharanthus roseus* cells. *J. Exp. Bot.* 58: 1025–1033.

Yin, Y., Katahira, R., and Ashihara, H. (2014). Metabolism of purine nucleosides and bases in suspension-cultured *Arabidopsis thaliana* cells. *Eur. Chem. Bull.* 3: 925–934.

Yin, Y., Katahira, R., and Ashihara, H. (2015). Metabolism of purine alkaloids and xanthine in leaves of mate (*Ilex paraguariensis*). *Nat. Prod. Commun.* 10: 707–712.

Zheng, X.-Q. and Ashihara, H. (2004). Distribution, biosynthesis and function of purine and pyridine alkaloids in *Coffea arabica* seedlings. *Plant Sci.* 166: 807–813.

Zheng, X.-Q., Koyama, Y., Nagai, C., and Ashihara, H. (2004). Biosynthesis, accumulation and degradation of theobromine in developing *Theobroma cacao* fruits. *J. Plant Physiol.* 161: 363–369.

Zrenner, R. and Ashihara, H. (2011). Nucleotide metabolism. In: *Plant Metabolism and Biotechnology* (eds. H. Ashihara, A. Crozier and A. Komamine), 135–162. Chichester, UK: Wiley.

Zrenner, R., Stitt, M., Sonnewald, U., and Boldt, R. (2006). Pyrimidine and purine biosynthesis and degradation in plants. *Annu. Rev. Plant Biol.* 57: 805–836.

17

Metabolism of Purine Alkaloids and Biotechnology

17.1 Introduction

Caffeine is the end product of purine alkaloid biosynthesis in most methylxanthine-producing plants. In a limited number of species, caffeine is either further metabolized to methyluric acids or rapidly catabolized via the conventional purine catabolic pathway following sequential cleavage of the three methyl groups on the purine ring.

Often cell and tissue culture models are used to produce phytochemicals that would otherwise be difficult to synthesize. Caffeine does not fall into this category as the substantial amounts required for medicinal products, and as an additive to soft drinks can be obtained by either large-scale chemical synthesis or as a byproduct of procedures used to decaffeinate tea and coffee. Caffeine production by tissue and cell cultures of coffee and tea has, therefore, had little economic impact and these test systems have been used principally for early research on caffeine biosynthesis. The topic has been reviewed by Ashihara and Crozier (1999b). More recent work on the metabolism of purine alkaloids, including biotechnology-based studies on caffeine, are discussed in this chapter.

17.2 Metabolism of Purine Alkaloids

17.2.1 Methylurate Biosynthesis

Theacrine (1,3,7,9-tetramethyluric acid), was discovered by Johnson (1937) as a minor component of *Camellia sinensis* leaves. It accumulates in substantial amounts in kucha (*Camellia assamica* var. *kucha*) (Ye et al. 1999). The biosynthesis of theacrine was investigated using ^{14}C-labelled adenosine, caffeine, and *S*-adenosyl-L-methionine (SAM). These feeding experiments established that caffeine is synthesized in kucha leaves by the same pathway that is utilized in *C. sinensis*, and it is then converted to theacrine, by successive oxidation and methylation steps with 1,3,7-trimethyluric acid serving as an intermediate (Zheng et al. 2002) (Figure 17.1).

Methyluric acid synthesis in leaves of *Coffea dewevrei*, *Coffea liberica*, and *Coffea abeokutae* was studied in feeding experiments with ^{14}C-labelled theobromine, caffeine, and theacrine (Petermann and Baumann 1983). In all three species, theobromine was converted to caffeine, which, in turn, was metabolized to libertine via methylliberine (Figure 17.1).

Plant Nucleotide Metabolism – Biosynthesis, Degradation, and Alkaloid Formation, First Edition.
Hiroshi Ashihara, Iziar A. Ludwig and Alan Crozier.

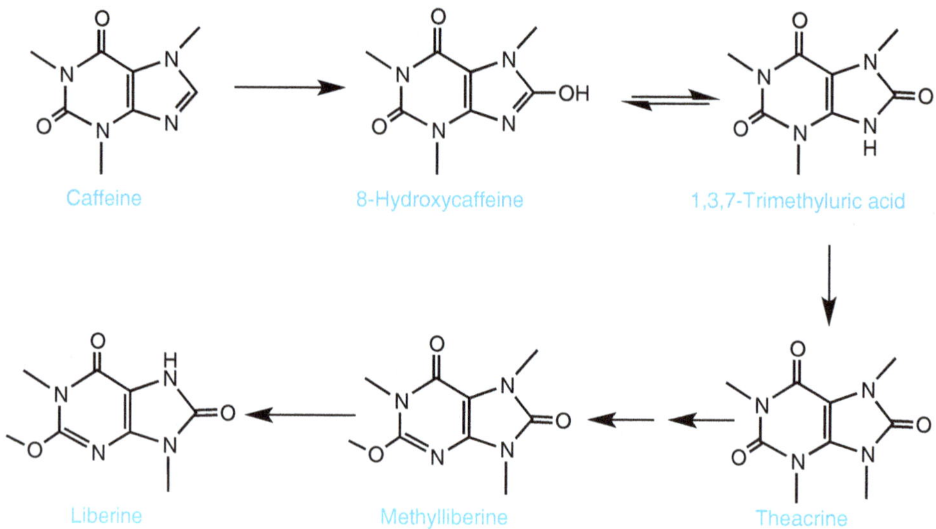

Figure 17.1 Biosynthesis pathway of methyluric acids in plants. Enzymes participating in the reactions have not yet been identified.

Enzyme activity that converts caffeine to methyluric acids has yet to be isolated from plants. It is of note that the bacterium *Pseudomonas* contains a novel caffeine dehydrogenase. This enzyme oxidizes caffeine to trimethyluric acid stoichiometrically and hydrolytically, without producing hydrogen peroxide. The enzyme is not $NAD(P)^+$-dependent but coenzyme Q_0 is the preferred electron acceptor (Yu et al. 2008). Arguably, a similar enzyme could participate in the formation of methyluric acids in plants.

17.2.2 The Major Pathway of Caffeine Degradation

In contrast to research on caffeine biosynthesis, there have been relatively few studies on the catabolism of caffeine. The participating enzymes have not been detected in cell-free preparations and the available data were obtained by feeding radiolabelled substrates to leaves.

Biodegradation of caffeine to xanthine was demonstrated initially in *Coffea arabica* leaves (Kalberer 1964, 1965). It was found that radioactivity from [^{14}C]caffeine applied to aged coffee leaves was incorporated into urate and CO_2. In the 1980s, Suzuki and Waller analysed radiolabelled products using paper chromatography and reported that both immature and mature coffee fruits degraded [2-^{14}C]caffeine to theobromine, theophylline, 3-methylxanthine, 7-methylxanthine, allantoin, allantoate, and urea (Suzuki and Waller 1984b). They suggested that catabolism of caffeine occurs in mature, ripened coffee fruits with theobromine and theophylline as the first biodegradation products, and that theobromine was, therefore, involved in both biosynthesis and degradation of caffeine, whereas theophylline is associated primarily with degradation (Suzuki and Waller 1984a).

Subsequently, Ashihara et al. (1996) performed analysis with high performance liquid chromatography (HPLC) and on-line radioactivity detection with the identity of

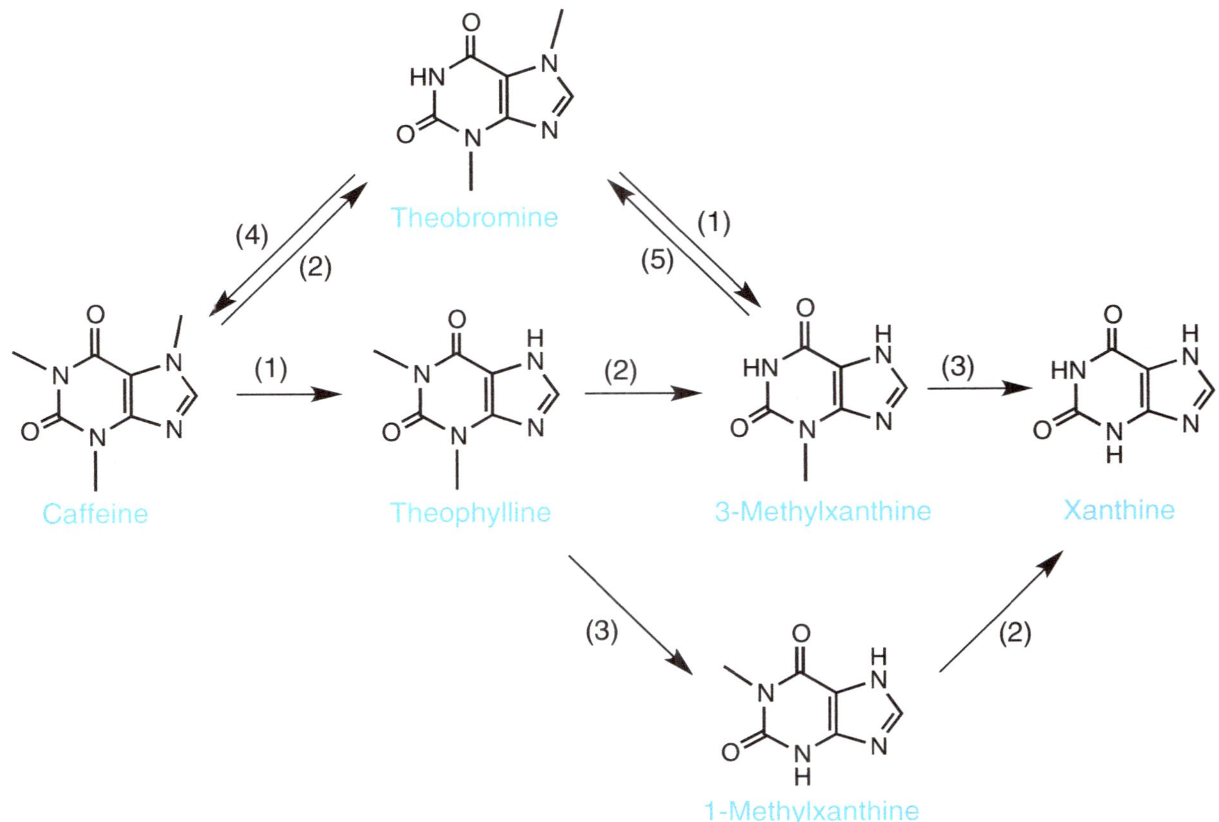

Figure 17.2 Interconversion of xanthine alkaloids in plants. Possible enzymes: (1) 7*N*-demethylase, (2) 1*N*-demethylase, (3) 3*N*-demethylase, (4) caffeine synthase (1*N*-methyltransferase), and (5) 7*N*-methyltransfease.

some metabolites being confirmed by gas chromatography–mass spectrometry. This research established that degradation via theophylline is the major route for caffeine catabolism in leaves of *C. arabica*. The study also demonstrated that theobromine is not a direct catabolite of caffeine and that catabolism proceeds via a three step caffeine → theophylline → 3-methylxanthine → xanthine pathway (Figure 17.2).

Catabolism of caffeine via theophylline has also been demonstrated in *C. sinensis* (Ashihara et al. 1997), but not in maté leaves (Yin et al. 2015). Several uncharacterized demethylases participate in the sequential demethylation reactions. Xanthine, formed by demethylation of caffeine, is converted to urate after which the purine ring is further degraded via allantoin (Figure 7.3).

17.2.3 Purine Catabolic Pathways in Alkaloid Plants

Although detailed enzymatic studies on xanthine catabolism have not been carried out with caffeine-accumulating plants, degradation of xanthine appears to occur in all plant

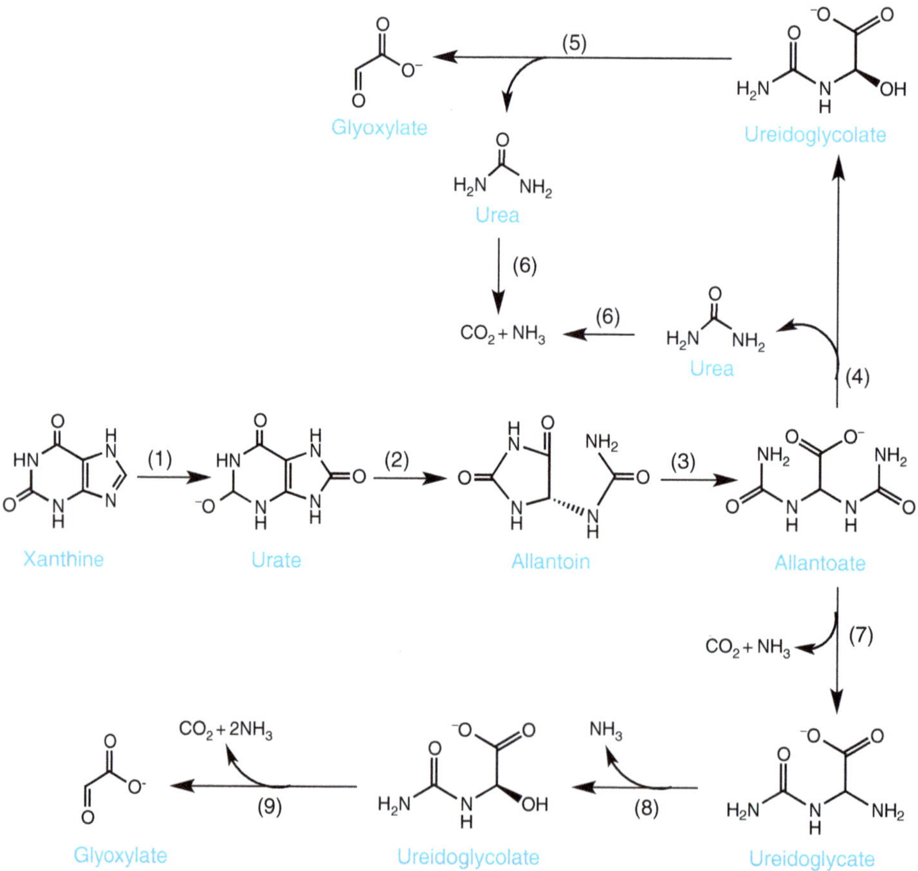

Figure 17.3 Catabolic pathways of xanthine in plants. Enzymes: (1) xanthine dehydrogenase (1.1.1.204), (2) uricase (1.7.3.3), (3) allantoinase (3.5.2.4), (4) allantoicase (3.5.3.4), (5) ureidoglycolate lyase (4.3.2.3), (6) urease (3.5.1.5), (7) allantoin deaminase (3.5.3.9), (8) ureidoglycine amidohydrolase (3.5.3.26), and (9) ureidoglycolate amidohydrolase (3.5.1.116). The allantoicase pathway (steps 4–6) and the allantoic acid amidohydrolase pathway (steps 7–9) have been proposed (see text).

species via what is termed the conventional purine catabolic pathway in which allantoate is produced via a xanthine → urate → hydroxyisourate → 2-oxy-4-hydroxy-4-carboxy-5-ureidoimidazozoline → (*S*)-allantoin → allantoate pathway (see Chapter 7).

As described in Chapter 7, it has been proposed that there are two different pathways for the metabolism of allantoate in plants. In the allantoicase pathway allantoate is degraded to CO_2, NH_3, and glyoxylate via urea and ureidoglycolate (Muñoz et al. 2006). An alternative route, the allantoate amidohydrolase pathway, has been proposed by Winkler et al. (1988), in which allantoate is initially converted to ureidoglycine, CO_2, and NH_3. The NH_3 is released directly and urea formation is not involved in the subsequent catabolism to glyoxylate (Figure 17.3). Since ^{14}C-urea was found when [8-^{14}C]xanthine was administered to leaves of *C. sinensis* (Deng and Ashihara 2010) and *C. arabica* (Nazario and Lovatt 1993), the Muñoz pathway appears to operate in tea and coffee plants.

17.3 Diversity of Purine Alkaloid Metabolism in Plants

17.3.1 *Coffea* Species

Caffeine levels decrease as *C. arabica* tissues age, although substantial quantities remain in aged leaves and fully ripened fruits. Ashihara et al. (1996) reported a detailed investigation of caffeine catabolism in which *C. arabica* leaves were incubated with a number of ^{14}C-labelled purine alkaloids. The release of $^{14}CO_2$ from [1-methyl-^{14}C]-, [3-methyl-^{14}C]-, and [2-^{14}C]caffeine by mature leaves was only 1.0%, 0.1%, and 0.03%, respectively, of the total radioactivity taken up by the leaves, when they were incubated for an 18-hour pulse followed by a 24-hour chase. This demonstrates that the catabolism of applied caffeine was very slow.

Catabolism of caffeine in leaves of three low caffeine-containing coffee species, *Coffea salvatrix*, *Coffea eugenioides*, and *Coffea bengalensis* has been compared (Ashihara and Crozier 1999a). As with *C. arabica*, little or no catabolism was observed in leaf segments of *C. salvatrix* and *C. bengalensis*. However, very rapid catabolism occurred in *C. eugenioides* with more than 75% of [8-^{14}C]caffeine being catabolized in both young and mature leaves, with radioactivity recovered as theophylline, 3-methylxanthine, 1-methylxanthine, xanthine, allantoin, allantoate, urea, and CO_2. This indicates that [8-^{14}C]caffeine undergoes demethylation to xanthine via theophylline and 3-methylxanthine, and that 1-methylxanthine may also be involved (Figure 17.2). Of the three low caffeine-containing *Coffea* species, only *C. eugenioides* possessed a strong capacity for caffeine catabolism. Degradation of [8-^{14}C]caffeine was slow in *C. salvatrix* and *C. bengalensis* indicating that the low caffeine content of these *Coffea* species is due primarily to a reduced biosynthetic activity.

Silvarolla et al. (2004) reported that a natural caffeine-deficient *C. arabica* from Ethiopia (AC1) degraded radiolabelled caffeine, at a rate similar to that which occurring in *C. arabica* cv. Mundo Novo. Therefore, the low caffeine endogenous content in the AC1 plants is probably due to a reduced rate of biosynthesis rather than enhanced degradation of caffeine.

In contrast to the slow metabolism of caffeine in *C. arabica* leaves, [8-^{14}C]theophylline is catabolized very rapidly (Ashihara et al. 1996). Most of the theophylline taken up by leaf discs of young, mature, and aged *C. arabica* leaves was recovered as $^{14}CO_2$.

Radioactivity was also associated with 3-methylxanthine, 7-methylxanthine, xanthine, allantoin, and allantoate. Incorporation of radioactivity into 3-methylxanthine was high in young and mature leaves while the accumulation of and 7-methylxanthine and xanthine was greatest in aged leaves. The inclusion of 5 mM allopurinol in the incubation medium had major effects on [8-^{14}C]theophylline catabolism. The production of $^{14}CO_2$ declined dramatically, as a consequence of the degradation of xanthine being blocked, and there were concomitant increases in the incorporation of label into xanthine, 7-methylxanthine, and 3-methylxanthine. It is noteworthy that >70% of the radioactivity recovered from aged leaves treated with allopurinol was incorporated into 7-methylxanthine. While the presence of 3-methylxanthine and xanthine as catabolites of theophylline is to be anticipated, the detection of 7-methylxanthine, especially in such large amounts, was unexpected, as it cannot be formed directly from theophylline as *N*3- and *N*1-demethylation and *N*7-methylation steps are involved. The most likely route to 7-methylxanthine would appear to involve *N*7-methylation of xanthine.

In contrast to theophylline, theobromine was preferentially utilized for caffeine biosynthesis and only small quantities were catabolized. In mature *C. arabica* leaves, catabolism of theobromine was higher than that of caffeine; the release rates of $^{14}CO_2$ from [2-^{14}C]theobromine and [2-^{14}C]caffeine were, respectively, 5% and 0.03% of total radioactivity after incubation of an 18-hour pulse followed by a 24-hour chase (Ashihara et al. 1996). This implies that a portion of the [2-^{14}C]theobromine was subjected to demethylation, yielding xanthine, which was further catabolized to $^{14}CO_2$. Catabolism of theobromine to xanthine would be via 3-methylxanthine, although radioactivity was not incorporated into the methylxanthine in detectable quantities during the experimental period. An outline of theobromine and caffeine metabolism in *C. arabica* is illustrated in Figure 17.4. When allopurinol inhibits the entry of xanthine into the purine catabolism pathway, xanthine is converted to 7-methylxanthine, which accumulates.

17.3.2 *Camellia* Species

Ashihara et al. (1997) found that only a small amount of $^{14}CO_2$ was released from [8-^{14}C]caffeine when young, mature, and aged leaves of *C. sinensis* were incubated over a 48-hour period. This finding indicates that the rate of catabolism of caffeine, as in *C. arabica*, is extremely low.

Caffeine metabolism in high and low caffeine-containing cultivars of *C. sinensis* have been compared (Ashihara et al. 1995). The results suggest that different levels of caffeine accumulation is mainly a consequence of caffeine biosynthetic activity, although catabolic activity, namely, the rate of release of $^{14}CO_2$ from [2-^{14}C]caffeine, was about two times higher in the low-caffeine cultivars. Therefore, catabolic activity of caffeine may also impact on the content of caffeine in tea (Ashihara et al. 1995).

In contrast to caffeine, theophylline was rapidly metabolized to a variety of compounds. There were marked differences in the metabolic fate of [8-^{14}C]theophylline in young leaves compared to mature and aged leaves (Table 17.1). Similar to coffee leaves (Ashihara et al. 1996), the catabolism of [8-^{14}C]theophylline by leaf discs from mature and aged leaves was rapid with more than 75% of the total radioactivity being recovered as 3-methylxanthine, xanthine, and allantoin and CO_2 with small amounts of label also being incorporated into theobromine and caffeine (Table 17.1). The addition of

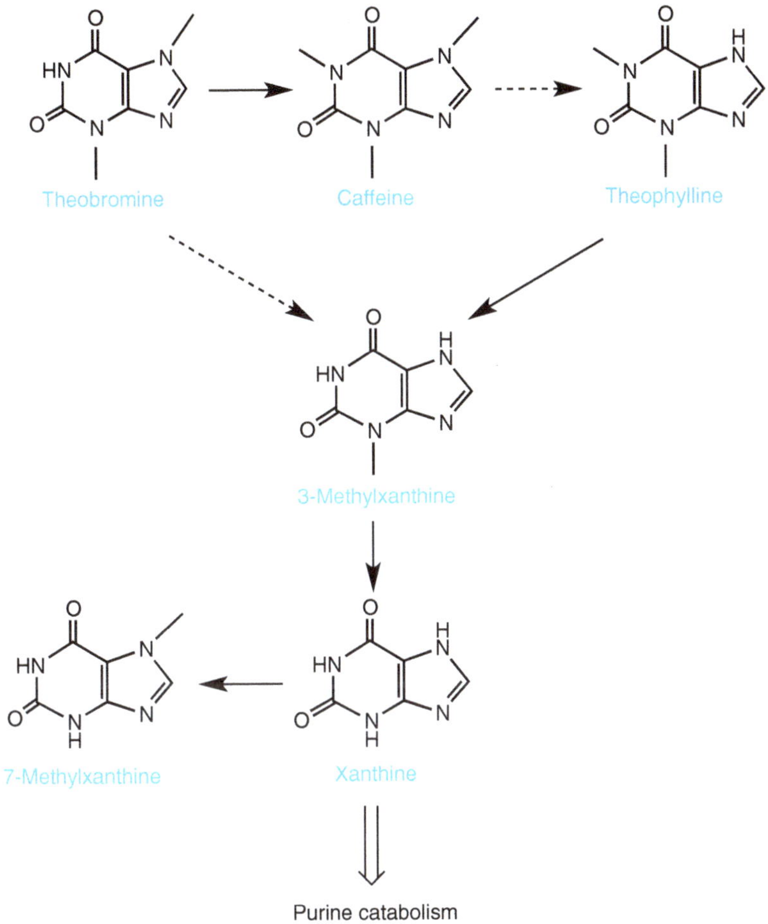

Figure 17.4 Diversity of xanthine alkaloid interconversion in *Coffea arabica* (Ashihara et al. 1996).

5 mM allopurinol resulted in a marked decline in purine catabolism and a concomitant increase in the accumulation of [^{14}C]xanthine. However, conversion of theophylline to caffeine was not influenced by allopurinol and, unlike *C. arabica*, there was no evidence of a conversion of xanthine to 7-methylxanthine (Table 17.1). These results demonstrate that theophylline is principally a precursor of methylxanthine catabolism and that it is not involved in caffeine biosynthesis to any extent in mature and aged tea leaves.

In young tea leaves, catabolism of [8-^{14}C]theophylline was slower than in mature and aged leaves. A marked reduction in $^{14}CO_2$ output suggests that the purine catabolism is less active in young leaves than in older leaves. In young leaves ~20% of radioactivity was incorporated into caffeine. Since radioactivity from [8-^{14}C]theophylline is also incorporated into 3-methylxanthine and theobromine, a theophylline → 3-methylxanthine → theobromine → caffeine pathway (Figure 17.2), appears to be functioning in young tea leaves.

Theobromine is utilized predominantly for caffeine biosynthesis in *C. sinensis* leaves (Ashihara et al. 1997). The highest rate of [^{14}C]caffeine production from

Table 17.1 Summary of the metabolism of [8-^{14}C]theophylline by young, mature, and aged leaves of *Camellia sinensis* in the presence and absence of 5 mM allopurinol for 24 h.

	Young leaves		Mature leaves		Aged leaves	
	Control	Allopurinol	Control	Allopurinol	Control	Allopurinol
Theophylline	47.3	48.1	17.2	22.1	15.7	26.3
3-Methylxanthine	11.6	17.5	8.4	10.9	8.3	12.5
Xanthine	3.1	11.4	1.7	52.5	3.7	50.7
Allantoin	5.0	1.0	1.4	1.6	2.7	1.9
CO_2	10.8	4.6	67.8	3.4	68.2	6.5
Theobromine	4.0	5.5	1.9	4.3	0.6	0.9
Caffeine	18.2	11.9	1.7	1.3	0.4	0.8

Levels of residual [8-^{14}C]theophylline and radiolabelled metabolites are expressed as a percentage of the total radioactivity recovered at the end of the incubation period.
Source: Data are from Ashihara et al. (1997).

[2-^{14}C]theobromine was observed in the mature leaves. Although the results do not agree with the finding that caffeine biosynthesis is most active in young tea leaves (Ashihara and Kubota 1986), it may be a consequence of the conversion of theobromine to caffeine, the last step of caffeine biosynthesis, being more rapid in mature leaves than in young leaves. The rate-limiting step in the overall caffeine biosynthesis pathway may be earlier, arguably the 7-methylxanthosine synthase reaction, so that the overall capacity for the biosynthesis of caffeine is highest in young leaves. No detectable incorporation of radioactivity from [2-^{14}C]theobromine into CO_2 or any other metabolites was detected in young leaves (Ashihara et al. 1997). An outline of caffeine metabolism in *C. sinensis* is illustrated in Figure 17.5.

Metabolism of xanthine alkaloids has been examined in three other *Camellia* species; theobromine-accumulating *Camellia ptilophylla* (cacao tea) and *Camellia irrawadiensis,* and theacrine-accumulating *C. assamica* var. *kucha* (kucha). In *C. ptilophylla* leaf sections, only a limited amount of [8-^{14}C]caffeine was converted to theobromine (6–8% of total radioactivity) and CO_2 (0.2–2%). Most of the [2-^{14}C]theobromine (>93%) was recovered unchanged after a 21-hour incubation (Ashihara et al. 1998). The metabolism of [8-^{14}C]theophylline in *C. irrawadiensis,* was the same as in *C. sinensis* (Figure 17.5), with metabolism to 3-methylxanthine, xanthine, allantoin, and CO_2 with a small amount of radioactivity also being incorporated into theobromine and caffeine (Table 17.2). However, in contrast to *C. sinensis,* complete degradation of the theophylline purine ring declined significantly with leaf age and this was accompanied by a relatively high recovery of unmetabolized [8-^{14}C]theophylline from aged leaves (Table 17.2).

In kucha, caffeine was metabolized to theacrine by expanding buds, and young and mature leaves; the conversion to methyluric acid was highest in young tissues (Zheng et al. 2002). However, most of the [8-^{14}C]caffeine remained unmetabolized. Small amounts of radioactivity, <3% of total activity, were recovered as theobromine, ureides (allantoin and allantoate), CO_2, and an unknown metabolite that arguably could be an intermediate between caffeine and theacrine.

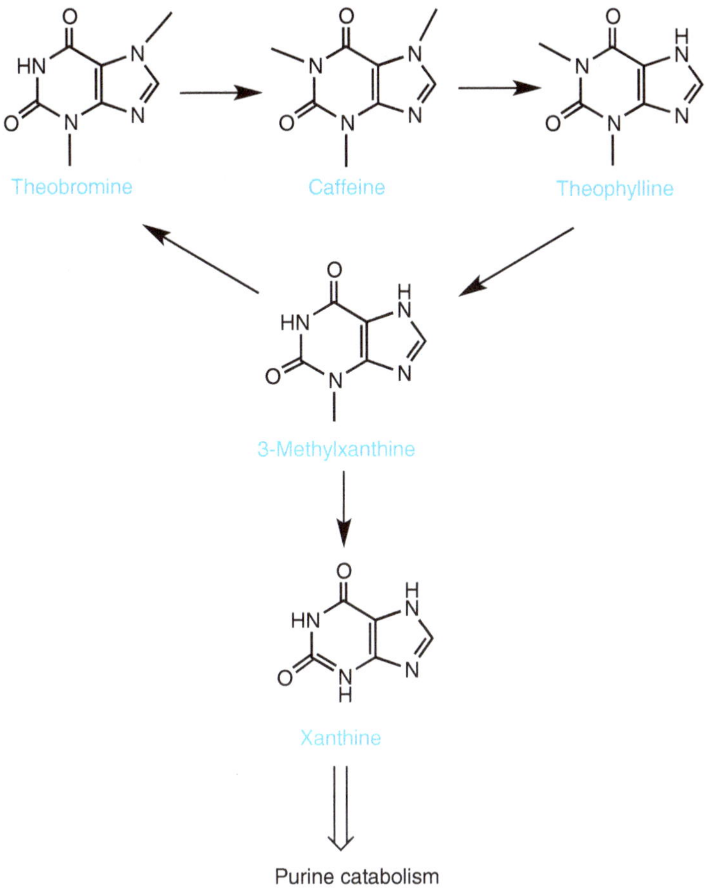

Figure 17.5 Diversity of xanthine alkaloid interconversion in *Camellia sinensis* (Ashihara et al. 1997).

Table 17.2 Summary of the metabolism of [8-^{14}C]theophylline by young, mature, and aged leaves of *Camellia irrawadiensis* in the presence and absence of 5 mM allopurinol for 24 h.

Tea plant	Young leaves		Mature leaves		Aged leaves	
Treatment	Control	Allopurinol	Control	Allopurinol	Control	Allopurinol
Theophylline	54.1	79.3	43.4	60.5	75.7	81.7
3-Methylxanthine	10.0	13.3	19.0	20.6	12.5	11.5
Xanthine	0.4	1.1	1.8	12.9	1.1	5.5
Ureides	2.2	0.2	17.6	0.7	1.9	0.3
CO_2	30.0	1.3	16.8	3.5	7.7	0.1
Theobromine	2.1	2.9	1.2	1.4	0.9	0.8
Caffeine	1.3	1.9	0.3	0.4	0.2	0.2

Levels of residual [8-^{14}C]theophylline and radiolabelled metabolites are expressed as a percentage of the total radioactivity recovered at the end of the incubation period.
Source: Data are from Ito et al. (1997).

17.3.3 Maté Species

In young leaves of maté (*Ilex paraguariensis*), more than 30% of the radioactivity from [8-^{14}C]theobromine was incorporated into metabolites, with the most heavily labelled being caffeine, followed by 3-methylxanthine. In mature leaves, a small proportion of [8-^{14}C]theobromine is converted to 3-methylxanthine and CO_2. In addition to theobromine, exogenous theophylline is utilized for caffeine biosynthesis in maté leaves. However, it has been shown that theophylline is not an intermediate in the main caffeine biosynthesis pathway from purine nucleotides (Ashihara 1993).

17.3.4 Cacao Species

In contrast to coffee and tea leaves, [8-^{14}C]theobromine is degraded by leaves of cacao (*Theobroma cacao*) to CO_2 via 3-methylxanthine, xanthine, and allantoate (Koyama et al. 2003). In young leaves, 12% of the total radioactivity taken up by leaf segments was found in caffeine and less than 4% in 3-methylxanthine. However, in developed and mature leaves, theobromine was not utilized for caffeine biosynthesis but was degraded to CO_2 via 3-methylxanthine. Thus, theobromine is degraded by the conventional purine catabolic pathway. [8-^{14}C]Caffeine was converted to theobromine and theophylline. In young cacao leaves, metabolism of caffeine to theobromine exceeded conversion to theophylline.

In fully developed leaves, where the catabolic activity of xanthine alkaloids was the highest, radioactivity from [8-^{14}C]caffeine was incorporated principally in theophylline and CO_2. In the oldest leaves, the metabolism of caffeine was reduced (Koyama et al. 2003). Although conversion of caffeine to theobromine was detected in younger cacao leaves, caffeine is mainly catabolized to CO_2 via theophylline in the later stages of growth (Figure 17.6) as occurs in coffee and tea leaves (Figures 17.4 and 17.5).

The metabolic fate of [8-^{14}C]theobromine was also examined in cacao fruits at different stages of development (Zheng et al. 2004). Theobromine metabolism was only detected in the pericarp of mature fruits. The radioactivity was recovered in 3-methylxanthine (2%), ureides (3%), and CO_2 (29%). The degradation of [8-^{14}C]caffeine in the pericarp was also detected with radioactivity being incorporated into theophylline (2%), theobromine (1%), and CO_2 (6%). An outline of caffeine metabolism in cacao leaves and pericarp is illustrated in Figure 17.6.

17.3.5 Other Plant Species

Plants, as well as other organisms, metabolize exogenously supplied compounds as xenobiotics, which may be converted to non-toxic compounds. Ito et al. (1997) reported [8-^{14}C]theophylline metabolism in three non-purine alkaloid containing species. It is noteworthy that compared to species that synthesize purine alkaloids, the uptake and metabolism of [8-^{14}C]theophylline by leaf segments of *Avena sativa*, roots of *Vigna mungo*, and suspension cell cultures of *Catharanthus roseus* was extremely low. Relatively small amounts of radioactivity were associated with 3-methylxanthine (1–5% of total activity), xanthine (0–2%), and CO_2 (0–4%). The effects of allopurinol treatment were minimal, confirming that relatively minor amounts of the theophylline were being catabolized to xanthine and entering the purine catabolism pathway.

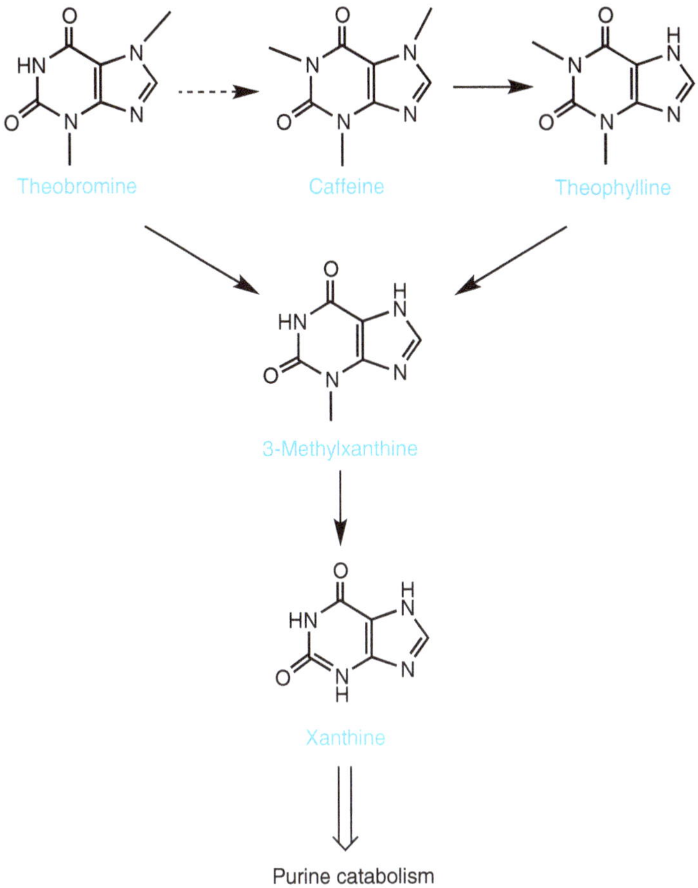

Figure 17.6 Diversity of xanthine alkaloid interconversion in *Theobroma cacao* (Koyama et al. 2003; Zheng et al. 2004).

17.3.6 Bacteria

Purine alkaloid biosynthesis from purine nucleotides does not occur in bacteria and animals, nonetheless they have the ability to metabolize exogenously supplied xanthine alkaloids. Comprehensive details of this topic are outside the scope of this book, but a summary is provided below.

Although high concentrations of caffeine seem to be toxic for microorganisms, several bacteria including *Pseudomonas putida* and *Serratia marcescens* and fungi, such as *Aspergillus niger* and *Penicillium roqueforti*, are able to break down caffeine and its catabolites and use them as a nitrogen source for growth (Mazzafera 2004). Catabolism of caffeine has been studied extensively in bacteria including *P. putida* and *S. marcescens*. Two distinct pathways of caffeine metabolism occur in bacteria: (i) oxidative and (ii) *N*-demethylation pathways. Oxidation of caffeine to form 1,3,7-trimethyluric acid (Figure 17.7) occurred in a mixed culture of *Rhodococcus* sp. and *Klebsiella* sp. (Madyastha and Sridhar 1998). The enzyme involved in the oxidation appears to be a non-NAD(P)$^+$-dependent caffeine dehydrogenase that was observed

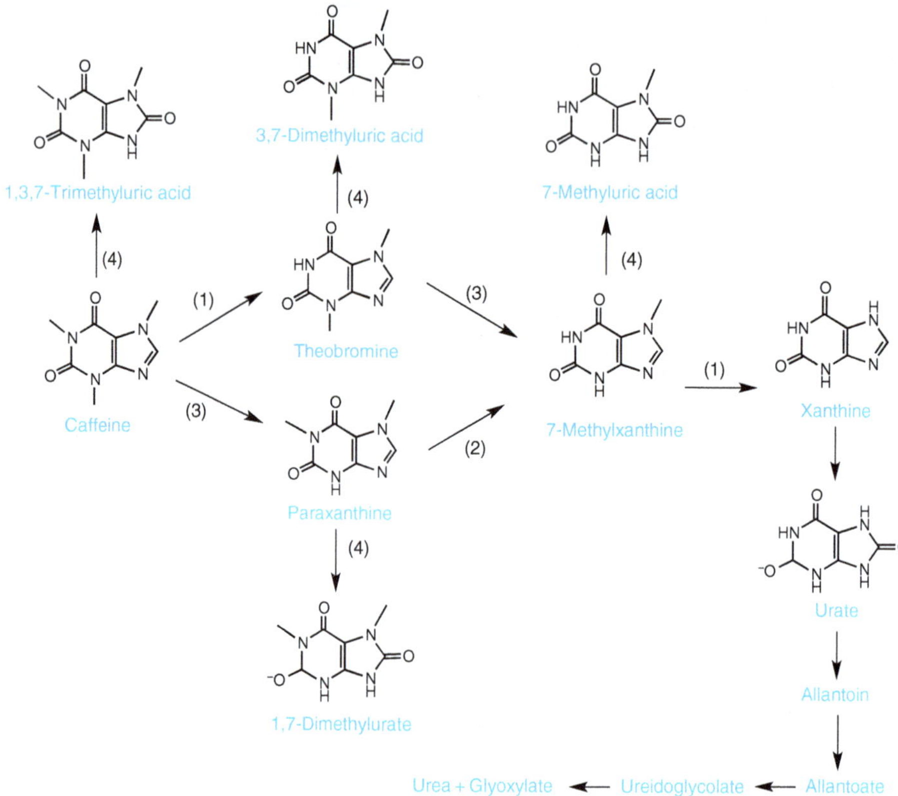

Figure 17.7 Caffeine metabolism proposed in bacteria (*Pseudomonas putida*). Possible enzymes: (1) 7*N*-demethylase; (2) 1*N*-demethylase; (3) 3*N*-demethylase; and (4) caffeine dehydrogenase, caffeine oxidase, or related enzymes. Source: Adapted from Yu et al. (2009) and Mazzafera (2004).

in *Pseudomonas* sp. (Yu et al. 2008). Several caffeine-degrading bacteria metabolize caffeine via the *N*-demethylation pathway and produce theobromine and minor amounts of paraxanthine as the initial products (steps 2 and 3 in Figure 17.7). Subsequently, *N*-demethylation of theobromine or paraxanthine to xanthine occurs via 7-methyxanthine. Xanthine is further oxidized to urate by xanthine dehydrogenase/oxidase as shown in higher plants (Yu et al. 2009). In contrast to plants, theophylline has not been reported to be a bacterial metabolite of caffeine. Yu et al. (2009) reported that *P. putida* strain CBB5 was also capable of utilizing theophylline as the sole carbon and nitrogen source. This bacterium *N*-demethylated theophylline to 1-methylxanthine and 3-methylxanthine, which were both further demethylated forming xanthine. Two distinct pathways for metabolism of caffeine and theophylline operate in *P. putida*.

A novel methylxanthine *N*-demethylase with broad substrate specificity was purified from *P. putida* (Summers et al. 2011). It was characterized as a soluble oxygenase composed of two subunits (A and B). The *N*-demethylation activity was dependent on a specific reductase, which oxidized NAD(P)H and presumably transferred electrons to the *N*-demethylase for *N*-demethylation. Genes encoding the two *N*-demethylase subunits and their reductase were identified within a gene cluster. Functional expression of these genes with the recombinant enzymes established *N*-demethylase A and B as individual Rieske oxygenases with, respectively, highly specific *N*1- and *N*3-demethylation activities (Summers et al. 2012).

17.4 Biotechnology of Purine Alkaloids

Using the gene sequences of the *N*-methyltransferases involved in caffeine biosynthesis, two types of transgenic plants have been produced. One is the construction of genetically modified coffee and tea plants with a reduced caffeine content, in which caffeine production is suppressed. The other is the introduction of caffeine biosynthesis into non-caffeine producing plants in order to confer resistance against herbivores. Examples of transgenic plants using caffeine synthase-related genes are shown in Table 17.3.

17.4.1 Decaffeinated Coffee Plants

As discussed in more detail in Chapter 23, caffeine often induces insomnia in humans. Furthermore, occasional heavy coffee drinkers, especially the elderly, sometimes suffer unpleasant side effects including palpitations, gastrointestinal disturbances, anxiety, and tremor. A restricted daily coffee intake is also advised for pregnant women and those with liver disease (Crozier et al. 2012). It is against this background that there has been an increasing demand for decaffeinated coffee, with 'decaf' now accounting for ~10% of coffee consumption worldwide (Crozier and Ashihara 2006). The flavour and aroma of decaffeinated coffees have improved greatly in recent years as more sophisticated decaffeination methods have been introduced. The latest procedure involves the use of supercritical fluid extraction with liquid CO_2. Such methods require considerable investment in machinery and expenditure of energy. In the long-term, the demand for decaffeinated coffee could be better met by the use of a traditional breeding programme. Unfortunately, this would require a long time to establish new

Table 17.3 Transgenic plants using genes involved in caffeine biosynthesis.

Plants	Coffee	Tea	Tobacco	Chrysanthemum
Species	*Coffea canephora* *Coffea arabica*	*Camellia sinensis*	*Nicotiana tabacum*	*Dendranthema grandiflorum* cv. Shinba
Transferred gene	*CaMXMT1*	Caffeine synthase (*FJ554589*)	*CaMXMT1, CaMXMTMT1/2 CaDXM1*	*CaMXMT CaDXM1*
Method	Antisense, RNAi	RNAi	Sense	Sense
Xanthine alkaloid content	3–10 $mg\,g^{-1}$ f.w.	15–40 $mg\,g^{-1}$ d.w.	3–5 $\mu g\,g^{-1}$ f.w.	3–4 $\mu g\,g^{-1}$ d.w.
Purpose	Decaffeinated coffee	Decaffeinated tea	Endogenous pesticides	Fungal resistance
Reference	Ogita et al. (2003)	Mohanpuria et al. (2011)	Uefuji et al. (2005)	Kim et al. (2011a, b)

hybrid coffee plants on a commercial scale. The use of genetically modified plants could overcome this problem. Production of decaffeinated transgenic plants that were obtained by the introduction of antisense or double-stranded RNA interference (RNAi) constructs of the *N*-methyltransferases related genes has been reported for both coffee (Ashihara et al. 2011; Ogita et al. 2003, 2004, 2005) and tea (Mohanpuria et al. 2011).

Studies on transgenic decaf coffee plants were initiated by Sano's group at the Nara Institute of Science and Technology in Japan. The first approach was to construct transgenic coffee with a reduced caffeine content by the RNA interference method, in which mRNA of the target gene is selectively degraded by small double-stranded RNA species (Ogita et al. 2003, 2004, 2005). The 30-untranslated region and the coding region of *CaMXMT* cDNA were selected to design the RNAi constructs.

Two different RNAi constructs, *RNAi-S* having a short insert with 150 bp, and *RNAi-L* with a long insert of 360 bp, were inserted into the *pBIH1-IG* vector, which was introduced into the EHA101 strain of *Agrobacterium tumefaciens* to transform *C. arabica* and *Coffea canephora* plants. The resulting transformed lines were assayed for expression of *N*-methyltransferase genes by reverse transcription-polymerase chain reaction (RT-PCR), and it was found that the *CaMXMT*-RNAi construct suppressed transcripts for not only *CaMXMT* but also *CaXMT* and *CaDXMT*. The homogeneity between *CaMXMT*, *CaXMT*, and *CaDXMT* is over 90% in the coding region (Uefuji et al. 2003), suggesting that the primary small double-stranded *CaMXMT*-RNA progressively produces many secondary small double-stranded RNAs spanning its coding region to the adjacent sequence of the initiator region, which, in turn, destroys mRNAs for CaXMT and CaDXMT. The reduced level of transcripts indicated decreased activities of the corresponding enzymes, and this was confirmed by measuring their products, theobromine and caffeine. The caffeine content in the controls was ~8.4 mg g^{-1} f.w. while that in both RNAi-S and RNAi-L was ~4.0 mg g^{-1} f.w., showing an average 50% reduction. However, the amount was variable depending on the line, with one notable example of *C. canephora* showing up to a 70% reduction (Ogita et al. 2003).

The caffeine biosynthesis activity in leaves of these low caffeine transgenic plants is reduced. Metabolic studies indicate that substrates of caffeine biosynthesis are catabolized via the conventional purine catabolic pathway and, as a consequence, other than a reduced caffeine content, these plants have the same profile of cellular metabolites as wild type tissues (Ashihara et al. 2006). Currently, caffeine-deficient beans from transgenic *C. arabica* plants have not been produced. When this is achieved with a >90% reduction in the content of caffeine, because of the substantial market for decaffeinated coffee, it will at least have the potential to have major commercial and eco-friendly implications.

17.4.2 Decaffeinated Tea Plants

Tea is a woody perennial crop with a long life cycle, self-incompatibility, and high inbreeding depression. These characteristics usually limit its genetic improvements by conventional breeding methods. To produce low-caffeine tea, conventional breeding may take 25 years or more (Mohanpuria et al. 2010). Yadav's group at the Institute of Himalayan Bioresource Technology in India attempted to produce a selected cultivar of tea with a low-caffeine content using RNAi technology. To suppress the caffeine synthesis in the selected tea, *C. sinensis* cv. Kangra jat, they isolated a partial gene fragment of caffeine synthase (CS), which catalysed the conversion of 7-methylxanthine to caffeine via theobromine, from the same cultivar and used it to design a RNAi construct (*pFGC1008-CS*). Somatic embryos were transformed with the developed construct using a biolistic method.

Transformed somatic embryos showed reduction in the levels of CS transcript expression as well as a lowered caffeine content. Plants were regenerated from the transformed somatic embryos. Transgenic plants showed a significant suppression of CS transcript expression and also showed a reduction of 44–61% in caffeine and 46–67% in theobromine compared to the controls. These results suggest that the RNAi construct using a single partial fragment of the *CS* gene significantly reduced the expression of the targeted endogenous gene. The resultant transgenic plants were morphologically similar at maturity (Mohanpuria et al. 2011), but it is necessary to confirm that, except for the low purine alkaloid content, the chemical constituents are essentially the same as a normal tea. The degree of reduction in the content of caffeine and their effects on the tea production in terms of quality and yield need to be evaluated. The stability and heritability pattern of this low caffeine and theobromine trait should also be studied to establish commercial decaf tea products.

17.5 Caffeine-Producing Transgenic Plants

Caffeine may constitute a part of the chemical defence system against pathogens and herbivores (Ashihara et al. 2008). Sano's group successfully constructed caffeine-producing transgenic plants, with species that would not normally synthesize the purine alkaloid, and examined tolerance traits against biotic attack (Uefuji et al. 2005). A multi-gene transfer vector (pBINNMT777) was constructed using *CaXMT1*, *CaMXMT1*, and *CaDXMT1*, and all were driven by the cauliflower mosaic virus 35S RNA promoter and nopaline synthase (NOS) terminator. The T-DNA region of the plasmid also contained genes for neomycin phosphotransferase (NPT II) and hygromycin phosphotransferase (HPT) as the antibiotic resistance marker. The pBIN-NMT777 was introduced into tobacco (Uefuji et al. 2005) and chrysanthemum (Kim et al. 2011b).

In transgenic tobacco leaves, the caffeine concentration varied from 0 to 5 $\mu g\,g^{-1}$ f.w. The highest concentration was found in aged leaves when plants entered the reproductive stage and formed flower buds. Immature fruits also contained caffeine but at a lower level (<1.3 $\mu g\,g^{-1}$ f.w.). In transgenic chrysanthemum, the caffeine content of fully matured leaves was 3 $\mu g\,g^{-1}$ f.w. Transgenic plants were tested for their tolerant traits against various biotic stresses including herbivore and pathogen attacks.

17.5.1 Antiherbivore Activity

Use was made of larvae of the tobacco cutworm (*Spodoptera litura*) and beet armyworm (*Spodoptera exigua*), which are major pests for many crop plants. When presented with fresh tobacco leaves containing caffeine at 5 $\mu g\,g^{-1}$ and wild type leaves, caterpillars positively avoided the transgenic leaves (Uefuji et al. 2005). The repellent effect was also observed with low caffeine content leaves (0.4 $\mu g\,g^{-1}$ tissue), with larvae eating only 4% of the transgenic leaves and up to 32% of the wild type leaves (Uefuji et al.

2005). A similar result was obtained with caterpillars of the beet armyworm feeding on transgenic chrysanthemum plants producing caffeine at a concentration of 3 μg g^{-1} f.w. (Kim et al. 2011a).

17.5.2 Antipathogen Activity

More than 70% of known plant diseases are caused by fungi, and 30% by bacteria, viruses, and other pathogens (Sano et al. 2013). Transgenic tobacco and chrysanthemum plants that produce caffeine were examined for their response to pathogens. Chrysanthemum plants were infected with a necrotrophic fungus grey mould (*Botrytis cinerea*), which causes death of the flowers, leaves, buds, and fruits of many plant species (Kim et al. 2011b). In the wild type plant, lesions appeared 72 hours after inoculation and rapidly spread from the infected site to the rest of the leaf. The lesion size exceeded 16 mm in diameter five days after infection. In the transgenic lines, the lesions appeared 90 hours after inoculation, and were smaller than the controls, ranging from 1 to 9 mm in diameter. To investigate resistance against microbial pathogens, use was made of *Pseudomonas syringae* pv. *glycinea*, which causes wild-fire disease. In wild type plants, distinct lesions were formed 24 hours after infection and developed into severe necrosis after 48 hours. In contrast, lesion development was strongly inhibited in the transgenic line even 48 hours after inoculation (Kim and Sano 2008). The number of propagated bacteria was lower in the transgenic line than in the control.

Resistance against viral pathogens was also examined (Kim and Sano 2008; Sano et al. 2013). Tobacco mosaic virus (TMV) has a wide host range of over 120 plant species, and causes mottled patterns on leaves, ultimately leading to plant death. When healthy leaves of wild-type and transgenic caffeine-producing tobacco plants were inoculated with TMV and kept at 30 °C, they do not recognize infection and the virus particles propagated. Upon a temperature shift to 23 °C, the hypersensitive response takes place, and a series of defence systems begin to operate. Physiologically, these responses can be assessed visibly by the formation and development of necrotic lesions. With wild-type plants, lesions appeared 48 hours after the temperature shift, and developed further over a subsequent 48 hour period. In transgenic lines, lesions were similarly formed 48 hours after a temperature shift but did not further develop. In addition to slow lesion formation, the size of lesions on transgenic lines was much smaller. The total number of lesions was also reduced in transgenic lines, with only 15–30% of the number that developed on the leaves of control plants. These results point to the effectiveness of caffeine in conferring tolerance against a wide range of pathogens.

Exogenously applied caffeine is toxic for a diversity of organisms at an average concentration ranging from 100 μg to 3 mg g^{-1} f.w. For example, cabbage leaves sprayed with 0.1% caffeine solution were toxic for snails and slugs. In contrast, the amount of caffeine produced in transgenic plants was at most 5 μg g^{-1} f.w., which is two to three orders of magnitude lower. Arguably, these low internal levels of endogenous caffeine activate the defence responses, possibly via salicylic acid-induced expression of pathogenesis-related proteins. The higher concentrations of exogenous caffeine are themselves toxic to bacteria and fungi on the leaf surface while much smaller amounts may be absorbed and act in the same way as the endogenous caffeine.

Despite the relatively low internal concentration of caffeine, the transgenic plants showed a tolerance against pests and pathogens. This apparent discrepancy in effective

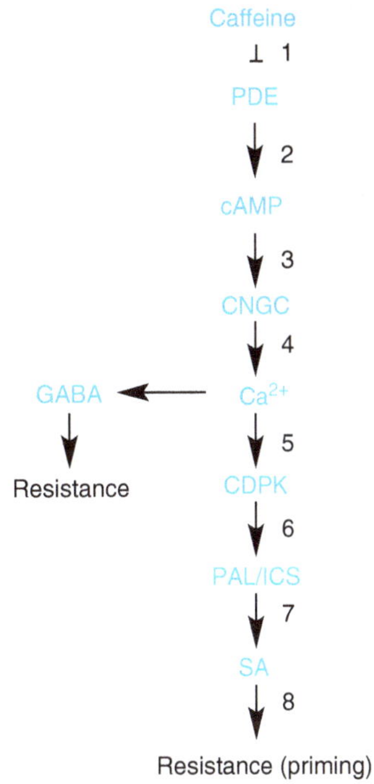

Figure 17.8 Caffeine signal cascade hypothesis in plants proposed by Sano et al. (2013). The molecular cascade of caffeine signal is predicted based on available literature and some experimental data. See text. PDE, phosphodiesterase; CNGC, cyclic nucleotide-gated channel; CDPK, calcium dependent protein kinase (CDPK) activity; PAL, phenylalanine ammonia lyase; ICS, isochorismic synthase; SA, salicylic acid.

concentration raised a question as to whether or not caffeine in transgenic plants is directly toxic for organisms. It is conceivable that endogenously produced caffeine induced some chemical changes in the leaves, thereby indirectly affecting the plant defence responses by a different mechanism (Sano et al. 2013).

The concept of the activation of plant defence systems by caffeine was proposed by Sano et al. (2013). Their simplest model for the action of caffeine, the caffeine signal pathway that enhances the defence response, is summarized in Figure 17.8.

The molecular cascade of the caffeine signal may be described as follows:

Step 1. Caffeine directly blocks phosphodiesterase.
Step 2. Degradation of cyclic adenosine monophosphate (cAMP) is inhibited and its level increases.
Step 3. Increased cAMP activates the cyclic nucleotide-gated channel (CNGC).
Step 4. Activated CNGC increases cytosolic Ca^{2+} level.
Step 5. Increased Ca^{2+} activates calcium-dependent protein kinase (CDPK) activity.
Step 6. Activated CDPK directly phosphorylates phenylalanine ammonia lyase (PAL) and/or isochorismic synthase (ICS).
Step 7. Activities of phosphorylated PAL and ICS increase, resulting in acceleration of salicylic acid (SA) production.
Step 8. The increased level of SA primes the defence response.

$GABA_A$ receptors and adenosine receptors have not been identified in plants. However, increased Ca^{2+} induces γ-aminobutyric acid accumulation, which enhances the defence reaction.

Although transgenic tobacco and chrysanthemum plants contain only a low amount of caffeine they nonetheless exhibited a strong tolerance against herbivores and pathogens.

The defence system was autonomously activated in the absence of external stresses. This can be regarded as the priming of defence responses, by which host plants go on standby to cope with a broad range of biotic stresses. This feature resembles mammalian immunization or vaccination, and it was proposed that plants can also be protected by expressing a mildly toxic 'antigenic' chemical, such as caffeine, *in planta*. This hypothesis was proposed by Sano et al. (2013) and while fascinating, direct evidence has yet to be obtained.

17.6 Summary

Caffeine is the end product of the four step purine alkaloid biosynthesis in most methylxanthine-producing plants. Metabolism of purine alkaloids, diversity of its alkaloid metabolism in plants and bacteria and finally biotechnology of purine alkaloids are discussed.

References

Ashihara, H. (1993). Purine metabolism and the biosynthesis of caffeine in maté leaves. *Phytochemistry* 33: 1427–1430.

Ashihara, H. and Crozier, A. (1999a). Biosynthesis and catabolism of caffeine in low-caffeine-containing species of *Coffea*. *J. Agric. Food Chem.* 47: 3425–3431.

Ashihara, H. and Crozier, A. (1999b). Biosynthesis and metabolism of caffeine and related purine alkaloids in plants. *Adv. Bot. Res.* 30: 117–205.

Ashihara, H. and Kubota, H. (1986). Patterns of adenine metabolism and caffeine biosynthesis in different parts of tea seedlings. *Physiol. Plant.* 68: 275–281.

Ashihara, H., Shimizu, H., Takeda, Y. et al. (1995). Caffeine metabolism in high and low caffeine containing cultivars of *Camellia sinensis*. *Z. Naturforsch.* 50c: 602–607.

Ashihara, H., Monteiro, A.M., Moritz, T. et al. (1996). Catabolism of caffeine and related purine alkaloids in leaves of *Coffea arabica* L. *Planta* 198: 334–339.

Ashihara, H., Gillies, F.M., and Crozier, A. (1997). Metabolism of caffeine and related purine alkaloids in leaves of tea (*Camellia sinensis* L.). *Plant Cell Physiol.* 38: 413–419.

Ashihara, H., Kato, M., and Ye, C.X. (1998). Biosynthesis and metabolism of purine alkaloids in leaves of cocoa tea (*Camellia ptilophylla*). *J. Plant Res.* 111: 599–604.

Ashihara, H., Zheng, X.Q., Katahira, R. et al. (2006). Caffeine biosynthesis and adenine metabolism in transgenic *Coffea canephora* plants with reduced expression of *N*-methyltransferase genes. *Phytochemistry* 67: 882–886.

Ashihara, H., Sano, H., and Crozier, A. (2008). Caffeine and related purine alkaloids: biosynthesis, catabolism, function and genetic engineering. *Phytochemistry* 69: 841–856.

Ashihara, H., Ogita, S., and Crozier, A. (2011). Purine alkaloid metabolism. In: *Plant Metabolism and Biotechnology* (eds. H. Ashihara, A. Crozier and A. Komamine), 163–189. Chichester, UK: Wiley.

Crozier, A. and Ashihara, H. (2006). The cup that cheers. Caffeine biosynthesis: biochemistry and molecular biology. *Biochemist* 28: 23–26.

Crozier, T.W.M., Stalmach, A., Lean, M.E.J., and Crozier, A. (2012). Espresso coffees, caffeine and chlorogenic acid intake: potential health implications. *Food Funct.* 3: 30–33.

Deng, W.-W. and Ashihara, H. (2010). Profiles of purine metabolism in leaves and roots of *Camellia sinensis* seedlings. *Plant Cell Physiol.* 51: 2105–2118.

Ito, E., Crozier, A., and Ashihara, H. (1997). Theophylline metabolism in higher plants. *Biochim. Biophys. Acta* 1336: 323–330.

Johnson, T.B. (1937). Purines in the plant kingdom: the discovery of a new purine in tea. *J. Am. Chem. Soc.* 59: 1261–1263.

Kalberer, P. (1964). Untersuchungen zum Abbau des Kaffeins in den Blattern von *Coffea arabica*. *Ber. Schwieiz. Bot. Ges.* 74: 62–107.

Kalberer, P. (1965). Breakdown of caffeine in the leaves of *Coffea arabica* L. *Nature* 205: 597–598.

Kim, Y.S. and Sano, H. (2008). Pathogen resistance of transgenic tobacco plants producing caffeine. *Phytochemistry* 69: 882–888.

Kim, Y.-S., Lim, S., Kang, K.-K. et al. (2011a). Resistance against beet armyworms and cotton aphids in caffeine-producing transgenic chrysanthemum. *Plant Biotechnol.* 28: 393–395.

Kim, Y.S., Lim, S., Yoda, H. et al. (2011b). Simultaneous activation of salicylate production and fungal resistance in transgenic chrysanthemum producing caffeine. *Plant Signal. Behav.* 6: 409–412.

Koyama, Y., Tomoda, Y., Kato, M., and Ashihara, H. (2003). Metabolism of purine bases, nucleosides and alkaloids in theobromine-forming *Theobroma cacao* leaves. *Plant Physiol. Biochem.* 41: 977–984.

Madyastha, K.M. and Sridhar, G.R. (1998). A novel pathway for the metabolism of caffeine by a mixed culture consortium. *Biochem. Biophys. Res. Commun.* 249: 178–181.

Mazzafera, P. (2004). Catabolism of caffeine in plants and microorganisms. *Front. Biosci.* 9: 1348–1359.

Mohanpuria, P., Kumar, V., and Yadav, S. (2010). Tea caffeine: metabolism, functions, and reduction strategies. *Food Sci. Biotechnol.* 19: 275–287.

Mohanpuria, P., Kumar, V., Ahuja, P., and Yadav, S. (2011). Producing low-caffeine tea through post-transcriptional silencing of caffeine synthase mRNA. *Plant Mol. Biol.* 76: 523–534.

Muñoz, A., Raso, M., Pineda, M., and Piedras, P. (2006). Degradation of ureidoglycolate in French bean (*Phaseolus vulgaris*) is catalysed by a ubiquitous ureidoglycolate urea-lyase. *Planta* 224: 175–184.

Nazario, G.M. and Lovatt, C.J. (1993). Regulation of purine metabolism in intact leaves of *Coffea arabica*. *Plant Physiol.* 103: 1195–1201.

Ogita, S., Uefuji, H., Yamaguchi, Y. et al. (2003). Producing decaffeinated coffee plants. *Nature* 423: 823.

Ogita, S., Uefuji, H., Morimoto, M., and Sano, H. (2004). Application of RNAi to confirm theobromine as the major intermediate for caffeine biosynthesis in coffee plants with potential for construction of decaffeinated varieties. *Plant Mol. Biol.* 54: 931–941.

Ogita, S., Uefuji, H., Morimoto, M., and Sano, H. (2005). Metabolic engineering of caffeine production. *Plant Biotechnol.* 22: 461–468.

Petermann, J. and Baumann, T.W. (1983). Metabolic relations between methylxanthines and methyluric acids in Coffea. *Plant Physiol.* 73: 961–964.

Sano, H., Kim, Y.-S., and Choi, Y.-E. (2013). Like cures like: caffeine immunizes plants against biotic stresses. *Adv. Bot. Res.* 682: 73–300.

Silvarolla, M.B., Mazzafera, P., and Fazuoli, L.C. (2004). Plant biochemistry. A naturally decaffeinated arabica coffee. *Nature* 429: 826–826.

Summers, R.M., Louie, T.M., Yu, C.L., and Subramanian, M. (2011). Characterization of a broad-specificity non-haem iron *N*-demethylase from *Pseudomonas putida* CBB5 capable of utilizing several purine alkaloids as sole carbon and nitrogen source. *Microbiology* 157: 583–592.

Summers, R.M., Louie, T.M., Yu, C.-L. et al. (2012). Novel, highly specific *N*-demethylases enable bacteria to live on caffeine and related purine alkaloids. *J. Bacteriol.* 194: 2041–2049.

Suzuki, T. and Waller, G.R. (1984a). Biodegradation of caffeine: formation of theophylline and caffeine in mature *Coffea arabica* fruits. *J. Sci. Food Agric.* 35: 66–70.

Suzuki, T. and Waller, G.R. (1984b). Biosynthesis and biodegradation of caffeine, theobromine, and theophylline in *Coffea arabica* L. fruits. *J. Agric. Food Chem.* 32: 845–848.

Uefuji, H., Ogita, S., Yamaguchi, Y. et al. (2003). Molecular cloning and functional characterization of three distinct *N*-methyltransferases involved in the caffeine biosynthetic pathway in coffee plants. *Plant Physiol.* 132: 372–380.

Uefuji, H., Tatsumi, Y., Morimoto, M. et al. (2005). Caffeine production in tobacco plants by simultaneous expression of three coffee *N*-methyltrasferases and its potential as a pest repellant. *Plant Mol. Biol.* 59: 221–227.

Winkler, R.G., Blevins, D.G., Polacco, J.C., and Randall, D.D. (1988). Ureide catabolism in nitrogen-fixing legumes. *Trends Biochem. Sci.* 13: 97–100.

Ye, C., Lin, Y., Su, J., and Zhang, H. (1999). Purine alkaloids in *Camellia assamica* var. kucha Chang et Wang. *Acta Sci. Nat. Univ. Sunyatseni* 38: 82–86.

Yin, Y., Katahira, R., and Ashihara, H. (2015). Metabolism of purine alkaloids and xanthine in leaves of mate (*Ilex paraguariensis*). *Nat. Prod. Commun.* 10: 707–712.

Yu, C.L., Kale, Y., Gopishetty, S. et al. (2008). A novel caffeine dehydrogenase in *Pseudomonas* sp. Strain CBB1 oxidizes caffeine to trimethyluric acid. *J. Bacteriol.* 190: 772–776.

Yu, C.L., Louie, T.M., Summers, R. et al. (2009). Two distinct pathways for metabolism of theophylline and caffeine are coexpressed in *Pseudomonas putida* CBB5. *J. Bacteriol.* 191: 4624–4632.

Zheng, X.Q., Ye, C.X., Kato, M. et al. (2002). Theacrine (1,3,7,9-tetramethyluric acid) synthesis in leaves of a Chinese tea, kucha (*Camellia assamica* var. kucha). *Phytochemistry* 60: 129–134.

Zheng, X.-Q., Koyama, Y., Nagai, C., and Ashihara, H. (2004). Biosynthesis, accumulation and degradation of theobromine in developing *Theobroma cacao* fruits. *J. Plant Physiol.* 161: 363–369.

Part VI

Pyridine Nucleotide Metabolism

18

Pyridine (Nicotinamide Adenine) Nucleotide Biosynthesis *De Novo*

18.1 Introduction

Nicotinamide adenine dinucleotide (NAD) and nicotinamide adenine dinucleotide phosphate (NADP) are coenzymes found in all living cells. NAD was discovered by Harden and Young (1906). It consists of two nucleotides, namely adenine nucleotide and a pyridine nucleotide joined through their phosphate groups. Warburg and Christian (1936) showed the function of the nucleotide coenzyme in hydride transfer and identified the nicotinamide portion as the site of redox reactions. They used the term, DPN (diphosphopyridine nucleotide) instead of NAD. Kornberg (1948) was the first to detect an enzyme that used the purine and pyridine nucleotides for the synthesis of NAD, namely, nicotinamide mononucleotide (NMN) + ATP → NAD + PPi. The NMN is a salvage product of nicotinamide, and this enzyme (EC 2.7.7.1) is also involved in the formation of nicotinate mononucleotide (NaMN) in the *de novo* pathway.

Many nucleotide reviews have described only purine and pyrimidine nucleotides without mention of pyridine nucleotides. However, there are reviews on pyridine nucleotides in plants (Ashihara et al. 2015; Noctor et al. 2011; Noctor et al. 2006; Wagner and Backer 1992; Zrenner and Ashihara 2011).

18.2 Two Distinct Pathways of *De Novo* Nicotinate Mononucleotide Biosynthesis

Two distinct routes, (i) the aspartate and (ii) the kynurenine (or tryptophan) pathways have been reported to operate in various organisms. The aspartate pathway consists of only three reactions leading to NaMN (Figure 18.1). Most prokaryotes utilize this *de novo* pathway (Begley et al. 2001).

In contrast, in eukaryotes from *Saccharomyces cerevisiae* to mammals, the *de novo* pathway starts with tryptophan (Belenky et al. 2007; Panozzo et al. 2002). This pathway is more complex than the aspartate pathway with seven steps involved in NaMN formation (Figure 18.2). Historically, the role of tryptophan as a precursor in mammalian NAD biosynthesis was first suggested by nutritional studies in which humans suffering from pellagra, a disease of nicotinamide deficiency, recovered after addition of tryptophan to their diet (Krehl et al. 1945).

There has been debate on the participating pathways in plants. Initially, Gustafson (1949) reported that tryptophan was an intermediate of nicotinate biosynthesis and

Plant Nucleotide Metabolism – Biosynthesis, Degradation, and Alkaloid Formation, First Edition.
Hiroshi Ashihara, Iziar A. Ludwig and Alan Crozier.

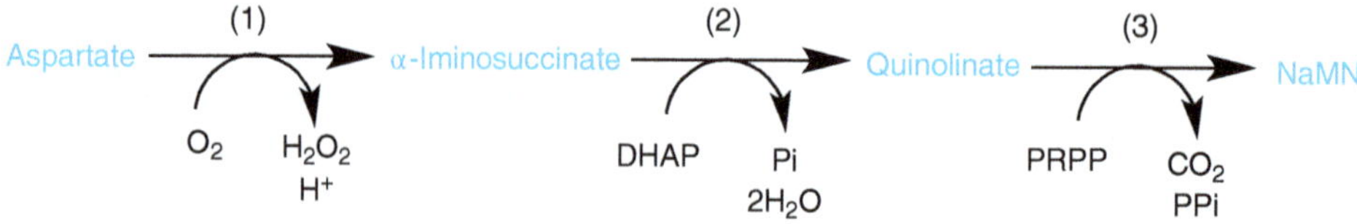

Figure 18.1 The aspartate pathway of nicotinate mononucleotide (NaMN) synthesis. (1) L-Aspartate oxidase; (2) quinolinate synthase; (3) quinolinate phosphoribosyltransferase. Quinolinate is also known as pyridine-2,3-dicarboxylate. DHAP, dihydroxyacetone phosphate; PRPP, 5-phosphoribosyl-1-pyrophosphate.

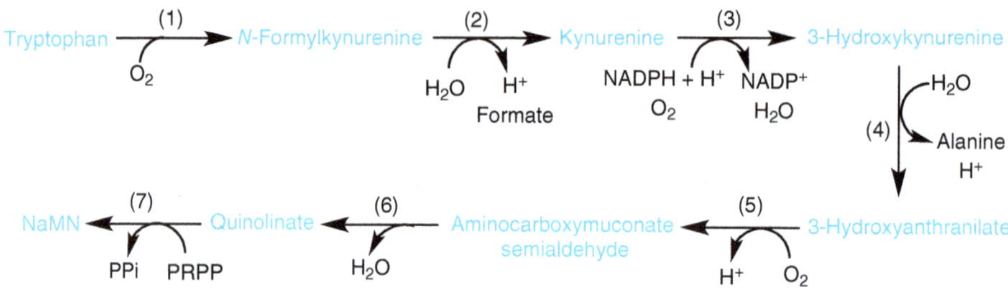

Figure 18.2 The kynurenine pathway of nicotinate mononucleotide (NaMN) synthesis. (1) Indoleamine 2,3-dioxygenase/tryptophan 2,3-dioxygenase; (2) kynurenine formamidase; (3) kynurenine 3-hydroxylase; (4) kynureninase; (5) 3-hydroxyanthranilate 3,4-dioxygenase; (6) non-enzymatic; (7) quinolinate phosphoribosyltransferase. In step (1), two enzymes catalyse the formation of *N*-formylkynurenine by cleavage of the pyrrole ring of tryptophan and the concomitant insertion of molecular oxygen. Source: Adapted from Magni et al. (2004).

several subsequent papers supported this proposal (Arditti and Tarr 1979; Tarr and Arditti 1982). However, Griffith et al. (1962) produced data in keeping with the operation of the aspartate pathway, which does not involve tryptophan as an intermediate. Incorporation of aspartate into the pyridine ring of pyridine alkaloids has also been demonstrated in plants (Jackanicz and Byerrum 1966; Yang and Waller 1965).

Genomic studies indicate the aspartate pathway is operative in plants (Katoh et al. 2006) while the kynurenine pathway occurs in animals, fungi, and some bacteria (Kurnasov et al. 2003; Rongvaux et al. 2003). Katoh and Hashimoto (2004) suggested the genome sequences of rice (*Oryza sativa*) contain the homologues of putative enzymes for all steps in the kynurenine pathway. However, most of the sequences shown by these investigators are not found in the current rice genome databases. Hence, it appears likely that monocotyledonous plants, such as rice, synthesize NAD from aspartate rather than tryptophan.

18.3 The Outline of the *De Novo* Pathway of NAD Biosynthesis in Plants

In the first step of the aspartate pathway, aspartate is oxidized to α-iminosuccinate by aspartate oxidase (EC 1.4.3.16) (Figure 18.3).

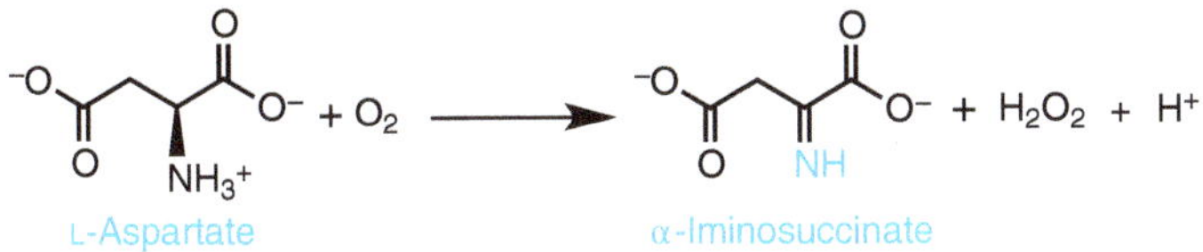

Figure 18.3 Step 1 in Table 18.1.

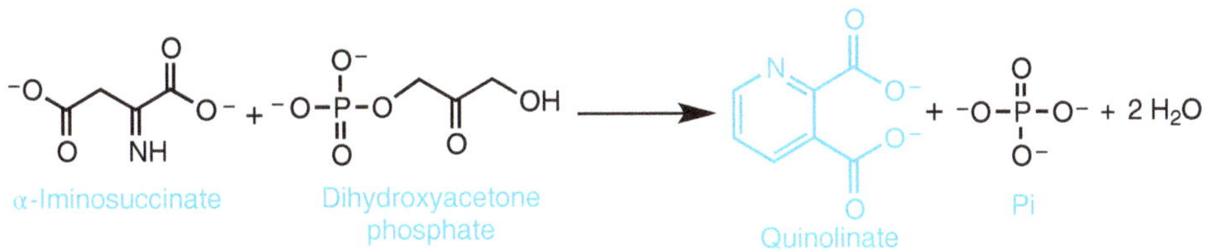

Figure 18.4 Step 2 in Table 18.1.

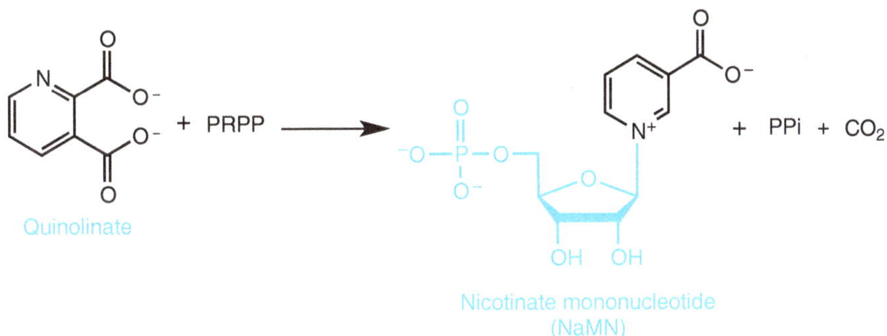

Figure 18.5 Step 3 in Table 18.1.

In the next step, α-iminosuccinate is condensed with dihydroxyacetone phosphate, or glyceraldehyde-3-phosphate, and cyclised to produce quinolinate by quinolinate synthase (EC 2.5.1.72) (Figure 18.4).

The third step is catalysed by quinolinate phosphoribosyltransferase (QPRT) (EC 2.4.2.19), which forms NaMN from quinolinate and 5-phosphoribosyl-1-pyrophostate (PRPP) (Figure 18.5). Quinolinate is produced both by the aspartate and kynurenine pathways so, three enzymatic steps are common in all organisms.

The steps between NaMN and NAD were discovered by Preiss and Handler (1958) and are referred to as the 'Preiss–Handler pathway' in older literature (Moat and Foster 1987). NaMN is also synthesized from nicotinate via the pyridine nucleotide cycle (see Chapter 13). NaMN reacts with ATP to form NaAD in a step catalysed by NaMN adenylyltransferase (EC 2.7.7.18) (Figure 18.6).

Finally, in a conversion catalysed by NAD synthase (glutamine-hydrolysing) (EC 6.3.5.1) the nicotinate moiety in NaAD reacts with glutamine and NAD is formed (Figure 18.7).

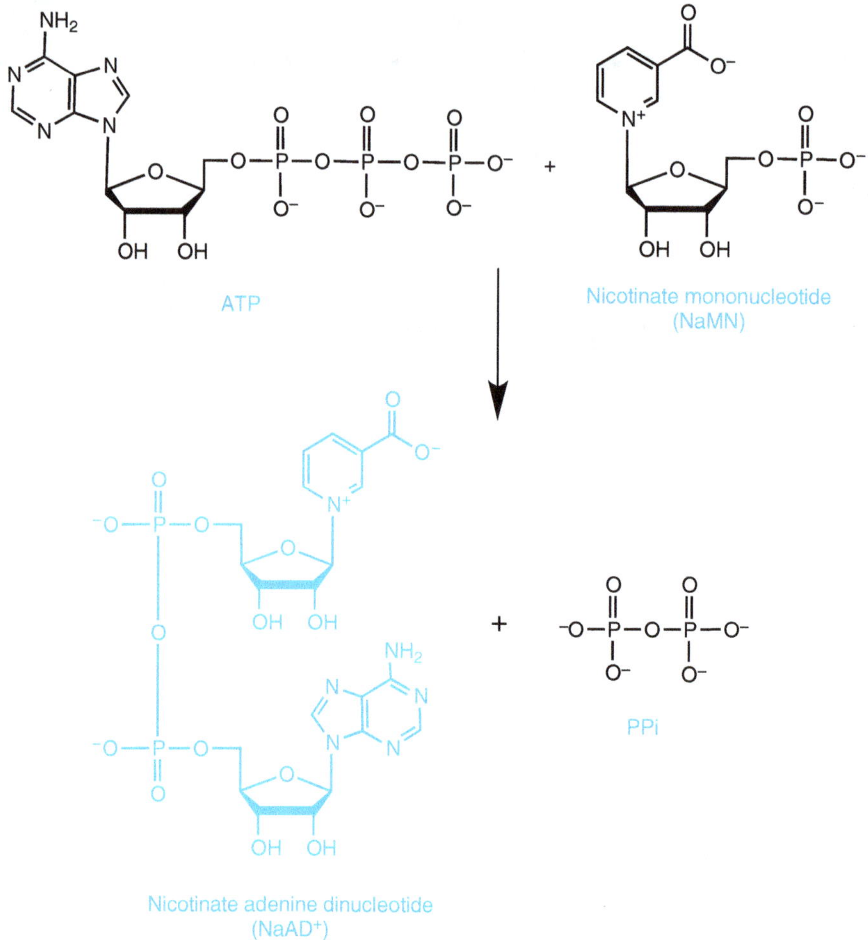

Figure 18.6 Step 4 in Table 18.1.

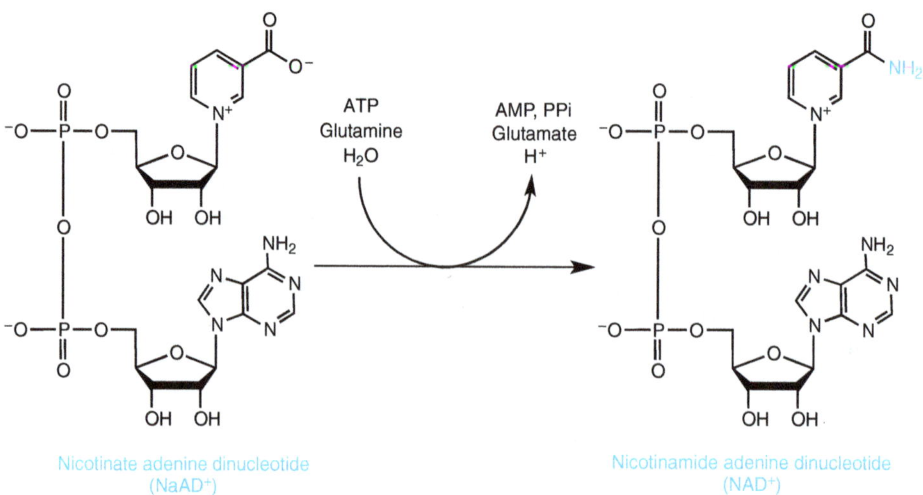

Figure 18.7 Step 5 in Table 18.1.

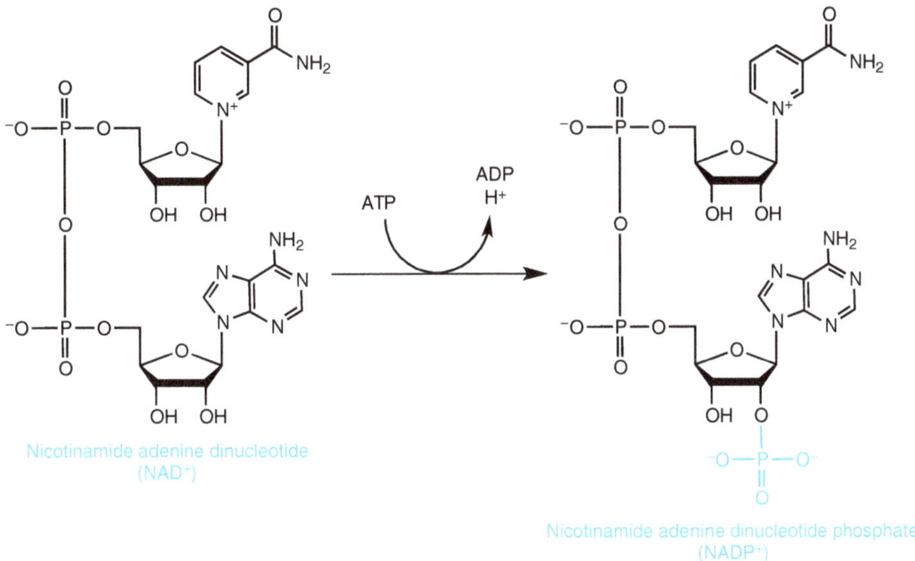

Figure 18.8 Step 6 in Table 18.1.

Figure 18.9 Structures of oxidized (NAD^+) and reduced (NADH) forms of NAD.

In metabolism, NAD (NAD^+ and NADH forms) is involved in redox reactions, carrying electrons from one reaction to another. NAD^+ is an oxidizing agent – it accepts electrons from other molecules and is reduced. The NADH form can be used as a reducing agent to donate electrons. These electron transfer reactions are the main function of NAD and NADP. $NADP^+$ is formed from NAD^+ by NAD^+ kinase (EC 2.7.1.23), and in mitochondria, NADPH is formed from NADH by NADH kinase (EC 2.7.1.86) (Figure 18.8). The structures of oxidized form (NAD^+) and the reduced form (NADH) are shown in Figure 18.9.

18.4 Enzymes Involved in *De Novo* NAD Synthesis in Plants

Summary of pyridine biosynthetic pathway, enzyme names, EC, genes, the TAIR locus is presented in Table 18.1.

Table 18.1 Summary of enzymes involved in pyrimidine nucleotide biosynthesis *de novo* in plants.

Step	Enzyme name	EC	Genes	TAIR Locus
1	L-Aspartate oxidase	1.4.3.16	*AO*	AT5G14760
2	Quinolinate synthase	2.5.1.72	*QS*	AT5G50210
3	Quinolinate phosphoribosyl transferase	2.4.2.19	*QPT*	AT2G01350
4	Nicotinate mononucleotide (NaMN) adenylyltransferase	2.7.7.18	*NaMNAT*	AT5G55810
5	NAD synthetase	6.3.5.1	*NADS*	AT1G55090
6	NAD kinase	2.7.1.23	*NADK1*	AT3G21070
	NAD kinase	2.7.1.23	*NADK2*	AT1G21640
	NADH kinase	2.7.1.86	*NADK3*	AT1G78590

18.4.1 L-Aspartate Oxidase and Quinolinate Synthase

Native L-aspartate oxidase (EC 1.4.3.16) (step 1 in Table 18.1) was partially purified from callus cells of cotton (*Gossypium hirsutum*) by Hosokawa et al. (1983). The cotton enzyme uses L-aspartate as a substrate, but not D-aspartate. Some of the chemical and kinetic properties of the cotton enzyme, namely a molecular weight of 83 000 and a Km for aspartate of 660 μM, are similar to an *E. coli* enzyme, which participates in the first step of pyridine nucleotide biosynthesis (Nasu et al. 1982).

No native enzymatic studies were performed on quinolinate synthase (EC 2.5.1.72) (step 2 in Table 10.1) in plants prior to computer searches of the *Arabidopsis thaliana* genome database. This revealed that a protein encoded by *At5g14760* has significant homology to *E. coli* L-aspartate oxidase and a protein encoded by *At5g50210* is a homologue of *E. coli* quinolinate synthase. To test whether the *A. thaliana* genes encode these enzymes, they were introduced and expressed in *E. coli* mutants deficient in L-aspartate oxidase (*nadB2*) or quinolinate synthase (*nadA2*). The results suggest that *At5g14760* and *At5g50210* encode these two enzymes, respectively, and both are single genes in the *A. thaliana* genome (Katoh et al. 2006).

The *A. thaliana* mutant *old5* (onset of leaf death 5) exhibits early leaf senescence. Schippers et al. (2008) reported that *OLD5* encodes quinolinate synthase. The *A. thaliana* quinolinate synthase was shown to carry a Cys desulfurase domain that stimulates reconstitution of the oxygen-sensitive Fe–S cluster that is required for quinolinate synthase activity (Narayana Murthy et al. 2007). The old5 lesion in this enzyme does not affect quinolinate synthase activity but it decreases its Cys desulfurase activity and, thereby, the long-term catalytic competence of the enzyme (Schippers et al. 2008).

L-Aspartate oxidase and quinolinate synthase of *A. thaliana* contain extra amino acid sequences at their N termini than their *E. coli* counterparts. It has been predicted that *A. thaliana* L-aspartate oxidase and quinolinate synthase are localized in plastids (Katoh et al. 2006).

18.4.2 Quinolinate Phosphoribosyltransferase

Native quinolinate phosphoribosyltransferase (QPRT, EC 2.4.2.19) (step 3 in Table 18.1) has been purified from endosperm of etiolated castor bean (*Ricinus communis*) seedlings (Mann and Byerrum 1974a, 1974b). The enzyme transfers the phosphoribosyl group from 5-phosphoribosyl-1-pyrophosphate to quinolinate forming NaMN with the release of CO_2 and PPi. The estimated molecular weight of the enzyme is ~70 000 and it consists of two subunits each with a molecular weight of ~35 000. Inhibition of QPRT by several substrate analogues suggested that both the 2- and the 3-carboxyl groups on the pyridine ring are needed for substrate binding to the enzyme (Mann and Byerrum 1974a; Mann and Byerrum 1974b).

Sinclair et al. (2000) characterized the cDNAs for QPRT from *Nicotiana rustica* and *Nicotiana tabacum*. The deduced proteins from both cDNAs are almost identical and contain a 24 amino acid *N*-terminal extension that has characteristics of a mitochondrial targeting sequence. In *N. tabacum* which contains nicotine as the major pyridine alkaloid, a QPRT transcript was detected in roots, the site of nicotine synthesis, but it was absent in leaves. Southern blot analysis of genomic DNA from several *Nicotiana* species suggests that QPRT is encoded by a small gene family.

The cDNA encoding QPRT has been isolated in *A. thaliana*. Disruption of QPRT causes embryo-lethal phenotypes of Arabidopsis. The transformation of heterozygous plants with a constitutively expressed wild-type QPRT vector complemented the embryo-lethal phenotype of the homozygous allele (Katoh and Hashimoto 2004; Katoh et al. 2006).

18.4.3 Nicotinate Mononucleotide Adenylyltransferase

Nicotinate mononucleotide adenylyltransferase (NaMNAT, EC 2.7.7.18) activity (step 4 in Table 18.1) has been detected in extracts from plant tissues including roots and callus cultures of tobacco (*N. tabacum)* (Wagner et al. 1986a, 1986b) and potato tubers (Katahira and Ashihara 2009). Native nicotinamide adenylyltransferase (NMNAT, EC 2.7.7.1) activity has also been detected in root extracts of tobacco, although the activity was less than 20% of that of NaMNT (Wagner and Wagner 1985). As mentioned below, molecular genetic analysis has revealed that NaMNAT and NMNAT are the same protein. Therefore, this enzyme is involved in the conversion of NaMN to NaAD in NAD biosynthesis, and also a NAD → NMN → NAD pathway in the pyridine nucleotide cycle, (see Chapter 13).

There are three human NMNAT isoforms (Zhang et al. 2003), but searches of the TAIR database suggested that *A. thaliana* possesses a single copy of a putative NMNAT-encoding gene. This gene encodes a bi-functional enzyme that expresses both NaMNAT and NMNAT activity (Hunt et al. 2004). The enzymatic activity of the AtNMNAT-encoded protein was assessed using a recombinant GST-AtNMNAT, i.e. AtNMNAT protein fused at the *C*-terminus of glutathione-*S*-transferase (GST), and the results obtained suggest that the *AtNMNAT* gene encodes a functional NMNAT enzyme (Hashida et al. 2007).

18.4.4 NAD Synthetase

Activity of glutamine-dependent NAD synthetase (EC 6.3.5.1) (step 5 in Table 18.1) has been detected in roots and callus cultures of tobacco (Wagner et al. 1986a; Wagner et al. 1986b). The *A. thaliana* database contains a homologue of the putative glutamine-dependent NAD synthetase gene (Table 18.1) (Katoh and Hashimoto 2004), but to date a recombinant enzyme has not been obtained.

18.4.5 NAD Kinase

NADP is produced from NAD by NAD kinases. The *de novo* biosynthesis of the NADP is catalysed by the ubiquitous NAD kinase family. The *A. thaliana* genome contains two genes encoding NAD^+ kinases (NADK, EC 2.7.1.23) (step 6 in Table 18.1), annotated as *NADK1, NADK2,* and one gene, *NADK3* encoding a NADH kinase (NADHK, EC 2.7.1.86) (step 7 in Table 10.1), the latter isoform preferring NADH as a substrate.

NAD kinase is often investigated because of its potential importance when plants are subjected to stress (Noctor et al. 2006). For example, NAD kinase is a calmodulin-binding protein (Harding et al. 1997) and its activity is subject to regulation by thiol/disulfide exchange (Delumeau et al. 2000). Two NAD^+-specific NAD kinases, named NADK1 and NADK2, have been cloned and characterized in *A. thaliana* (Turner et al. 2005; Turner et al. 2004). Compared with NADK1, NADK2 has a large *N*-terminal extension (Turner et al. 2004), probably involved in regulation by calmodulin (Harding et al. 1997). In contrast to NADK1 and NADK2, a third characterized *A. thaliana* isoform, NADK3, shows a strong preference for NADH, over NAD, with the Km value for NADH being 40 μM, while for NAD^+ it is 2400 μM, and the specificity constant (Vmax/Km) for NADH being 100-fold greater than for NAD^+. The yeast homologue of Arabidopsis NADK3 is a mitochondrial enzyme, whereas the plant enzyme appears to be located in the cytosol (Turner et al. 2005). The involvement of a putative cytoplasmic NADH kinase (NADK3) in several abiotic stress responses has been observed in plants. The cytoplasmic NADPH is provided primarily by the product of the *NADK3* gene in *A. thaliana* (Mao-Feng et al. 2006).

18.5 Summary

The *de novo* biosynthetic pathways of pyridine nucleotides in plants and animals are different. The plant route from aspartate to NAD is an aspartate → α-iminosuccinate → quinolinate → nicotinate mononucleotide → nicotinate adenine dinucleotide → NAD → NADP pathway.

References

Arditti, J. and Tarr, J.B. (1979). Niacin biosynthesis in plants. *Am. J. Bot.* 66: 1105–1113.

Ashihara, H., Ludwig, I., Katahira, R. et al. (2015). Trigonelline and related nicotinic acid metabolites: occurrence, biosynthesis, taxonomic considerations, and their roles in planta and in human health. *Phytochem. Rev.* 14: 765–798.

Begley, T.P., Kinsland, C., Mehl, R.A. et al. (2001). The biosynthesis of nicotinamide adenine dinucleotides in bacteria. *Vitam. Horm.* 61: 103–119.

Belenky, P., Bogan, K.L., and Brenner, C. (2007). NAD^+ metabolism in health and disease. *Trends Biochem. Sci.* 32: 12–19.

Delumeau, O., Paven, M.-C.M.-L., Montrichard, F., and Laval-Martin, D.L. (2000). Effects of short-term NaCl stress on calmodulin transcript levels and calmodulin-dependent NAD kinase activity in two species of tomato. *Plant Cell Environ.* 23: 329–336.

Griffith, T., Hellman, K.P., and Byerrum, R.U. (1962). Studies on the biosynthesis of the pyridine ring of nicotine. *Biochemistry* 1: 336–340.

Gustafson, F.G. (1949). Tryptophane as an intermediate in the synthesis of nicotinic acid by green plants. *Science* 110: 279–280.

Harden, A. and Young, W.J. (1906). The alcoholic ferment of yeast-juice. Part II. The coferment of yeast-juice. *Proc. Royal Soc. London. Ser. B* 78: 369–375.

Harding, S.A., Oh, S.H., and Roberts, D.M. (1997). Transgenic tobacco expressing a foreign calmodulin gene shows an enhanced production of active oxygen species. *The EMBO J.* 16: 1137–1144.

Hashida, S.n., Takahashi, H., Kawai-Yamada, M., and Uchimiya, H. (2007). *Arabidopsis thaliana* nicotinate/nicotinamide mononucleotide adenyltransferase (AtNMNAT) is required for pollen tube growth. *Plant J.* 49: 694–703.

Hosokawa, Y., Mitchell, E., and Gholson, R.K. (1983). Higher plants contain L-asparate oxidase, the first enzyme of the *Escherichia coli* quinolinate synthetase system. *Biochem. Biophys. Res. Commun.* 111: 188–193.

Hunt, L., Lerner, F., and Ziegler, M. (2004). NAD – new roles in signalling and gene regulation in plants. *New Phytol.* 163: 31–44.

Jackanicz, T.M. and Byerrum, R.U. (1966). Incorporation of aspartate and malate into the pyridine ring of nicotine. *J. Biol. Chem.* 241: 1296–1299.

Katahira, R. and Ashihara, H. (2009). Profiles of the biosynthesis and metabolism of pyridine nucleotides in potatoes (*Solanum tuberosum* L.). *Planta* 231: 35–42.

Katoh, A. and Hashimoto, T. (2004). Molecular biology of pyridine nucleotide and nicotine biosynthesis. *Front. Biosci.* 9: 1577–1586.

Katoh, A., Uenohara, K., Akita, M., and Hashimoto, T. (2006). Early steps in the biosynthesis of NAD in Arabidopsis start with aspartate and occur in the plastid. *Plant Physiol.* 141: 851–857.

Kornberg, A. (1948). The participation of inorganic pyrophosphate in the reversible enzymatic synthesis of diphosphopyridine nucleotide. *J. Biol. Chem.* 176: 1475–1476.

Krehl, W.A., Teply, L.J., Sarma, P.S., and Elvehjem, C.A. (1945). Growth-retarding effect of corn in nicotinic acid-low rations and its counteraction by tryptophane. *Science* 101: 489–490.

Kurnasov, O., Goral, V., Colabroy, K. et al. (2003). NAD biosynthesis: identification of the tryptophan to quinolinate pathway in bacteria. *Chem. Biol.* 10: 1195–1204.

Magni, G., Amici, A., Emanuelli, M. et al. (2004). Enzymology of NAD^+ homeostasis in man. *Cell. Mol. Life Sci.* 61: 19–34.

Mann, D.F. and Byerrum, R.U. (1974a). Avtivation of the *de novo* pathway for pyridine nucleotide biosynthesis prior to ricinine biosynthesis in caster beans. *Plant Physiol.* 53: 603–609.

Mann, D.F. and Byerrum, R.U. (1974b). Quinolinic acid phosphoribosyltransferase from castor bean endosperm. I. Purification and characterization. *J. Biol. Chem.* 249: 6817–6823.

Mao-Feng, C., Peng-Cheng, W., Qi-Jun, C. et al. (2006). NADK3, a novel cytoplasmic source of NADPH, is required under conditions of oxidative stress and modulates abscisic acid responses in Arabidopsis. *Plant J.* 47: 665–674.

Moat, A.G. and Foster, J.W. (1987). Biosynthesis and salvage pathways of pyridine nucleotides. In: *Pyridine Nucleotide Coenzymes. Chemical, Biochemical, and Medical Aspects, Part B* (eds. D. Dolphin, O. Avramovic and R. Poulson), 1–24. New York: Wiley.

Narayana Murthy, U.M., Ollagnier-de-Choudens, S., Sanakis, Y. et al. (2007). Characterization of *Arabidopsis thaliana* SufE2 and SufE3: functions in chloroplast iron-sulfur cluster assembly and NAD synthesis. *J. Biol. Chem.* 282: 18254–18264.

Nasu, S., Wicks, F.D., and Gholson, R.K. (1982). L-aspartate oxidase, a newly discovered enzyme of *Escherichia coli*, is the B protein of quinolinate synthetase. *J. Biol. Chem.* 257: 626–632.

Noctor, G., Queval, G., and Gakiere, B. (2006). NAD(P) synthesis and pyridine nucleotide cycling in plants and their potential importance in stress conditions. *J. Exp. Bot.* 57: 1603–1620.

Noctor, G., Hager, J., and Li, S. (2011). Biosynthesis of NAD and its manipulation in plants. *Adv. Bot. Res.*: 153–201.

Panozzo, C., Nawara, M., Suski, C. et al. (2002). Aerobic and anaerobic NAD^+ metabolism in *Saccharomyces cerevisiae*. *FEBS Lett.* 517: 97–102.

Preiss, J. and Handler, P. (1958). Biosynthesis of diphosphopyridine nucleotide. II. Enzymatic aspects. *J. Biol. Chem.* 233: 493–500.

Rongvaux, A., Andris, F., Van Gool, F., and Leo, O. (2003). Reconstructing eukaryotic NAD metabolism. *BioEssays* 25: 683–690.

Schippers, J.H.M., Nunes-Nesi, A., Apetrei, R. et al. (2008). The Arabidopsis onset of leaf *death5* mutation of quinolinate synthase affects nicotinamide adenine dinucleotide biosynthesis and causes early ageing. *Plant Cell* 20: 2909–2925.

Sinclair, S., Murphy, K., Birch, C., and Hamill, J. (2000). Molecular characterization of quinolinate phosphoribosyltransferase (QPRTase) in *Nicotiana*. *Plant Mol. Biol.* 44: 603–617.

Tarr, J.B. and Arditti, J. (1982). Niacin biosynthesis in seedlings of *Zea mays*. *Plant Physiol.* 69: 553–556.

Turner, W.L., Waller, J.C., Vanderbeld, B., and Snedden, W.A. (2004). Cloning and characterization of two NAD kinases from Arabidopsis. Identification of a calmodulin binding isoform. *Plant Physiol.* 135: 1243–1255.

Turner, W.L., Waller, J.C., and Snedden, W.A. (2005). Identification, molecular cloning and functional characterization of a novel NADH kinase from *Arabidopsis thaliana* (Thale cress). *Biochem. J.* 385: 217–223.

Wagner, K.G. and Backer, A.I. (1992). Dynamics of nucleotides in plants studied on a cellular basis. In: *International Review of Cytology*, vol. 134 (eds. K.W. Jeon and M. Friedlander), 1–84. San Diego: Academic Press.

Wagner, R. and Wagner, K.G. (1985). The pyridine nucleotide cycle in tobacco. Enzyme activities for the *de novo* synthesis of NAD. *Planta* 165: 532–537.

Wagner, R., Feth, F., and Wagner, K.G. (1986a). Regulation in tobacco callus of enzyme activities of the nicotine pathway II. The pyridine-nucleotide cycle. *Planta* 168: 408–413.

Wagner, R., Feth, F., and Wagner, K.G. (1986b). The regulation of enzyme activities of the nicotine pathway in tobacco. *Physiol. Plant.* 68: 667–672.

Warburg, O. and Christian, W. (1936). Pyridin, der wasserstoffübertragende Bestandteil von Gärungsfermenten. *Helv. Chim. Acta* 19: E79–E88.

Yang, K.S. and Waller, G.R. (1965). Biosynthesis of the pyridine ring of ricinine from quinolinic acid glycerol and aspartic acid. *Phytochemistry* 4: 881–889.

Zhang, X., Kurnasov, O.V., Karthikeyan, S. et al. (2003). Structural characterization of a human cytosolic NMN/NaMN adenylyltransferase and implication in human NAD biosynthesis. *J. Biol. Chem.* 278: 13503–13511.

Zrenner, R. and Ashihara, H. (2011). Nucleotide Metabolism. In: *Plant Metabolism and Biotechnology* (eds. H. Ashihara, A. Crozier and A. Komamine), 135–162. Chichester, UK: Wiley.

19

Pyridine Nucleotide Cycle

19.1 Introduction

Like purine and pyrimidine metabolism, in addition to *de novo* biosynthesis of pyridine nucleotides, a salvage synthesis pathway also occurs. However, the pyridine ring is more efficiently recycled than is the case with degradation of purine and pyrimidine. The pyridine nucleotide cycle (PNC), which includes pyridine nucleotide catabolism and regeneration of pyridine nucleotides, was first postulated to occur in microbial cells. Evidence for the formation and degradation of the pyridine nucleotides in plants, animals, and microorganisms was summarized by Gholson (1966, 1968). The diversity of PNCs in various organisms was reviewed by Moat and Foster (1987). In plants, nicotinate is salvaged to nicotinate mononucleotide (NaMN) by the PNC. Alternatively it can be utilized for pyridine alkaloid synthesis (Waller et al. 1966). Until the early 1990s, various routes of the PNC had been found and were referred to according to the number of enzymatic steps required for degradation through to the reformation of nicotinamide dinucleotide (NAD). In plants, a schematic presentation of the possible routes called PNC III, IV, V, and VI was proposed (Wagner and Backer 1992; Wagner et al. 1986a). Subsequently some routes were revised and different PNCs have been found in plants where the major cycles are PNC V and PNC VII (Ashihara et al. 2015; Noctor et al. 2006). The structures of the metabolites involved in the PNC are illustrated in Figure 19.1 together with their commonly used abbreviations.

19.2 Pyridine Nucleotide Cycle

PNCs show wide variations between species and even organs (Ashihara et al. 2005; Moat and Foster 1987; Noctor et al. 2006; Wagner and Backer 1992). The simplest cycle (PNC II) appears to be a NAD → NMN → NAD pathway. Wagner and Wagner (1985) found NAD pyrophosphatase (EC 3.6.1.22) and NMN adenyltransferase (EC 2.7.7.1) activity in *Nicotiana tabacum*, so this cycle might operate in tobacco. However, a more complicated cycle appears to occur in most plants as experiments using radioactive precursors, such as [^{3}H]quinolinate, have shown the incorporation of radioactivity into catabolites of NAD (Ashihara et al. 2005).

Further studies on the profile of enzyme activities involved in PNC indicate the operation of some minor routes (Ashihara and Deng 2012; Katahira and Ashihara 2009).

Plant Nucleotide Metabolism – Biosynthesis, Degradation, and Alkaloid Formation, First Edition.
Hiroshi Ashihara, Iziar A. Ludwig and Alan Crozier.

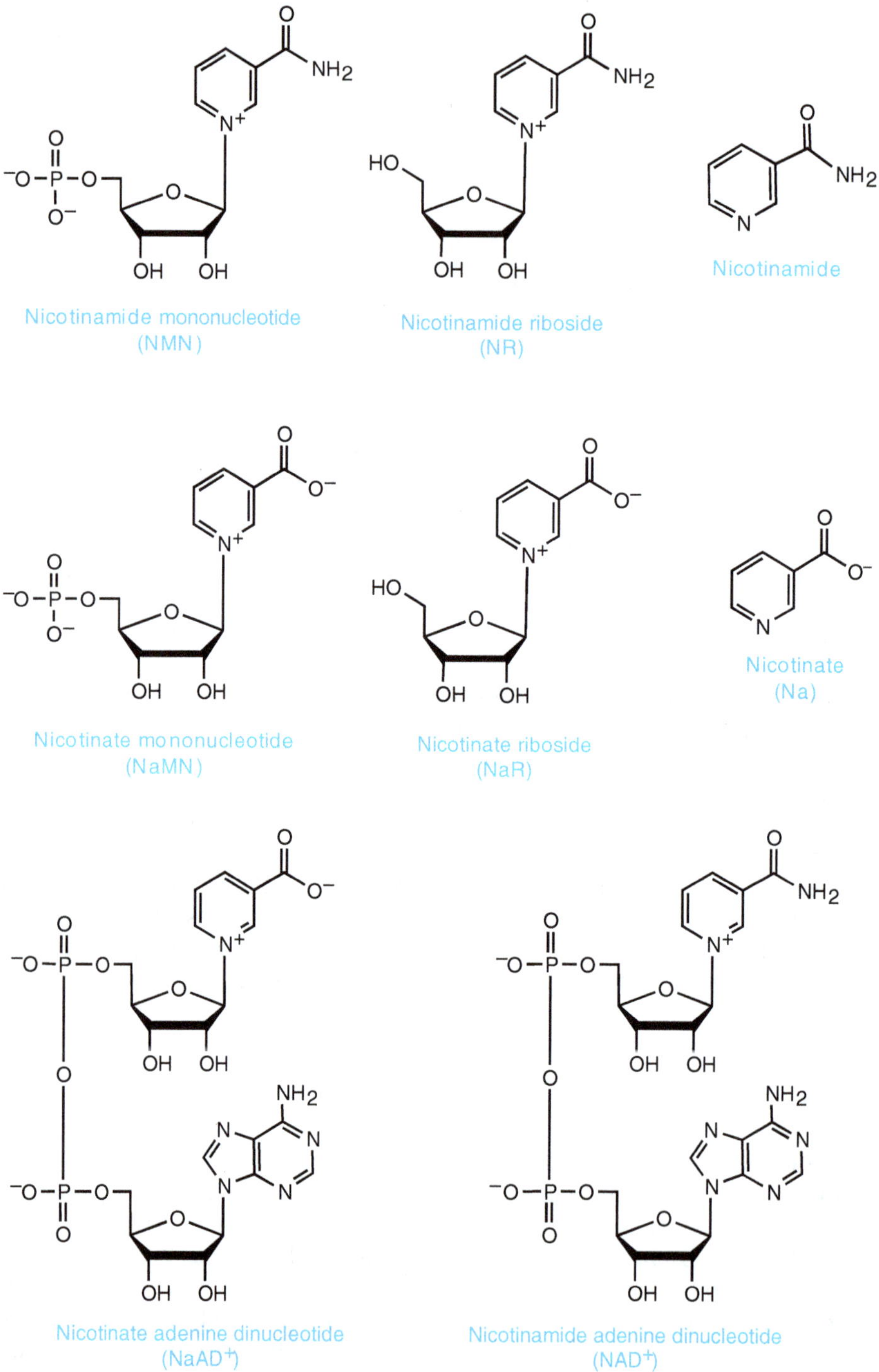

Figure 19.1 Structures of metabolites involved in the pyridine nucleotide cycles in plants. Commonly used abbreviations of metabolites that appeared in this text are also shown.

Table 19.1 Profiles of activities of enzymes involved in the pyridine nucleotide cycle and related pathways in growing potato tubers.

Step[a)]	Enzyme (EC)	Reaction	Activity	
			Potato	Tea
1	NAD^+ ADP-ribosyltransferase (EC 2.4.2.30)	NAD^+ + (ADP-ribosyl)$_n$-acceptor → nicotinamide + (ADP-ribosyl)$_{n+1}$- acceptor	—	—
2	NAD pyrophosphatase (3.6.1.22)	NAD → NMN + AMP	44.1	3.4
3	5′-Nucleotidase (3.1.3.5)	NMN → NR + Pi	1.5	120
3a	NMN phosphoribohydrolase (3.2.2.14)	NMN → N + Ribose-5-P	nd	nd
3b	Nicotnamide phophoribosyltransferase (2.4.2.12)	N + PRPP → NMN + PPi	nd	nd
4	NR nucleosidase (3.2.2-)	NR → N + ribose	1.3	2880
5	Nicotinamidase (3.5.1.19)	N → Na + NH_3	5.7	6.5
5a	NR deaminase (3.5.1.-)	NR → NaR + NH_3	12.5	nd
5b	NMN deamidase (3.5.1.42)	NMN → NaMN + NH_3	nd	nd
6	Nicotinate phosphoribosyltransferase (6.3.4.21)	Na + PRPP → NaMN + PPi	0.4	0.2
7	NaMN adenylyltransferase (2.7.7.18)	NaMN + ATP → NaAD + PPi	23.1	5.3
9	NaR phosphorylase (2.4.2.-)	Na + Ribose-1-P → NaR + Pi	nd	nd
10	NaR kinase (2.7.1.173)	NaR + ATP → NaMN + ADP	0.7	0.1
10a	NR kinase (2.7.1.22)	NR + ATP → NMN + ADP	—	—
11	5′-Nucleotidase (3.1.3.5)	NaMN → NaR + Pi	6.6	8.6
12	NaR nucleosidase (3.2.2.-)	NaR → Na + ribose	1.4	4.2
13	Nicotinate *N*-glucosyltransfease (2.4.1.196)	Na + UDP-glucose → Nicotinate-*N*-glucoside + UDP	23.8	0.3
14	Trigonelline synthase (2.1.1.7)	Na + SAM → Trigonelline + SAH	nd	nd

The activities are expressed as pkat mg^{-1} protein.

a) NAD, nicotinamide dinucleotide; N, nicotinamide; Na, nicotinate; NMN, nicotinate adenine dinucleotide; NR, nicotinamide riboside; NaR, nicotinate riboside; NaMN, nicotinate mononucleotide; PRPP, 5-phosphoribosy-1-pyrophosphate; UDP, uridine diphosphate. nd, No activity was detected. —, Not measured.

Source: Potato tuber, Katahira and Ashihara (2009); tea leaves, Ashihara and Deng (2012).

As an example, comprehensive profiles of enzyme activities of pyridine nucleotide-related metabolism in potato tubers and tea leaves are presented in Table 19.1.

19.2.1 Major Pyridine Nucleotide Cycles in Plants

Two major PNCs in plants are illustrated in Figure 19.2. An eight-component pyridine nucleotide cycle (PNC VII), NAD → NMN → nicotinamide riboside (NR) → nicotinamide → nicotinate → nicotinate mononucleotide (NaMN) → nicotinate

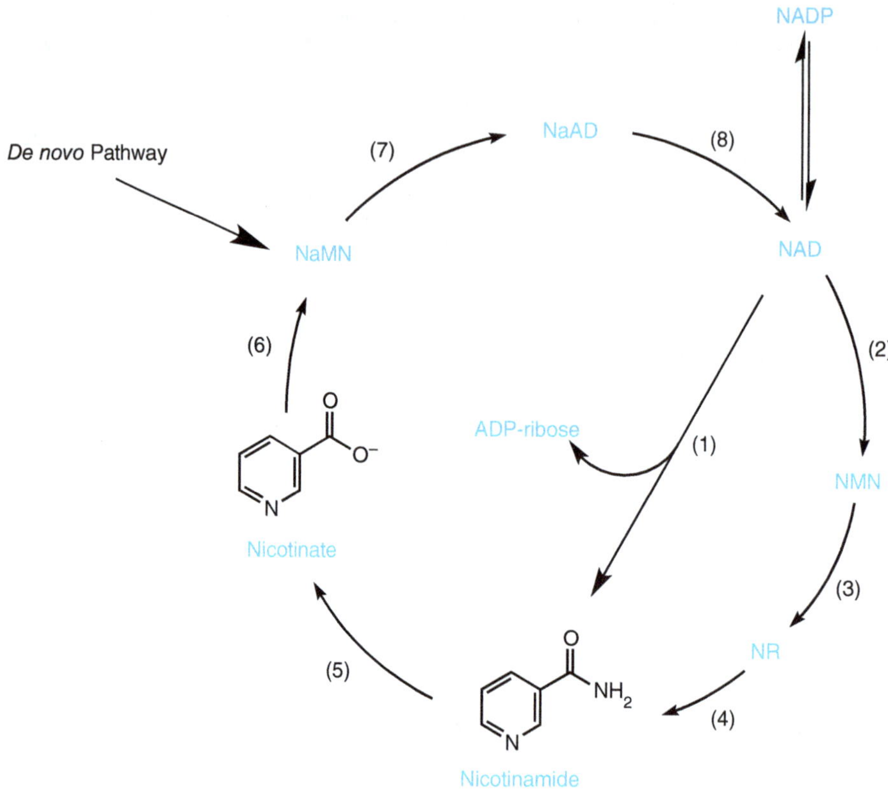

Figure 19.2 Typical two pyridine nucleotide cycles in plants. The PNC VII contains steps 2–8 and PNC V contains steps 1 and 5–8. NaMN, nicotinate mononucleotide; NaAD, nicotinamide adenosine dinucleotide; NADP, nicotinamide dinucleotide phosphate; NAD, nicotinamide dinucleotide; NMN, nicotinamide mononucleotide; NR, nicotinamide riboside; NaR, nicotinate riboside.

adenine dinucleotide (NaAD) → NAD route (steps 2–8 in Figure 19.2) and a six-component pyridine nucleotide cycle (PNC V), NAD → nicotinamide → nicotinate → NaMN → NaAD → NAD route (steps 1 and 5–8 in Figure 19.2) have been proposed as the main pyridine cycles in plants (see Zrenner and Ashihara 2011).

19.2.2 Alternative Pyridine Nucleotide Cycles in Plants

Detailed pathways of NAD metabolism including minor routes and other pathways observed in animals are illustrated in Figure 19.3. NR deaminase and nicotinate *N*-riboside (NaR) kinase are unique plant enzymes. NaR kinase occurs in mung bean (*Vigna radiate*) (Matsui and Ashihara 2008) and NR deaminase in potato (Katahira and Ashihara 2009). Low activity of NaR kinase was found in tea leaves but NR deaminase was undetectable (Ashihara and Deng 2012). Enzyme profiles involved in the PNCs in potato tubers and tea leaves are shown in Table 19.1. These findings imply that a new six-component pyridine nucleotide cycle (PNC VI), namely a NAD → NMN → NR → nicotinate riboside (NaR) → NaMN → NaAD → NAD

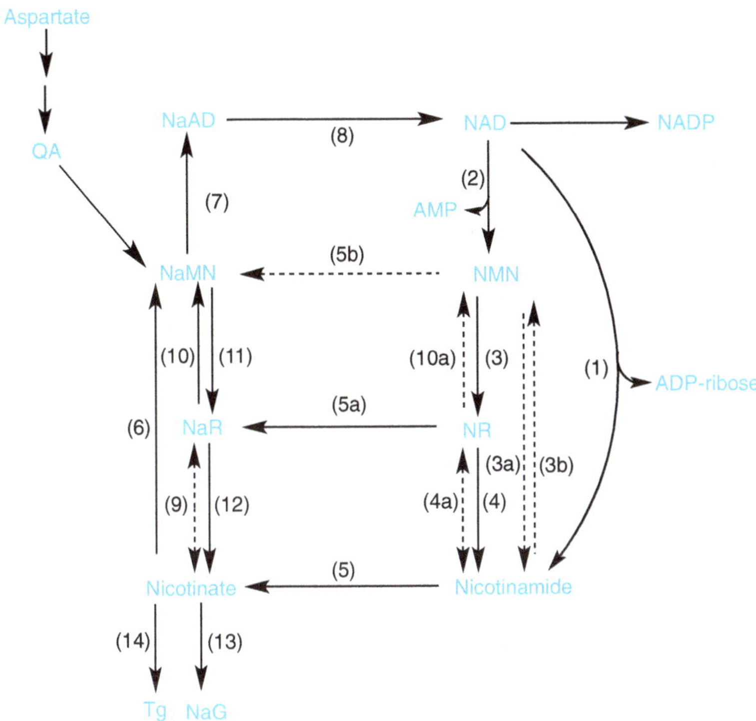

Figure 19.3 Detailed metabolic pathways involved in the pyridine nucleotide cycles. Steps 1–8 is the same in Figure 19.2. Steps 9–14, 3a-3b, 4a, 5a-5b, and 12a are included in this figure. The dotted arrows are steps which are not found in plants. Steps 19 and 14 are nicotinate derivatives, nicotinate-*N*-glucoside and trigonelline, respectively (see part VII, pyridine alkaloids). NaMN, nicotinate mononucleotide; NaAD, nicotinate adenosine dinucleotide; NaR, nicotinate riboside; NADP, nicotinamide dinucleotide phosphate; NAD, nicotinamide dinucleotide; NMN, nicotinamide mononucleotide; NR, nicotinamide riboside; Tg, trigonelline; NaG, nicotinate-*N*-glucoside.

route (steps 2–3, 5a, 12, and 7–8 in Figure 19.3) which together with the PNC VII, NAD → NMN → NR → NaR → nicotinate → NaMN → NaAD → NAD pathway (steps 2–3, 5a, 12, and 6–8 in Figure 19.3) can function in some plants. The major steps 1–8 are also shown in Figure 19.2.

19.2.3 Rate-Limiting Step of the Pyridine Cycle

The most likely rate-limiting step of the pyridine cycle appears to be that catalysed by nicotinate phosphoribosyltransferase (NaPRT) (step 6 in Figures 19.2 and 19.3), an enzyme that requires ATP for activity (Zheng et al. 2005). Hence, availability of ATP, as well as 5-phosphoribosyl-1-pyrophosphate (PRPP), regulates the formation of NaMN. In contrast, other conversions, including those catalysed by nicotinamidase (step 5 in Figures 19.2 and 19.3) and NaMN adenylyltransferase (step 7 in Figures 19.2 and 19.3), seem not to be the rate limiting steps in the PNC, since there is little accumulation of the substrates nicotinamide and NaMN in most plants (Katahira and Ashihara 2009; Matsui et al. 2007). An adequate supply of ATP would also appear to be required for NaMN adenylyltransferase to form the adenine nucleotide moiety of NAD.

19.3 Catabolism of NAD

19.3.1 Reactions from NAD to Nicotinate

In plants NAD appears to be catabolized to nicotinamide via at least two routes. The first is direct formation from NAD which releases ADP-ribose and is catalysed by poly(ADP-ribose) polymerases, mono(ADP ribosyl)transferase or NAD-dependent deacetylase (step 1 in Figure 19.2). For example, NAD^+ ADP-ribosyltransferase (EC 2.4.2.30) catalyses the reaction: NAD^+ + (ADP-D-ribosyl)$_n$-acceptor → nicotinamide + (ADP-D-ribosyl)$_{n+1}$-acceptor + H^+ (Ashihara et al. 2015).

The second pathway is the three-step formation of nicotinamide, namely, NAD → NMN → NR → nicotinamide (Ashihara et al. 2005; Noctor et al. 2006). NAD is hydrolysed to NMN and AMP by NAD pyrophosphatase (EC 3.6.1.22) (step 2 in Figure 19.2). NMN is further hydrolysed to NR and Pi. This reaction may be catalysed by a non-specific 5′-nucleotidase (EC 3.1.3.5) (step 3 in Figure 19.2). NR is hydrolysed by nicotinamide riboside nucleosidase (EC 3.2.2.-) (step 4 in Figure 19.2).

The activity of NMN nucleosidase (NMN glycohydrolase, EC 3.2.2.14) has been detected in *N. tabacum* (Wagner et al. 1986a, 1986b, 1986c), but it is absent in potato tubers and tea leaves (Ashihara and Deng 2012; Katahira and Ashihara 2009) (Table 19.1). The reaction catalysed by this enzyme is: NMN + H_2O → ribose-5-phosphate + nicotinamide (step 3a in Figure 19.3). The enzyme assay system used in the Wagner et al. studies did not include the detection of ribose 5-phosphate. It is, therefore, possible that they measured the combined activity of 5′-nucleotidase and nicotinamide riboside nucleosidase (steps 3 and 4 in Figure 19.2) in their crude enzyme preparations.

Nicotinamide is readily converted to nicotinate by a plant-specific nicotinamidase (step 5 in Figure 19.2). In addition to nicotinamidase, NR deaminase (step 5a in Figure 19.3) has been detected in potato tubers (Table 19.1) (Katahira and Ashihara 2009). NR is converted to NaR by this enzyme. Hence, the end product of NAD catabolism in plants is nicotinate, and in some instances, NaR.

19.3.2 Degradation of Pyrimidine Ring

Although nicotinate is a major end product of NAD catabolism in plants, it has been postulated that it is further degraded in parsley (*Petroselinum hortense*) (Barz 1985) and tea (Ashihara and Deng 2012). Degradation of the pyridine ring of nicotinate appears to take place through nicotinate-*N*-glucoside (NaG) (step 13 in Figure 19.2). Radioactivity from [2-^{14}C]nicotinate was incorporated into NaG and glutaric acid (pentanedioic acid) by tea leaf disks. It was not possible to determine the detailed degradation route of the ring structure of pyridine, because of the very low radioactivity recovered in intermediate metabolites (Ashihara and Deng 2012).

Aerobic nicotin degradation has been reported to occur in microorganisms, such as *Pseudomonas putida* (Jiménez et al. 2008). The catabolic sequence, known as the maleamic acid pathway, is nicotinate → 6-hydroxynicotinate → 2,5- dihydroxypyridine → maleamic acid → maleic acid → fumaric acid. However, in tea leaves, radioactivity from [2-^{14}C]nicotinate was not incorporated into any of these metabolites (Ashihara and Deng 2012). Consequently, a novel catabolic pathway, which is distinct from the pathway in microorganisms, may be operative in plants. However, it is worth noting that pyridine ring catabolism occurs in only restricted plant species and the activity is very low.

19.3.3 Nicotinate Conversion to Nicotinate-*N*-Glucoside and *N*-Methylnicotinate

Although plants do not have a strong capacity for pyridine ring degradation, the nicotinate metabolites that are produced, namely are nicotinate-*N*-glucoside (**1**) and *N*-methylnicotinate (**2**, aka trigonelline), represent an important reservoir of nicotinate (Neuhann et al. 1979). This topic is described in Chapter 20.

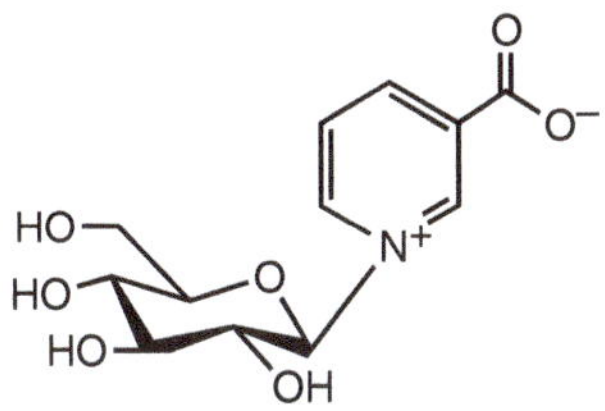

Nicotinate-*N*-glucoside (**1**)

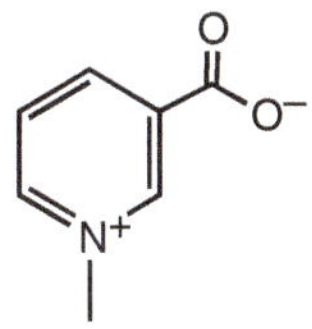

Trigonelline (**2**)

19.4 Enzymes Involved in NAD Catabolism

19.4.1 Direct NAD Cleavage Enzymes

NAD is cleaved by three classes of cleavage enzymes that produce nicotinamide and an ADP-ribosyl product (step 1 in Figure 19.2). The three classes are (i) poly(ADP-ribose) polymerases or monoADP-ribose transferases, (ii) ADP-ribosyl cyclase, and (iii) sirtuins (type III protein lysine deacetylases) (Ashihara et al. 2015; Belenky et al. 2007). Poly(ADP-ribose) polymerases are the main enzymes responsible for the post-translational modification known as polyADP-ribosylation. These enzymes play important roles in genotoxic stress tolerance, DNA repair, programmed cell death, transcription, and cell cycle control in animals. Similar effects are being discovered in plants, as well as roles in plant-specific processes (Briggs and Bent 2011). While these findings are of note, the topics are beyond the scope of this review. The Sir 2 (silent information regulator 2) family of proteins, which are also known as sirtuins, are NAD-dependent protein deacetylases. They contain a 200-amino acid domain conserved from bacteria to humans (Frye 2000). Plant genomes appear to contain fewer *SIR2* homologues than other eukaryotes. In *Arabidopsis thaliana*, only two *SIR2* family gene sequences have been identified. Very limited research has been carried out on the function of plant Sir2-related proteins.

19.4.2 NAD Pyrophosphatase

Activity of NAD pyrophosphatase (EC 3.6.1.22) (step 2 in Figure 19.3) has been detected in extracts from tobacco (Wagner et al. 1986a), potato (Katahira and Ashihara 2009), and tea (Ashihara and Deng 2012) (Table 19.1). The hydrolysis of a variety of nucleoside diphosphate derivatives is catalysed by a family of proteins called Nudix (nucleoside diphosphates linked to a moiety X) hydrolases (Bessman et al. 1996). Twenty-four genes of the Nudix hydrolase homologues (AtNUDTs) with predicted localizations in the cytosol, chloroplasts, and mitochondria, exist in *A. thaliana*. Dobrzanska et al. (2002) described the cloning and characterization of an *A. thaliana* cDNA encoding a Nudix hydrolase that degrades NADH. The deduced sequence of AtNUDT1 contains 147 amino acids. The recombinant AtNUDT1 was expressed in *E. coli* and purified. In the presence of Mn^{2+} and the optimal pH of 7.0, the recombinant AtNUDT1 catalysed the hydrolysis of NADH. The catalytic properties of NADH pyrophosphatase from *A. thaliana* resemble those described for *E. coli* enzymes. The substrate specificity of the enzyme has a marked preference for NADH over NAD^+. The catalytic velocity with NADH is 20-fold faster than with NAD^+. Ogawa et al. (2005) reported a comprehensive analysis of nine types of cytosolic AtNUDT proteins. The recombinant proteins of AtNUDT-2, -6, -7, and -10 showed both NADH pyrophosphatase and ADP-ribose pyrophosphatase activities. Jambunathan and Mahalingam (2006) demonstrated that the *A. thaliana Growth Factor Gene 1* (*GFG1; At4g12720*) which encodes a Nudix hydrolase, is an NADH pyrophosphatase and ADP-ribose pyrophosphatase.

19.4.3 5′-Nucleotidase and Nicotinamide Riboside Nucleosidase

NMN is probably converted to nicotinamide releasing Pi and ribose by two enzymatic steps (steps 3 and 4 in Figure 19.3). 5′-Nucleotidase and nucleosidase appear to participate in the conversion (Ashihara and Deng 2012; Katahira and Ashihara 2009) (Table 19.1).

The catabolic enzymes converting NMN to nicotinamide vary in different organisms. In bacteria, such as *Azotobacter vinelandii*, a glycohydrolase catalyses the conversion of NMN to nicotinamide and ribose-5-phosphate (Imai 1979) (step 3a in Figure 19.3). In contrast, two enzymes, 5′-nucleotidase and NR phosphorylase (steps 3 and 4a in Figure 19.3), are responsible for NMN catabolism in animals. No specific NMN glycohydrolase (step 3a in Figure 19.3) have been found in animal tissues (Imai and Anderson 1987). In plants, NR nucleosidase (step 4 in Figure 19.2), but not NR phosphorylase (step 4a in Figure 19.3), participates in this conversion. Especially high NR nucleosidase activity is found in young tea leaves (Table 19.1) (Ashihara and Deng 2012; Katahira and Ashihara 2009).

19.4.4 Nicotinamidase and Nicotinamide Riboside Deaminase

High activity of nicotinamidase (EC 3.5.1.19, step 5 in Figures 19.2 and 19.3) has been detected in the extracts of several plants (Ashihara and Deng 2012; Ashihara et al. 2005; Katahira and Ashihara 2009; Yin et al. 2008; Zheng et al. 2004, 2005). Wang and Pichersky (2007) reported that the *A. thaliana* gene *At2g22570* encodes a nicotinamidase, which is expressed in all tissues, with the highest levels occurring in roots and stems. The 244-residue protein, designated AtNIC1, converts nicotinamide to nicotinate and has a Km of 118 μM and a Kcat of 0.93 s^{-1}. Plants homozygous for a null *AtNIC1* allele, *nic1-1*, have lower levels of NAD and NADP under normal growth conditions.

Katahira and Ashihara (2009) detected NR deaminase activity (step 5a in Figure 19.3) in crude extracts of potato tubers (Table 19.1). The presence of the deaminase had not previously been reported in any organism. Although NMN deaminase (EC 3.5.1.42) (step 5b in Figure 19.3) from *A. vinelandii* is able to catalyse the deamination of NR *in vitro* (Imai 1973), its activity was fivefold lower than that of NMN cleavage. Since NMN deaminase activity could not been detected in potato tubers and in tea leaves (Table 19.1), the NR deaminase detected by Katahira and Ashihara (2009) may be a new NR specific deaminase, distinct from bacterial NMN deaminase.

19.5 Salvage of Nicotinamide and Nicotinate

In contrast to animals (Revollo et al. 2004; van der Veer et al. 2007), nicotinamide is not directly salvaged to NMN in plants because they lack nicotinamide phosphoribosyltransferase activity (step 3b in Figure 19.3, Table 19.1) (Ashihara and Deng 2012; Ashihara et al. 2005; Katahira and Ashihara 2009; Wang and Pichersky 2007; Zheng et al. 2005). Instead, plants salvage nicotinate for NaMN synthesis via NaPRT (step 6 in Figures 19.2 and 19.3). In addition to the widely distributed nicotinate salvage by NaPRT, salvage of NaR for NAD synthesis by NaR kinase may operate in cotyledons of mung bean (*Vigna radiata*) seedlings. When applied to cotyledons [carboxyl-^{14}C]NaR is readily converted to pyridine nucleotides. This conversion was also found in embryonic axes, but the rate was slower than in cotyledons (Matsui and Ashihara 2008). This pathway, therefore, would appear to function principally in the utilization of stored pyridine compounds.

19.5.1 Nicotinate Phosphoribosyltransferase

Nicotinate phosphoribosyltransferase activity (NaPRT, EC 6.3.4.21, step 6 in Figures 19.2 and 19.3) has been found in several plant extracts including mung bean (Zheng et al. 2005), potato tubers, and young tea leaves shown in Table 19.1. The IUBMB Enzyme Nomenclature old EC number of this enzyme, EC 2.4.2.11, which was created 1961, was deleted in 2013. The enzyme can form NaMN from nicotinate and PRPP in the absence of ATP. However, when ATP is available the enzyme is phosphorylated resulting in a much lower Km for nicotinate. In *Salmonella typhimurium*, the phospho-enzyme is hydrolysed during the transferase reaction, regenerating the low affinity form. The presence of ATP shifts the products/substrates equilibrium from 0.67 to 1100 (Vinitsky and Grubmeyer 1993). The ATP is required for its activity in plants (Katahira and Ashihara 2009; Yin et al. 2008) (Figure 19.4).

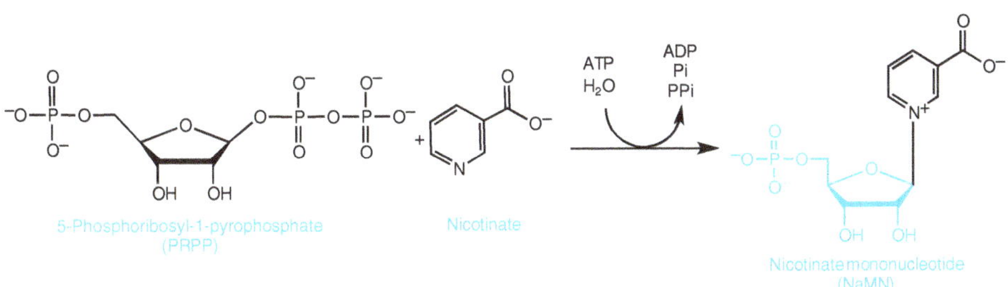

Figure 19.4 Step 6 in Figures 19.2 and 19.3 catalysed by nicotinate phosphoribosyltransferase.

Arabidopsis thaliana genes *At2g23420* and *At4g36940* encode NaPRT, but characterization of the recombinant enzymes has not yet been achieved. In suspension culture of *Lotus japonicus* cells, the expression profile of NaPRT was examined using expression sequence tags obtained from the *L. japonicus* database. The transcript level of NaPRT, encoding salvage enzymes, was highest in the lag phase of cell growth, corresponding to the activity of NaPRT (Yin et al. 2008). In microorganisms, feedback inhibition by NAD of NaPRT purified from *Brevibacterium* has been reported (Dulyaninova et al. 2000).

19.5.2 Nicotinate Riboside Kinase

As described in Section 19.5, NaR kinase activity (step 12 in Figure 19.3) was first detected in extracts from mung bean seedlings (Matsui and Ashihara 2008) and relatively high activity is also found in potato tubers (Katahira and Ashihara 2009) (Table 19.1) (Figure 19.5).

Although detailed kinetic studies could not be performed, the reaction of native NaR kinase activity has been measured in cotyledons of mung bean seedlings. Activity was 0.15 pkat mg^{-1} protein in cotyledons, but was much lower, <0.01 pkat mg^{-1} protein, in the embryonic axes. Activity of another NaR phosphotransferase, which catalyses NaMN synthesis from NaR and AMP, was detectable (0.02 pkat mg^{-1} protein). The activity of NaPRT, which is a conventional pyridine salvage enzyme, is higher than NaR kinase activity in mung bean seedlings especially in the cotyledons and embryonic axes, hence, NaR kinase may make a limited contribution to salvage for pyridine nucleotide synthesis. The operation of the NaR salvage pathway in cotyledons was demonstrated from the *in situ* metabolism of [carboxyl-^{14}C]NaR which showed the conversion of NaR to NaMN, NMN, and NAD (Matsui and Ashihara 2008).

The plant NaR kinase lacks nicotinamide riboside (NR) kinase activity. In contrast, salvage of NR by NR kinase (step 12a in Figure 19.3) has been demonstrated in fungi and humans (Bieganowski and Brenner 2004). This is interesting because plants salvage nicotinate and NaR, while fungi and animals salvage nicotinamide and NR (Ashihara et al. 2015).

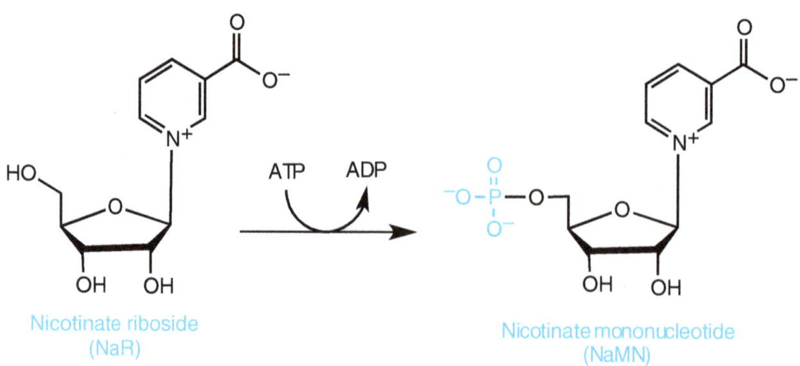

Figure 19.5 Step 10 in Figure 19.3 catalysed by nicotinamide riboside kinase.

19.6 Summary

There are two major PNCs in plants. The seven step pyridine nucleotide cycle (PNC VII) is a NAD → nicotinamide mononucleotide → nicotinamide riboside → nicotinamide → nicotinate → nicotinate mononucleotide → nicotinate adenine dinucleotide → NAD route. Another pathway is a five step cycle (PNC V), NAD → nicotinamide → nicotinate → nicotinate mononucleotide → nicotinate adenosine dinucleotide → NAD route. Further PNCs occur in different organs and tissues of plants.

References

Ashihara, H. and Deng, W.-W. (2012). Pyridine metabolism in tea plants: salvage, conjugate formation and catabolism. *J. Plant Res.* 125: 781–791.

Ashihara, H., Stasolla, C., Yin, Y. et al. (2005). *De novo* and salvage biosynthetic pathways of pyridine nucleotides and nicotinic acid conjugates in cultured plant cells. *Plant Sci.* 169: 107–114.

Ashihara, H., Ludwig, I., Katahira, R. et al. (2015). Trigonelline and related nicotinic acid metabolites: occurrence, biosynthesis, taxonomic considerations, and their roles in planta and in human health. *Phytochem. Rev.* 14: 765–798.

Barz, W. (1985). Metabolism and degradation of nicotinic acid in plant cell cultures. In: *Primary and Secondary Metabolism of Plant Cell Cultures* (eds. K.H. Neumann, W. Barz and E. Reinhard), 186–195. Berlin: Springer-Verlag.

Belenky, P., Bogan, K.L., and Brenner, C. (2007). NAD^+ metabolism in health and disease. *Trends Biochem. Sci.* 32: 12–19.

Bessman, M.J., Frick, D.N., and O'Handley, S.F. (1996). The MutT proteins or "Nudix" hydrolases, a family of versatile, widely distributed, "housecleaning" enzymes. *J. Biol. Chem.* 271: 25059–25062.

Bieganowski, P. and Brenner, C. (2004). Discoveries of nicotinamide riboside as a nutrient and conserved NRK genes establish a preiss-handler independent route to NAD^+ in fungi and humans. *Cell* 117: 495–502.

Briggs, A.G. and Bent, A.F. (2011). Poly(ADP-ribosyl)ation in plants. *Trends Plant Sci.* 16: 372–380.

Dobrzanska, M., Szurmak, B., Wyslouch-Cieszynska, A., and Kraszewska, E. (2002). Cloning and characterization of the first member of the Nudix family from *Arabidopsis thaliana*. *J. Biol. Chem.* 277: 50482–50486.

Dulyaninova, N.G., Podlepa, E.M., Toulokhonova, L.V., and Bykhovsky, V.Y. (2000). Salvage pathway for NAD biosynthesis in Brevibacterium ammoniagenes: regulatory properties of triphosphate-dependent nicotinate phosphoribosyltransferase. *Biochim. Biophys. Acta* 1478: 211–220.

Frye, R.A. (2000). Phylogenetic classification of prokaryotic and eukaryotic Sir2-like proteins. *Biochem. Biophys. Res. Commun.* 273: 793–798.

Gholson, H.R. (1966). The pyridine nucleotide cycle. *Nature* 212: 933–935.

Gholson, R.K. (1968). Regulation of NAD biosynthesis *via* the pyridine nucleotide cycle. *J. Vitaminol.* 14: 114–122.

Imai, T. (1973). Purification and properties of nicotinamide mononucleotide amidohydrolase from *Azotobacter vinelandii*. *J. Biochem.* 73: 139–153.

Imai, T. (1979). Isolation and properties of a glycohydrolase specific for nicotinamide mononucleotide from *Azotobacter vinelandii*. *J. Biochem.* 85: 887–899.

Imai, T. and Anderson, B.M. (1987). Metabolism of nicotinamide mononucleotide in beef liver. *Arch. Biochem. Biophys.* 254: 241–252.

Jambunathan, N. and Mahalingam, R. (2006). Analysis of Arabidopsis growth factor gene 1 (GFG1) encoding a nudix hydrolase during oxidative signaling. *Planta* 224: 1–11.

Jiménez, J.I., Canales, Á., Jiménez-Barbero, J. et al. (2008). Deciphering the genetic determinants for aerobic nicotinic acid degradation: the nic cluster from *Pseudomonas putida* KT2440. *Proc. Natl. Acad. Sci. USA* 105: 11329–11334.

Katahira, R. and Ashihara, H. (2009). Profiles of the biosynthesis and metabolism of pyridine nucleotides in potatoes (*Solanum tuberosum* L.). *Planta* 231: 35–42.

Matsui, A. and Ashihara, H. (2008). Nicotinate riboside salvage in plants: presence of nicotinate riboside kinase in mungbean seedlings. *Plant Physiol. Biochem.* 46: 104–108.

Matsui, A., Yin, Y., Yamanaka, K. et al. (2007). Metabolic fate of nicotinamide in higher plants. *Physiol. Plant.* 131: 191–200.

Moat, A.G. and Foster, J.W. (1987). Biosynthesis and salvage pathways of pyridine nucleotides. In: *Pyridine Nucleotide Coenzymes. Chemical, Biochemical, and Medical Aspects, Part B* (eds. D. Dolphin, O. Avramovic and R. Poulson), 1–24. New York: John Wiley & Sons.

Neuhann, H., Leienbach, K.W., and Barz, W. (1979). Degradation of nicotinamide adenine dinucleotide in cell suspension cultures. *Phytochemistry* 18: 61–64.

Noctor, G., Queval, G., and Gakiere, B. (2006). NAD(P) synthesis and pyridine nucleotide cycling in plants and their potential importance in stress conditions. *J. Exp. Bot.* 57: 1603–1620.

Ogawa, T., Ueda, Y., Yoshimura, K., and Shigeoka, S. (2005). Comprehensive analysis of cytosolic Nudix hydrolases in *Arabidopsis thaliana*. *J. Biol. Chem.* 280: 25277–25283.

Revollo, J.R., Grimm, A.A., and Imai, S. (2004). The NAD biosynthesis pathway mediated by nicotinamide phosphoribosyltransferase regulates Sir2 activity in mammalian cells. *J. Biol. Chem.* 279: 50754–50763.

van der Veer, E., Ho, C., O'Neil, C. et al. (2007). Extension of human cell lifespan by nicotinamide phosphoribosyltransferase. *J. Biol. Chem.* 282: 10841–10845.

Vinitsky, A. and Grubmeyer, C. (1993). A new paradigm for biochemical energy coupling. *Salmonella typhimurium* nicotinate phosphoribosyltransferase. *J. Biol. Chem.* 268: 26004–26010.

Wagner, K.G. and Backer, A.I. (1992). Dynamics of nucleotides in plants studied on a cellular basis. In: *International Review of Cytology*, vol. 134 (eds. K.W. Jeon and M. Friedlander), 1–84. San Diego: Academic Press.

Wagner, R. and Wagner, K.G. (1985). The pyridine-nucleotide cycle in tobacco: enzyme activities for the *de novo* synthesis of NAD. *Planta* 165: 532–537.

Wagner, R., Feth, F., and Wagner, K.G. (1986a). The pyridine-nucleotide cycle in tabacco: enzyme activities for the recycling of NAD. *Planta* 167: 226–232.

Wagner, R., Feth, F., and Wagner, K.G. (1986b). Regulation in tobacco callus of enzyme activities of the nicotine pathway II. The pyridine-nucleotide cycle. *Planta* 168: 408–413.

Wagner, R., Feth, F., and Wagner, K.G. (1986c). The regulation of enzyme activities of the nicotine pathway in tobacco. *Physiol. Plant.* 68: 667–672.

Waller, G.R., Yang, K.S., Gholson, R.K. et al. (1966). The pyridine nucleotide cycle and its role in the biosynthesis of ricinine by *Ricinus communis* L. *J. Biol. Chem.* 241: 4411–4418.

Wang, G. and Pichersky, E. (2007). Nicotinamidase participates in the salvage pathway of NAD biosynthesis in Arabidopsis. *Plant J.* 49: 1020–1029.

Yin, Y., Matsui, A., Sakuta, M., and Ashihara, H. (2008). Changes in pyridine metabolism profile during growth of trigonelline-forming *Lotus japonicus* cell cultures. *Phytochemistry* 69: 2891–2898.

Zheng, X.Q., Nagai, C., and Ashihara, H. (2004). Pyridine nucleotide cycle and trigonelline (*N*-methylnicotinic acid) synthesis in developing leaves and fruits of *Coffea arabica*. *Physiol. Plant.* 122: 404–411.

Zheng, X.Q., Hayashibe, E., and Ashihara, H. (2005). Changes in trigonelline (*N*-methylnicotinic acid) content and nicotinic acid metabolism during germination of mungbean (*Phaseolus aureus*) seeds. *J. Exp. Bot.* 56: 1615–1623.

Zrenner, R. and Ashihara, H. (2011). Nucleotide Metabolism. In: *Plant Metabolism and Biotechnology* (eds. H. Ashihara, A. Crozier and A. Komamine), 135–162. Chichester, UK: Wiley.

Part VII

Pyridine Alkaloids

20

Occurrence and Biosynthesis of Pyridine Alkaloids

20.1 Introduction

In this chapter the occurrence and biosynthesis of trigonelline and related nicotinate metabolites are described. High concentrations of trigonelline are found in coffee seeds and some members of the Fabaceae, while trace amounts occur in many other plant species. In contrast, the occurrence of pyridine alkaloids derived from nicotinate is limited. Nicotinate, a precursor of the secondary pyridine metabolites, is derived from pyridine nucleotides. *In planta*, pyridine nucleotide biosynthesis *de novo* is initiated from aspartic acid (see Chapter 12).

Trigonelline (1-*N*-methylnicotinate, nicotinate *N*-methylbetaine) (**1** in Figure 20.1) was originally isolated from fenugreek (*Trigonella foenum-graecum*) by Johns (1885). This was the first isolation of a pyridine derivative from a living organism.

Trigonelline is synthesized from nicotinate which is a catabolite of pyridine nucleotides. The biosynthesis of trigonelline is, therefore, analogous to that of caffeine (1,3,7-*N*-trimethylxanthine), which is derived from purine nucleotides (Ashihara 2008; Ashihara et al. 2013) (see Chapter 15). Both the caffeine and trigonelline biosynthetic pathways involve reactions catalysed by *S*-adenosyl-L-methionine (SAM)-dependent-*N*-methyltransferases. Trigonelline has a more widespread distribution in the plant kingdom than caffeine (Ashihara et al. 2013). Several hypotheses have been proposed concerning the physiological functions of trigonelline in plants. Minorsky (2002) summarized the state of knowledge regarding *in planta* functions of trigonelline in his column in the journal *Plant Physiology*. Trigonelline first gained the attention of plant biologists because it proved to be very effective in inducing G2 arrest in the root apices of numerous plant species (Evans and Tramontano 1981). Subsequently, evidence has been gradually accumulating indicating that the effects of trigonelline are not limited to cell cycle regulation. Although at the time little was known about trigonelline metabolism, Minorsky's short article in 2002 provided a fascinating insight into the potential roles of trigonelline *in planta*. Studies with coffee (*Coffea arabica* and *Coffea canephora*), mung bean (*Vigna radiata*), and potato (*Solanum tuberosum*) have elucidated the metabolic pathways which produce trigonelline (Ashihara 2008; Ashihara et al. 2012; Katahira and Ashihara 2009; Zheng et al. 2005). Furthermore, trigonelline in coffee has attracted interest as a potentially protective dietary agent in human health (Ashihara 2015).

Plant Nucleotide Metabolism – Biosynthesis, Degradation, and Alkaloid Formation, First Edition.
Hiroshi Ashihara, Iziar A. Ludwig and Alan Crozier.

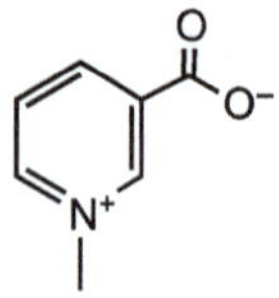

Trigonelline (**1**)

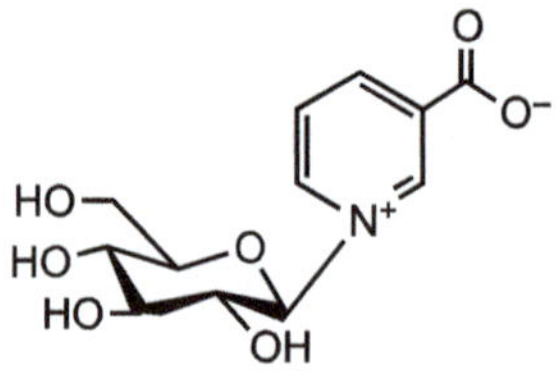

Nicotinate-*N*-glucoside (**2**)

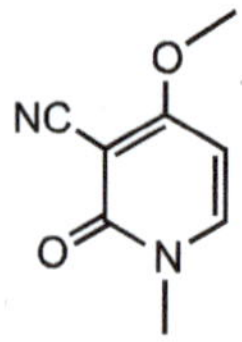

Ricinine (**3**)

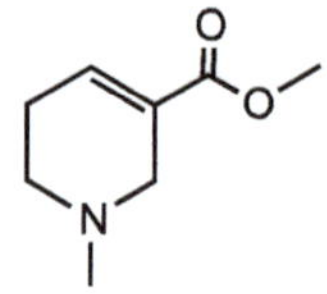

Arecoline (**4**)

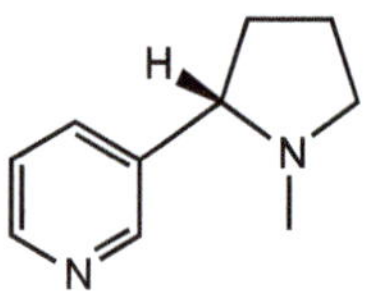

Nicotine (**5**)

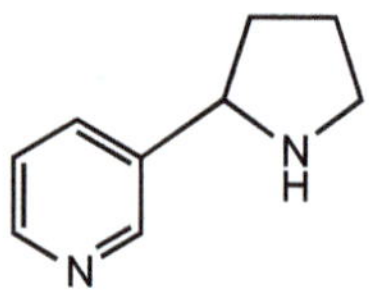

Nornicotine (**6**)

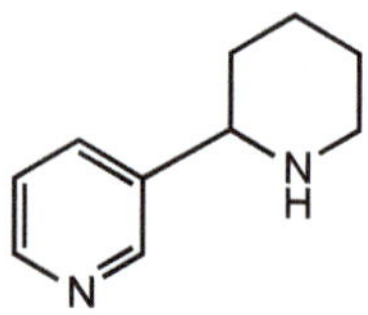

Anabasine (**7**)

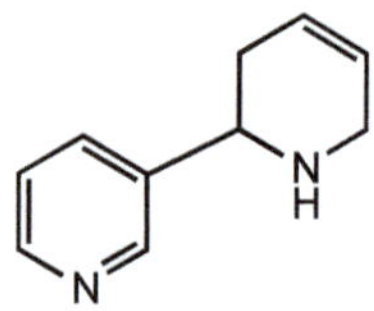

Anatabine (**8**)

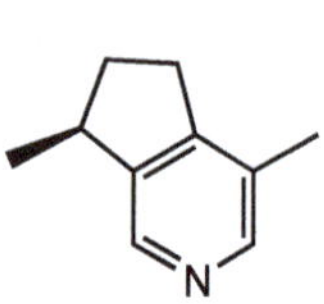

Actinidine (**9**)

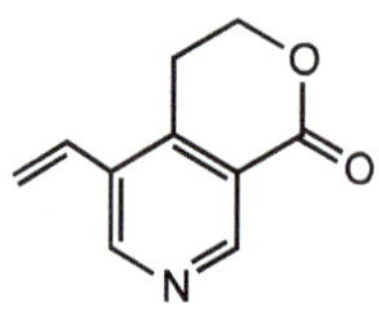

Gentianine (**10**)

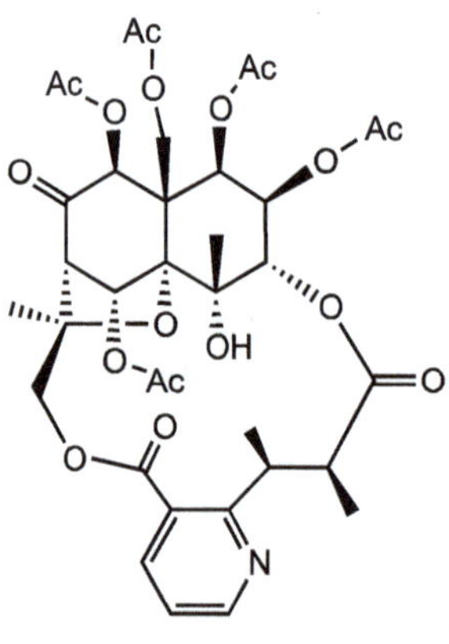

Evonine (**11**)
Ac = CH_3CO

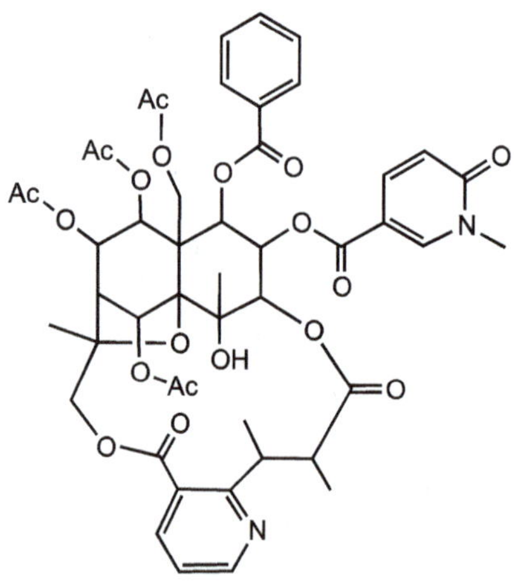

Hippocrateine (**12**)
Ac = CH_3CO

Figure 20.1 Chemical structures of pyridine alkaloids and related compounds.

This chapter first describes the occurrence of trigonelline in plants and for comparative purposes, the occurrence of other pyridine alkaloids derived from nicotinate is also summarized. *In planta* and eco-chemical functions of trigonelline are discussed in Chapter 21. The involvement of trigonelline in human nutrition and health is a topic covered in Section 23.4.2.

20.2 Occurrence of Pyridine Alkaloids

20.2.1 Trigonelline in Plants

The available literature indicates that trigonelline accumulates in seeds of legumes and coffee. Tramontano et al. (1986) studied the concentration of trigonelline in seeds of representatives of most super orders (21 of a total 24) of the Dicotyledoneae, investigating whether phylogenetic relationships are reflected in the relative amounts of trigonelline. With the exception of the Fabiflorae, the quantities of trigonelline were variable within certain super orders, whereas other super orders, such as the Violiflorae, Primuliflorae, Rosiflorae, Rutiflorae, Araliiflorae, and Gentianiflorae, yielded levels that were consistently low. Trace amounts of trigonelline were present in all of the super orders when compared with the relatively high levels in the Fabiflorae in which the highest concentration, 9.5 $\mu mol\,g^{-1}$ f.w., was detected in a leguminous species, cork bush (*Mundulea sericea*) (Tramontano et al. 1986). Subsequently, higher concentrations have been found to occur in seeds of clover (*Trifolium incarnatum*, 58 $\mu mol\,g^{-1}$ f.w.), coffee (*C. arabica*, 53 $\mu mol\,g^{-1}$ f.w.), and alfalfa (*Medicago sativa*, 12 $\mu mol\,g^{-1}$ f.w.) (Matsui et al. 2007).

Trigonelline occurs not just in seeds but other organs, including leaves and roots. High levels of trigonelline in these tissues have been found only in a few species including coffee (*C. arabica*), orange jessamine (*Murraya paniculata*), and four o'clock flower (*Mirabilis jalapa*). Trigonelline has been detected in young leaves and roots of *C. arabica* at concentrations of 19.5 and 6.4 $\mu mol\,g^{-1}$ f.w., respectively, while young stems contained 47.7 $\mu mol\,g^{-1}$ f.w. The seeds and pericarp of mature *Coffea fruits* contained 25.5 and 12.0 $\mu mol\,g^{-1}$ f.w., respectively (Zheng et al. 2004).

Trigonelline concentrations in young leaves and roots of orange jessamine were 13.4 and 4.1 $\mu mol\,g^{-1}$ f.w., respectively. In the four o'clock flower the trigonelline concentration in young leaves was 2.6 $\mu mol\,g^{-1}$ f.w. and this decreased with age, with mature leaves containing 0.4 $\mu mol\,g^{-1}$ f.w. A similar trend was observed in roots with the radicle of seven-day-old seedlings containing 3.5 $\mu mol\,g^{-1}$ f.w. of the pyridine alkaloid with levels in the roots of four-month-old plants declining to 0.8 $\mu mol\,g^{-1}$ f.w. In the three species noted above, trigonelline occurred in all plant parts with the concentrations being higher in younger than in mature tissues (Ashihara and Watanabe 2014). This is in contrast to the distribution of caffeine in *C. arabica* where it is located principally in older leaves and stems (Zheng and Ashihara 2004). Trace amounts of trigonelline have been detected in the shoots of 96 of the 143 angiosperm species examined by Blunden et al. (2005).

20.2.2 Other Pyridine Alkaloids in Plants

In addition to trigonelline, several secondary metabolites related to pyridine have been detected in plants. They vary from simple nicotinate metabolites to complicated sesquiterpene pyridine derivatives that occur in only a few species (Table 20.1).

Nicotinate *N*-glucoside (NaG, **2**), ricinine (**3**), and arecoline (**4**) are simple nicotinate derivatives. Accumulation of NaG occurs infrequently, although it is readily synthesized from nicotinate in ferns and some angiosperms as summarized below. Nishitani et al. (1995) reported that NaG was detected in cultured tobacco cells but only when nicotinate or nicotinamide was included in the culture medium. Ricinine, a cyano-substituted pyridine compound, is a toxic alkaloid produced by caster bean (*Ricinus communis*). It is present in all parts of the young caster bean plant at levels of ~6 μmol g^{-1} f.w. (Waller et al. 1966b). Arecoline, which was first isolated from seeds of the betel nut palm (*Areca catechu*) by Jahns (1891), is a cholinergic alkaloid.

Nicotine (**5**) and related compounds, such as nornicotine (**6**), anabasine (**7**), and anatabine (**8**), are polycyclic non-condensing pyridine derivatives (Table 20.1). Nicotine

Table 20.1 Occurrence of pyridine alkaloids in plants.

Compound	Species	Common name	Organ	Concentration (μmol g^{-1} d.w.)	Reference
Trigonelline (**1**)	*Trifolium incarnatum*	Clover (Crimson)	Seed	58	Matsui et al. (2007)
	Coffea arabica	Coffee	Seed	53	
	Glycine max	Soybean	Seed	1.2	
	Oryza sativa	Rice	Seed	0.04	
Nicotinate *N*-glucoside (**2**)	*Nicotiana tabacum*	Tobacco	Cells	0	Nishitani et al. (1995)
Ricinine (**3**)	*Ricinus communis*	Caster bean	Leaf	23	Waller et al. (1966b)
Arecoline (**4**)	*Areca catechu*	Areca palm	Seed		Jahns (1891)
Nicotine (**5**)	*Nicotiana tabacum*		Leaf	67	Saitoh et al. (1985)
Nornicotine (**6**)	*Nicotiana glutinosa*		Leaf	47	
Anabasine (**7**)	*Nicotiana glauca*		Leaf	57	
Anatabine (**8**)	*Nicotiana otophora*		Leaf	1.4	
Actinidine (**9**)	*Actinidia polygama*	Silver vine			Sakan et al. (1959)
Gentianine (**10**)	*Enicostema littorale*				Iyer et al. (1956)
	Halenia corniculata				Hesse (2002)
Evonine (**11**)	*Euonymus europaeus*	European spindle			Pailer and Libiseller (1962)
	Euonymus sieboldiana	Mayumi			Wada et al. (1971)
Hippocrateine (**12**)	*Hippocratea excelsa*	Cancerina	Root bark		Mata et al. (1990)

is the most abundant alkaloid in tobacco comprising more than 90% of the total alkaloid content. Minor alkaloids in tobacco include nornicotine, anabasine, and anatabine. The distribution of alkaloids in leaves and roots of 60 *Nicotiana* species has been reported (Saitoh et al. 1985). Nicotine was the dominant constituent in leaves of 33 species including *Nicotiana tabacum*, while nornicotine was a major component in 24 species, anabasine in 2 species, *Nicotiana glauca* and *Nicotiana debneyi*, and anatabine in 1 species, *Nicotiana otophora*. In roots, nicotine was the major pyridine derivative in 51 species, nornicotine in 2 species, *Nepenthes alata* and *Nicotiana africana*, and anabasine in 7 species, *N. glauca*, *Nicotiana solanifolia*, *Nicotiana benavidesii*, *Nicotiana cordgolia*, *Nicotiana debneyr*, *Nicotiana maritlma*, and *Nicotiana hesperis*. In most species the concentrations of anatabine and anabasine are higher in roots than in leaves, but the reverse occurs in *N. glauca*, *N. tabacum*, *Nicotiana repanda*, and *N. hesperis*, *N. alata*, *Nicotiana corymbosa*, and *N. hesperis*.

Actinidine (**9**) is found in silver vine (*Actinidia polygama*) and the essential oil of roots of valerian (*Valeriana officinalis*) (Sakan et al. 1959). Gentianine (**10**), a monoterpene alkaloid, has been isolated from a number of plants of the families Gentianaceae and Loganiaceae including *Halenia corniculata* and *Enicostema littorale* (Hesse 2002).

There are many sesquiterpene pyridine derivatives. For example, evonine (**11**) was isolated from *Euonymus* species (Pailer and Libiseller 1962; Wada et al. 1971) while hippocrateine (**12**) occurs in *Hippocratea excels* (Mata et al. 1990). Furukawa et al. (2002) reported the isolation of 19 sesquiterpene pyridine alkaloids from the stem bark of *Hippocratea excels*.

20.3 Biosynthesis of Pyridine Alkaloids

In plants, there are potentially two metabolic routes leading from nicotinate: trigonelline formation catalysed by trigonelline synthase, and NaG production mediated by nicotinate glucosyltransferase. In a limited number of plants, however, nicotinate serves as a precursor for the synthesis of pyridine alkaloids, such as nicotine and ricinine (Shoji and Hashimoto 2011; Waller et al. 1966b). Although detailed pathways have not yet been elucidated, pyridine rings of complex pyridine alkaloids originate from nicotinate derived from pyridine nucleotides.

20.3.1 Trigonelline Biosynthesis

In a narrow sense, trigonelline synthesis refers to trigonelline formation from nicotinate. The reaction is: SAM + nicotinate → *S*-adenosyl-L-homocysteine (SAH) + trigonelline (Figure 20.2).

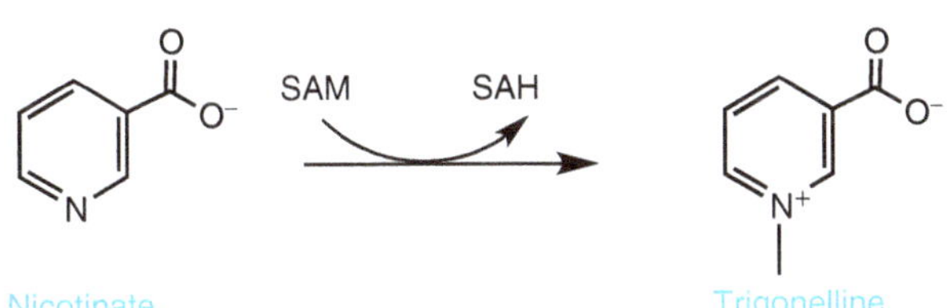

Figure 20.2 Production of trigonelline from nicotinate.

Trigonelline synthase (EC 2.1.1.7, nicotinate *N*-methyltransferase) has been studied in extracts from pea (*Pisum sativum*) (Joshi and Handler 1960), duckweed (*Lemna paucicostata*) (Taguchi et al. 1989), soybean (*Glycine max*) (Chen and Wood 2004; Hohl et al. 1988; Upmeier et al. 1988a), and mung bean (*V. radiata*) (Zheng et al. 2005) using native enzymes and recombinant *Escherichia coli* enzymes from *C. arabica* genes (Mizuno et al. 2014).

The estimated molecular weight of the partially purified soybean enzyme was 85 000–90 000 by gel filtration and ~42 000 by SDS-PAGE (Upmeier et al. 1988a; Chen and Wood 2004) suggesting that it is a dimer consisting of two identical subunits. Trigonelline synthase activity, in recombinant proteins from unidentified clones, which were tentatively named caffeine synthases (*CtCS3* and *CtCS4*), and reported as coffee trigonelline genes, *CTgS1* and *CTgS2*. *CTgS1* and *CTgS2* are highly homologous (>80%) with coffee caffeine synthases (*CCS1, CTS1, CTS2* and *CmXRS1*) at the nucleotide level. The amino acid sequences for CTgS1 and CTgS2 are similar to the coffee caffeine synthases, with the sequence identity between CCS1 and CTgS1 being more than 82%. These recombinant proteins catalyse the methylation of nicotinate, but do not use xanthine as a substrate. The Km values of CTgS1 and CTgS2 were 121 and 184 μM for nicotinate, and 68 and 120 μM for SAM, respectively (Mizuno et al. 2014).

Although nicotinate utilized for trigonelline synthesis is provided by the pyridine nucleotide cycle (see Chapter 20), trigonelline is also derived from nicotinate originating from nicotinamide mononucleotide (NaMN), which is produced by the *de novo* pathway. The pathway is NaMN → nicotinate riboside (NaR) → nicotinate (steps 11–13 in Figure 19.3) in trigonelline accumulating legumes. Zheng et al. (2008) reported that radioactivity from [^{14}C]NaMN was incorporated into trigonelline in the presence of SAM in crude cell-free extracts of the embryonic axes of germinating mung bean seeds. The cell-free extracts also possessed NaMN nucleotidase, NaR nucleosidase, and trigonelline synthase activities. *In situ* tracer experiments using embryonic axes segments showed that [^{14}C]NaR was readily converted to nicotinate and trigonelline. The NaMN → NaR → nicotinate → trigonelline pathway, therefore, operates in the embryonic axes of mung bean seedlings. These results indicate that trigonelline is synthesized from NaMN, an intermediate of the *de novo* biosynthetic pathway of pyridine nucleotides. However, this direct formation of trigonelline does not occur widely: a similar study with coffee fruits has shown that trigonelline is synthesized principally from nicotinate produced by the degradation of nicotinamide dinucleotide (NAD) (Ashihara et al. 2011).

20.3.2 Nicotinate *N*-Glucoside Biosynthesis

Similar to trigonelline, biosynthesis of NaG is also widely distributed in the plant kingdom. The reaction is: nicotinate + UDP-D-glucose + H^+ → NaG + UDP (Figure 20.3).

NaG synthase (nicotinate *N*-glucosyltransferase, EC 2.4.1.196) activity has been reported in the extracts of various organs of potato (Katahira and Ashihara 2009) and cell cultures of parsley (Upmeier et al. 1988b) and tobacco (Taguchi et al. 1997). Some of the properties of the partially purified enzymes from parsley and tobacco have been characterized. The Km values are similar: 170 and 220 μM for nicotinate, and 1.2 and 4.3 mM for UDP-glucose, respectively. In contrast, they have different pH optima, 7.8–8.2 for parsley enzyme and 6.1 for the tobacco enzyme. The estimated molecular

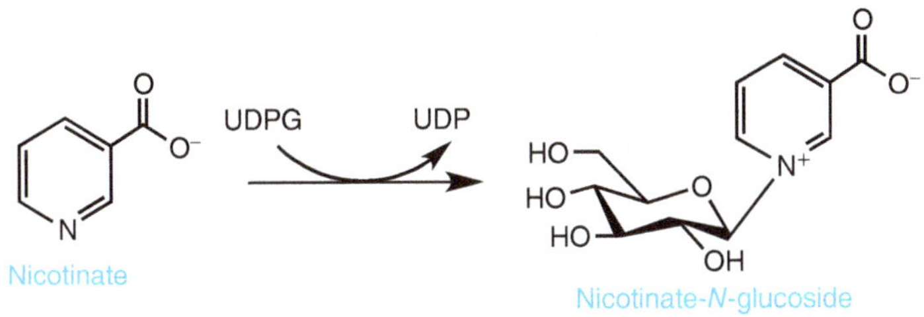

Figure 20.3 Production of nicotinate-*N*-glucoside from nicotinate.

weight of the parsley enzyme by gel filtration column chromatography was 46 000. Highly purified enzyme preparations and cloning of the NaG synthase gene have, to date, not been reported.

20.3.3 The Diversity of Biosynthetic Reactions

The diversity of the biosynthetic activity of trigonelline and NaG in different species and organs has been examined by several researchers. Willeke et al. (1979) reported that when nicotinate is added to suspension cultures of higher plants, one of two alternative pathways comes into operation. It is either converted via methylation to trigonelline or conjugated with arabinose. Only when conjugated as the *N*-arabinoside, does the molecule then undergo further metabolism and degradation. The metabolic fate of [carboxyl-^{14}C]nicotinate was investigated using cell cultures of ~50 plant species. It was found that the arabinoside was, with few exceptions, formed exclusively by members of the subclass Asteridae, while other plant groups generally followed the trigonelline route. This metabolic feature, thus, appears to be a marker at the level of order or subclass.

The production of the arabinoside by a few members of the Rosidae and Dilleniidae supports the view of some taxonomists of the presence of a phylogenetic line linking these orders with the Asteridae. The fact that there are these little-known taxa makes the character more interesting, since its distribution pattern questions the validity of the existing system of plant classification above the family level (Harborne and Turner 1984).

Nicotinate *N*-arabinoside described by Willeke et al. (1979) has now been identified as NaG (Upmeier et al. 1988c). Therefore, the diversity of the formation of nicotinate *N*-arabinoside in the original paper by Willeke et al. (1979) should be corrected to that of NaG. Furthermore, it was originally reported that nicotinate *N*-arabinoside, now known to be NaG, was a unique metabolite of cultured cells. This is incorrect as NaG is found in many intact plant tissues including the leaves of *N. tabacum* (Mizusaki et al. 1970).

The investigation of Willeke et al. (1979) using cell cultures is an example of a novel biochemical approach being utilized to reveal the chemosystematics of the plant kingdom (Harborne and Turner 1984). Their speculation is fascinating, but some questions remain unanswered. In their studies, NaG formation was found only in cell cultures of the subclass Asteridae and in the higher orders of the subclasses Rosidae and Dilleniidae. Ashihara et al. (2011), however, found that NaG was also formed from [carbony-^{14}C]nicotinamide by all 10 species of ferns that were investigated. Furthermore, secondary metabolism, such as alkaloid formation, depends strongly on the culture conditions, and alkaloids present in plant tissues are often absent in cultured cells (Barz 1985). Suspension-cultured potato cells, for instance, produce only NaG (Willeke et al. 1979) whereas the leaves of potato plants synthesize both trigonelline and NaG (Katahira and Ashihara 2009).

The diversity in nicotinate metabolite formation was investigated with leaf discs of ~50 plant species using carbonyl-^{14}C-labelled nicotinamide rather than [carboxyl-^{14}C]nicotinate as a precursor. This is because exogenous nicotinamide is readily absorbed by the leaf discs and converted to nicotinate while uptake of exogenous nicotinate is relatively low (Ashihara and Deng 2012). The results are illustrated in Figure 20.4. The order of the plant species is according to Smith et al. (2006) for ferns and APG III for angiospermae (The Angiosperm Phylogeny Group 2009). In ferns, the major nicotinate conjugate formed from nicotinate is NaG.

20.3.3.1 Ferns

The metabolic fate of [carbonyl-^{14}C]nicotinamide was investigated in 10 ferns, *Psilotum nudum* (Psilotaceae) (1 in Figure 20.4), *Angiopteris evecta* (Marattiaceae) (2), *Lygodium japonicum* (Lygodiaceae) (3), *Cyathea lepifera* (Cyatheaceae) (4), *Acrostichum aureum* (Pteridaceae) (5), *Asplenium antiquum* (Aspleniaceae) (6), *Thelypteris acuminate* (Thelypteridaceae) (7), *Diplazium subsinuatum* (Woodsiaceae) (8), *Blechnum orientale* (Blechnaceae) (9), and *Crytomium fortune* (Dryopteridaceae) (10) (Ashihara et al. 2011, 2012).

As well as eight ferns belonging to the class Pteridopsida (3–10), *P. nudum* (Psilotaceae) (1) from the class Psilotopsida was also examined. This fern lacks true leaves and roots and, traditionally, has not been viewed as a true fern. However, molecular evidence suggests that it is a fern, which has lost a number of pteridophytic characteristics (Qiu and Palmer 1999). *A. evecta* (Marattiaceae) of the class Marattiopsida (2) is a huge fern and is considered to be primitive as very similar fossilized fronds have been found in rocks almost three hundred million years old.

With one exception, all fern species produced a large quantity of NaG from [^{14}C]nicotinamide. The exception was *A. antiquum* (Aspleniaceae) which yielded only a very small amount of trigonelline after a long-term, 18-hour incubation. Increases in the release of $^{14}CO_2$ with incubation time were accompanied by decreases in [carboxyl-^{14}C]NaG. Turnover of NaG in ferns therefore appears to be rapid, as also occurs in parsley cell cultures (Upmeier et al. 1988c) and tea leaves (Ashihara and Deng 2012). Formation of NaG is the principal characteristic of nicotinate metabolism in ferns (Figure 20.4).

20.3.3.2 Gymnosperms

In gymnosperms the major nicotinate metabolite is trigonelline, which was detected in five species, namely: *Cycas revoluta* (Cycadaceae) (11), *Ginkgo biloba* (Ginkgoaceae) (12), *Picea glauca* (Pinaceae) (13), *Pinus luchuensis* (Pinaceae) (14), and *Casuarina equisetifolia* (Casuarinaceae) (15). *Cycas revolute* (11) was an exception as similar amounts of NaG and trigonelline were formed. It is possible that *C. revoluta* (11) is a key plant in understanding the diversity of nicotinate metabolite formation.

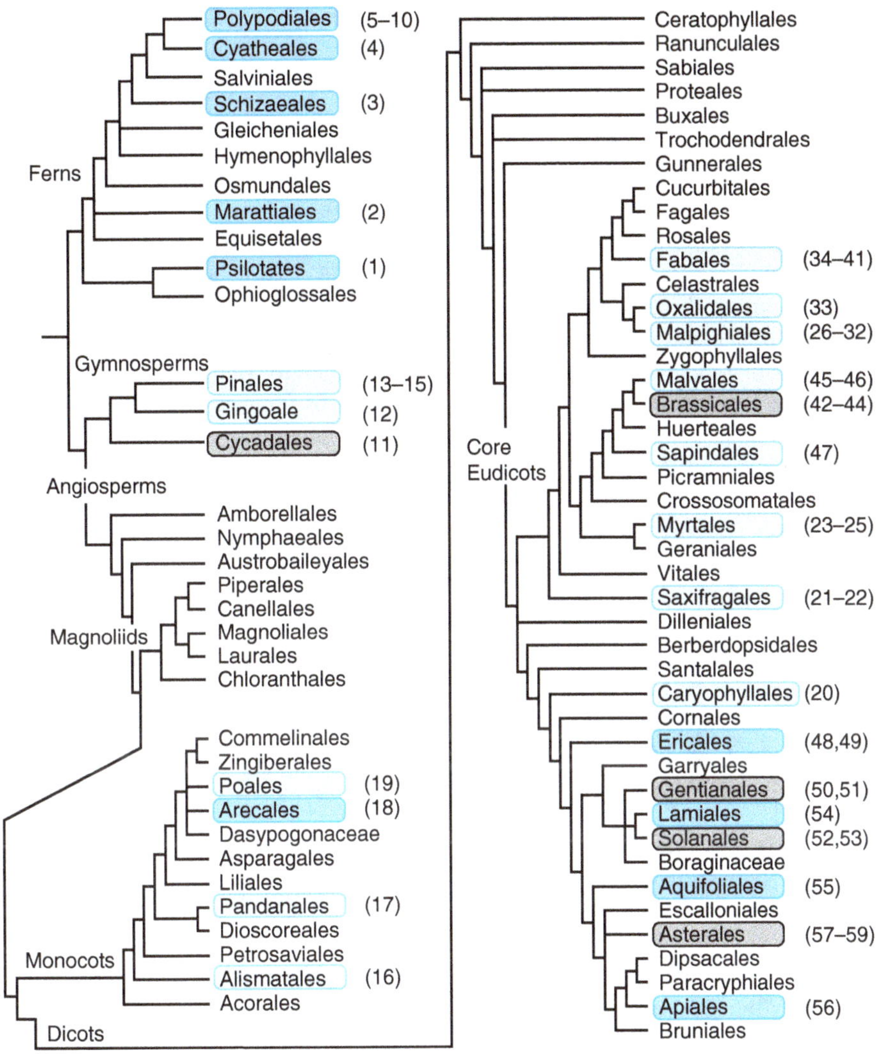

Figure 20.4 Phylogenetic relationship between trigonelline- and nicotinate-*N*-glucoside (NaG)-synthesizing plant orders. The red boxes are trigonelline-forming orders and blue boxes are NaG-forming plants. The grey boxes show the orders that include species that produce both metabolites or include more than two species that produce either trigonelline or NaG (see text). Numbers shown are species that were investigated (see text).

20.3.3.3 Angiosperms

In angiosperm monocots, three of four species, *Alocasia odora* (Araceae) (16), *Pandanus odoratissimus* (Pandanaceae) (17), and *Oryza sativa* (Poaceae) (19), produced trigonelline, while NaG accumulated in *Arenga trewmula* (Arecaceae) (18). In dicots (20–59), 25 of 39 species examined synthesized trigonelline and 12 species produced NaG. Both NaG and trigonelline are synthesized as major products in *S. tuberosum* (Solanaceae) (53) and *Chrysanthemum coronarium* (Asteraceae) (59).

In general, trigonelline- and NaG-forming angiosperms can be separated around the border between the orders of Fabales and Brassicales, although there are clear exceptions to this rule. All species in Fabales (34–41) produce trigonelline. In Brassicales, *Raphanus sativus* var. *sativus* (44) produces trigonelline; *Brassica napus* (43) and *Arabidopsis thaliana* (42) produce NaG while *A. thaliana* (42) also produces limited amounts of trigonelline. Usually species in the same orders produce the same nicotinate conjugates, but some exceptions are found. In Gentianales, *C. arabica* (Rubiaceae) (50) produces trigonelline, but *Catharanthus roseus* (Apocynaceae) (51) produces NaG. In Asterales (57–59), *Scaevola taccada* (Goodeniaceae) (57) and *Farfugium japonicum* (Asteraceae) (58) produce NaG, while *C. coronarium* (Asteraceae) (59) produces both metabolites. These results indicate that all the plants, probably without exception, have the ability to convert exogenous nicotinamide or nicotinate to trigonelline and/or NaG. NaG formation is restricted to ferns and selected orders of angiosperms, whereas other plants produce trigonelline. In most cases the formation of NaG and trigonelline is mutually exclusive, but there are several exceptions.

20.3.3.4 Nicotinate Conjugate Formation

Different nicotinate conjugate formation in different plant species may depend mainly on the activities of nicotinate *N*-glucosyltransferase and *N*-methyltransferase (trigonelline synthase). As the substrates of these enzymes, UDP-glucose and SAM, are ubiquitous in plants, their supply would appear not to impact qualitatively on the conversion of nicotinate to NaG and trigonelline. The expression of genes encoding these two enzymes inherently control the formation of nicotinate metabolites. These genes have yet to be identified in plants and further studies are needed to clarify the situation.

A comprehensive survey of the nicotinate metabolite formation was carried out using cultured cells or leaf discs and it has been reported that different metabolites are produced in different organs of the same species (Katahira and Ashihara 2009; Koster et al. 1989; Matsui et al. 2007). The biosynthetic activity of the formation of NaG and trigonelline is more or-less organ-specific. Katahira and Ashihara (2009) compared the metabolism of [carboxyl-^{14}C]nicotinate with activities of nicotinate *N*-glucosyltransferase and trigonelline synthase in leaves, the green, photosynthesizing above ground stems and etiolated below ground stems, stolons, tubers, and illuminated greenish tubers of potato plants. Significant amounts of [carboxyl-^{14}C]nicotinate were converted to NaG in all instances, but trigonelline synthesis occurred predominantly in green photosynthesizing tissues. Trigonelline synthesis was low in etiolated stems, but was more active in the illuminated tubers. The greater activity of trigonelline synthase, reflected in the *in situ* biosynthesis of trigonelline from [carboxyl-^{14}C]nicotinate in green organs compared to non-illuminated tubers, stolons, and etiolated stems, suggests that trigonelline synthesis is associated with chloroplasts in potato. In this context, it is of interested that Matsui et al. (2007) reported that leaves of *Oryza sativa, Lotus japonicus,* and *Theobroma cacao* possess higher trigonelline biosynthetic activity than other organs.

Although many plants possess trigonelline biosynthetic activity, accumulation of the pyridine alkaloid, as discussed above, is found in only limited species including coffee and orange jessamine. This suggests that continuous supply of nicotinate, a substrate of trigonelline synthase, may be an important factor for the accumulation of trigonelline. However, clear mechanisms underlying accumulation of trigonelline remain to be elucidated.

20.3.4 Biosynthesis of Ricinine

Historically, the pyridine nucleotide cycle in plants was studied using caster bean which produces ricinine. During the late 1960s and 1970s many papers were published on the ricinine biosynthesis and pyridine metabolism, and these studies were reviewed by Waller and Dermer (1981).

Interdependency between the pyridine nucleotide cycle and ricinine biosynthesis was shown by use of ^{14}C- and ^{15}N-labelled compounds (Waller et al. 1966a; Yang and Waller 1965) and inhibitors (Johnson and Waller 1974). It was reported that the pyridine moieties of quiniolinate, nicotinate, and nicotinamide, NaMN, NAD, and NADP were incorporated into ricinine. All of these pyridine compounds except quinolinate are members of the pyridine nucleotide cycle, and they could function as a source of metabolic intermediates for the biosynthesis of ricinine.

Mann and Byerrum (1974) provided evidence for the activation of the *de novo* pathway for pyridine nucleotide biosynthesis prior to ricinine biosynthesis in castor beans. The relationship was estimated by comparison of enzyme activities and ricinine content in germinating caster beans (Mann and Byerrum 1974). A sixfold increase in the activity of quinolinate phosphoribosyltransferase (QPRT), an enzyme involved in *de novo* pyridine nucleotide synthesis, was found in etiolated seedlings four days after imbibition. This was preceded by an increase in ricinine content and thereafter both QPRT and ricinine decreased.

However, the level of a key enzyme in the pyridine nucleotide cycle, nicotinamidase, was unusually low. The activity of another enzyme of the cycle, nicotinate phosphoribosyltransferase, declined during germination to the level found in mature plants. Therefore, a positive correlation exists only between the activity of QPRT and the biosynthesis of ricinine. In contrast, activity of the pyridine nucleotide cycle was not closely related to the ricinine biosynthesis.

Calculations based on QPRT levels in young cotyledons and endosperm of caster bean indicate that the ricinine is formed from pyridine nucleotide cycle components in the cotyledons, such as nicotinate, which is synthesized in the endosperm (Mann and Byerrum 1974). The intermediates between nicotinate or nicotinamide and ricinine have been tentatively identified (Figure 20.5).

Robinson (1965) obtained a crude enzyme preparation from castor beans that catalysed the oxidation of *N*-methylnicotinonitrile to the corresponding 4- and 6-pyridones. However, Waller et al. (1966b) reported that *N*-methylnicotinonitrile was not incorporated into ricinine *in vivo*. Since ricinine formation requires pyridinium oxidase activity, enzyme compartmentalization might have occurred or alternatively *N*-methylnicotinonitrile may be unable to be transported through the membrane defining the subcellular location of the pyridinium oxidase.

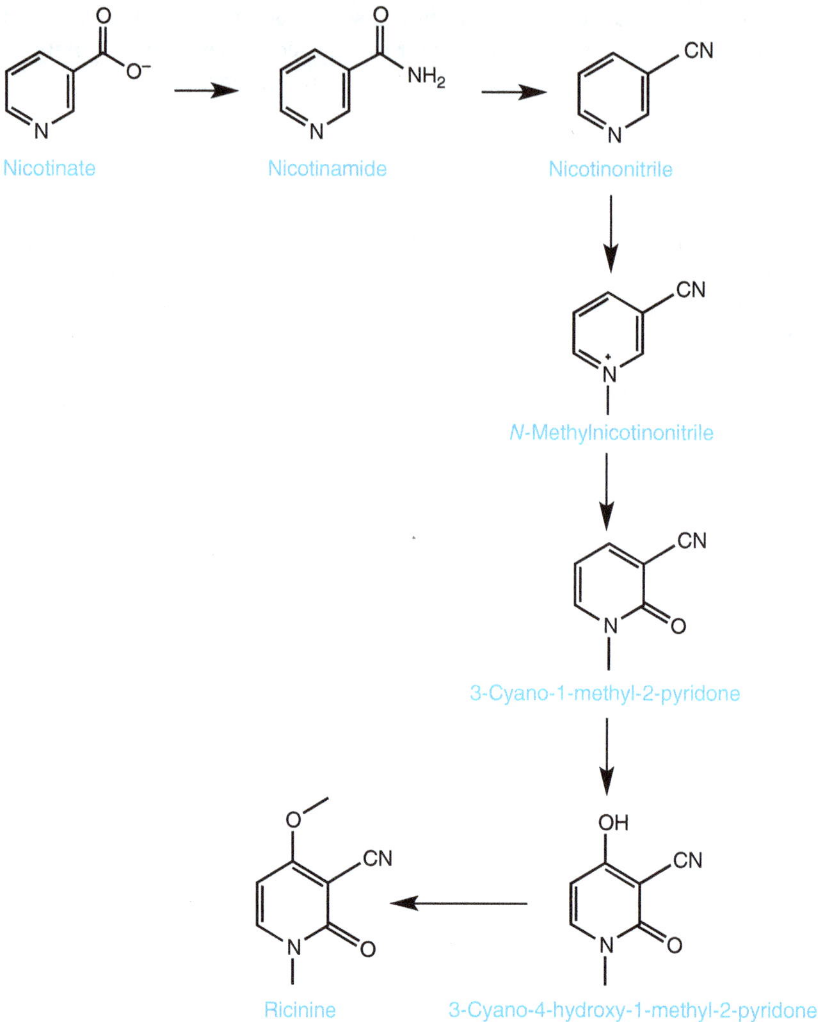

Figure 20.5 Proposed pathway of ricinine biosynthesis in *Ricinus communis* (adapted from Waller and Dermer 1981). To-date there are no reports of enzymatic studies being carried out.

Fu and Robinson (1970) found that seven species out of 11 members of the family Euphorbiaceae exhibited pyridinium oxidase activity when incubated with *N*-methylnicotinonitrile. Three isoforms of pyridinium oxidase, A, B, and C, which catalysed formation of the 4- and 6-pyridones, but exhibited different behaviour on DEAE-cellulose columns, were found in *R. communis*. The pyridinium oxidase B was purified from *R. communis* and characterized (Fu and Robinson 1972). Plants belonging to other families, including *N. tabacum*, showed no such oxidase activity (Waller and Dermer 1981), suggesting that the enzyme system is phylogenetically specific.

Detailed studies on the ricinine biosynthetic pathway at the molecular level have, as yet, not been carried out.

20.3.5 Biosynthesis of Nicotine (Pyridine Ring)

Nicotine (**1**) and related tobacco alkaloids are composed of pyridine and pyrrolidine rings (Figure 20.1). Anabasine (**7**) comprises a pyridine and a piperidine ring. The pyridine and pyrrolidine rings are derived, respectively, from nicotinate or its derivatives, and ornithine and possibly from arginine via the symmetric diamine putrescine pathway (Figure 20.6). On the other hand, the piperidine ring of anabasine is synthesized from lysine through decarboxylation by lysine decarboxylase to produce cadaverine, and then oxidation by diamine oxidase to yield 5-aminopentanal followed by spontaneous intramolecular cyclisation resulting in Δ^1-piperidine (Figure 20.6) (Shoji and Hashimoto 2011). Anatabine (**8**), a minor alkaloid found in some species of *Nicotiana*, consists of two rings both of them are derived from nicotinate (Leete and Slattery 1976) (Figure 20.6).

The formation of the pyridine ring of nicotine will now be considered briefly. An excellent review on nicotine biosynthesis has been published by Shoji and Hashimoto (2011), and readers interested in an in depth commentary on the topic are referred to this article.

The biosynthetic pathway of the pyridine section of the nicotine biosynthesis pathway was investigated by several researchers in the 1960s. For example, Waller and co-workers demonstrated that [2,3,7,8-$^{14}C_4$]quinolinate, [2,3,7-$^{14}C_3$]nicotinate, and [2,3,7-$^{14}C_3$]NAD were incorporated with similar efficiencies into the pyridine structure of nicotine in *Nicotiana rustica* (Frost et al. 1967; Yang et al. 1965; Yang and Waller 1965). They postulated that in tobacco metabolites of the pyridine nucleotide cycle are incorporated into the pyridine ring for nicotine biosynthesis. Like Arabidopsis, the *de novo* biosynthetic pathway of pyridine biosynthesis in tobacco starts with quinolinic acid synthesis from aspartate (Figure 12.1) and NAD enters into the pyridine nucleotide cycle (Figure 13.2).

The expression pattern of three genes involved in *de novo* pyridine biosynthesis, namely, L-aspartate oxidase, quinolinate synthase and QPRT, is closely related to the activity of nicotine biosynthesis. Their transcripts are most abundant in nicotine-producing roots while being expressed at basal levels in leaves, and subjected to regulation by *NIC* loci and activation by jasmonate treatment as well as other genes involved in nicotine biosynthesis (Sinclair et al. 2000; Goossens et al. 2003; Shoji et al. 2010). Such co-regulation indicates the aspartate pathway of *de novo* pyridine biosynthesis is under the same transcriptional control as nicotine biosynthesis in tobacco.

It has been presumed that nicotinate or its derivatives couple with *N*-methyl-Δ^1-pyrrolinium at the final stage in the formation of nicotine (Figure 20.6). However, the coupling mechanism of the pyridine ring to the pyrrolidine ring has not been fully elucidated. Tracer studies using ^{14}C- and ^{3}H-labelled nicotinate have established that the pyridine moiety is incorporated into nicotine. In the biosynthesis of nicotine from nicotinate in *N. tabacum*, the C-7 carboxyl group of nicotinate is lost and presumably replaced by the pyrrolidine ring (Dawson et al. 1953). It appeared that a

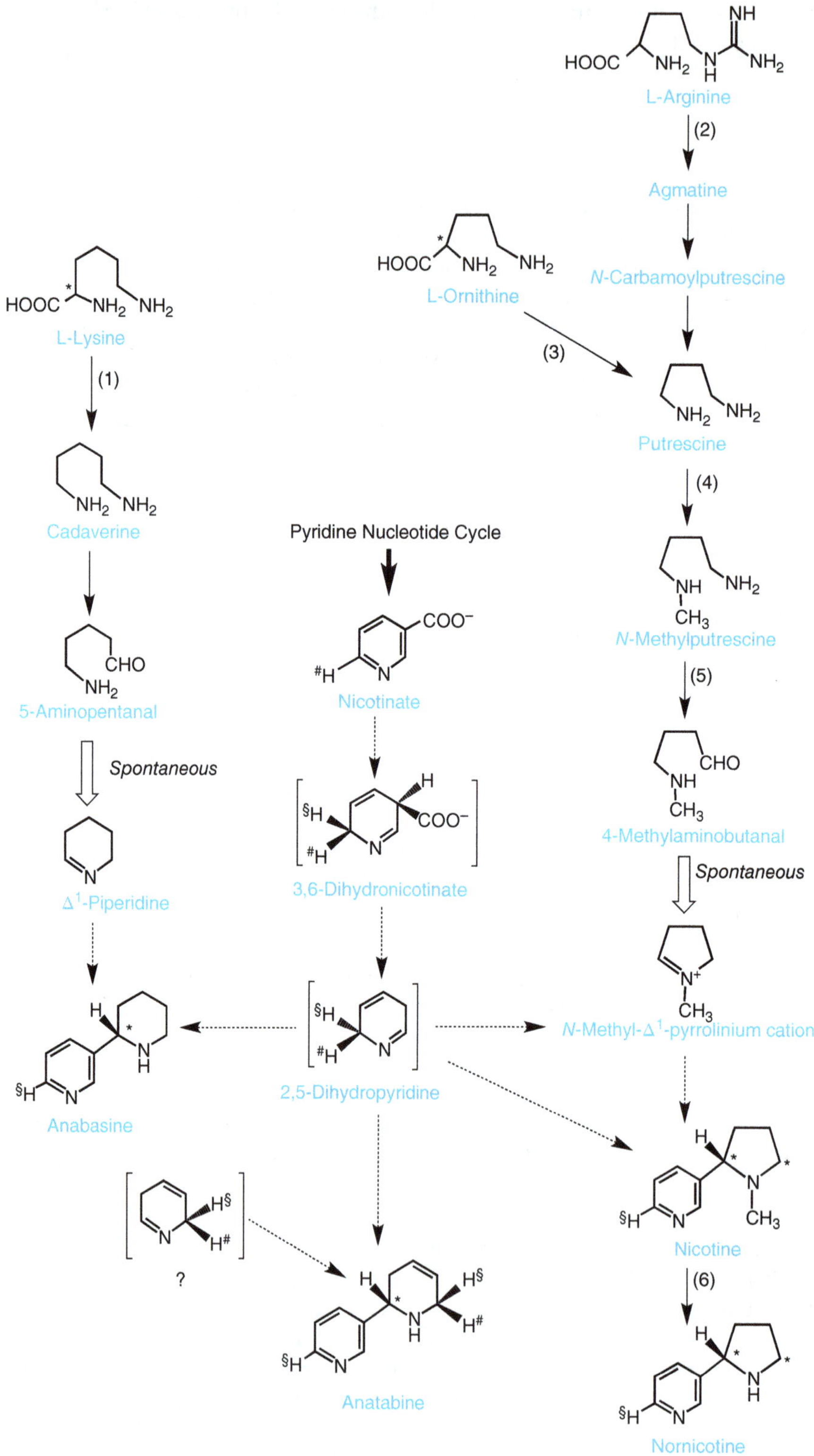
L-Arginine
(2)
Agmatine
N-Carbamoylputrescine
L-Ornithine
(3)
Putrescine
(4)
N-Methylputrescine
(5)
4-Methylaminobutanal
Spontaneous
N-Methyl-Δ1-pyrrolinium cation
L-Lysine
(1)
Cadaverine
5-Aminopentanal
Spontaneous
Δ1-Piperidine
Anabasine
Pyridine Nucleotide Cycle
Nicotinate
3,6-Dihydronicotinate
2,5-Dihydropyridine
?
Anatabine
Nicotine
(6)
Nornicotine

Figure 20.6 Proposed pathway of nicotine alkaloid biosynthesis (adapted from Shoji and Hashimoto 2011). The C-2 carbons of ornithine and lysine, the hydrogen at C-6 of nicotinate, and another hydrogen introduced at this position are indicated using symbols to show their fates. Hypothetical intermediates derived from nicotinate are included in brackets, and undefined steps including the ring couplings are shown with broken arrows. Enzymes (EC numbers) shown are (1), lysine decarboxylase (EC 4.1.1.18); (2), arginine decarboxylase (EC 4.1.1.19); (3), ornithine decarboxylase (EC 4.1.1.17); (4), putrescine *N*-methyltransferase (EC 2.1.1.53); (5), *N*-methylputrescine oxidase (EC 1.4.3.22); (6), nicotine *N*-demethylase (EC 1.14.13.B3, preliminary BRENDA-supplied EC number).

1,6-dihydropyridine intermediate is attacked by an *N*-methyl-Δ^1-pyrrolidinium cation at C-3 position, with simultaneous decarboxylation (Scott and Glynn 1967) (Figure 20.6). This mechanism also explains why tritium-labelled nicotine derived from [6-^{3}H]nicotinate is not labelled specifically at C-6 (Leete and Liu 1973).

In the case of anatabine, the reduction of nicotinate by the introduction of hydrogen at C-6 in the *pro-(R)* position was demonstrated by Leete (1978) who postulated that 3,6-dihydronicotinate and its decarboxylated form 2,5-dihydropyridine were potential intermediates derived from nicotinate (Figure 20.6). Nicotine synthase activity, which catalyses the formation of (*S*)-nicotine from nicotinate and 1-methy-Δ^1-pyrrolidinium chloride, was detected in crude root extracts of *Nicotiana glutinosa* and *N. tabacum* by Friesen and Leete (1990). However, the results have not been replicated by other researchers (Hibi et al. 1994). Instead, a novel reductase, A622, encoding gene, isolated initially as a gene repressed in low nicotine tobacco mutants (*nic1* and *nic2*) (Hibi et al. 1994), might be required for the coupling step. Since the production of all the pyridine alkaloids was inhibited by A622 suppression (DeBoer et al. 2009) (Kajikawa et al. 2009), A622 could be the enzyme, or one of the enzymes, acting in the coupling step(s). However, the enzymatic reaction of A622 has yet to be defined.

20.4 Summary

Several pyridine alkaloids occur in plants. The biosynthesis of trigonelline and/or nicotinate *N*-glucoside are well established in all plant species. The alkaloid ricinine and half of the pyridine ring of nicotine are produced from nicotinate.

References

Ashihara, H. (2008). Trigonelline (*N*-methylnicotinic acid) biosynthesis and its biological role in plants. *Nat. Prod. Commun.* 3: 1423–1428.

Ashihara, H. (2015). Plant biochemistry: Trigonelline biosynthesis in *Coffea arabica* and *Coffea canephora*. In: *Coffee in Health and Disease Prevention* (ed. V.R. Preedy), 19–28. San Diego: Academic Press.

Ashihara, H. and Deng, W.-W. (2012). Pyridine metabolism in tea plants: salvage, conjugate formation and catabolism. *J. Plant Res.* 125: 781–791.

Ashihara, H. and Watanabe, S. (2014). Accumulation and function of trigonelline in non-leguminous plants. *Nat. Prod. Commun.* 9: 795–798.

Ashihara, H., Yin, Y., and Watanabe, S. (2011). Nicotinamide metabolism in ferns: formation of nicotinic acid glucoside. *Plant Physiol. Biochem.* 49: 275–279.

Ashihara, H., Yin, Y., Katahira, R. et al. (2012). Comparison of the formation of nicotinic acid conjugates in leaves of different plant species. *Plant Physiol. Biochem.* 60: 190–195.

Ashihara, H., Yokota, T., and Crozier, A. (2013). Biosynthesis and catabolism of purine alkaloids. In: *Adv. Bot. Res*, vol. 68, 111–138.

Barz, W. (1985). Metabolism and degradation of nicotinic acid in plant cell cultures. In: *Primary and Secondary Metabolism of Plant Cell Cultures* (eds. K.H. Neumann, W. Barz and E. Reinhard), 186–195. Berlin: Springer-Verlag.

Blunden, G., Patel, A.V., Armstrong, N. et al. (2005). Betaine distribution in angiosperms. *Biochem. Syst. Ecol.* 33: 904–920.

Chen, X. and Wood, A.J. (2004). Purification and characterization of *S*-adenosyl-L-methionine nicotinic acid-*N*-methyltransferase from leaves of *Glycine max*. *Biol. Plant.* 48: 531–535.

Dawson, R.F., Christman, D.R., and Anderson, R.C. (1953). Alkaloid biogenesis. IV. The non-availability of nicotinic acid-carboxyl-C14 and its ethyl ester for nicotine biosynthesis. *J. Am. Chem. Soc.* 75: 5114–5116.

DeBoer, K., Lye, J., Aitken, C. et al. (2009). The A622 gene in Nicotiana glauca (tree tobacco): evidence for a functional role in pyridine alkaloid synthesis. *Plant Mol. Biol.* 69: 299–312.

Evans, L.S. and Tramontano, W.A. (1981). Is trigonelline a plant hormone pea seedlings? *Am. J.Bot.* 68: 1282–1289.

Friesen, J.B. and Leete, E. (1990). Nicotine synthase - an enzyme from nicotiana species which catalyzes the formation of (S)-nicotine from nicotinic acid and 1-methyl-δ'pyrrolinium chloride. *Tetrahedron Lett.* 31: 6295–6298.

Frost, G.M., Yang, K.S., and Waller, G.R. (1967). Nicotinamide adenine dinucleotide as a precursor of *Nicotine in Nicotiana* rustica L. *J. Biol. Chem.* 242: 887–888.

Fu, P. and Robinson, T. (1970). Pyridinium oxidases in the family euphorbiaceae. *Phytochemistry* 9: 2443–2446.

Fu, P. and Robinson, T. (1972). Isolation and characterization of pyridinium oxidase B. *Phytochemistry* 11: 95–103.

Furukawa, M., Makino, M., Uchiyama, T. et al. (2002). Sesquiterpene pyridine alkaloids from *Hippocratea excelsa*. *Phytochemistry* 59: 767–777.

Goossens, A., Häkkinen, S.T., Laakso, I. et al. (2003). A functional genomics approach toward the understanding of secondary metabolism in plant cells. *Proc. Natl. Acad. Sci.* 100: 8595–8600.

Harborne, J.B. and Turner, B.L. (1984). *Plant Chemosystematics*. London: Academic Press.

Hesse, M. (2002). *Alkaloids nature's Curse or Blessing*. New York: Wiley-VCH.

Hibi, N., Higashiguchi, S., Hashimoto, T., and Yamada, Y. (1994). Gene expression in tobacco low-nicotine mutants. *Plant Cell* 6: 723–735.

Hohl, U., Upmeier, B., and Barz, W. (1988). Growth and nicotinic acid biotransformation in batch cultured and airlift fermenter grown soybean cell suspension cultures. *Appl. Microbiol. Biotechnol.* 28: 319–323.

Iyer, R.S., Pathak, B., and Bose, A. (1956). Gentianine from an Indian medicinal plant. *Naturwissenschaften* 43: 251–252.

Jahns, E. (1891). Über die Alkaloide der Arekanuß. *Arch. Pharm.* 229: 669–707.

Johns, E. (1885). Ueber die Alkaloide des Bockshornsamens. *Ber. Dtsch. Chem. Ges.* 18: 2518–2523.

Johnson, R.D. and Waller, G.R. (1974). The relationship of the pyridine nucleotide cycle to ricinine biosynthesis in *Ricinus Communis*. *Phytochemistry* 13: 1493–1500.

Joshi, J.G. and Handler, P. (1960). Biosynthesis of Trigonelline. *J. Biol. Chem.* 235: 2981–2983.

Kajikawa, M., Hirai, N., and Hashimoto, T. (2009). A PIP-family protein is required for biosynthesis of tobacco alkaloids. *Plant Mol. Biol.* 69: 287–298.

Katahira, R. and Ashihara, H. (2009). Profiles of the biosynthesis and metabolism of pyridine nucleotides in potatoes (*Solanum tuberosum* L.). *Planta* 231: 35–42.

Koster, S., Upmeier, B., Komossa, D., and Barz, W. (1989). Nicotinic acid conjugation in plants and plant cell cultures of potato (*Solanum tuberosum*). *Z. Naturforsch.* 44c: 623–628.

Leete, E. (1978). Stereochemistry of the reduction of nicotinic acid when it serves as a precursor of anatabine. *J. Chem. Soc. Chem. Commun.*: 610–611.

Leete, E. and Liu, Y.Y. (1973). Metabolism of [2-^{3}H]- and [6-^{3}H]-nicotinic acid in intact *Nicotiana tabacum* plants. *Phytochemistry* 12: 593–596.

Leete, E. and Slattery, S.A. (1976). Incorporation of [2-^{14}C]- and [6-^{14}C]nicotinic acid into the tobacco alkaloids. Biosynthesis of anatabine and alpha,beta-dipyridyl. *J. Am. Chem. Soc.* 98: 6326–6330.

Mann, D.F. and Byerrum, R.U. (1974). Avtivation of the *de novo* pathway for pyridine nucleotide biosynthesis prior to ricinine biosynthesis in caster beans. *Plant Physiol.* 53: 603–609.

Mata, R., Calzada, F., Díaz, E., and Toscano, R.A. (1990). Chemical studies on Mexican plants used in traditional medicine, XV. Sesquiterpene evoninoate alkaloids from *Hippocratea excelsa*. *J. Nat. Prod.* 53: 1212–1219.

Matsui, A., Yin, Y., Yamanaka, K. et al. (2007). Metabolic fate of nicotinamide in higher plants. *Physiol. Plant.* 131: 191–200.

Minorsky, P.V. (2002). Trigonelline: A diverse regulator in plants. *Plant Physiol.* 128: 7–8.

Mizuno, K., Matsuzaki, M., Kanazawa, S. et al. (2014). Conversion of nicotinic acid to trigonelline is catalyzed by *N*-methyltransferase belonged to motif B′ methyltransferase family in *Coffea arabica*. *Biochem. Biophys. Res. Commun.* 452: 1060–1066.

Mizusaki, S., Tanabe, Y., Kisaki, T., and Tamaki, E. (1970). Metabolism of nicotinic acid in tobacco plants. *Phytochemistry* 9: 549–554.

Nishitani, H., Yamada, Y., Ohshima, N. et al. (1995). Identification of *N*-(β-D-glucopyranosyl)nicotinic acid as a major metabolite from niacin in cultured tobacco cells. *Biosci. Biotechnol. Biochem.* 59: 1336–1338.

Pailer, M. and Libiseller, R. (1962). Über Evonymus-Alkaloide, 1. Mitt.: Zur Reindarstellung und Konstitution des Evonins, (aus *Evonymus europaea* L.). *Monatsh. Chem.* 93: 403–416.

Qiu, Y.L. and Palmer, J.D. (1999). Phylogeny of early land plants: insights from genes and genomes. *Trends Plant Sci.* 4: 26–30.

Robinson, T. (1965). Enzymes of alkaloid biosynthesis—I. : enzymatic oxidation of 1-methylnicotinonitrile. *Phytochemistry* 4: 67–74.

Saitoh, F., Noma, M., and Kawashima, N. (1985). The alkaloid contents of sixty Nicotiana species. *Phytochemistry* 24: 477–480.

Sakan, T., Fujino, A., Murai, F. et al. (1959). On the structure of actinidine and matatabilactone, the effective components of *Actinidia polygama*. *Bull. Chem. Soc. Jpn.* 32: 315–316.

Scott, T.A. and Glynn, J.P. (1967). The incorporation of [2,3,7-^{14}C] nicotinic acid into nicotine by *Nicotiana tabacum*. *Phytochemistry* 6: 505–510.
Shoji, T. and Hashimoto, T. (2011). Nicotine biosynthesis. In: *Plant Metabolism and Biotechnology*, 191–216. Wiley.
Shoji, T., Kajikawa, M., and Hashimoto, T. (2010). Clustered transcription factor genes regulate nicotine biosynthesis in tobacco. *Plant Cell* 22: 3390–3409.
Sinclair, S., Murphy, K., Birch, C., and Hamill, J. (2000). Molecular characterization of quinolinate phosphoribosyltransferase (QPRTase) in *Nicotiana*. *Plant Mol. Biol.* 44: 603–617.
Smith, A.R., Pryer, K.M., Schuettpelz, E. et al. (2006). A classification for extant ferns. *Taxon* 55: 705–731.
Taguchi, H., Nishitani, H., Okumura, K. et al. (1989). Biosynthesis and metabolism of trigonelline in *Lemna paucicostata* 151. *Agric. Biol. Chem.* 53: 2867–2871.
Taguchi, H., Sasatani, K., Nishitani, H., and Okumura, K. (1997). Finding of UDP-glucose : nicotinic acid-*N*-glucosyltransferase activity in cultured tobacco cells and its properties. *Biosci. Biotechnol. Biochem.* 61: 720–722.
The Angiosperm Phylogeny Group (2009). An update of the Angiosperm Phylogeny group classification for the orders and families of flowering plants: APG III. *Bot. J. Linn. Soc.* 161: 105–121.
Tramontano, W.A., McGinley, P.A., Ciancaglini, E.F., and Evans, L.S. (1986). A survey of trigonelline concentrations in dry seeds of the dicotyledoneae. *Environ. Exp. Bot.* 26: 197–205.
Upmeier, B., Gross, W., Koster, S., and Barz, W. (1988a). Purification and properties of *S*-adenosyl-L-methionine:nicotinic acid-*N*-methyltransferase from cell suspension cultures of *Glycine max* L. *Arch. Biochem. Biophys.* 262: 445–454.
Upmeier, B., Thomzik, J.E., and Barz, W. (1988b). Enzymatic studies on the reversible synthesis of nicotinic acid-*N*-glucoside in heterotrophic parsley cell suspension cultures. *Z. Naturforsch.* 43c: 835–842.
Upmeier, B., Thomzik, J.E., and Barz, W. (1988c). Nicotinic acid-*N* -glucoside in heterotrophic parsley cell suspension cultures. *Phytochemistry* 27: 3489–3493.
Wada, H., Shizuri, Y., Yamada, K., and Hirata, Y. (1971). Evonine, an alkaloid obtained from *Euonymus sieboldiana* blume. I. Relationship of functional groups and partial structure. *Tetrahedron Lett.* 12: 2655–2658.
Waller, G.R. and Dermer, O.C. (1981). Enzymology of alkaloid metabolism in plants and microorganisms. In: *The Biochemistry of Plants, a Comprehensive Treatise*, vol. 7 (Secondary Plant Products) (ed. E.E. Conn), 317–402. New York: Academic Press.
Waller, G.R., Ryhage, R., and Meyerson, S. (1966a). Mass spectrometry of biosynthetically labeled ricinine. *Anal. Biochem.* 16: 277–286.
Waller, G.R., Yang, K.S., Gholson, R.K. et al. (1966b). The pyridine nucleotide cycle and its role in the biosynthesis of ricinine by *Ricinus communis* L. *J. Biol. Chem.* 241: 4411–4418.
Willeke, U., Heeger, V., Meise, M. et al. (1979). Mutually exclusive occurrence and metabolism of trigonelline and nicotinic acid arabinoside in plant cell cultures. *Phytochemistry* 18: 105–110.
Yang, K.S. and Waller, G.R. (1965). Biosynthesis of the pyridine ring of ricinine from quinolinic acid glycerol and aspartic acid. *Phytochemistry* 4: 881–889.
Yang, K.S., Gholson, R.K., and Waller, G.R. (1965). Studies on nicotine biosynthesis. *J. Am. Chem. Soc.* 87: 4184–4188.

Zheng, X.Q. and Ashihara, H. (2004). Distribution, biosynthesis and function of purine and pyridine alkaloids in Coffea arabica seedlings. *Plant Sci.* 166: 807–813.

Zheng, X.Q., Nagai, C., and Ashihara, H. (2004). Pyridine nucleotide cycle and trigonelline (*N*-methylnicotinic acid) synthesis in developing leaves and fruits of *Coffea arabica*. *Physiol. Plant.* 122: 404–411.

Zheng, X.Q., Hayashibe, E., and Ashihara, H. (2005). Changes in trigonelline (*N*-methylnicotinic acid) content and nicotinic acid metabolism during germination of mungbean (*Phaseolus aureus*) seeds. *J. Exp. Bot.* 56: 1615–1623.

Zheng, X.Q., Matsui, A., and Ashihara, H. (2008). Biosynthesis of trigonelline from nicotinic acid mononucleotide in mungbean seedlings. *Phytochemistry* 69: 390–395.

21

Physiological Aspect and Biotechnology of Trigonelline

21.1 Introduction

Trigonelline (**1**) can accumulate in high concentrations in some plants, such as coffee and legumes. In most species, the level of trigonelline, and that of nicotinate *N*-glucoside (**2**) (NaG), both of which are produced from nicotinate (**3**), are much lower. In this chapter, the biosynthesis of trigonelline, and its various roles in plants, and plant biotechnology are described.

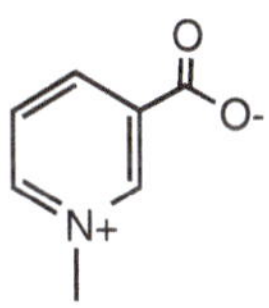

Trigonelline (**1**)

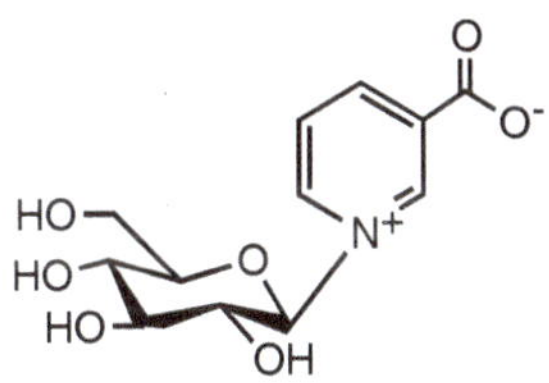

Nicotinate-*N*-glucoside (**2**)

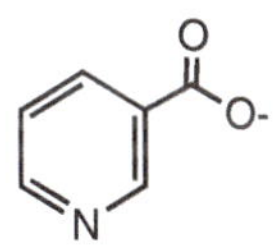

Nicotinate (**3**)

21.2 Physiological Aspect of Trigonelline Biosynthesis

21.2.1 Coffee

Changes in trigonelline biosynthesis occur during the growth of leaves and fruits of coffee plants. Trigonelline is present in all parts (leaves, cotyledons, stems, and roots) of six-month-old *Coffea arabica* seedlings (Zheng and Ashihara 2004). The concentration is highest in the upper part of the stems, including buds (180 $\mu mol\, g^{-1}$ d.w.), and lowest in roots (25 $\mu mol\, g^{-1}$ d.w.). The trigonelline concentration in leaves is 60–80 $\mu mol\, g^{-1}$ d.w., with the level in young leaves higher than in older leaves.

Biosynthetic activity of trigonelline was estimated from the conversion of [carboxyl-^{14}C]nicotinate to trigonelline (Figure 21.1). Trigonelline synthesis was found in all parts of coffee seedlings. Metabolism studies indicated that a large fraction of the radioactivity from [^{14}C]nicotinate (59% in leaves, 53% in cotyledons, 36% in stems, and 29% in roots) was incorporated into trigonelline during a four-hour incubation period. Thus, trigonelline biosynthetic activity is found in all parts of coffee seedlings, even in mature tissues. Since the presence of trigonelline demethylase was confirmed (Taguchi et al. 1987), it has been considered that trigonelline may be a

Plant Nucleotide Metabolism – Biosynthesis, Degradation, and Alkaloid Formation, First Edition.
Hiroshi Ashihara, Iziar A. Ludwig and Alan Crozier.

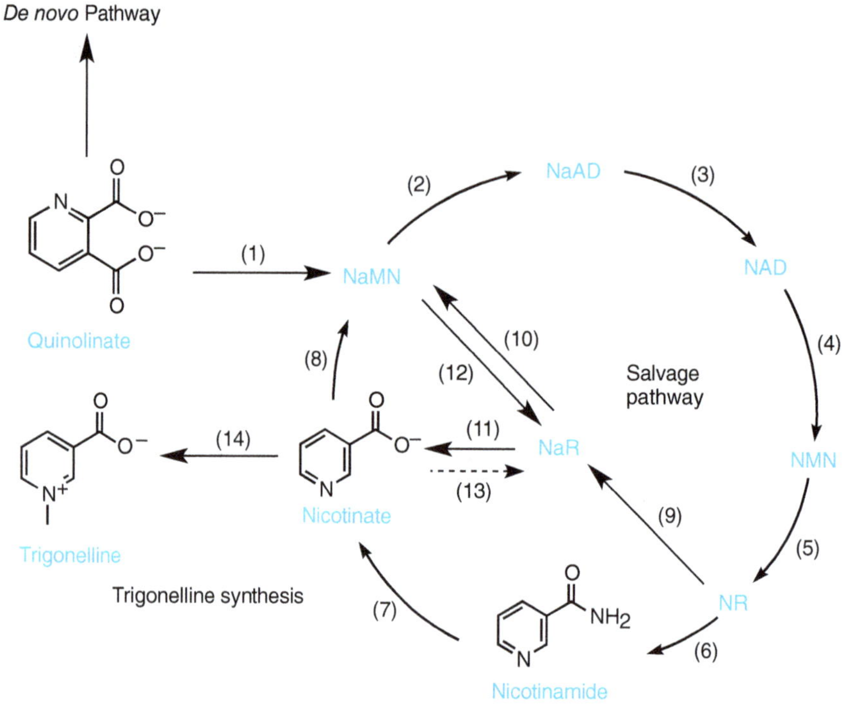

Figure 21.1 Possible pyridine nucleotide cycle and the trigonelline synthesis pathways in coffee and legumes. The numbers shown correspond to: (1) quinolinate phosphoribosyltransferase (EC 2.4.2.19); (2) nicotinate mononucleotide adenylyltransferase (2.7.7.18); (3) NAD synthase (6.3.5.1); (4) NAD pyrophosphatase (3.6.1.22); (5) 5′-nucleotidase (3.1.3.5); (6) nicotinamide riboside nucleosidase (3.2.2.-); (7) nicotinamidase (3.5.1.19); (8) nicotinate phosphoribosyltransferase (2.4.2.11); (9) nicotinamide riboside deaminase; (10) nicotinate riboside kinase; (11) nicotinate riboside nucleosidase; (12) nicotinate mononucleotide nucleotidase; (13) nicotinate phosphorylase; (14) trigonelline synthase (EC 2.1.1.7). The direct pathway for nicotinamide formation from NAD by poly (ADP-ribose) polymerases (2.4.2.30), mono (ADP-ribosyl) transferase (2.4.2.31), or NAD dependent deacetylase (Sir 2) (3.5.1.-) is omitted from this map. Metabolites: NAD, nicotinamide adenine dinucleotide; NaAD, nicotinate adenine dinucleotide; NaMN, nicotinate mononucleotide; NaR, nicotinate riboside; NMN, nicotinamide mononucleotide; NR, nicotinamide riboside. Source: Adapted from Ashihara et al. (2011).

reusable form of nicotinate. However, no specific function has yet been proposed for the trigonelline that accumulates in coffee seedlings.

Biosynthesis and accumulation of trigonelline during growth and ripening of fruits of *C. arabica* and *Coffea canephora* have also been investigated (Koshiro et al. 2006). In addition to caffeine, trigonelline is a major alkaloid in coffee seeds. Accumulation of these alkaloids in *C. arabica* and *C. canephora* seeds have been monitored. Growth stages are specified by letters, A to G. They correspond to the pinhead and small (A), rapid expansion and pericarp growth (B), endosperm formation (C), early dry matter accumulation (D), mature (green) (E), ripening (pink) (F), and fully ripened (red) (G) stages. Trigonelline content increased at stages D and E (Table 21.1).

Concentrations of trigonelline in ripe seeds (stage G) of *C. arabica* and *C. canephora* were ~1.3% of dry weight. High biosynthetic activity of trigonelline, derived from

Table 21.1 Changes in trigonelline (Tg) content (μmol part^{-1}) and its synthesis from [carbonyl-^{14}C]nicotinamide (kBq [100 mg]$^{-1}$ f.w.) and in parentheses as a % of total radioactivity taken up during development of coffee fruits.

Growth stage	Part of fruits	*Coffea arabica*		*Coffea canephora*	
		Tg content	Tg synthesis	Tg content	Tg synthesis
A	Fruit (whole)	0.22	2.86 (50.3)	0.42	5.93 (79.7)
B	Fruit (whole)	0.68	3.20 (52.4)	1.0	4.64 (72.9)
C	Fruit (whole)	3.1	2.54 (55.4)	3.8	4.46 (75.1)
D	Fruit (whole)	10.3	1.41 (51.7)	6.9	2.26 (60.1)
E	Pericarp	2.2	3.01 (55.7)	3.0	5.47 (70.6)
	Seed	15.4	2.26 (51.2)	21.9	1.52 (54.4)
F	Pericarp	8.6	1.37 (31.0)	10.4	2.83 (45.9)
	Seed	25.1	0.04 (2.5)	20.1	0.13 (6.9)
G	Pericarp	4.9	0.35 (11.0)	7.4	0.48 (8.1)
	Seed	34.0	nd	29.1	nd

nd – not detected.
Source: Adapted from Koshiro et al. (2006).

[carbonyl-^{14}C]nicotinamide, was detected in young fruits (stages A–C) and in the pericarp of developing fruits (stage E) (Table 21.1). Trigonelline accumulates in seeds, but biosynthetic activity is higher in the pericarp than in seeds of the coffee fruits (Table 21.1). At least some trigonelline, therefore, seems to be transported from the pericarp to the seeds.

Gene expression studies on the two trigonelline synthases in *C. arabica* have been performed by Mizuno et al. (2014). The trigonelline synthases, *CTgS1* and *CTgS2*, are highly homologous with coffee caffeine synthases at the nucleotide level. The expression of *CTgS1* and *CTgS2* was evaluated by semi-quantitative reverse transcription-polymerase chain reaction (RT-PCR) designed for specific amplification and detection of these transcripts. Transcripts of CTgS1 and CTgS2 were detected in all organs, namely, flower buds, closed leaves, and young leaves. RT-PCR of total RNAs from young fruits at various stages of development, designated as days after flowering (DAF) was performed. The transcript accumulation of CTgS1 was maximum in DAF five fruits, and then tended to decrease slowly. On the other hand, *CTgS1* was expressed in mature rather than in young leaves. The signals for CTgS2 transcripts were extremely weak in leaves.

Trigonelline in coffee seeds may contribute to the bitter taste of the resultant beverage. To determine the trigonelline biosynthetic pathway in coffee fruits, Ashihara et al. (2011) investigated the metabolic fate of [carboxyl-^{14}C]nicotinate riboside (NaR) and *in situ* activity of related enzymes. The radiolabelled NaR was rapidly converted to nicotinate mononucleotide (NaMN) which was utilized for nicotinamide adenine dinucleotide (NAD) synthesis. NaR was also converted to trigonelline, but this process took longer than NAD synthesis. These results indicate that an efficient nicotinate riboside salvage system functions in coffee fruits, and that trigonelline is synthesized mainly from nicotinate produced by the degradation of NAD (Figure 21.1).

21.2.2 Leguminous Plants

Tramontano et al. (1983) reported fluctuations in the levels of trigonelline, nicotinate, and nicotinamide during germination of *Pisum sativum* seeds. Trigonelline in cotyledons was transported to the epicotyl during germination. Smaller amounts migrated to the roots, although the concentration in root tips decreased during the 10 day period after imbibition. In contrast to trigonelline, only small amounts of nicotinate and nicotinamide were detected in the seeds.

Zheng et al. (2005) reported on the trigonelline content and biosynthetic activity in cotyledons and embryonic axes of germinating etiolated mung bean (*Vigna radiata*) seedlings. Trigonelline in the cotyledons of dry seeds decreased during the three days after imbibition. In contrast, the trigonelline content of the embryonic axes increased with seedling growth and reached a peak at day five (Table 21.2).

The highest trigonelline biosynthetic activity was in the embryonic axis although the alkaloid was also transported from the cotyledons to the embryonic axis (Table 21.3). More than half of [carboxyl-^{14}C]nicotinate taken up by cotyledons was converted to pyridine nucleotides and less than a quarter was incorporated into trigonelline. Nicotinate salvage activity for pyridine nucleotide synthesis was therefore higher than the biosynthetic activity of trigonelline. No significant metabolic changes were observed during germination. Interestingly, trigonelline synthesis still occurred even in the later stage of germination when the trigonelline content was markedly reduced (Table 21.2). In the embryonic axes, there was much more trigonelline synthetic activity than pyridine nucleotide synthesis during the entire germination period. Trigonelline synthesis gradually declined after three days of germination. This suggests that trigonelline synthesis is highest in young rather than older cells.

In addition to being converted to nicotinate, degradation of NAD is utilized for trigonelline synthesis (steps 4–7 and 14 in Figure 21.1). Trigonelline is also derived from nicotinate originating from NaMN, which is produced by the *de novo* pathway (steps 1, 12, 11, and 14 in Figure 21.1) in legumes. The activity of both nicotinamidase (EC 3.5.1.19) and quinolinate phosphoribosyltransferase (EC 2.4.2.19) in the cotyledons

Table 21.2 Changes in fresh weight (mg) and trigonelline content (nmol) in cotyledons and embryonic axes of mungbean seeds during germination.

	Fresh weight		Trigonelline content	
Day	Cotyledon	Embryonic axis	Cotyledon	Embryonic axis
0	57	1	243	9
1	113	27	261	55
2	106	116	259	149
3	75	249	107	192
4	34	568	30	283
5	31	699	33	381
6	30	753	40	316
7	28	757	17	292

Source: Adapted from Zheng et al. (2005).

Table 21.3 Metabolic fate of [carboxyl-^{14}C]nicotinate (10 μM) in cotyledons and embryonic axes of 1-, 3-, and 5-day-old mung bean seedlings. The duration of incubation was 4 h.

	Day 1		Day 3		Day 5	
	Cotyledon	Embryonic axis	Cotyledon	Embryonic axis	Cotyledon	Embryonic axis
Trigonelline	25.5	65.1	19.8	65.1	16.8	51.7
Nucleotides	54.1	29.4	72.0	24.2	71.1	36.4
CO_2	0.4	0.2	0.3	0.2	0.1	0.1
Nicotinate	19.9	5.3	0.8	5.3	12.0	11.8

The rates of incorporation of radioactivity into trigonelline, nucleotides, CO_2, and nicotinate are expressed as a percentage of the total radioactivity taken up by the segments. The nucleotides are NAD, NADP, and very small amounts of nicotinate mononucleotide and nicotinamide mononucleotide.
Source: Adapted from data of Zheng et al. (2005).

Table 21.4 Changes in the activities of trigonelline synthase and nicotinamidase in germinating mung bean seeds. Enzyme activities in cotyledons and embryonic axes are expressed as pkat per pair of cotyledons or pkat per axis.

	Trigonelline synthase		Nicotinamidase	
Day	Cotyledon	Embryonic axis	Cotyledon	Embryonic axis
0	1.4	0	0.5	0
1	1.6	0.3	1.1	0.4
2	0.7	0.3	0.8	0.8
3	0.7	1.9	2.1	3.8
4	0.7	2.2	3.8	4.4
5	0.5	1.5	4.4	3.8
6	0.4	2.9	6.1	10.6

Source: Adapted from data of Zheng et al. (2005).

and embryonic axes of germinating mung bean seeds has been investigated (Ashihara et al. 2011). The activities of nicotinamidase were 0.92 and 3.47 and quinolinate phosphoribosyltransferase were 0.60 and 5.39 pkat mg^{-1} protein, respectively, in cotyledons and embryo axes. Changes of trigonelline synthase (EC 2.1.1.7) and nicotinamidase activity were also measured during germination. Trigonelline synthase increased in the embryonic axes, but decreased in cotyledons while nicotinamide activity increased in both organs (Table 21.4).

Yin et al. (2008) investigated changes in the profile of pyridine metabolism during cell growth using trigonelline-forming suspension-cultured cells of the legume *Lotus japonicus*, (birdsfoot trefoil). Activity of the *de novo* and salvage pathways of NAD biosynthesis was estimated from the *in situ* metabolism of [^{3}H]quinolinate and [^{14}C]nicotinamide. Maximum activity of the *de novo* pathway for NAD synthesis occurred during the exponential phase of cell growth, whereas activity of the salvage pathway increased in the lag phase. Transcript levels of nicotinate phosphoribosyltransferase (NaPRT) and nicotinamidase-encoding salvage enzymes, were enhanced in the lag phase, whereas the maximum expression of NAD synthase was found in the exponential phase of growth. In keeping with these events, the activities of the salvage enzymes, NaPRT and NIC, increased one day after transfer of stationary phase cells to fresh medium. Highest *in situ* trigonelline synthesis, both from [^{3}H]quinolinate and [^{14}C]nicotinamide, was found in the stationary phase of cell growth. Trigonelline was produced principally when pyridine nucleotide synthesis declined. A sizable portion of trigonelline formed in the cells was released into the culture medium. It was suggested that trigonelline may act as a reservoir for nicotinate in legumes, especially in nodules (Yin et al. 2008).

Matsui et al. (2007) reported the metabolic fate of [^{14}C]nicotinamide in roots and leaves of one-month-old *L. japonicus* seedlings. Almost all nicotinamide was utilized for the synthesis of pyridine nucleotides in root nodules, whereas formation of nucleotides in root and leaf tissues was limited. In leaves nicotinamide was converted almost exclusively to trigonelline.

21.3 Physiological Aspects of Nicotinate *N*-Glucoside Biosynthesis

In contrast to trigonelline, NaG levels in plant cells are very low. Upmeier et al. (1988) measured the pool size of NaG in parsley cells cultivated *in vitro* in a nicotinate-free medium. The concentration of NaG remained almost constant (35–40 nmol g^{-1} f.w.) during the culture period, arguably, because NaG was being converted to nicotinate and utilized for NAD synthesis. The half-life of NaG, estimated using [U-^{14}C]aspartate, was ~24 hours.

The metabolic fate of [carbonyl-^{14}C]nicotinamide was examined in microspore-derived embryos of NaG-forming canola (*Brassica napus*) at different developmental stages: globular stage, early cotyledonary, late cotyledonary, and fully developed stages (Ashihara et al. 2008). The radiolabel in the culture medium was taken up readily by the embryos, especially in the globular and early cotyledonary stages of development, and rapidly converted to nicotinate, which was further metabolized. There was a ~10% incorporation of radioactivity from [^{14}C]nicotinamide into NAD/NADP in the globular and early cotyledonary stage embryos, compared with 6–8% in embryos at the late cotyledonary and fully developed stages. There was a 34% conversion of [^{14}C]nicotinamide into NaG by embryos at the globular stage of development but this declined to 3.8% in the fully developed embryos. High NaG formation in embryos at the globular stage may be due to NaG serving as a source of nicotinate during the later stages of embryo development that are characterized by intense growth and a requirement for high levels of NAD.

21.4 The Role of Trigonelline in Plants

Several *in planta* roles for trigonelline have been proposed, namely: as a nutrient source, a compatible solute, a bioactive substance for nyctinasty, cell cycle regulation, signal transduction, detoxification of nicotinate, and an eco-chemical function such as host selection by herbivores.

21.4.1 Role of Trigonelline as a Nutrient Source

Trigonelline appears to be an end product of NAD metabolism. If trigonelline is converted to nicotinate, which is salvaged by NaPRT, and utilized for the resynthesis of NAD, trigonelline in seeds may act as a storage form of nicotinate. The activity of trigonelline demethylase, which catalyses the conversion of trigonelline to nicotinate, has been detected in various organisms including *C. arabica* seeds (Shimizu and Mazzafera 2000; Taguchi and Shimabayashi 1983). The enzyme occurs in relatively low amounts in seeds of leguminous species, such as *V. radiata* (Zheng et al. 2005). Bacteria, such as *Rhizobium meliloti*, in leguminous root nodules, are able to catabolize trigonelline in symbiotic conditions (Boivin et al. 1990, Boivin et al. 1991; Goldmann et al. 1991). It follows that trigonelline in cotyledons of legumes is translocated to root nodules and degraded for use as a nutrient source for bacteroid growth. Very high salvage activity of nicotinate for NAD synthesis has been detected in nodules of *L. japonicus* (Matsui et al. 2007).

Trigonelline released from roots of *Medicago sativa* (alfalfa) acts as a symbiont signal. To investigate the role of trigonelline catabolism in the *Rhizobium*–legume interaction, Boivin et al. (1991) studied the regulation of *trc* gene expression in free-living and endosymbiotic bacteria using *Escherichia coli lacZ* as a reporter gene. Experiments performed with free-living bacteria indicated that *trc* genes were organized in at least four transcription units of which three were specifically induced by trigonelline in the rhizosphere, infection threads, and bacteroids of alfalfa. This suggests that trigonelline is a nutrient source throughout the *Rhizobium*–legume association.

21.4.2 Role of Trigonelline as a Compatible Solute

Several reports suggest the accumulation of trigonelline as a compatible solute. Like proline and glycine betaine, trigonelline is synthesized as an osmoregulator when plants are subjected to drought or salt stress.

Tramontano and Jouve (1997) investigated the effect of NaCl on the trigonelline content of leaves of soybean (*Glycine max*) by treating nine-week-old plants with 100 ml of 100 mM NaCl three times weekly for a period of six weeks. Foliage trigonelline in control plants was 0.2–0.3 $\mu mol\,g^{-1}$ d.w. while in NaCl-treated plants it was higher at 0.4–1.2 $\mu mol\,g^{-1}$ d.w. throughout the exposure period until the plants began to senesce.

Cho et al. (1999) determined the concentration of trigonelline and the relative water content of non-stressed and salt-stressed leaves of 30-day-old seedlings of 17 cultivars of soybean seedlings. The seedlings were treated with NaCl in a step-wise manner over nine days (three days 30 mM, three days 70 mM, and three days 100 mM). Trigonelline concentrations ranged 0.46–1.2 $\mu mol\,g^{-1}$ d.w. in non-stressed leaf tissue and 0.55–1.6 $\mu mol\,g^{-1}$ d.w. in salt stressed leaf tissue. During salt treatment, trigonelline concentrations increased in ten cultivars, declined in five and remained unchanged in 2 cultivars.

Bray et al. (1991) examined effects of NaCl treatment on the glycine betaine content of a micropropagated *in vitro* clone of poplar (*Populus trichocarpa* X *deltoids*). The trigonelline levels in leaves and roots of the control plants were 8.5 and

1.7 μmol g^{-1} d.w., respectively, and when treated with 100 mM NaCl the level increased to 15.3 and 4.6 μmol g^{-1} d.w. Trigonelline can be considered as a sensitive indicator of stress as no accumulation of glycine betaine or proline was observed

Mangrove plants are a group of trees and shrubs growing along seashores and estuaries in tropical and subtropical areas that have special morphological and physiological characteristics to enable them to withstand high levels of salt stress (Tomlinson 1994). Yin et al. (2011) reported that the endogenous trigonelline content in leaves of *Derris indica*, a mangrove belonging to the Leguminosae, was 6.1 μmol g^{-1} d.w. This value is 5–12 times greater than in leaves of soybean. The concentration of trigonelline in leaf discs was increased by incubation with 500 mM NaCl (equivalent to ~100% sea water). The net increase was ~150 nmol h^{-1} g^{-1} d.w.

Ashihara et al. (2010) examined the effect of 250 mM NaCl, equivalent to ~50% sea water, on the synthesis of trigonelline from [carbonyl-^{14}C]nicotinamide by leaf discs of mangrove species. Statistically significant, short-term (three hours) NaCl-induced stimulation of trigonelline synthesis was observed in *Bruguiera gymnorrhiza*, *Rhizophora stylosa*, and *Sonneratia alba*. Trigonelline synthesis from labelled quinolinate, nicotinamide, and nicotinate in cultured cells of the mangrove, species, *Bruguiera sexangula*, has also been shown to be enhanced by salt stress (Suzuki-Yamamoto et al. 2006). Salt stress-mediated increase in trigonelline synthesis, therefore, appears to be closely related to the synthesis of osmoprotective compounds, to prevent water loss when the mangrove cells are exposed to strong saline conditions. Elevation of trigonelline levels induced by salt stress has also been reported to occur in tomato (Rajasekaran et al. 2001) and seashore paspalum (*Paspalum vaginatum*) (Lee et al. 2008).

Increases in trigonelline induced by salt- or drought stress may, therefore, act as a compatible solute. However, the concentration of trigonelline is much lower than that of other compatible solutes, such as glycine betaine, which can increase from 50 to 250 to 400 μmol g^{-1} d.w. when plants are subject to severe water deficient or salinity (Hanson et al. 1995). Trigonelline, therefore, does not act alone but may be one of several constituents functioning as a compatible solute.

21.4.3 Trigonelline and Nyctinasty

Leguminous species, such as *Mimosa pudica*, are well known for the involvement of a circadian rhythm in the movement of their leaves. This nyctinastic leaf movement, which is a feature of almost all legumes, is characterized by leaf closure at night and opening the following morning. It appears not to be regulated by a common signal molecule but by compounds that can vary depending upon the species (Ueda et al. 2000).

Trigonelline was isolated as a bioactive substance for nyctinasty in *Aeschynomene indica* (Ueda et al. 1995) being effective at inducing day-time leaf closure at a concentration of 0.1 μM. In contrast, trigonelline was without effect in other nyctinastic species, *Cassia mimosoides* and *M. pudica*.

21.4.4 Cell Cycle Regulation

Trigonelline was isolated from the cotyledons of pea (*P. sativum*) and shown to act as a factor that arrests the growth of cells in roots and shoots in the G2 stage of the cell cycle (Evans et al. 1979). Trigonelline promotes preferential cell arrest in the G2 stage for about 40% of the cell population in pea root meristems. The alkaloid is present in mature seeds and is transported from cotyledons to developing tissues during the early stages of germination. In the meristems of young embryos 3–10 days after imbibition, the concentration of trigonelline decreases in parallel with the fall in cells arrested at G2. Trigonelline has been isolated from excised cotyledons and when applied to decotyledonized seedlings or excised root meristems inhibits growth to the rate found in intact young seedlings. Predominant cell arrest in G2 occurs in roots of some plant species, although other species show most cell arrest in G1.

The proportion of cells arrested in G2 is highly correlated with the concentration of trigonelline within the root meristem. In pea, two naturally occurring cytokinins, the ribosides of isopentenyl adenine and zeatin, both at 0.1 μM, were antagonistic to the effect of simultaneous treatment with 1 μM trigonelline. In contrast, the synthetic cytokinins, kinetin and benzyl adenine, did not impact on the inhibitory effect of trigonelline. In soybean, isopentenyl adenine and zeatin, at 0.1 and 0.01 μM, promoted cell arrest in G2 in the absence of trigonelline. However, the proportion of cells arrested in G2 was higher in cultured roots co-treated with trigonelline and a cytokinin. In both legumes, cytokinin treatment did not affect trigonelline concentrations in the roots. These results imply an interaction between naturally occurring cytokinins and trigonelline in control of the cell cycle in pea and soybean (Tramontano et al. 1985; Tramontano et al. 1982).

21.4.5 Detoxification of Nicotinate

Exogenously supplied nicotinate and nicotinamide, but not trigonelline, both inhibit the growth of roots of mung bean seedlings (Zheng et al. 2005), and the division and colony formation of lettuce cells derived from protoplasts (Sasamoto and Ashihara 2014). One role of trigonelline synthesis could, therefore, be detoxification of excess nicotinate and nicotinamide released from the pyridine nucleotide cycle in plant cells. A similar role in detoxification of nicotinate has been attributed to the formation of NaG (Nishitani et al. 1995).

These findings are seemingly not consistent with the view that trigonelline is a cell cycle regulator that causes cell arrest in G2 and prevents ligation of replicons in the S-phase in lettuce root meristems (Mazzuca et al. 2000). The reasons for this inconsistency are unclear, but Mazzuca et al. used a higher concentration (3 mM) of trigonelline and different plant materials (root meristems) than those employed in the lettuce study. Francis and Sorrell (2001) stated that trigonelline did not arrest all cells in G2, so the meristematic cells appear to be differentially sensitive to cell cycle regulators in G2.

In addition to the endogenously-generated nicotinate by pyridine nucleotide cycles (see Figure 21.1), plants may have the ability to form nicotinate metabolites (trigonelline and/or NaG) for detoxification if nicotinate or nicotinamide is taken up externally. Formation of trigonelline and NaG is, thus, involved in xenobiotic metabolism, the biochemical modification of toxic compounds. Detoxification metabolism is often divided into three phases: modification, conjugation, and excretion. One of the most common modifications is hydroxylation catalysed by the cytochrome P-450-dependent oxidase system. Trigonelline and NaG formation seems to be categorized as a second phase metabolism in which reactions are catalysed by a large group of transferases.

Nicotinate is methylated by SAM-dependent methyltransferase and glucosylated by UDP-glucosyltransferase (Jakoby and Ziegler 1990).

21.4.6 Signal Transduction

It has been proposed that nicotinamide participates in an important part of the signal transduction chain involved in the response of plant cells to conditions that cause DNA strand breakage, especially in connection with oxidative stress. It was suggested that nicotinamide and trigonelline serve as potent inducers of defensive metabolism in plants, including glutathione metabolism, and the accumulation of secondary defence compounds (Berglund 1994; Berglund et al. 1996).

Kalbin et al. (1997) reported that strong UV-B irradiation of *P. sativum* plants caused 3.5- and 9.5-fold increases in the respective levels of nicotinamide and trigonelline, with a 60-fold increase in oxidized glutathione in leaves. These events were not paralleled by changes in the profiles of the mRNA transcript levels of chalcone synthase (CHS) and phenylalanine ammonia-lyase (PAL) genes. It was, therefore, concluded that neither nicotinamide nor trigonelline function as signalling molecules for CHS and PAL gene expression in response to UV-B exposures.

21.4.7 Role of Host Selection by Herbivores

It has been shown that some phytochemicals exert an effect on host selection by herbivores. Common Mormon (*Papilio polytes*) is a species of swallowtail butterfly and a Rutaceae feeder that utilizes *Toddalia asiatica* as a major host plant (Nakayama and Honda 2004). Another rutaceous plant, orange jessamine (*Murraya paniculata*) is not exploited by the butterfly. Nakayama and Honda reported that trigonelline, which is found in orange jessamine, exhibits moderate oviposition deterrence towards the female butterfly.

21.5 Biotechnology of Trigonelline

Trigonelline has potential protective effects in humans and manipulation of the trigonelline content in foods has been carried out. In the case of coffee, two methods are used for increase in trigonelline of the beverage. The first is the use of a low roasting temperature, which reduces the degradation of trigonelline while the second approach is the addition of trigonelline extracted from green beans to the roasted coffee powders. Trigonelline is also important as a source of vitamin B_3 (niacin, i.e. nicotinate and nicotinamide).

Trigonelline in coffee beans is demethylated to form nicotinate during the roasting process (Casal et al. 2000; Taguchi et al. 1985). Coffee provides 1–3 mg of nicotinate per cup (Adrian and Frangne 1991) which is 6–18% of the daily recommended 16 mg d^{-1} intake of vitamin B_3 for adult men (Allred et al. 2009). Genetic engineering of trigonelline formation could in the future be a useful means for creating trigonelline- or vitamin B_3-rich foods. As shown in the metabolism section, there is a direct pathway of trigonelline synthesis from quinolinate, which is formed from the *de novo* pathway (steps 1, 12, 11, and 14 in Figure 21.1). This pathway bypasses the complicated pyridine nucleotide cycle.

To obtain vitamin B_3-enriched foods, Noctor et al. (2011) tried to manipulate the pathway of pyridine nucleotide synthesis to achieve increases in NAD. Although vitamin B_3-enriched plants have not yet been obtained, in *E. coli*, the enhancement of the pyridine nucleotide pool has been achieved by genetic manipulation of the NAD biosynthetic pathways (Heuser et al. 2007). After simultaneous overexpression of the genes encoding NaPRT and NAD synthetase, intracellular concentrations of NADP and NAD increased two- and sevenfold, respectively. NAD is an important metabolite in primary metabolism, so its accumulation appears to be limited compared with that of secondary metabolites such as trigonelline. Trigonelline-rich genetically-modified plants are, therefore, better candidates for enhancing vitamin B_3 levels.

Trigonelline demethylating enzyme activity has been found in animals – with especially high activity in pig liver (Taguchi and Shimabayashi 1983). Stimulation of activities of the first three enzymes of *de novo* pyridine nucleotide biosynthesis (steps 1–3 in Figure 21.2), and three enzymes involved in the conversion of NaMN to trigonelline (steps 4–6 in Figure 21.2) would be required to produce trigonelline-enriched plants.

Addition of a trigonelline demethylase gene from other organisms is necessary to obtain vitamin B_3-rich plants, although accumulation of nicotinate in vacuoles may be required, because a high concentration of nicotinate in cytoplasm inhibits plant growth (Sasamoto and Ashihara 2014; Zheng et al. 2005). In addition to potential benefits to human health, trigonelline-rich genetically-modified plants could be of agricultural and environmental value. Several environmental factors including salinity and drought adversely affect plant growth and development and final yield of a crop. It is estimated that less than 10% of the world's arable lands are free of major environmental stresses with drought and salinity stresses being the most widespread (Ashraf and Foolad 2007).

Research is in progress to establish stress-tolerant genetically-modified plants that are capable of producing elevated amounts of compatible solutes in response to environmental stresses. At present, glycine betaine and proline are target molecules, but trigonelline may also be a possible candidate compatible solute.

As noted above there is a high concentration of trigonelline in orange jessamine, which does not act as a host for the eggs of the swallowtail butterfly (Nakayama and Honda 2004). Thus, further possible use of trigonelline in agriculture is as a natural pesticide to reduce insect infestations.

The ability to manipulate internal cellular concentrations of trigonelline would also be of value in clarifying the *in planta* roles of the pyridine alkaloid that have been proposed on the basis of data obtained with exogenous applications of trigonelline. Before any of this can be achieved information is required on the structure of trigonelline synthase and trigonelline demethylase genes and the mechanism of accumulation of trigonelline and its metabolites in plant tissues.

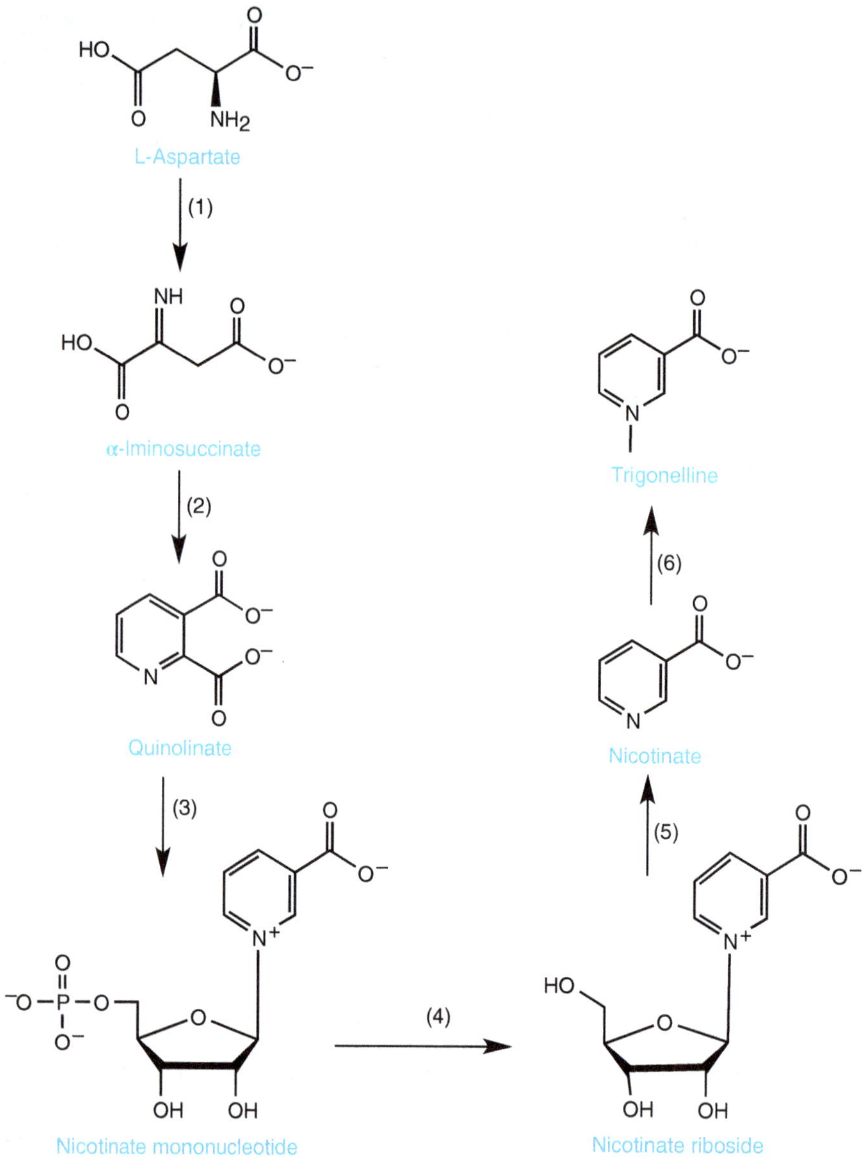

Figure 21.2 A possible trigonelline pathway from aspartate which is required to produce trigonelline-enriched plants. Enzymes shown are: (1) L-aspartate oxidase; (2) quinolinate synthase; (3) quinolinate phosphoribosyltransferase (decarboxylating); (4) nicotinate mononucleotide nucleotidase; (5) nicotinate riboside nucleosidase; (6) trigonelline synthase.

21.6 Summary

Trigonelline, produced from nicotinate, accumulates in substantial amounts in some plants, especially coffee and leguminous species. Enzyme activity and gene expression of trigonelline synthase in *C. arabica* have been investigated. Several *in planta* roles for trigonelline have been proposed including as a nutrient source, a compatible solute, and an involvement in cell cycle regulation. Genetic manipulation of trigonelline levels has potential agricultural uses, and could also enhance positive effects on health of trigonelline-containing foods.

References

Adrian, J. and Frangne, R. (1991). Synthesis and availability of niacin in roasted coffee. *Adv. Exp. Med. Biol.* 289: 49–59.

Allred, K.F., Yackley, K.M., Vanamala, J., and Allred, C.D. (2009). Trigonelline is a novel phytoestrogen in coffee beans. *J. Nutr.* 139: 1833–1838.

Ashihara, H., Luit, B., Belmonte, M., and Stasolla, C. (2008). Metabolism of nicotinamide, adenine and inosine in developing microspore-derived canola (*Brassica napus*) embryos. *Plant Physiol. Biochem.* 46: 752–759.

Ashihara, H., Yin, Y., Deng, W.W., and Watanabe, S. (2010). Pyridine salvage and nicotinic acid conjugate synthesis in leaves of mangrove species. *Phytochemistry* 71: 47–53.

Ashihara, H., Deng, W.-W., and Nagai, C. (2011). Trigonelline biosynthesis and the pyridine nucleotide cycle in *Coffea arabica* fruits: metabolic fate of [carboxyl-^{14}C]nicotinic acid riboside. *Phytochem. Lett.* 4: 235–239.

Ashraf, M. and Foolad, M.R. (2007). Roles of glycine betaine and proline in improving plant abiotic stress resistance. *Environ. Exp. Bot.* 59: 206–216.

Berglund, T. (1994). Nicotinamide, a missing link in the early stress response in eukaryotic cells: a hypothesis with special reference to oxidative stress in plants. *FEBS Lett.* 351: 145–149.

Berglund, T., Kalbin, G., Strid, A. et al. (1996). UV-B- and oxidative stress-induced increase in nicotinamide and trigonelline and inhibition of defensive metabolism induction by poly(ADP-ribose)polymerase inhibitor in plant tissue. *FEBS Lett.* 380: 188–193.

Boivin, C., Camut, S., Malpica, C.A. et al. (1990). *Rhizobium meliloti* genes encoding catabolism of trigonelline are induced under symbiotic conditions. *The Plant Cell* 2: 1157–1170.

Boivin, C., Barran, L.R., Malpica, C.A., and Rosenberg, C. (1991). Genetic analysis of a region of the *Rhizobium meliloti* pSym plasmid specifying catabolism of trigonelline, a secondary metabolite present in legumes. *J. Bacteriol.* 173: 2809–2817.

Bray, L., Chriqui, D., Gloux, K. et al. (1991). Betaines and free amino acids in salt stressed vitroplants and winter resting buds of *Populus trichocarpa* x *deltoides*. *Physiol. Plant.* 83: 136–143.

Casal, S., Beatriz Oliveira, M., and Ferreira, M.A. (2000). HPLC/diode-array applied to the thermal degradation of trigonelline, nicotinic acid and caffeine in coffee. *Food Chem.* 68: 481–485.

Cho, Y., Lightfoot, D.A., and Wood, A.J. (1999). Trigonelline concentrations in salt stressed leaves of cultivated *Glycine max*. *Phytochemistry* 52: 1235–1238.

Evans, L.S., Almeida, M.S., Lynn, D.G., and Nakanishi, K. (1979). Chemical characterization of a hormone that promotes cell arrest in G2 in complex tissues. *Science* 203: 1122–1123.

Francis, D. and Sorrell, D. (2001). The interface between the cell cycle and plant growth regulators: a mini review. *Plant Growth Regul.* 33: 1–12.

Goldmann, A., Boivin, C., Fleury, V. et al. (1991). Betaine use by rhizosphere bacteria: genes essential for trigonelline, stachydrine, and carnitine catabolism in *Rhizobium meliloti* are located on *pSym* in the symbiotic region. *Mol. Plant Microbe Interact.* 4: 571–578.

Hanson, A.D., Rivoal, J., Burnet, M., and Rathinasabapathi, B. (1995). Biosynthesis of quaternary ammonium and tertiary sulphonium compounds in response to water deficit. In: *Environment and Plant Metabolism Flexibility and Acclimation* (ed. N. Smirnoff), 189–198. Oxford: BIOS Scientific Publishers.

Heuser, F., Schroer, K., Lütz, S. et al. (2007). Enhancement of the NAD(P)(H) pool in *Escherichia coli* for biotransformation. *Eng. Life Sci.* 7: 343–353.

Jakoby, W.B. and Ziegler, D.M. (1990). The enzymes of detoxication. *J. Biol. Chem.* 265: 20715–20718.

Kalbin, G., Ohlsson, A.B., Berglund, T. et al. (1997). Ultraviolet-B- radiation-induced changes in nicotinamide and glutathione metabolism and gene expression in plants. *Eur. J. Biochem.* 249: 465–472.

Koshiro, Y., Zheng, X.Q., Wang, M. et al. (2006). Changes in content and biosynthetic activity of caffeine and trigonelline during growth and ripening of *Coffea arabica* and *Coffea canephora* fruits. *Plant Sci.* 171: 242–250.

Lee, G., Carrow, R.N., Duncan, R.R. et al. (2008). Synthesis of organic osmolytes and salt tolerance mechanisms in *Paspalum vaginatum*. *Environ. Exp. Bot.* 63: 19–27.

Matsui, A., Yin, Y., Yamanaka, K. et al. (2007). Metabolic fate of nicotinamide in higher plants. *Physiol. Plant.* 131: 19–200.

Mazzuca, S., Bitonti, M.B., Innocenti, A.M., and Francis, D. (2000). Inactivation of DNA replication origins by the cell cycle regulator, trigonelline, in root meristems of *Lactuca sativa*. *Planta* 211: 127–132.

Mizuno, K., Matsuzaki, M., Kanazawa, S. et al. (2014). Conversion of nicotinic acid to trigonelline is catalyzed by *N*-methyltransferase belonged to motif B′ methyltransferase family in *Coffea arabica*. *Biochem. Biophys. Res. Commun.* 452: 1060–1066.

Nakayama, T. and Honda, K. (2004). Chemical basis for differential acceptance of two sympatric rutaceous plants by ovipositing females of a swallowtail butterfly, *Papilio polytes* (Lepidoptera, Papilionidae). *Chemoecology* 14: 199–205.

Nishitani, H., Yamada, Y., Ohshima, N. et al. (1995). Identification of *N*-(β-D-Glucopyranosyl)nicotinic acid as a major metabolite from niacin in cultured tobacco cells. *Biosci. Biotechnol. Biochem.* 59: 1336–1338.

Noctor, G., Hager, J., and Li, S. (2011). Biosynthesis of NAD and its manipulation in plants. *Adv. Bot. Res.* 58: 153–201.

Rajasekaran, L.R., Aspinall, D., Jones, G.P., and Paleg, L.G. (2001). Stress metabolism. IX. Effect of salt stress on trigonelline accumulation in tomato. *Can. J. Plant Sci.* 81: 487–498.

Sasamoto, H. and Ashihara, H. (2014). Effect of nicotinic acid, nicotinamide and trigonelline on the proliferation of lettuce cells derived from protoplasts. *Phytochem. Lett.* 7: 38–41.

Shimizu, M.M. and Mazzafera, P. (2000). A role for trigonelline during imbibition and germination of coffee seeds. *Plant Biol.* 2: 605–611.

Suzuki-Yamamoto, M., Mimura, T., and Ashihara, H. (2006). Effect of short-term salt stress on the metabolic profiles of pyrimidine, purine and pyridine nucleotides in cultured cells of the mangrove tree, *Bruguiera sexangula*. *Physiol. Plant.* 128: 405–414.

Taguchi, H. and Shimabayashi, Y. (1983). Findings of trigonelline demethylating enzyme activity in various organisms and some properties of the enzyme from hog liver. *Biochem. Biophys. Res. Commun.* 113: 569–574.

Taguchi, H., Sakaguchi, M., and Shimabayashi, Y. (1985). Trigonelline content in coffee beans and the thermal conversion of trigonelline into nicotinic acid during the roasting of coffee beans. *Agric. Biol. Chem.* 49: 3467–3471.

Taguchi, H., Sakaguchi, M., Yamaki, K., and Shimabayashi, Y. (1987). Biosynthesis of trigonelline in the coffee plant. *Nippon Nogeikagaku Kaishi, Jpn.* 61: 183–189.

Tomlinson, P.B. (1994). *The Botany of Mangroves*. Cambridge, U.K: Cambridge University Press.

Tramontano, W.A. and Jouve, D. (1997). Trigonelline accumulation in salt-stressed legumes and the role of other osmoregulators as cell cycle control agents. *Phytochemistry* 44: 1037–1040.

Tramontano, W.A., Hartnett, C.M., Lynn, D.G., and Evans, L.S. (1982). Relationship between trigonelline concentration and promotion of cell arrest in G_2 in cultured roots of *Pisum sativum*. *Phytochemistry* 21: 1201–1206.

Tramontano, W.A., Lynn, D.G., and Evans, L.S. (1983). Trigonelline, nicotinic acid and nicotinamide in seedlings of *Pisum sativum*. *Phytochemistry* 22: 673–678.

Tramontano, W.A., Evans, L.S., and McGinley, P.A. (1985). Effects of cytokinins on promotion of cell arrest in G2 by trigonelline and trigonelline concentrations in cultured roots of *Pisum sativum* and *Glycine max*. *Environ. Exp. Bot.* 25: 83–88.

Ueda, M., Yamamura, S., and Niwa, M. (1995). Trigonelline, a leaf-closing factor of the nyctinastic plant, *Aeschynomene indica*. *Phytochemistry* 39: 817–819.

Ueda, M., Shigemori, H., Sata, N., and Yamamura, S. (2000). The diversity of chemical substances controlling the nyctinastic leaf-movement in plants. *Phytochemistry* 53: 39–44.

Upmeier, B., Thomzik, J.E., and Barz, W. (1988). Nicotinic acid-*N*-glucoside in heterotrophic parsley cell suspension cultures. *Phytochemistry* 27: 3489–3493.

Yin, Y., Matsui, A., Sakuta, M., and Ashihara, H. (2008). Changes in pyridine metabolism profile during growth of trigonelline-forming *Lotus japonicus* cell cultures. *Phytochemistry* 69: 2891–2898.

Yin, Y., Sasamoto, H., and Ashihara, H. (2011). Pyridine metabolism and trigonelline synthesis in leaves of the mangrove legume trees *Derris indica* (*Millettia pinnata*) and *Caesalpinia crista*. *Nat. Prod. Commun.* 6: 1835–1838.

Zheng, X.Q. and Ashihara, H. (2004). Distribution, biosynthesis and function of purine and pyridine alkaloids in *Coffea arabica* seedlings. *Plant Sci.* 166: 807–813.

Zheng, X.Q., Hayashibe, E., and Ashihara, H. (2005). Changes in trigonelline (*N*-methylnicotinic acid) content and nicotinic acid metabolism during germination of mung bean (*Phaseolus aureus*) seeds. *J. Exp. Bot.* 56: 1615–1623.

Part VIII

Other Nucleotide-Related Metabolites

22

Sugar Nucleotides

22.1 Introduction

Nucleotide sugars are monosaccharides that have a nucleotide, usually uridine triphosphate (UTP), bound to the carbohydrate chain through an energy rich linkage. UTP is synthesized by the *de novo* pyrimidine biosynthesis (see Figure 8.1 in Chapter 8).

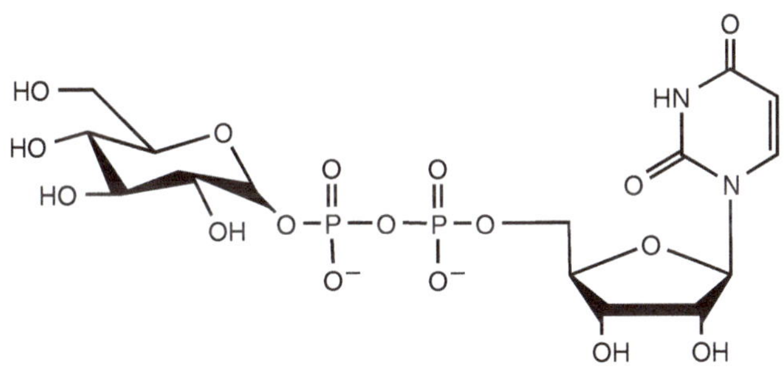

UDP-glucose (**1**)

Among these compounds, uridine diphosphate-glucose (UDP-glucose) (**1**) is one of the more important (Geisler et al. 2004). It is the major glucosyl donor for carbohydrates in higher plants, serving as a direct precursor for the synthesis of sucrose, which is a major product of photosynthesis in green leaves, accounting for much of the CO_2 fixed during photosynthesis. Sucrose serves as a long-distance transport molecule in most plants and it also has an important storage role. In addition, UDP-glucose is the direct precursor for cellulose and callose synthesis occurring at the plasmalemma (Amor et al. 1995; Schlupmann et al. 1994). Cellulose, the principal scaffolding component of all plant cell walls, is a polymer derived from β-1,4 linked glucose units. Callose differs from cellulose in consisting of β-1,3-glucan chains. It is produced in response to wounding, and in some types of cell it is made at specific stages of cell wall development, such as in growing pollen tubes and in the cell plates of dividing cells.

Sugar nucleotides are essential building blocks for metabolism of all living organisms. In plant sugar metabolism, sugar nucleotides are mainly involved in maintenance of carbohydrates. Nucleotide sugars are the activated forms of monosaccharides serving as glycose donors in glycosylation reactions. Those reactions are catalysed by glycosyltransferases. In plants, nucleotide sugars are the universal sugar donors for the formation of polysaccharides, glycoproteins, proteoglycans, glycolipids, and

Plant Nucleotide Metabolism – Biosynthesis, Degradation, and Alkaloid Formation, First Edition.
Hiroshi Ashihara, Iziar A. Ludwig and Alan Crozier.

glycosylated secondary metabolites. At least 100 genes that encode enzymes involved in the formation of nucleotide sugars have been identified. Nucleotide sugars are formed using (i) carbohydrates derived from photosynthesis, (ii) the sugar generated by hydrolysis of translocated sucrose, (iii) sugars released from storage carbohydrates, (iv) the salvage of sugars from glycoproteins and glycolipids, (v) the recycling of sugars released during primary and secondary cell wall restructuring, and (vi) sugar generated during plant–microbe interactions (Bar-Peled and O'Neill 2011).

22.2 The Sugar Nucleotide Moiety

The sugar nucleotide moiety comprises the sugar derivatives of three of the four ribonucleoside diphosphates (ADP, GDP, CDP, and UDP) and one deoxyribonucleoside diphosphate (TDP), and related compounds (Table 22.1). UDP-glucose was first identified in *Beta vulgaris* (beet) plants and later was detected in extracts of mung bean (*Vigna radiata*). Four UDP nucleotides, UDP-glucose, UDP-galactose, UDP-xylose, and UDP-arabinose (66 : 28 : 4 : 2) have been discovered, as well as UDP-*N*-acetyl-galactosamine and UDP-glucuronate. The topic has been reviewed in depth by Frank (2013). Currently, there are 15 UDP-sugars, 8 ADP-sugars, 8 GDP-sugars, and 2 TDP-sugars (Table 22.1). CDP-sugars do not occur in plants, although animal CDP-ribitol has been discovered in skeletal muscles (Gerin et al. 2016).

Historically, using mainly HPLC and/or ion-exchange chromatography, sugar nucleotides have been quantified in many plant materials (Frank 2013). In addition to UDP-glucose, ADP-glucose is found in some plant materials, although other kinds of sugar nucleotides are rare. One of the few sources are cell suspension cultures where they occur in low concentrations (Table 22.2). In most cells, the principal sugar nucleotide to accumulate is UDP-glucose. The levels of other sugar nucleotides are relatively low. The concentrations of UDP-glucose and ADP-glucose are, respectively, 199 and 6.6 nmol g^{-1} f.w. in *Datura innoxia* (Mitsui and Ashihara 1988), 102 and

Table 22.1 Sugar nucleotides in plants.

Nucleosides	D-Sugars
UDP	D-Glucose, D-galactose, D-fructose, D-rhamnose, D-mannose, D-ribose, D-xylose, L-arabinose, D-xylose, D-fucose, D-apiose, D-glucuronate, D-galacturonate, *N*-acetyl-D-glucosamine, *N*-acetyl-D-galactosamine
ADP	D-Glucose, D-galactose, D-fructose, D-mannose, D-ribose, L-arabinose, D-xylose, D-glucuronate
GDP	D-Glucose, D-galactose, D-mannose, D-rhamnose, D-ribose, D-xylose, L-arabinose, D-fucose
CDP	None
TDP	D-Glucose, D-galacturonate

This group comprises the sugar derivatives of three of ribonucleoside diphosphate (UDP, ADP, and GDP) and one deoxyribonucleoside diphosphate (TDP) and related sugar compounds.
Source: Adapted from Frank (2013).

Table 22.2 UDP-glucose and other sugar nucleotides in plants.

Plant	Sample	Sugar nucleotide contents (nmol g^{-1} f.w.)
Catharanthus roseus	Cells	UDPGlu (95); UDPGal (4); UDPXyl (0.3); UDPAra (3); UDPGlcA(18)
Datura inoxia	Cells	UDPGlu (199); UDPGlcNAc (4.1); UDPGlcA (4.8); UDPGalA (0.6); ADPGlc (6.6)
Glycine max	Cells	UDPGlu (102); UDPGal (34); UDPXyl (2.4); UDPAra (2.4); UDPRha (1.9); ADPGlc (<0.4)
Oryza sativa	Cells	UDPGlu (32); UDPGal (10); UDPXyl (3.1); UDPAra (4.0); UDPFru (2.8); UDPRha (2.9); UDPRib (3.0); UDPMan (2.2); UDPGlcA (0.6); UDPGalA (0.2); ADPGlc (6.2); ADPPRib (8.5); ADPXyl (0.3); ADPAra (0.4); GDPGlc (13.2); GDPXyl (14.3); GDPFuc (1.0); GDPRha (2.0); GDP (4.3)
Arabidopsis thaliana	Seedlings	UDPGlu (30); UDPGal (4); UDPXyl (0.7); UDPAra (0.7); UDPGlcA (0.6)
Phaseolus vulgaris	Seedling	UDPGlu (51); ADPGlc (5.0); ADPRib (3.5)
Zea mays	Grain	UDPGlu + UDPGal (6.8); ADPGal (4.1); ADPMan (2.1); ADPGlcNAc (0.7); UDPGlcNAc (1.5); UDPGalNAc (2.4)

Abbreviations of sugars are as follows: Ara, L-Arabinose; Fru, D-fructose; Fuc, D-fucose; Gal, D-galactose; GalA, D-galacturonate; GalNac, *N*-acetyl-D-galactosamine; Glc, D-glucose; GlcA, glucuronate; GlcNAc, D- *N*-acetyl-D-glucosamine; Rha, D-rhamnose; Rib, D-ribose; Man, D-mannose; Xyl, D-xylose.
Source: Data for *Arabidopsis thaliana* adapted from Behmüller et al. (2014) and others are from Frank (2013).

<0.4 nmol g^{-1} f.w. in *Glycine max* (ap Rees et al. 1984) and 51 and 5 nmol g^{-1} f.w. in *Phaseolus vulgaris* (Mangat 1979). In studies with cell walls, several different UDP-sugars have been detected (Buchanan et al. 2015).

22.3 Enzymes of Sugar Nucleotide Biosynthesis

In plants the main sugar nucleotide, UDP-glucose, is essential for sucrose and cell wall biosynthesis and for secondary metabolism (Bar-Peled and O'Neill 2011; Payyavula et al. 2014). UDP-glucose synthesis is catalysed by UDP-glucose pyrophosphorylase, UDP-sugar pyrophosphorylase and sucrose synthase.

22.3.1 UDP-Glucose Pyrophosphorylase

The properties of UDP-glucose pyrophosphorylase have been studied since the 1950s when it was isolated from yeast (Munch-Petersen 1955). Plant UDP-glucose pyrophosphorylase was initially purified from mung bean seedlings (Ginsburg 1958). It was registered as UTP-glucose-1-phosphate uridylyltransferase. The IUBMB enzyme nomenclature is EC 2.7.7.9 and the catalysed reaction is UTP + α-D-glucose-1-phosphate $\leftrightharpoons$ UDP-glucose + PPi (Figure 22.1).

UDP-glucose pyrophosphorylase is responsible for the formation and degradation of UDP-glucose, the key precursor for sucrose formation and cell wall components.

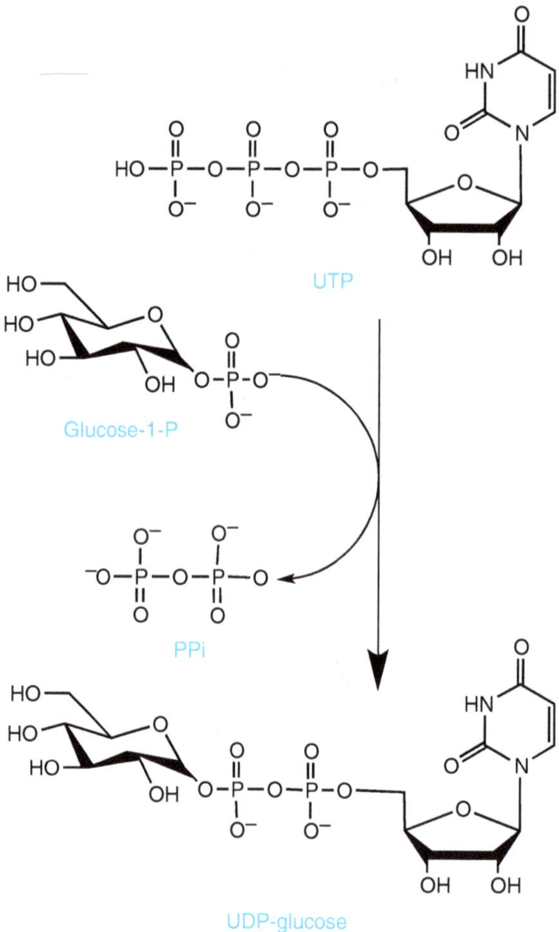

Figure 22.1 The conversion of UTP to UDP-glucose catalysed by UDP-glucose pyrophosphorylase (EC 2.7.7.9) or UDP-sugar pyrophosphorylase (EC 2.7.7.64) with glucose-1-phosphate acting as a substrate.

The properties of enzymes are shown in Table 22.3. In *Arabidopsis thaliana*, enzymes Km values for the formation substrates, UTP and glucose-1-P, are respectively, 0.03–0.17 and 0.14–0.24 mM. The Km of degradation-direction substrates of UDP-glucose pyrophosphorylase, UDP-glucose and PPi, are 0.04–0.06 and 0.03–0.07 mM, respectively (Meng et al. 2008; Okazaki et al. 2009). In melon, the Km of glucose-1-P is 0.24 mM and that of UDP-glucose is 0.11 mM (Dai et al. 2006). The Vmax values are slightly higher in degradation than in formation mode in both plant species.

UDP-glucose pyrophosphorylase is one of the key enzymes of the carbohydrate metabolic pathway and is found in microorganisms, plants, and animals. The reaction of the enzyme is channelled towards formation of glucose-1-P and/or UGP-glucose because of metabolic coupling with other pathways of sugar metabolism, involved either in synthesis or degradation of sucrose (Figure 22.2). In source tissues, such as photosynthetic leaves, where sucrose is being produced and stored in vacuoles or transported to sink tissues, UDP-glucose pyrophosphorylase is thought to be the primary enzyme for UDP-glucose synthesis from sucrose by being coupled to the reaction catalysed by sucrose phosphate synthase (EC 2.4.1.14). On the other hand,

Table 22.3 Substrates and Km values of UDP-glucose pyrophosphorylase.

Plant	Direction[a)]	Substrate	Km (mM)	Vmax	Reference
Arabidopsis thaliana (UGPase-1)	Formation	UTP	0.08	340	(Meng et al. 2008)
		G1P[b)]	0.14		
	Degradation	UDPG[c)]	0.04	510	
		PPi	0.03		
Arabidopsis thaliana (UGPase-2)	Formation	UTP	0.17	385	(Meng et al. 2008)
		G1P	0.36		
	Degradation	UDPG	0.06	715	
		PPi	0.07		
Arabidopsis thaliana (UGPase-3)	Formation	UTP	0.031		(Okazaki et al. 2009)
		G1P	0.24		
Cucumis melo (melon)	Formation	G1P	0.24	238	(Dai et al. 2006)
	Degradation	UDPG	0.11	277	

a) Direction: formation is UTP + G1P → UDPG + PPi, degradation is UDPG + PPi → UTP + G1P.
b) G1P, glucose-1-phosphate.
c) UDPG, UDP-glucose.

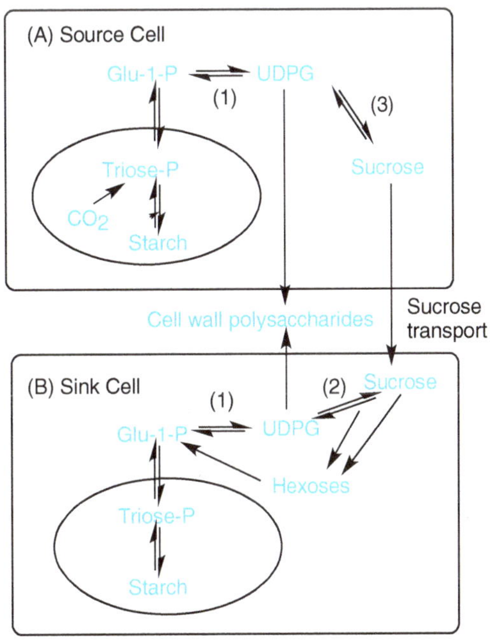

Figure 22.2 Possible involvement of UDP-glucose pyrophosphorylase (UGPase) in cell wall synthesis in source and sink tissues. UGPase in photosynthetic (source) tissues generates UDP-glucose (UDPG), and works in conjunction with sucrose-phosphate synthetase (SPS, EC 2.4.1.14) and sucrose-phosphate phosphatase (SPP, EC 3.1.3.24) to produce sucrose. In a sink (e.g. roots), UGPase can work in conjunction with sucrose synthase (SuSy) and/or invertases to breakdown sucrose, but can also remobilize glucose-1-P back to UDPG. Note that UDPG is used for cell wall polysaccharide synthesis in both source and sink tissues. Enzymes: (1), UGPase; (2), SuSy; (3) SPS and SPP. Source: Adapted from Kleczkowski et al. (2004).

in sink tissues, such as roots and developing tubers, where sucrose is hydrolysed, UDP-glucose pyrophosphorylase is involved in the metabolism of UDP-glucose formed by sucrose synthase-catalysed breakdown of sucrose (see Section 22.3.3). The glucose-1-P produced by this reaction can either be catabolized through the glycolysis pathway or converted into ADP-glucose by ADP-glucose pyrophosphorylase (EC 2.7.7.27) and finally stored as starch (Kleczkowski 1994).

22.3.2 UDP-Sugar Pyrophosphorylase

UDP-sugar pyrophosphorylase is registered as UTP-monosaccharide-1-phosphate uridylyltransferase. The IUBMB enzyme nomenclature is EC 2.7.7.64 and the reaction is as follows:

$$\text{UTP} + \text{a monosaccharide 1-phosphate} \leftrightharpoons \text{UDP-monosaccharide} + \text{PPi}$$

UDP-sugars, activated forms of monosaccharides, are synthesized through *de novo* and salvage pathways and serve as substrates for the synthesis of polysaccharides, glycolipids, and glycoproteins in higher plants.

UDP-sugar pyrophosphorylase, with an apparent molecular mass was 67 000 Da, was first purified from sprouts of pea (*Pisum sativum*) (Kotake et al. 2004). The enzyme catalysed the formation of UDP-glucose, UDP-galactose, UDP-glucuronate, UDP-arabinose, and UDP-xylose from respective monosaccharide 1-phosphates in the presence of UTP as a co-substrate. This demonstrated that the enzyme has broad substrate specificity towards monosaccharide 1-phosphates. Maximum activity of the enzyme occurred at pH 6.5–7.5, and at 45 °C in the presence of 2 mM Mg^{2+}. The Km values for glucose-1-P and arabinose-1-P were, respectively, 0.34 and 0.96 mM. The enzyme is a novel type of UDP-sugar pyrophosphorylase, which catalyses the formation of various UDP-sugars at the end of salvage pathways in higher plants (Table 22.4).

UDP-sugar pyrophosphorylase is studied in glycobiology investigating the structure, biosynthesis, and biology of saccharides (sugar chains or glycans) that are widely distributed in nature. It is probable that the relative amounts and the architecture of cell wall polysaccharides and of the sugar moieties of glycolipids and glycoproteins in higher plants are regulated by the amount of the available nucleotide sugars, as well as the levels of glycosyltransferases that incorporate monosaccharide units from respective nucleotide sugars into the polymers. Activated nucleotide sugars that serve as glycosyl donors for these polymers are generated through *de novo* pathways, in which various UDP- and GDP-sugars are produced through sequential interconversions from the starting substrates UDP-glucose and GDP-mannose (Reiter and Vanzin 2001). In the salvage pathways, alternative routes to synthesize nucleotide sugars are (i) free monosaccharides released during the degradation of polysaccharides, and (ii) glycoconjugates, which are first phosphorylated by the action of monosaccharide kinases, then converted into nucleotide sugars by the action of pyrophosphorylases in the presence of the respective nucleotide triphosphates as co-substrates. Pyrophosphorylases principally catalyse both the forward (synthesis of NDP-sugars) and reverse (pyrophospholysis) reactions: monosaccharide 1-phosphates + NTP $\leftrightharpoons$ NDP-sugars + PPi. UDP-sugar pyrophosphorylase is responsible for the formation and degradation of UDP-sugars, the key precursor for sucrose formation and for cell wall components.

Table 22.4 Substrates and Km values of UDP-sugar pyrophosphorylase.

Plant	Direction[a)]	Substrate	Km (mM)	Vmax	Reference
Pisum sativum	Formation	UTP	0.048	81[b)]	Kotake et al. (2004)
		Glucose-1-P	0.34	106	
		Galactose-1-P	0.58	161	
		Glucuronate-1-P	0.48	66	
		Arabinose-1-P	0.96	71	
		Xylose 1P	1.98	49	
	Degradation	UDP-Glucose	0.34	145	
		PPi	0.25	64	
Pisum sativum	Formation	UTP	1.15	127[b)]	Ohashi et al. (2006)
		Galacturonate-1-P	2.27	164	
	Degradation	UDP-Galacturonate	0.70	442	
		PPi	1.26	756	
Arabidopsis thaliana	Formation	UTP	0.077	—[c)]	Kotake et al. (2007)
		Glucose-1-P	0.23	84[b)]	
		Galactose-1-P	0.27	87	
		Glucuronate-1-P	0.094	51	
		Arabinose-1-P	1.50	45	
		Xylose-1-P	1.00	23	
	Degradation	UDP-glucose	0.28	—	
		PPi	0.16	—	
Arabidopsis thaliana	Formation	UTP	2.9	8.9[d)]	Liu et al. 2013
		dUTP	3.7	5.6	
		dTTP	8.1	1.2	

a) Direction: Formation is UTP + monosaccharide-1-P → UDP-monosaccharide + PPi; degradation is UDP-monosaccharide + PPi → UTP + monosaccharide-1-P.
b) $\mu mol\ min^{-1}\ mg^{-1}$ protein.
c) — Not measured.
d) mg^{-1} protein.

In addition to the first pea enzyme (Kotake et al. 2004), UDP-galacturonate formation and degradation has also been detected in pea (Ohashi et al. 2006) and the properties of the enzyme have also been studied in *A. thaliana* (Kotake et al. 2007; Liu et al. 2013). The properties of the enzymes are shown in Table 22.4. Both pea and *A. thaliana* UDP-sugar pyrophosphorylase catalyse the formation of UDP-glucose, UDP-galactose, UDP-glucuronate, UDP-arabinose, and UDP-xylose. The phosphorylase exhibits broad substrate specificity towards monosaccharide 1-phosphates. In contrast, mannose 1-phosphate, fructose 1-phosphate, and glucose 6-phosphate are not substrates. UTP can be replaced by dUTP and dTTP (Liu et al. 2013) but not ATP, CTP, GTP, or ITP (Kotake et al. 2004).

22.3.3 Sucrose Synthase

Sucrose synthase, which catalyses the reaction UDP + sucrose ⇋ UDP-glucose + D-fructose (Figure 22.3), is also called UDP-glucose-fructose glucosyltransferase (EC 2.4.1.13). Although UDP is generally considered to be the preferred nucleoside diphosphate for sucrose synthase, numerous studies have shown that ADP can also serve as an effective acceptor molecule to produce ADP-glucose. Sucrose synthase has a dual role in producing both UDP-glucose, necessary for cell wall and glycoprotein biosynthesis, and ADP-glucose, which is required for starch biosynthesis (Baroja-Fernández et al. 2003).

Sucrose synthase has been purified from etiolated *Phaseolus aureus* seedlings (Delmer 1972). Subunit analyses indicate that the enzyme is a tetramer composed of identical protein subunits with an approximate molecular weight of 94 000 Da. An equilibrium

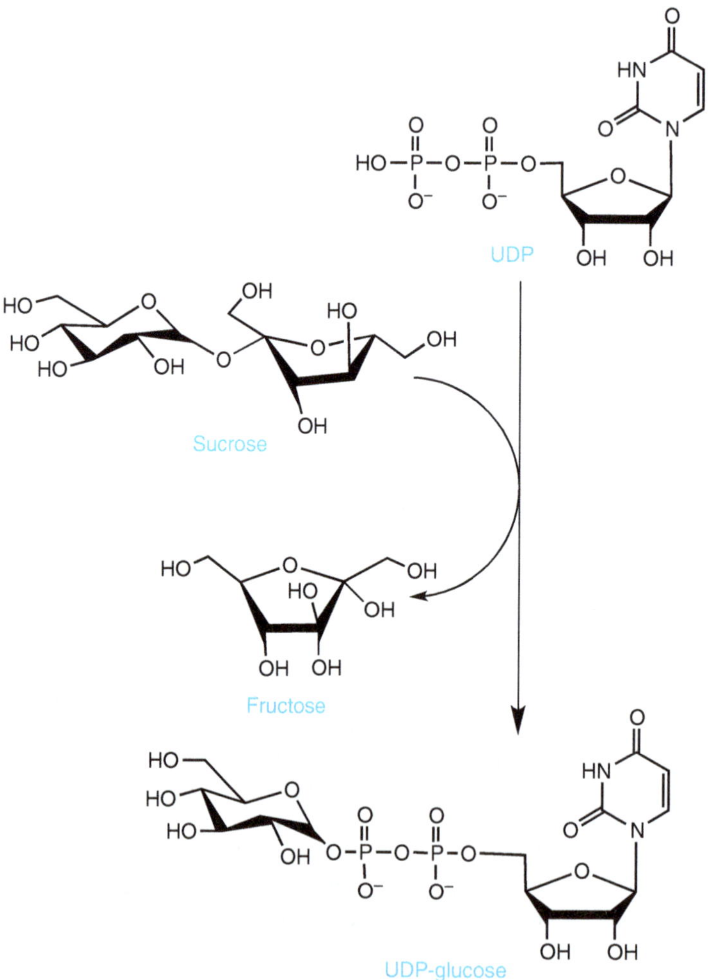

Figure 22.3 The conversion of UDP to UDP-glucose is catalysed by sucrose synthase (EC 2.4.1.13) with sucrose acting as a substrate.

constant of 0.15 was estimated for the reaction at pH 7.5 and 25 °C in the direction of UDP-glucose synthesis. When UDP and ADP are present at saturating levels, the respective Km figures for sucrose were 17 and 29 mM. When extensively purified, the enzyme retains activity for UDP, ADP, TDP, CDP, and GDP as substrates. Km and Vmax values are shown in Table 22.5.

Two isoforms of sucrose synthase have been purified from the Japanese pear (*Pyrus serotine)* fruit and their properties compared (Tanase and Yamaki 2000). Isoform I appeared mainly in young fruit and isoform II principally in mature fruit. The molecular weights of native and subunit proteins of isoforms I and II were almost the same and both enzymes appeared to be tetramers composed of an 83 kDa polypeptide. However, the ionic charges of the native proteins and the kinetic parameters of two isoforms were different. Specifically, the Km value for UDP-glucose with isoform I was the same as that for UDP, while the Km value for UDP-glucose with isoform II was less than that for UDP (Table 22.5). The isoform II reacted more readily for sucrose synthesis than sucrose cleavage compared with isoform I. Therefore, it is considered that both isoforms play different roles in the utilization of carbohydrate in young and mature fruit.

Kinetic properties of barley and potato sucrose synthase are shown in Table 22.5. The Km values for ADP and UDP were shown to be very similar in the presence of saturating sucrose, whereas the Vmax was 1.5-fold higher with UDP compared with ADP. The kinetic studies of the sucrose cleavage reaction in the presence of fixed nucleoside diphosphate revealed a notably higher affinity for sucrose in the presence of saturating UDP than in the presence of saturating ADP (Baroja-Fernández et al. 2003).

22.4 Localization of UDP-Glucose-Producing Enzymes

UDP-glucose pyrophosphorylase, UDP-sugar pyrophosphorylase, and sucrose synthase activities occur in several parts of plant cells, tissues, and organs (Table 22.6). UDP-glucose pyrophosphorylase has been found in all plant tissues that have been studied to date. Less is known about the distribution of UDP-sugar pyrophosphorylase within plants, and most data are based on gene expression studies. For example, in *A. thaliana*, UDP-sugar pyrophosphorylase is expressed in most cell types, especially in leaf and root vascular tissues, cotyledons, trichomes, cauline leaves, stem epidermis, and pollen (Kotake et al. 2007; Litterer et al. 2006). For sucrose synthase, a rule of thumb is that its activity and protein content are low or very low in mature leaves, but are prominent in sink tissues (Barratt et al. 2009; Koch 2004). With respect to intracellular localization, the bulk of UDP-glucose pyrophosphorylase is confined to the cytosol (Kleczkowski 1994) yet there have been reports of its presence in the Golgi apparatus, membrane fractions, chloroplasts, cell walls, and even in xylem sap.

22.5 UDP-Glucose-Interconversion

Major uridine nucleotide-sugar interconversion pathways in plants are shown in Figure 22.4. The *de novo* and salvage pathways are major routes for UDP-sugar production. The *de novo* pathway focuses on the interconversion of nucleotides to produce

Table 22.5 Substrates and Km and Vmax values of sucrose synthase.

Plant	Direction[a)]	Substrate	Km (mM)	Vmax	Reference
Phaseolus aureus	Formation	UDP	0.19	100[b)]	Delmer (1972)
		ADP	0.19	28.0	
		TDP	0.30	5.8	
		CDP	0.44	3.4	
		GDP	0.17	2.8	
		Sucrose (UDP)	17	—[c)]	
		Sucrose (ATP)	29	—	
Pyrus serotine	Formation	Sucrose (UDP)	37.6	3.6[d)]	Tanase and Yamaki (2000)
Isoform SSI		UDP	0.07	3.6	
		ADP	0.11	2.9	
	Degradation	Fructose	18.6	3.4	
		UDP-glucose	0.05	2.0	
		ADP-glucose	0.17	2.0	
Isoform SSII	Formation	Sucrose (UDP)	36.0	2.4[d)]	Tanase and Yamaki (2000)
		UDP	0.41	2.3	
		ADP	1.94	0.8	
	Degradation	Fructose	9.77	2.9	
		UDP-glucose	0.03	2.5	
		ADP-glucose	0.78	1.6	
Hordeum vulgare	Formation	UDP	0.22	290[d)]	Baroja-Fernández et al. (2003)
		ADP	0.15	190	
		Sucrose (UDP)	30	290	
		Sucrose (ADP)	210	190	
Solanum tuberosum	Formation	UDP	0.22	80[d)]	Baroja-Fernández et al. (2003)
		ADP	0.30	65	
		Sucrose (UDP)	35	80	
		Sucrose (ADP)	220	65	

a) Direction: formation, NDP + sucrose → NDP-glucose + fructose; degradation: NDP-glucose + fructose → NDP + sucrose.

b) Relative %.

c) — Not measured.

d) mg^{-1}.

Table 22.6 Subcellular location of plant UDPG producing three enzymes.

	(i) UGPase	(ii) USPase	(iii) SuSy
Cytosol	+	+	+
Plastid	+		+
Mitochondria			+
Nucleus			+
Vacuole			+
Golgi	+		+
Plasmalemma			+
Microsome membranes	+	+	
Cell wall	+		+
Xylem sap	+		+

(i) UDPG pyrophosphorylase (UGPase), (ii) UDP-sugar pyrophosphorylase, and (iii) sucrose synthase (SuSy).
Source: Adapted from Kleczkowski et al. (2010).

UDP sugars. The salvage pathway produces nucleotide sugars by phosphorylation and subsequent pyrophosphorylation of monosaccharides, which are released by cell wall turnover or spontaneous hydrolysis of energy-rich nucleotide sugars in cells. The phosphorylation is catalysed by sugar-specific monosaccharide-1-kinases, whereas the pyrophosphorylation involves a UDP-sugar pyrophosphorylase that accepts at least six different sugar-1-phosphates as a substrate (Kotake et al. 2010).

22.6 Other Metabolites

In addition to sugar nucleotides, cyclic nucleotides and diadenosine tetraphosphate are nucleotide-related metabolites.

22.6.1 Cyclic Nucleotides

A cyclic nucleotide is a single-phosphate nucleotide with a cyclic bond between the sugar and phosphate groups. Adenosine 3′,5′-cyclic monophosphate (**2**), commonly referred to as cAMP, was first described as a regulator of glycogen breakdown in the liver (Robison et al. 1971) and is now known to be a secondary messenger for a wide variety of cellular responses in living organisms ranging from a species of soil-living amoeba (*Dictyostelium*) to *Homo sapiens* (Robison et al. 1995). The question of whether cAMP is an authentic secondary messenger in signal transduction pathways in higher plants attracted the interest of plant scientists for decades (Assaman 1995; Bolwell 1995). The natural occurrence of cAMP, guanosine 3′,5′-cyclic monophosphate (**3**, cGMP) and other cyclic nucleotides in higher plants, at one time a topic of much debate (Martinez-Atiena et al. 2007), is now well established, as is the presence of the nucleotidyl cyclases and cyclic nucleotide phosphodiesterases involved in their synthesis and breakdown (Newton and Smith 2004). The presence of cyclic nucleotides

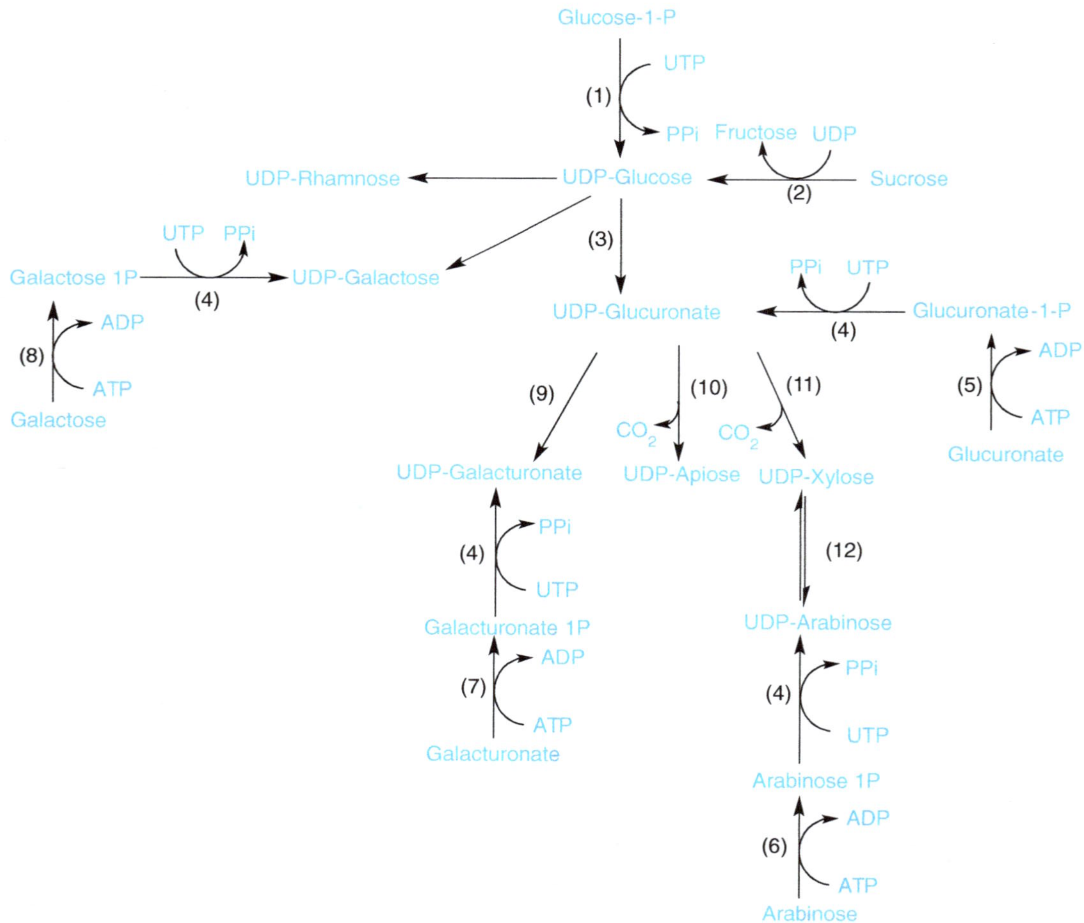

Figure 22.4 *De novo* synthesis and recycling pathways of UDP-sugars. Enzymes: (1) UDP-glucose pyrophosphorylase; (2) Sucrose synthase; (3) UDP-glucose dehydrogenase; (4) UDP-sugar pyrophosphorylase; (5) glucuronate-1-kinase; (6) 5 arabinose-1-kinase; (7) galacturonate-1-kinase; (8) galactose-1-P kinase; (9) UDP-glucuronate 4-epimerase; (10) UDP-apiose synthase; (11) UDP-xylose synthase; (12) UDP-xylose epimerase. Source: Adapted from Carpita and McCann (2000) and Behmüller et al. (2014).

in plants and their role in signal transduction, which is beyond the scope of this article, has been the topic of numerous reviews including those by Gehring (2010), Gehring and Turek 2017) and Isner and Maathuis (2016).

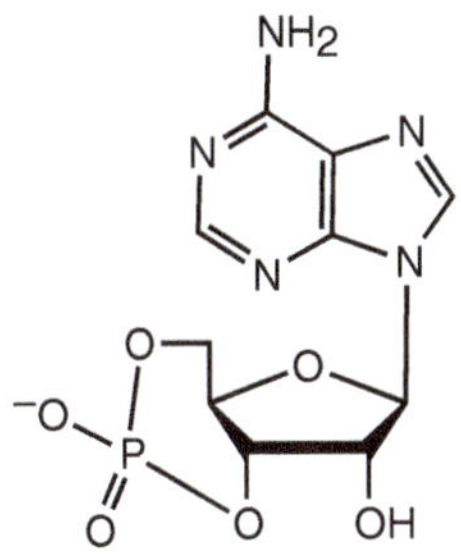

Adenosine 3′,5′-cyclic monophosphate (**2**)

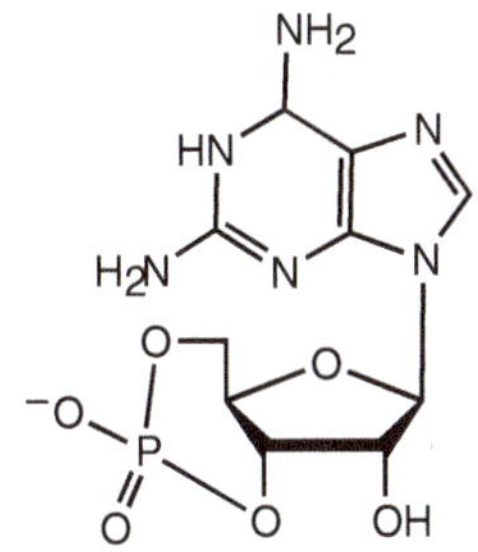

Guanosine 3′,5′-cyclic monophosphate (**3**)

22.6.2 Diadenosine Tetraphosphate

Diadenosine tetraphosphate (**4**, Ap4A) is a putative intracellular signal molecule, ubiquitous in nature being common to all organisms from bacteria to humans. Adenosine polyphosphates are capable of inducing multiple physiological effects (Luo et al. 2004). In plants, potential biochemical conversions of minor mononucleotides, such as adenosine-5′-tetraphosphate (**5**), adenosine-5′-pentaphosphate (**6**), and dinucleotides including Ap4A, has been examined. Unlike various bacteria, fungi, and animals, these nucleotides have not been identified in plants (Guranowski 2004). However, plant enzymes potentially involved in the synthesis and degradation of **5** and **6** have begun to be investigated (Pietrowska-Borek et al. 2014).

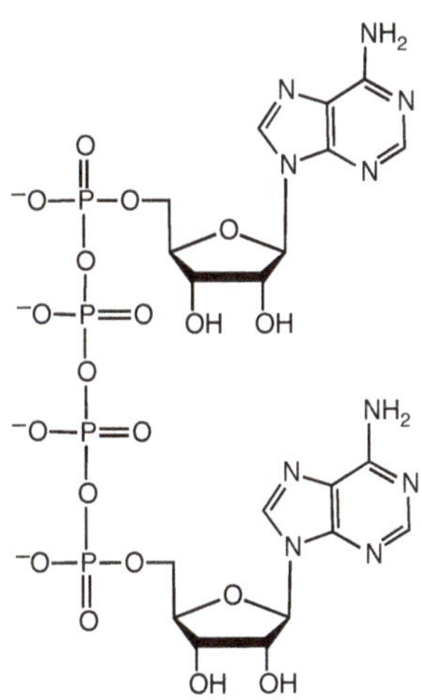

Diadenosine tetraphosphate (**4**)

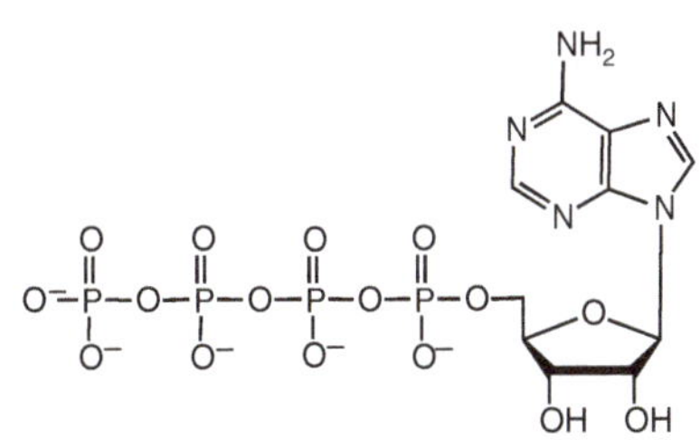

Adenosine-5′-tetraphosphate (**5**)

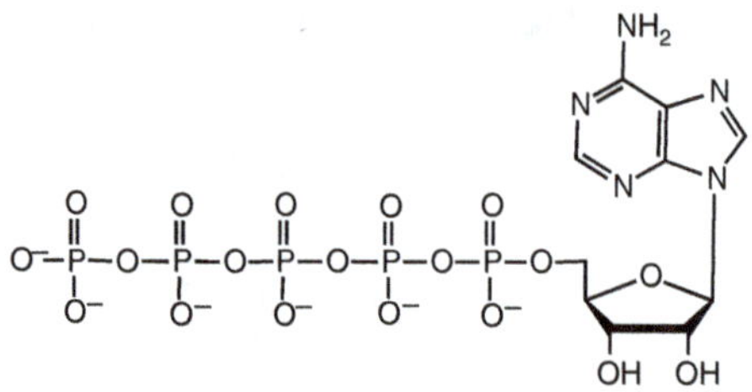

Adenosine-5′-pentaphosphate (**6**)

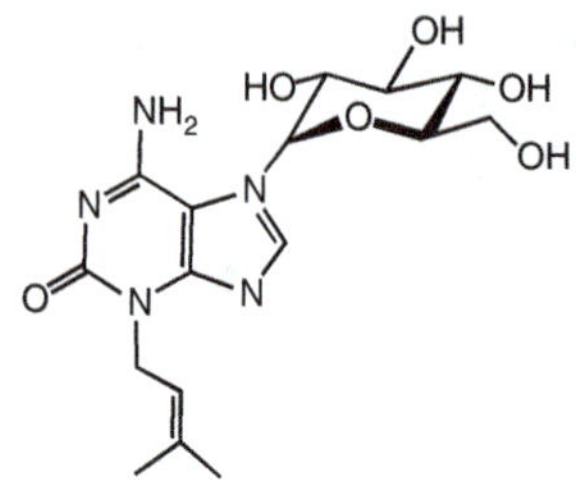

Locustoside A (**7**)

22.6.3 Purine Alkaloid Glucosides

Purine alkaloid glycosides are rare. One source is seeds of *Gleditsia japonica,* a member of the Fabaceae that is used as a traditional oriental medicine. Initially Kajimoto et al. (2010) identified N^3-prenylisoguanine-N^6-glucoside (7, locustoside A) (Kajimoto et al. 2010) and subsequently, Harauchi et al. (2017) isolated a number of related glycosides from the same source.

22.7 Summary

Sugar nucleotides (nucleotide sugars) are the activated forms of nucleotides and sugars. The UDP-glucose plays a key role in plant metabolism. Three enzymes, UDP-glucose pyrophosphorylase, UDP-sugar pyrophosphorylase and sucrose synthase, are involved in the synthesis of UDP-glucose and other UDP-sugars.

References

Amor, Y., Haigler, C.H., Johnson, S. et al. (1995). A membrane-associated form of sucrose synthase and its potential role in synthesis of cellulose and callose in plants. *Proc. Nat. Acad. Sci.* 92: 9353–9357.

ap Rees, T., Leja, M., Macdonald, F.D., and Green, J.H. (1984). Nucleotide sugars and starch synthesis in spadix of *Arum maculatum* and suspension cultures of *Glycine max. Phytochemistry* 23: 2463–2468.

Assaman, S.M. (1995). Cyclic AMP as a secondary messenger in higher plants. *Plant Physiol* 108: 885–889.

Baroja-Fernández, E., Muñoz, F.J., Pozueta-Romero, J. et al. (2003). Sucrose synthase catalyzes the de novo production of ADPglucose linked to starch biosynthesis in heterotrophic tissues of plants. *Plant Cell Physiol.* 44: 500–509.

Bar-Peled, M. and O'Neill, M.A. (2011). Plant nucleotide sugar formation, Interconversion, and salvage by sugar recycling. *Annu. Rev. Plant Biol.* 62: 127–155.

Barratt, D.H.P., Derbyshire, P., Findlay, K. et al. (2009). Normal growth of *Arabidopsis* requires cytosolic invertase but not sucrose synthase. *Proc. Nat. Acad. Sci.* 106: 13124–13129.

Behmüller, R., Forstenlehner, I.C., Tenhaken, R., and Huber, C.G. (2014). Quantitative HPLC-MS analysis of nucleotide sugars in plant cells following off-line SPE sample preparation. *Anal. Bioanal. Chem.* 406: 3229–3237.

Bolwell, G.P. (1995). Cyclic AMP, the reluctant messenger in plants. *Trends Biochem. Sci.* 20: 492–495.

Buchanan, B.B., Gruissen, W., and Jones, R.L. (2015). *Biochemistry and Molecular Biology of Plants*. Rockville: American Society of Plant Biologists.

Carpita, N. and McCann, M. (2000). The cell wall. In: *Biochemistry and Moleculer Biology of Plants* (eds. B.B. Buchanan, W. Gruissem and R.L. Jones), 52–108. Rockville: American Society of Plant Physiologists.

Dai, N., Petreikov, M., Portnoy, V. et al. (2006). Cloning and expression analysis of a UDP-galactose/glucose pyrophosphorylase from melon fruit provides evidence for the major metabolic pathway of galactose metabolism in raffinose oligosaccharide metabolizing plants. *Plant Physiol.* 142: 294–304.

Delmer, D.P. (1972). The purification and properties of sucrose synthetase from etiolated *Phaseolus aureus* seedlings. *J. Biol. Chem.* 247: 3822–3828.

Frank, A.W. (2013). Mononucleotides. In: *Chemistry of Plant Phosphorus Compounds* (ed. A.W. Frank), 305–403. Oxford: Elsevier.

Gehring, C. (2010). Adenylcyclases and cAMP in plant signaling – past and present. *Cell Commun. Signal* 8: 15. https://doi.org/10.1186/1478-811X-8-15.

Gehring, C. and Turek, I.S. (2017). Cyclic nucleotide monophosphates and their cyclases in plant signaling. *Front. Plant Sci.* 8: 1704. https://doi.org/10.3389/fpls.2017.01704.

Geisler, M., Wilczynska, M., Karpinski, S., and Kleczkowski, L.A. (2004). Toward a blueprint for UDP-glucose pyrophosphorylase structure/function properties: homology-modeling analyses. *Plant Mol. Biol.* 56: 783–794.

Gerin, I., Ury, B., Breloy, I. et al. (2016). ISPD produces CDP-ribitol used by FKTN and FKRP to transfer ribitol phosphate onto α-dystroglycan. *Nature Comm.* 7: 11534.

Ginsburg, V. (1958). Purification of uridinediphosphate glucose pyrophosphorylase from mung bean seedlings. *J. Biol. Chem.* 232: 55–61.

Guranowski, A. (2004). Metabolism of diadenosine tetraphosphate (Ap4A) and related nucleotides in plants; review with historical and general perspective. *Front. Biosci.* 9: 1398–1411.

Harauchi, Y., Kajimoto, T., Ohta, E. et al. (2017). Prenylated purine alkaloids from seeds of *Gleditsia japonica*. *Phytochemistry* 143: 145–150.

Isner, J.-C. and Maathuis, F.J.M. (2016). cGMP signalling in plants: from enigma to main stream. *Funct. Plant Biol.* 42: 93–101.

Kajimoto, T., Aoki, N., Ohta, E., and Ohta, S. (2010). Locustoside a – a new purine alkaloid glucoside from seeds of *Gleditsia japonica*. *Phytochem. Lett.* 3: 198–200.

Kleczkowski, L.A. (1994). Glucose activation and metabolism through UDP-glucose pyrophosphorylase in plants. *Phytochemistry* 37: 1507–1515.

Kleczkowski, L.A., Geisler, M., Ciereszko, I., and Johansson, H. (2004). UDP-glucose pyrophosphorylase. An old protein with new tricks. *Plant Physiol.* 134: 912–918.

Kleczkowski, L.A., Kunz, S., and Wilczynska, M. (2010). Mechanisms of UDP-glucose synthesis in plants. *Crit. Rev. Plant Sci.* 29: 191–203.

Koch, K. (2004). Sucrose metabolism: regulatory mechanisms and pivotal roles in sugar sensing and plant development. *Curr. Opin. Plant Biol.* 7: 235–246.

Kotake, T., Yamaguchi, D., Ohzono, H. et al. (2004). UDP-sugar pyrophosphorylase with broad substrate specificity toward various monosaccharide 1-phosphates from pea sprouts. *J. Biol. Chem.* 279: 45728–45736.

Kotake, T., Hojo, S., Yamaguchi, D. et al. (2007). Properties and physiological functions of UDP-sugar pyrophosphorylase in *Arabidopsis*. *Biosci. Biotechnol. Biochem.* 71: 761–771.

Kotake, T., Hirosawa, C., Ando, Y., and Tsumuraya, Y. (2010). Generation of nucleotide sugars for biomass formation in plants. *Plant Biotechnol.* 27: 231–236.

Litterer, L.A., Schnurr, J.A., Plaisance, K.L. et al. (2006). Characterization and expression of *Arabidopsis* UDP-sugar pyrophosphorylase. *Plant Physiol. Biochem.* 44: 171–180.

Liu, J., Zou, Y., Guan, W. et al. (2013). Biosynthesis of nucleotide sugars by a promiscuous UDP-sugar pyrophosphorylase from *Arabidopsis thaliana* (AtUSP). *Bioorg. Med. Chem. Lett.* 23: 3764–3768.

Luo, J., Jankowski, V., Güngär, N. et al. (2004). Endogenous diadenosine tetraphosphate, diadenosine pentaphosphate, and diadenosine hexaphosphate in human myocardial tissue. *Hypertension* 43: 1055–1059.

Mangat, B.S. (1979). The effect of Diuron on the soluble nucleotide pool and growth of bean and corn seedlings. *Pestic. Biochem. Physiol.* 10: 251–258.

Martinez-Atiena, J., Ingelgem, C., Roef, L., and Maathuis, F.J.M. (2007). Plant cyclic nucleotide signalling. Facts and fiction. *Plant Signal. Behav.* 2: 540–543.

Meng, M., Wilczynska, M., and Kleczkowski, L.A. (2008). Molecular and kinetic characterization of two UDP-glucose pyrophosphorylases, products of distinct genes, from *Arabidopsis*. *Biochim. Biophys. Acta* 1784: 967–972.

Mitsui, K. and Ashihara, H. (1988). Nucleotide pools and purine metabolism in tissue cultures of wild-type *Dature innoxia* and adenine-requiring auxotroph. *Plant Cell Physiol.* 29: 1171–1183.

Munch-Petersen, A. (1955). Investigations of the properties and mechanism of the uridine diphosphate glucose pyrophosphorylase reaction. *Acta Chem. Scand.* 9: 1523–1536.

Newton, R.P. and Smith, C.J. (2004). Cyclic nucleotides. *Phytochemistry* 65: 2423–2437.

Ohashi, T., Cramer, N., Ishimizu, T., and Hase, S. (2006). Preparation of UDP-galacturonic acid using UDP-sugar pyrophosphorylase. *Anal. Biochem.* 352: 182–187.

Okazaki, Y., Shimojima, M., Sawada, Y. et al. (2009). A chloroplastic UDP-glucose pyrophosphorylase from *Arabidopsis* is the committed enzyme for the first step of sulfolipid biosynthesis. *Plant Cell* 21: 892–909.

Payyavula, R.S., Tschaplinski, T.J., Jawdy, S.S. et al. (2014). Metabolic profiling reveals altered sugar and secondary metabolism in response to UGPase overexpression in *Populus*. *BMC Plant Biol.* 14: 265–265.

Pietrowska-Borek, M., Czekała, Ł., Belchí-Navarro, S. et al. (2014). Diadenosine triphosphate is a novel factor which in combination with cyclodextrins synergistically enhances the biosynthesis of trans-resveratrol in *Vitis vinifera* cv. Monastrell suspension cultured cells. *Plant Physiol. Biochem.* 84: 271–276.

Reiter, W.-D. and Vanzin, G.F. (2001). Molecular genetics of nucleotide sugar interconversion pathways in plants. *Plant Mol. Biol.* 47: 95–113.

Robison, G.A., Butcher, R.W., and Sutherland, E.W. (1971). *Cyclic AMP*. New York: Academic Press.

Robison, G.A., Butcher, R.W., and Assmann, S.M. (1995). Cyclic AMP as a second messenger in higher plants. Status and future prospects. *Plant Physiol.* 108: 885–889.
Schlupmann, H., Bacic, A., and Read, S.M. (1994). Uridine diphosphate glucose metabolism and callose synthesis in cultured pollen tubes of *Nicotiana alata* link et Otto. *Plant Physiol.* 105: 659–670.
Tanase, K. and Yamaki, S. (2000). Purification and characterization of two sucrose synthase isoforms from Japanese pear fruit. *Plant Cell Physiol.* 41: 408–414.

23

Cytokinins

23.1 Introduction

Cytokinins along with gibberellins, auxins, abscisic acid, and brassinosteroids are major classes of plant hormones. Among these naturally occurring phytohormones, cytokinins are unique in that they have an adenine-type structure. Their discovery goes back to the 1950s when Skoog at the University of Wisconsin-Madison found that yeast extracts stimulated the growth of parenchyma cells of tobacco (*Nicotiana tabacum*) cultured *in vitro*. Later they established that the undefined component(s) in yeast could be replaced by the N^6-furfuryl aminopurine, kinetin (**1**). Kinetin, in combination with auxin, promoted the initiation and maintenance of cell division of the tobacco cells. Kinetin is not found in living plants, however in 1963 a naturally-occurring kinetin-like substance, *trans*-zeatin (**2**), was isolated from the endosperm of immature seeds of corn (*Zea mays*) by Letham in New Zealand. Zeatin and related compounds are referred to as cytokinins. Many naturally-occurring cytokinins have now been isolated and they are defined structurally as adenine derivatives with an isoprene-based side attached at the N^6 amino group. Cytokinins with a benzyl or hydroxybenzyl group at N^6 also occur, but they are rare. N^6-(Δ^2-Isopentenyl)adenine (**3**), *cis*-zeatin (**4**) and *trans*-dihydrozeatin (**5**) are the major cytokinins with an isoprenoid side chain (Bishop et al. 2015; Crozier et al. 2000).

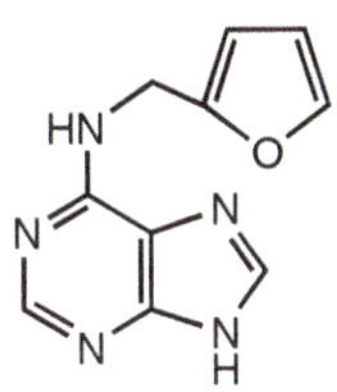
Kinetin (**1**)

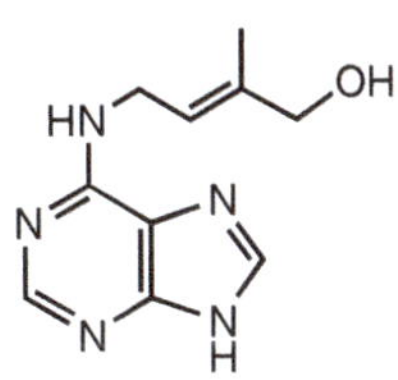
trans-Zeatin (**2**)

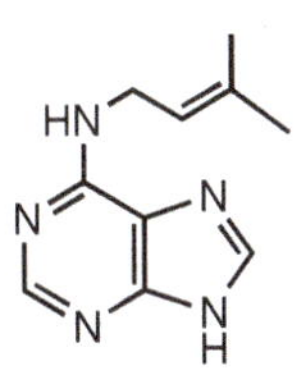
N^6-(Δ^2-Isopentenyl)adenine (**3**)

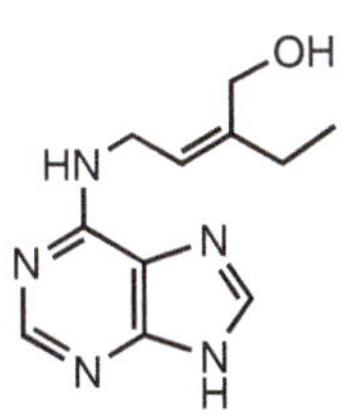
cis-Zeatin (**4**)

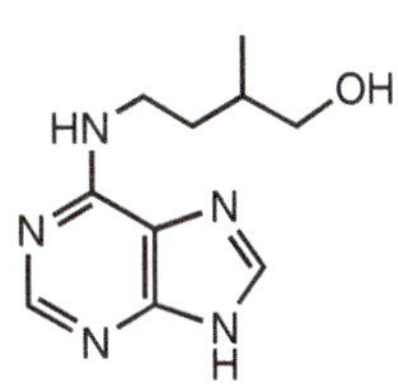
trans-Dihydrozeatin (**5**)

Plant Nucleotide Metabolism – Biosynthesis, Degradation, and Alkaloid Formation, First Edition.
Hiroshi Ashihara, Iziar A. Ludwig and Alan Crozier.

Major sites of cytokinin biosynthesis are roots, cambium, and other actively dividing tissues. Cytokinin biosynthesis, interconversions and degradation are involved in the regulation of cytokinin homeostasis during plant development (Ashihara et al. 2018; Åstot et al. 2000; Kurakawa et al. 2007; Kwade et al. 2005; Romanov et al. 2006; Schoor et al. 2011). Cytokinins are synthesized as ribonucleotides and three forms of cytokinins, namely cytokinin nucleotides, nucleosides, and free bases occur in plant cells (Sakakibara 2006, 2010).

Together with auxin, cytokinins promote plant cell division and influence differentiation of cells in culture. A high cytokinin/auxin ratio promotes shoot production, auxin alone initiates root growth while equimolar amounts of auxin and cytokinin result in the proliferation of undifferentiated callus cells. They also have multiple roles in plant growth and development including inducing stomatal opening, the regulation of apical dominance, nutrient allocation, responses to external biotic and abiotic signals, and retardation of leaf sensecence (Bishop et al. 2015; Crozier et al. 2000; Schoor et al. 2011; Zhang et al. 2013).

23.2 Adenosine Phosphate-Isopentenyl Formation

The first step of cytokinin biosynthesis is catalysed by adenosine phosphate-isopentenyltransferases (IPTs) and substrates of this enzymes are adenosine-phosphates (ATP, ADP, or AMP) and dimethylallyl diphosphate (DMAPP). The reaction is shown using ATP in Figure 23.1.

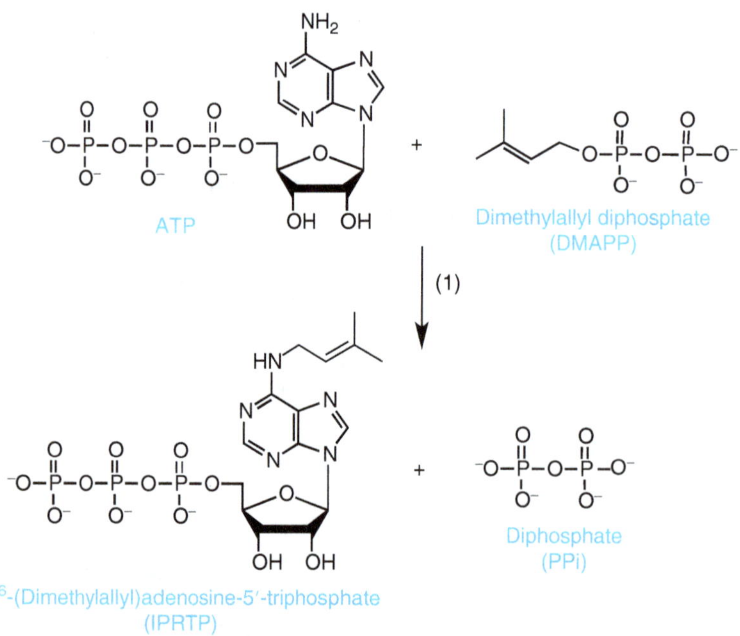

Figure 23.1 Structures of naturally occurring cytokinins in plants. The first reaction step (1) in the biosynthesis of isoprenoid cytokinins catalysed by adenosine phosphate-isopentenyltransferase (IPT). This enzyme is also registered as ADP/ATP-dependent adenylate dimethylallyltransferase (EC 2.5.1.112) by IUBMB Enzyme Nomenclature.

Two EC number enzymes are registered in the International Union of Biochemistry and Molecular Biology (IUBMB) for the adenosine phosphate-isopentenyltransferases. The enzyme (EC 2.5.1.112) is adenylate dimethylallyltransferase (ADP/ATP-dependent). The enzymes from *Arabidopsis thaliana* (Kakimoto 2001; Takei et al. 2001) and *Humulus lupulus* (Sakano et al. 2004) have similar properties. One of the isoforms (IPT4) from *A. thaliana* is specific for ADP and ATP and the Km of DMAPP and ATP are, respectively, 6.5 and 18 μM (Kakimoto 2001). The *A. thaliana* enzyme (EC 2.5.1.112) has very low specificity for AMP (Sakakibara and Takei 2002).

The AMP utilizing enzyme, adenylate dimethylallyltransferase (EC 2.5.1.27), is involved in the biosynthesis of cytokinins in a cell-free system from cytokinin-autotrophic tobacco tissue cultures (Chen and Melitz 1979).

23.3 *trans*-Zeatin Phosphate Synthesis

The dominant pathway of *trans*-zeatin phosphate synthesis, *trans*-hydroxylation of iP-type cytokinins, is catalysed by a P450 monooxygenase, CYP735A (EC 1.14.14.–) (Chen and Leisner 1984; Takei et al. 2004) (step 3 in Figure 23.2). In *A. thaliana*, there are two genes, *CYP735A1* and *CYP735A2* encoding the cytokinin *trans*-hydroxylase (Takei et al. 2004). In the rice genome, highly similar genes, *CYP735A3* and *CYP735A4*, also occur, suggesting that they encode enzymes involved in the biosynthesis of *trans*-zeatin (Nelson et al. 2004). *A. thaliana* CYP735As utilize isopentenyl-nucleotides but not the nucleoside and free-base forms. CYP735As preferentially catalyse isopentenyl-AMP or isopentenyl-ADP rather than isopentenyl-ATP (Takei et al. 2004). Arguably, one of the physiological roles of the cytokinin nucleotides is as a metabolic pool for substrates prior to side chain modifications (Hwang and Sakakibara 2006).

23.4 Formation of Cytokinin Bases

Cytokinins are effective long-distance signal molecules transported within the plant in xylem and phloem. The xylem transported form of cytokinins are predominantly ribosides, but the cytokinin receptors show a preference for the free base, therefore, conversion of nucleosides to bases is necessary (Hwang and Sakakibara 2006) (step 4 and 5 in Figure 23.2).

The cytokinin interconversion pathways are shown in Figure 23.3. In vegetative tissues of *A. thaliana*, the major route for the production of free base forms of cytokinins from ribonucleotides (step 3 in Figure 23.3) is mediated by cytokinin ribotide phosphoribohydrolase (named LONELY GUY proteins) (Sakakibara 2006). This conversion is very similar to that catalysed by AMP phosphoribohydrolase (EC 3.2.2.4), however, AMP is not hydrolysed by LONELY GUY (Kuroha et al. 2009). In addition to LONELY GUY, degradation enzymes of purine nucleotides and nucleosides also contribute to the production of the free base cytokinins. Chen and Kristopeit (1981) reported that two isoforms of 5′-nucleotidases from wheat germ have high affinity for isopentenyladenine riboside 5′-monophosphate. Interconversion of cytokinins are shown in Figure 23.3. Two purine salvage enzymes, adenine phosphoribosyltransferase and adenosine kinase, have been shown to contribute to the control of cytokinin activity as discussed below.

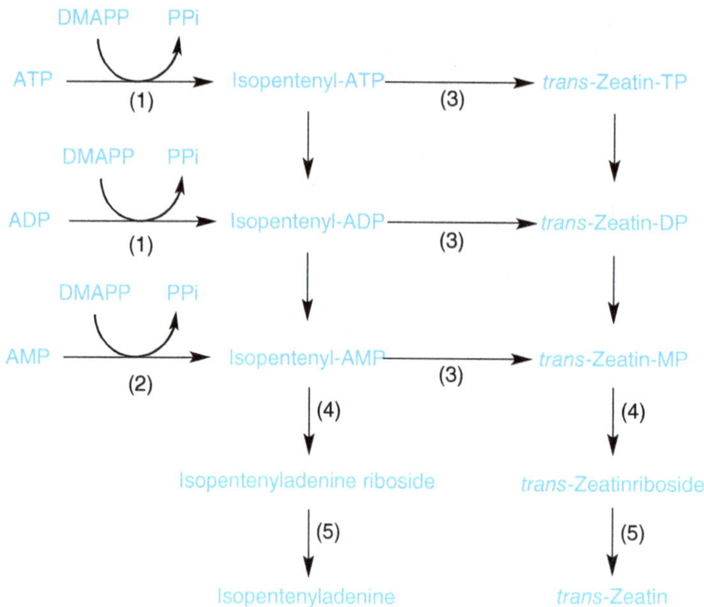

Figure 23.2 Current model of purine cytokinin biosynthesis pathway in plants. Interconversions among nucleotides, nucleosides, and nucleobases are catalysed by enzymes of the purine salvage pathway. (1) adenosine phosphate-isopentenyltransferase (IPT); (2) adenylate dimethylallyltransferase; (3) P450 monooxygenase, CYP735A; (4) 5′-nucleotidase; (5) riboside nucleosidase. DMAPP, dimethylallyl pyrophosphate. Source: Adapted from Hwang and Sakakibara (2006).

23.5 Effect of Nucleotide Enzymes in Cytokinins

23.5.1 Cytokinin Inactivation by Adenine Phosphoribosyltransferase

APT1 is the causal gene of the high-dose cytokinin-resistant mutants of *A. thaliana*. The APT1 protein is functionally predominant among the five members of the adenine phosphoribosyl transferase family (Zhang et al. 2013). Loss of APT1 activity in plants leads to excess accumulation of cytokinin bases, and evokes a myriad of cytokinin-regulated responses, such as delayed leaf senescence, anthocyanin accumulation, and downstream gene expression. These results indicated that APT1 is a key metabolic enzyme participating in cytokinin inactivation by phosphoribosylation.

Like adenine, exogenously applied cytokinin bases are rapidly salvaged to cytokinin ribotides by adenine phosphoribosyltransferase *in planta*. Thus, cytokinin bases transported to plant cells appear to be temporarily stored as cytokinin nucleotides, an inactive form, after which they are converted to the active form by the hydrolysing cytokinin activation enzymes noted above. The nucleotide formation by adenine phosphoribosyltransferase plays a key role in metabolizing exogenous cytokinins (Bishop et al. 2015).

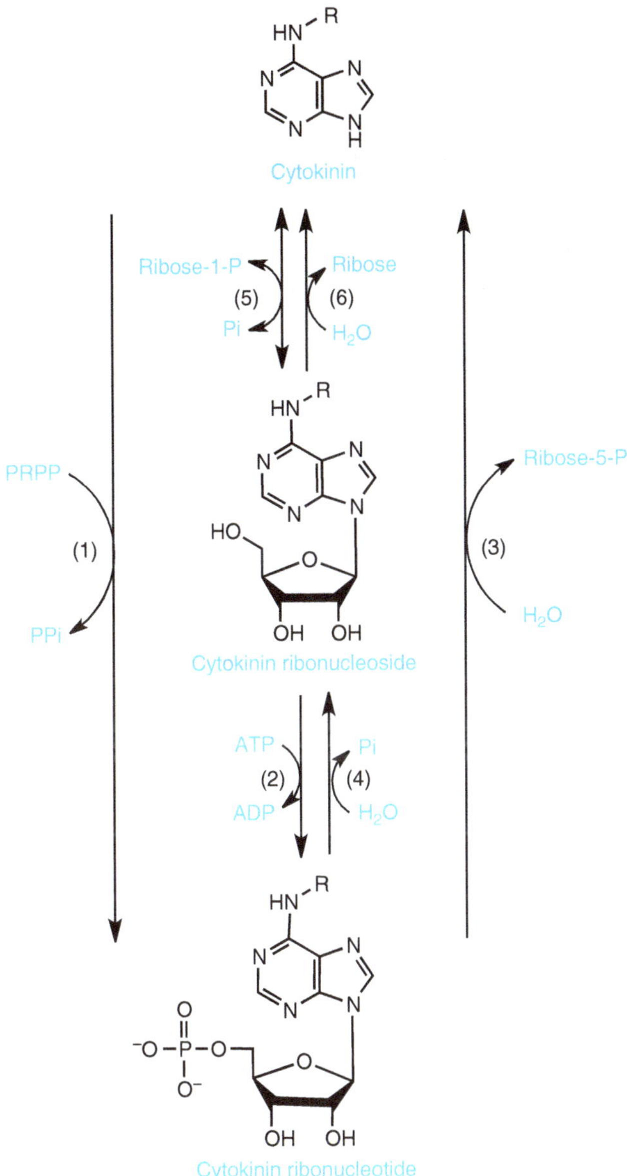

Figure 23.3 The cytokinin interconversion pathway. Numbers indicate the reaction steps catalysed by the enzymes shown below. EC numbers for cytokinin specific enzymes have not yet been allocated. (1) cytokinin base (adenine) phosphoribosyltransferase; (2) cytokinin riboside (adenosine) kinase; (3) cytokinin ribotide phosphoribohydrolase (LONELY GUY, LOG); (4) 50-nucleotidase; (5) cytokinin riboside (purine nucleoside) phosphorylase (StCKP1); and (6) cytokinin riboside (adenosine) nucleosidase. Cytokinin ribotide phosphoribohydrolase (LOG) is distinct from AMP phosphoribohydrolase (Kuroha et al. 2009). Cytokinin riboside phosphorylase (StCKP1) has a stronger affinity for iPR than for adenosine (Bromley et al. 2014). Source: Adapted from Bishop et al. (2015).

23.5.2 Homeostasis of Cytokinin by Adenosine Kinase

The contribution of adenosine kinase (ADK) to cytokinin interconversions was reported by Schoor et al. (2011). Use was made of lines of *A. thaliana* silenced in *adk* expression to understand the contribution of ADK activity to *in vivo* cytokinin metabolism. Both small interfering RNA- and artificial microRNA-mediating silencing of *adk* led to impaired root growth, small, crinkled rosette leaves, and reduced apical dominance. ADK-deficient roots and leaves exhibited irregular cell division. Root tips had an uneven arrangement of root cap cells, reduced meristem size, and enlarged cells in the elongation zone; rosette leaves exhibited decreased cell size but increased cell abundance. *In vivo* feeding of ^{3}H-isopentenyladenine riboside and ^{3}H-*trans*-zeatin riboside in ADK-deficient leaves showed a reduced incorporation of radioactivity into the isopentenyladenine ribotide and *trans*-zeatin ribotide, respectively. There were higher levels of cytokinin ribosides in *adk*-silenced plants. From these metabolic and phenotypic analyses of ADK deficient lines, it was concluded that ADK contributes to cytokinin homeostasis *in planta*.

23.5.3 Endodormancy of Potato and Purine Nucleoside Phosphorylase

Endodormancy of potato tubers is an internally imposed state of zero growth and minimal metabolism. Duration of endodormancy is shortened by chill treatment: a period of exposure to temperatures near to but not below freezing, mimicking the early stages of a cold season. Termination of endodormancy in response to the chill treatment allows buds of temperate perennials to resume growth whenever the cold season ends. It has been shown that the exit from endodormancy in potato tubers is linked to an increase in cytokinins and a higher cell sensitivity to cytokinins. These factors are amplified by chilling (Turnbull and Hanke 1985a,b).

Bromley et al. (2014) isolated a nucleoside phosphorylase superfamily protein (StCKP1) from potato shoot apices that interconverted cytokinin ribosides and free bases. The predicted polypeptide includes a cleavable signal peptide and motifs for purine nucleoside phosphorylase activity. N^6-(Δ^2-Isopentenyl)adenine (**3**), *trans*-zeatin (**2**), and *trans*-dihydrozeatin (**5**) were converted by StCKP1 into ribosides in the presence of ribose-1-phosphate. StCKP1 in the presence of Pi did not possess purine nucleoside activity and did not catalyse the reversed reaction, the conversion of cytokinin ribosides to cytokinin bases. Evidence was presented that StCKP1 is active in tubers as a negative regulator of cytokinins, prolonging dormancy by a chill-reversible mechanism.

23.6 New Purine-Related Plant Growth Regulators

Many people who cultivate manicured lawns know of the curse of 'fairy rings' and the rapid grass growth associated with them. The rings are arcs of fungus-stimulated growth. Choi et al. (2014) reported that the compounds responsible for the effects of the fairy-ring-forming fungus *Lepista sordida* are 2-azahypoxanthine (AHX) and its metabolite 2-aza-8-oxohypoxanthine (AOH). Treatment of rice (*Oryza sativa*), wheat (*Triticum aestivum*), and other crop species in greenhouse and/or field experiments with AHX and AOH stimulated growth in all instances (see Choi et al. 2018).

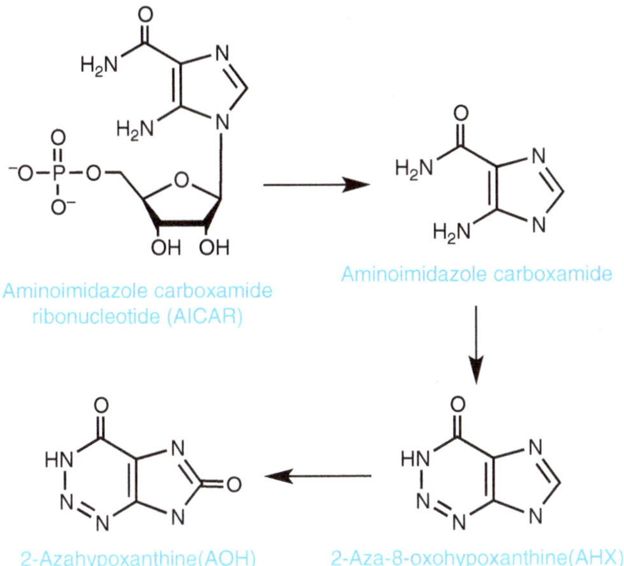

Figure 23.4 Biosynthesis of plant growth stimulators, 2-azahypoxanthine (AHX) and 2-aza-8-oxohypoxanthine (AOH) are formed from the intermediates of *de novo* biosynthesis of purine nucleotide, aminoimidazole carboxamide ribonucleotide (AICAR) via aminoimidazole carboxamide (AICA). Source: Based on the data of Choi et al. (2018).

Furthermore, it was shown that both AHX and AOH are produce naturally by higher plants, and are present in rice at levels similar to those of traditional plant hormones such as gibberellins. Data was obtained demonstrating that AHX is produced from the purine *de novo* biosynthesis pathway with aminoimidazole carboxamide ribonucleotide (AICAR) being converted to aminoimidazole carboxamide and then AHX and AOH (Figure 23.4).

In rice, as well as being metabolized to AOH, AHX is converted to several *N*-glycosides, which lack biological activity, suggesting that this is a means whereby rice, and potentially other plants, regulate the activity of AHX and AOH (Choi et al. 2018). Although this research is preliminary, it is tempting to draw parallels with Japanese studies on the fungus *Gibberella fujikuroi* and the bakanae disease of rice in the 1920s which led to the discovery of gibberellins as naturally occurring plant hormones (Phinney 1983). The effects of AOH and AHX on plant growth appear to resemble those of gibberellins, but their purine ring-like structure, which contains five rather than four nitrogen atoms (Figure 23.4), is closer to that of cytokinins than the diterpenoid gibberellins (see Bishop et al. 2015; Crozier et al. 2000).

23.7 Summary

Purine nucleotide, adenine-type cytokinins with an N^6-linked isoprenoid side chain are synthesized as ribonucleotides and occur in plant cells as nucleotides, nucleosides, and free bases. They have a major hormonal role and impact on many phases of plant growth and development. Recent research suggests that 2-azahypoxanthine and

2-aza-8-oxohypoxanthine, both of which have a purine ring-like structure but lack a side chain, may also promote the growth of higher plants.

References

Ashihara, H., Stasolla, C., Fujimura, T., and Crozier, A. (2018). Purine salvage in plants. *Phytochemistry* 147: 89–124.

Åstot, C., Dolezal, K., Nordström, A. et al. (2000). An alternative cytokinin biosynthesis pathway. *Proc. Natl. Acad. Sci. U. S. A.* 97: 14778–14783.

Bishop, G., Sakakibara, H., Seo, M., and Yamaguchi, S. (2015). Biosynthesis of hormones. In: *Biochemistry and Molecular Biology of Plants*, 2e (eds. B.B. Buchanan, W. Gruissem and R.L. Jones), 769–833. Chichester, UK: Wiley.

Bromley, J.R., Warnes, B.J., Newell, C.A. et al. (2014). A purine nucleoside phosphorylase in *Solanum tuberosum* L. (potato) with specificity for cytokinins contributes to the duration of tuber endodormancy. *Biochem. J.* 458: 225–237.

Chen, C.-M. and Kristopeit, S.M. (1981). Metabolism of cytokinin: dephosphorylation of cytokinin ribonucleotide by 5′-nucleotidases from wheat germ cytosol. *Plant Physiol.* 67: 494–498.

Chen, C.-M. and Leisner, S.M. (1984). Modification of cytokinins by cauliflower microsomal enzymes. *Plant Physiol.* 75: 442–446.

Chen, C.-M. and Melitz, D.K. (1979). Cytokinin biosynthesis in a cell-free system from cytokinin-autotrophic tobacco tissue cultures. *FEBS Lett.* 107: 15–20.

Choi, J.-H., Ohnishi, T., Yamakawa, Y. et al. (2014). The source of "Fairy rings": 2-azahypoxanthine and its metabolite found in a novel purine metabolic pathway in plants. *Angew. Chem. Int. Ed.* 53: 1552–1555.

Choi, J.-H., Wu, J., Sawada, A. et al. (2018). *N*-glucosides of fairy chemicals, 2-azahypoxanthine and 2-aza-8-oxohypoxanthine, in rice. *Org. Lett.* 20: 312–314.

Crozier, A., Kamita, Y., Bishop, G., and Yokota, T. (2000). Biosynthesis of hormones and elicitor molecules. In: *Biochemistry and Molecular Biology of Plants* (eds. B.B. Buchanan, W. Gruissen and R.L. Jones), 850–829. Rockville, MD: American Society of Plant Physiologists.

Hwang, I. and Sakakibara, H. (2006). Cytokinin biosynthesis and perception. *Physiol. Plant.* 126: 528–538.

Kakimoto, T. (2001). Identification of plant cytokinin biosynthetic enzymes as dimethylallyl diphosphate: ATP/ADP isopentenyltransferases. *Plant Cell Physiol.* 42: 677–685.

Kurakawa, T., Ueda, N., Maekawa, M. et al. (2007). Direct control of shoot meristem activity by a cytokinin – activating enzyme. *Nature* 445: 652–655.

Kuroha, T., Tokunaga, H., Kojima, M. et al. (2009). Functional analyses of LONELY GUY cytokinin – activating enzymes reveal the importance of the direct activation pathway in arabidopsis. *Plant Cell* 21: 3152–3169.

Kwade, Z., Świątek, A., Azmi, A. et al. (2005). Identification of four adenosine kinase isoforms in tobacco BY-2 cells and their putative role in the cell cycle-regulated cytokinin metabolism. *J. Biol. Chem.* 280: 17512–17519.

Nelson, D.R., Schuler, M.A., Paquette, S.M. et al. (2004). Comparative genomics of rice and *Arabidopsis*. Analysis of 727 cytochrome P450 genes and pseudogenes from a monocot and a dicot. *Plant Physiol.* 135: 756–772.

Phinney, B.O. (1983). The history of gibberellins. In: *The Biochemistry and Physiology of Gibberellins*, vol. 1 (ed. A. Crozier), 1–18. New York, NY: Praeger.

Romanov, G.A., Lomin, S.N., and Schmülling, T. (2006). Biochemical characteristics and ligand-binding properties of *Arabidopsis* cytokinin receptor AHK3 compared to CRE1/AHK4 as revealed by a direct binding assay. *J. Exp. Bot.* 57: 4051–4058.

Sakakibara, H. (2006). Cytokinins: activity, biosynthesis, and translocation. *Annu. Rev. Plant Biol.* 57: 431–449.

Sakakibara, H. (2010). Cytokinin biosynthesis and metabolism. In: *Plant Hormones* (ed. P.J. Davies), 95–114. Springer.

Sakakibara, H. and Takei, K. (2002). Identification of cytokinin biosynthesis genes in *Arabidopsis*: a breakthrough for understanding the metabolic pathway and the regulation in higher plants. *J. Plant Growth Regul.* 21: 17–23.

Sakano, Y., Okada, Y., Matsunaga, A. et al. (2004). Molecular cloning, expression, and characterization of adenylate isopentenyltransferase from hop (*Humulus lupulus* L.). *Phytochemistry* 65: 2439–2446.

Schoor, S., Farrow, S., Blaschke, H. et al. (2011). Adenosine kinase contributes to cytokinin interconversion in *Arabidopsis. Plant Physiol.* 157: 659–672.

Takei, K., Sakakibara, H., and Sugiyama, T. (2001). Identification of genes encoding adenylate isopentenyltransferase, a cytokinin biosynthesis enzyme, in *Arabidopsis thaliana. J. Biol. Chem.* 276: 26405–26410.

Takei, K., Yamaya, T., and Sakakibara, H. (2004). *Arabidopsis* CYP735A1 and CYP735A2 encode cytokinin hydroxylases that catalyze the biosynthesis of trans-Zeatin. *J. Biol. Chem.* 279: 41866–41872.

Turnbull, C.G.N. and Hanke, D.E. (1985a). The control of bud dormancy in potato tubers. *Planta* 165: 359–365.

Turnbull, C.G.N. and Hanke, D.E. (1985b). The control of bud dormancy in potato tubers. Measurement of the seasonal pattern of changing concentrations of zeatin-cytokinins. *Planta* 165: 366–376.

Zhang, X., Chen, Y., Lin, X. et al. (2013). Adenine phosphoribosyl transferase 1 is a key enzyme catalyzing cytokinin conversion from nucleobases to nucleotides in arabidopsis. *Mol. Plant* 6: 1661–1672.

Part IX

Dietary Plant Alkaloids, Their Bioavailability, and Potential Impact on Human Health

24

Bioavailability and Potential Impact on Human Health of Caffeine, Theobromine, and Trigonelline

24.1 Caffeine

24.1.1 Dietary Caffeine

Although consumption of carbonated beverages containing added caffeine (**1**) is increasing, for most consumers coffee and tea remain the principal dietary source of the purine alkaloid (Heckman et al. 2010; Mahoney et al. 2018). Coffee is the largest cash commodity in the United States, besides oil. Drinking an estimated 1.7 cups per capita daily, US consumption is around 400 million cups of coffee per day. In the United Kingdom, around 50 million cups of tea are drunk daily.

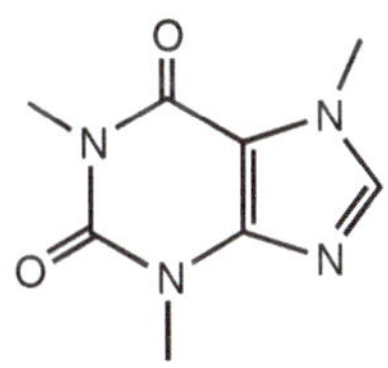

Caffeine (**1**)

The concentration of caffeine in green coffee beans varies between 0.9–1.3 and 1.5–2.5% dry matter for arabica and robusta coffees, respectively (Clifford et al. 2018). The caffeine content of beans is not markedly reduced by roasting, although losses, presumably due to sublimation, may occur if the heating process is carried out at higher temperatures (Crozier et al. 2012). Heckman et al. (2010) reported that caffeine levels in a cup of coffee vary between 50 and 100 mg. However, substantial variations occur within different coffee brews. The caffeine content of 20 specialty coffees purchased at different coffee shops in the United States varied from 58 to 259 mg per cup (McCusker et al. 2002). Similarly, analysis of the caffeine levels in commercial espresso coffees in the UK (Table 24.1), Spain, and Italy found that the amounts ranged between 51 and 322 mg per cup (Crozier et al. 2012; Ludwig et al. 2014). These findings show that depending on barista techniques, the amount of caffeine per serving can vary more than sixfold, which makes it very difficult to accurately assess caffeine intake on a population basis.

Plant Nucleotide Metabolism – Biosynthesis, Degradation, and Alkaloid Formation, First Edition.
Hiroshi Ashihara, Iziar A. Ludwig and Alan Crozier.

Table 24.1 Levels of caffeine in espresso coffees purchased from 20 coffee shops in Glasgow, United Kingdom.

Coffee shop	Volume of coffee (ml)	Caffeine per serving (mg)
Patisserie Francoise	52	322
S'mug	32	173
Costa Coffee	25	157
Little Italy	23	129
Paperino's	50	205
Peckhams	70	140
Chapter 1	26	140
University Cafe	49	230
Baguette Express	45	140
Kember & Jones	43	90
Heart Buchanan	24	156
Jellyhill	63	151
Coffee @ 491	49	98
Beanscene	48	77
Tinderbox	25	75
Café Cinnamon	59	242
Crepe á Croissant	34	95
Morton's	35	73
Antipasti	36	72
Starbucks	27	51
Range	**23–70**	**50–322**
Median	**43**	**140**

Source: After Crozier et al. (2012).

24.1.2 Bioavailability and Bioactivity of Caffeine

Caffeine is rapidly and almost completely absorbed in the stomach and small intestine and distributed to all tissues, including the brain. Peak plasma concentration (C_{max}) is reached within 60 minutes after intake and the half-life of caffeine is ~5 hours (Lang et al. 2013a). Caffeine metabolism occurs primarily in the liver by the cytochrome P450 isoform CYP1A1. The first metabolic step is the demethylation of caffeine leading to paraxanthine (1,7-dimethylxanthine), theobromine (3,7-dimethylxanthine), and theophylline (1,3-dimethylxanthine). These metabolites are then further demethylated in the liver to the corresponding monomethylxanthines before being oxidized to methyl uric acids and other products as illustrated in Figure 24.1.

Once absorbed, caffeine exhibits numerous and well-studied physiological effects (Heckman et al. 2010) and exerts most of them through the antagonism of the A_1 and A_2 subtypes of the adenosine receptor. The A_1 receptors are located in all parts of the brain with higher concentrations in the hippocampus, cerebral and cerebellar cortex, and certain thalamic nuclei, whereas the A_2 subtypes are located in the dopamine-rich

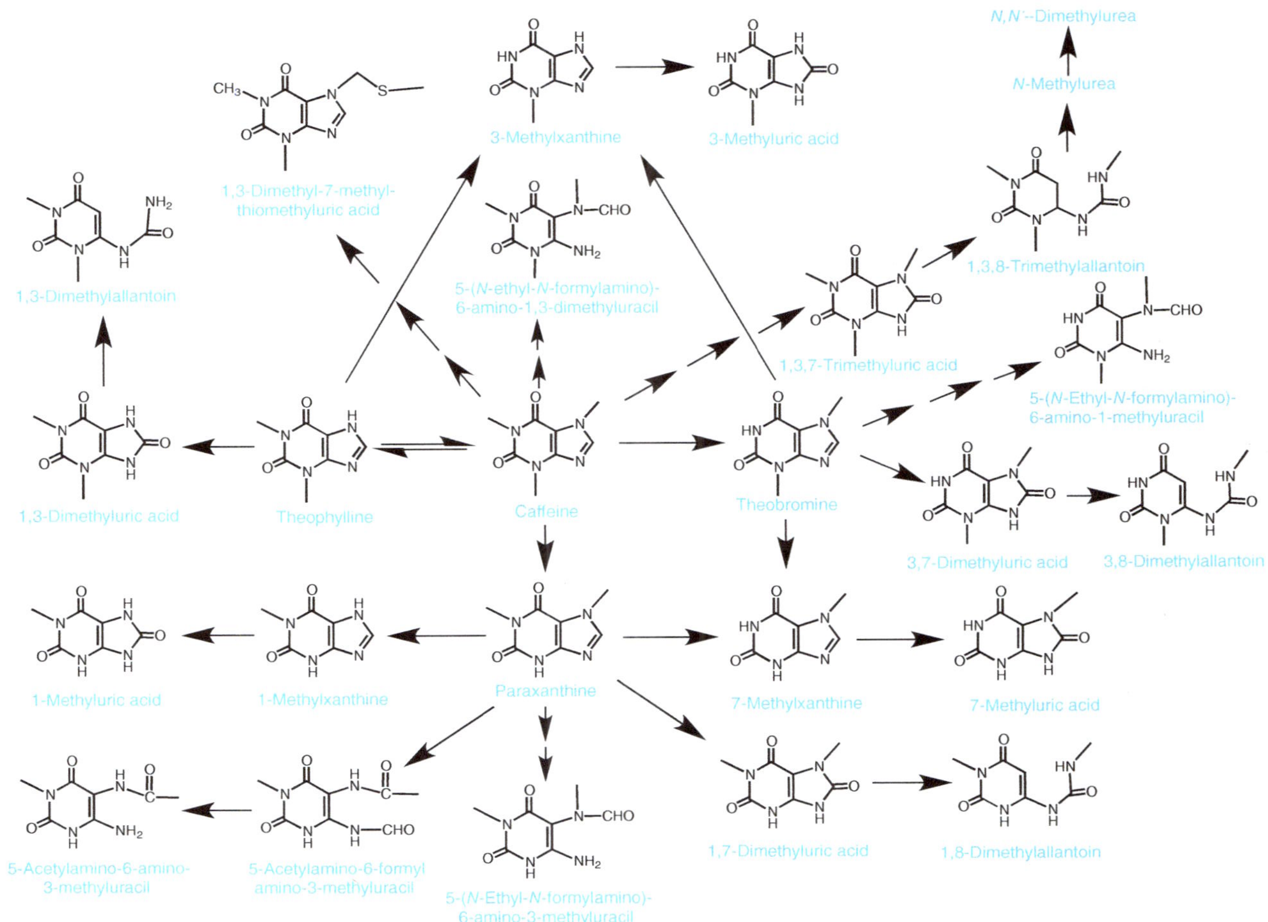

Figure 24.1 Pathways for human metabolism of caffeine to di- and monomethylxanthines, the corresponding tri-, di-, and monomethyl uric acids and three uracil derivatives by opening of the five membered ring. For further details and routes of caffeine metabolism in humans and selected mammals see Arnaud (2011).

regions of the brain (Fredholm et al. 1999). The antagonization of the action of adenosine elevates the level of dopamine which is responsible for many of the central nervous system stimulant and addictive properties of caffeine. Another mechanism of action of caffeine is the synergistic interaction with adrenalin and noradrenalin, main neurotransmitters for the sympathetic nervous system (Lean et al. 2012). The stimulatory effects of caffeine include enhanced perception, an increased capacity to remain awake for longer periods, and reduced fatigue. It also helps reduce symptoms associated with Parkinson's disease, and has been suggested as a potential contributor to reducing risk factors involved in metabolic syndrome, such as obesity.

Parkinson's disease leads to a progressive loss of dopaminergic neurons of the substantia nigra in the midbrain, which results in the deterioration of gross and small motor skills, and tremors. Caffeine improves the performance of the dopaminergic system by blocking the A_2 adenosine receptors and thereby stimulating the release of dopamine (Trevitt et al. 2009), and via this mechanism, is thought to reduce symptoms associated with Parkinson's disease, such as tremor, and freezing of gait, in animal models and humans (Trevitt et al. 2009; Kitagawa et al. 2007).

Caffeine consumption increases metabolic rate, energy expenditure, lipid oxidation, and lipolytic and thermogenic activities, and so has potential as an aid in weight loss and reducing the overall risk of developing metabolic syndrome (Heckman et al. 2010). Consumption of 300 mg caffeine per day has been found to increase energy expenditure by approximately 79 kcal d^{-1} (Rudelle et al. 2007). Although this increase is modest, the authors of this study indicate that such modifications are sufficient to prevent weight gain and, when consumed regularly as part of a healthy diet and exercise regime, caffeine containing beverages may provide benefits for weight control.

Caffeine might contribute to the overall antioxidant capacity of coffee. Devasagayam et al. (1996) studied caffeine as a protector agent of cell membranes against oxidative damage. The mM concentrations used in this study, however, were more than 40-fold higher than the average blood concentrations observed after consuming coffee in normal dietary amounts (MacDonald et al. 1991). Lee (2000) measured the protective effect of caffeine and its metabolites on LDL peroxidation at physiological, μM concentrations. Neither caffeine nor its dimethylxanthine metabolites exhibited antioxidant activity while, in contrast, 1-methylxanthine (**2**) and 1-methyluric acid (**3**), were highly effective antioxidants *in vitro* being able to prevent LDL oxidation.

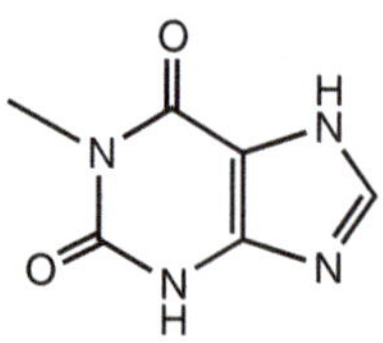

1-Methylxanthine (**2**)

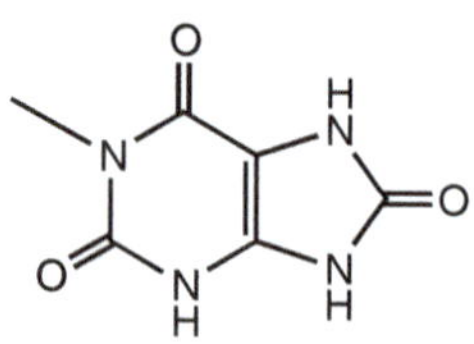

1-Methyluric acid (**3**)

There are also negative effects linked to caffeine intake that must be considered. The activation of adenosine receptors by caffeine has negative consequences on cardiovascular health, such as increased heart rate and blood pressure (Riksen et al. 2011). Caffeine in excess, can lead to a state of excitement and anxiety including adverse reactions like tachycardia, headache, palpitations, insomnia, restlessness, nervousness, and tremor. Dose-response varies between individuals, and even a single cup may cause sleeplessness with a racing mind for some people, while for others drinking 10 times this amount the effects can still be pleasant and not interfere with sleep. The substantial differences in the effects of caffeine between individuals are at least partly the result of genetic variations of susceptibility (Rogers et al. 2010). In addition, regular coffee drinkers develop a certain tolerance to the adverse impacts of caffeine, especially on adenosine antagonism and sleeplessness. The development of tolerance may have unpleasant consequences. People who drink coffee on a regular basis and decide to interrupt consumption can experience withdrawal effects. The absence of the adenosine antagonist may cause headaches due to intracranial vasodilatation (Guieu et al. 1998), and this coupled with reduced catecholamine and serotonin activity may lead to fatigue, mild depression and impaired cognitive function (Juliano et al. 2009; Juliano and Griffith 2004).

The typical 2.5–4.5 hours elimination half-life ($T_{½}$) of caffeine in adults can be extended up to 30 hours in the body by women taking an oral contraceptive, pregnant women, the developing fetus, young children, and those with liver disease. These groups are, therefore, much more susceptible to the effects of caffeine toxicity. Acute intoxication is characterized by dehydration, tachycardia, extreme anxiety, hallucinations, flushing, and delirium. Further fatal consequences are epileptic seizure, cardiac arrhythmia and ultimately ventricular fibrillation.

The fatal dose for humans depends on individual sensitivity but in rare instances can be as little as 2 g (Lean et al. 2012). In general, fatal caffeine overdose involves the ingestion of caffeine-containing medications rather than caffeinated foods or beverages (Higdon and Frei 2006). However, some cases of death after ingestion of high amounts of energy drinks and caffeinated energy mints have been attributed to caffeine (BBC 2001; Metro 2008; Laville 2000; BBC 2013). Although the amounts present in coffee make it unlikely to be a source for caffeine intoxication, isolated cases of caffeine toxicity have been attributed to excessive coffee intake (BBC 2007).

Pregnant women have a slower caffeine metabolism with an extended half-life of up to 10 hours. In addition, caffeine freely crosses the placenta and enzymes involved in caffeine metabolism are lacking in both the placenta and fetus (Heckman et al. 2010). The effect of caffeine consumption on pregnancy complications, such as spontaneous abortion and impaired fetal growth, has been studied extensively but epidemiological data are conflicting. While some studies did not detect a link between caffeine intake and the risk of spontaneous abortion (Fenster et al. 1997; Mills et al. 1993; Savitz et al. 2008), other investigations found a significant association at intake levels of >200 mg d^{-1} (Giannelli et al. 2003; Rasch 2003; Wen et al. 2001; Weng et al. 2008). Maternal caffeine intake has also been associated with a higher risk of low birth weight in some (Sengpiel et al. 2013; Vlajinac et al. 1997) but not all epidemiological studies (Bech et al. 2007; Fortier et al. 1993). Although the threshold at which the risk is significantly higher is not well characterized, results from the CARE study group suggest a reduced fetal growth for women consuming >100 mg d^{-1} (Boylan et al. 2008). Sengpiel et al. (2013) reported similar findings with a decrease in birth weight of 21–28 g for each daily 100 mg of maternal caffeine intake.

Based on a literature review concerning the effects of caffeine on health, the European Food Safety Agency (2015) estimated safe levels of caffeine consumption in the population. They concluded that habitual caffeine consumption up to 200 mg per day by pregnant women does not give rise to safety concerns for the fetus. This is in line with earlier advice from the UK Food Standards Agency that pregnant women should restrict caffeine to below 200 mg per day (Wadge 2009).

24.2 Theobromine

24.2.1 Interactions with Flavan-3-ols

The main source of theobromine (**4**) in the diet is cocoa which also contains much smaller amounts of caffeine and substantial quantities of flavan-3-ols, the main components of which are the monomer (−)-epicatechin (**5**) along with oligomeric and polymeric proanthocyanidins (Crozier et al. 2006). Consumption of cocoa results in improvements in endothelial function and other markers of vascular function, including blood pressure, pulse wave velocity, and circulating angiogenic cells (Heiss et al. 2003, 2010, 2015). These effects have been attributed to (−)-epicatechin rather the proanthocyanidins (Rodrigues-Mateos et al. 2018; Schroeter et al. 2006).

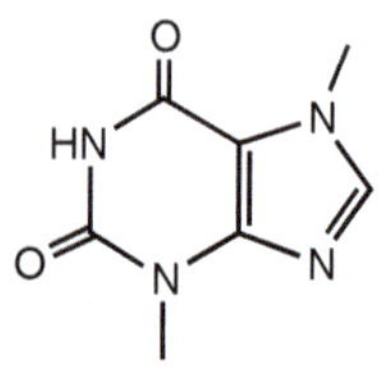

Theobromine (**4**)

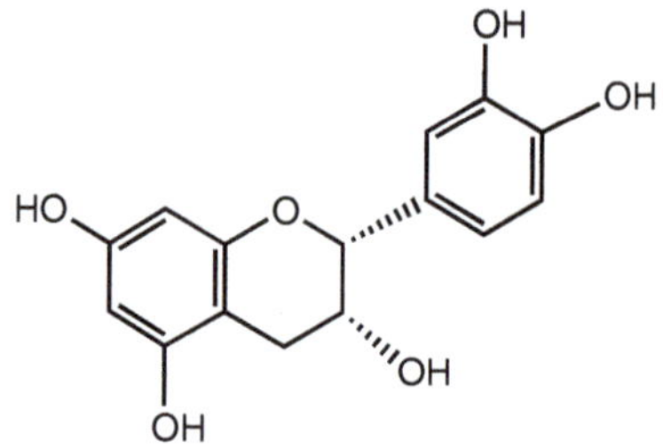

(−)-Epicatechin (**5**)

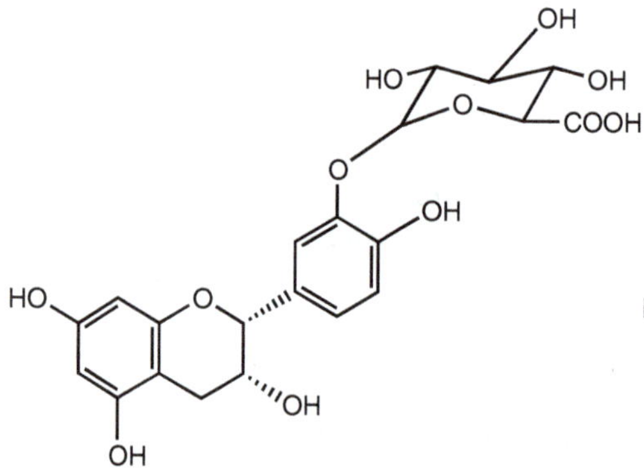

(−)-Epicatechin-3′-*O*-glucuronide (**6**)

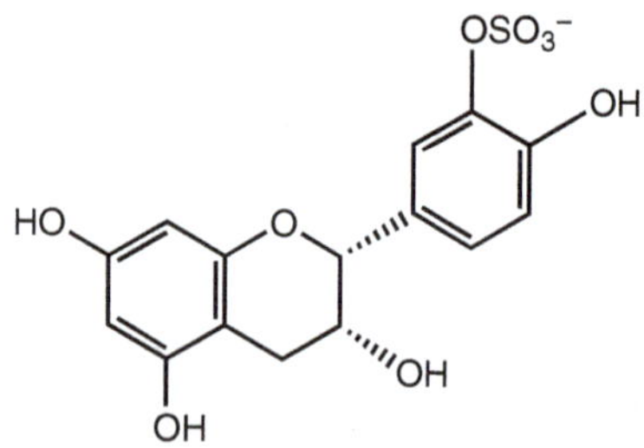

(−)-Epicatechin-3′-sulfate (**7**)

In a study by Sansone et al. (2017) the impact of ingestion of two cocoa beverages containing (i) 112 mg of (−)-epicatechin, 111 mg of theobromine and 11 mg of caffeine, and ii) 112 mg of (−)-epicatechin, 1.6 mg of theobromine and 1.6 mg of caffeine was compared. The effects of on vascular status was assessed by monitoring a number of biomarkers including flow-mediated vasodilation (FMD). The data obtained showed that FMD, which peaked 2 hours after cocoa intake, was

Figure 24.2 Time course of (a) flow mediated dilation (FMD) and (b) structurally-related (−)-epicatechin metabolites (SREMs) in plasma after the ingestion of cocoa containing (i) 112 mg of (−)-epicatechin, 111 mg of theobromine, and 11 mg of caffeine (CF + MX); and (ii) 112 mg of (−)-epicatechin, 1.6 mg of theobromine, and 1.6 mg of caffeine (CF) by human volunteers. Symbols are mean values and error bars are standard deviation ($n = 5$). Source: Sansone et al. (2017).

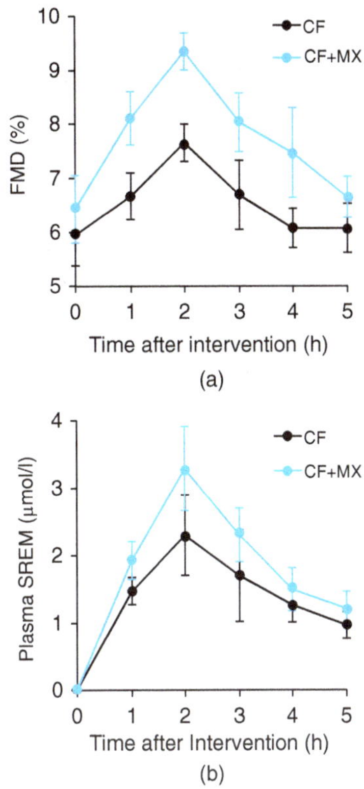

enhanced significantly by the presence of the higher level of theobromine in the ingested drink and this was associated with a parallel and significant increase in the concentration of structurally-related (−)-epicatechin metabolites, such as (−)-epicatechin-3′-*O*-glucuronide (**6**) and (−)-epicatechin-3′-sulfate (7) (Borges et al. 2018), in the circulatory system (Figure 24.2). This is in accordance with results obtained in a study with rats which showed that theobromine intake increases the absorption of (−)-epicatechin (Yamamoto et al. 2014).

Potential mechanisms by which theobromine may increase (−)-epicatechin absorption and thereby enhance vascular function are unknown but may include specific effects on gastrointestinal physiology (e.g. motility and/or secretion, blood flow, gastric pH) or even modulation of transporters regulating the efflux of flavan-3-ols in the gut. However, more work is needed to understand the theobromine/(−)-epicatechin interactions and the potential molecular events that mediate the synergism on vascular effects between these dietary compounds.

As noted above, the purine alkaloids in cocoa are comprised predominantly of theobromine with ~10-fold lower amounts of caffeine. However, in other dietary sources of flavan-3-ols, such as tea, caffeine predominates (Crozier et al. 2006). In this context, it is pertinent to ask whether or not the findings of Sansome et al. (2017) can be extrapolated to situations in which flavan-3-ols are co-ingested with substantial amounts caffeine. Previous studies suggest that theobromine and caffeine may have different potencies to mediate a given effect or even mediate completely different activities (Mitchell et al. 2011; Usmani et al. 2005). Blood pressure lowering and blood pressure increasing effects have, respectively, been reported with theobromine and caffeine. In healthy volunteers, acute supplementation with 700 mg theobromine decreased blood pressure, whereas 120 mg of caffeine had the opposite effect (Mitchell et al. 2011). Other studies suggest that caffeine may adversely affect arterial stiffness (Karatzis et al. 2005; Vlachopoulos et al. 2003), whereas consumption of theobromine-containing chocolate did not show any effect on arterial stiffness (Vlachopoulos et al. 2005, 2007). Thus, the extrapolation of the findings of Sansome et al. (2017) regarding the synergism between purine alkaloids and flavan-3-ols to foods other than cocoa, such as green and black tea, requires further investigation.

24.2.2 Toxicity of Theobromine

A number of species, including dogs, foxes, badgers, horses, and the New Zealand kea, have very limited capacity to metabolize theobromine, which tends to accumulate and becomes toxic with potentially fatal consequences (Jannsson et al. 2001; Gartrell and Reid 2007). Depending on the size of dog, as little as 100–200 mg of theobromine, the content of 200 g of milk chocolate or 20 g of dark chocolate, can be fatal. Caffeine consumption by horses leads to theophylline (**8**) accumulation and hyperactivity (Vickroy et al. 2008).

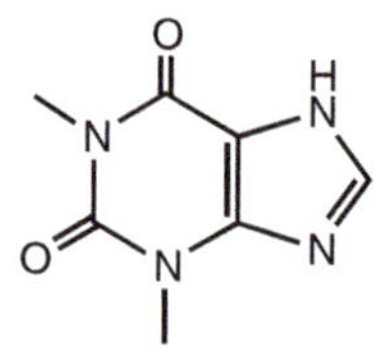

Theophylline (**8**)

24.3 Trigonelline

24.3.1 Dietary Trigonelline

Shoots and seeds of fenugreek (*Trigonella foenum-graecum*), used as a vegetable and a spice in South and Central Asia, contain respectively, ~0.12% and ~4.2%, trigonelline (**9**) (Radwan and Kokate 1980). Taguchi et al. (1985) surveyed trigonelline content in 104 foods and it was detected, not only in plant products but in fish, shellfish, and the liver of several animals. The highest concentration, 0.23%, occurred in turban shell (*Turbo cornutus*). Trigonelline containing plants included pea seeds (0.09%), celery (0.07%), and green asparagus (0.05%). Corol et al. (2012) examined the trigonelline content of ~200 cereal genotypes. They found low levels in milled cereal grains with the highest concentration being detected in oats at 0.003%. However, the most important commercially available trigonelline-rich dietary product is coffee. Trigonelline accumulates in the green coffee beans in similar concentrations as caffeine at ~1–3% of d.w. (Clifford 1985; Mazzafera 1991; Ky et al. 2001; Koshiro et al. 2006). Mazzafera (1991) surveyed the trigonelline content of 16 cultivars of *Coffea arabica* and found values varying from 1.5% (cv. Laurina) to 2.9% (cv. Semperflorens). Trigonelline in *Coffea canephora* and other *Coffea* species was also analysed and an especially high level, 3.1%, was detected in *C. canephora* cv. Guarini.

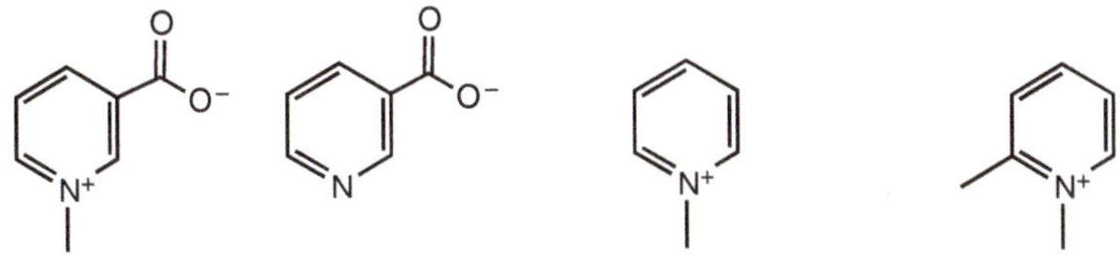

Trigonelline (**9**) Nicotinate (**10**) 1-Methylpyridinium (**11**) 1,2-Dimethylpyridinium (**12**)

During the roasting of green coffee beans trigonelline is partially degraded to nicotinate (**10**) (aka vitamin B_3 and niacin) and several pyridine derivatives, including 1-methylpyridinium (**11**) and 1,2-dimethylpyridinium (**12**). (Viani and Horman 1974; Stadler et al. 2002a,b). Depending on coffee variety and degree of roast trigonelline levels in roasted beans vary between 0.2 and 0.9 mg g^{-1} (Casal et al. 2000; Perrone et al. 2008). The amount in coffee brew depends largely on the coffee making procedure and range between 40 and 110 mg per serving (Farah 2012; Maeztu et al. 2001; Lang et al. 2013b). Coffee has been reported to provide 1–3 mg of nicotinate per cup (Higdon and Frei 2006), contributing to 6–18% of the recommended daily UK vitamin B_3 allowance of 16 mg d^{-1} for adult men (Lean et al. 2012).

24.3.2 Bioavailability and Bioactivity of Trigonelline

Several studies have investigated the bioavailability of trigonelline in humans and animals. Results obtained from a bioavailability study of trigonelline with humans who consumed one cup of coffee, containing 808 μmol of trigonelline, showed that the pyridine alkaloid attained a C_{max} of ~6 μM after ~3 hours and had a $T_{½}$ of ~5 hours (Lang et al. 2010). Urinary excretion of trigonelline 0–8 hours post-ingestion of coffee was 57.4% of intake for males and 46.2% for female volunteers, showing that substantial amounts had passed through the circulatory without being subjected to phase II metabolism (Table 24.2). Trigonelline was detected 15 minutes after ingestion of coffee, pointing towards absorption occurring initially in the stomach. The coffee also contained 172 mol of 1-methylpyridinium (**11**) which reached a C_{max} of ~0.8 μM with a more rapid T_{max} than that of trigonelline, ranging from 1.0–1.7 hours. The $T_{½}$ of 1-methylpyridinium was also more rapid than that of trigonelline at ~2.2 hours while urinary excretion of 1-methylpyridinium by males was more than 60% of intake both for males and females (Lang et al. 2010) (Table 24.2). This level of urinary excretion indicates that both trigonelline and 1-methypyridinium pass through the body without undergoing substantial phase II metabolism. In keeping with this possibility, a study on the absorption of trigonelline from the small intestine of germ-free and specific pathogen-free rats demonstrated that the pyridine alkaloid is not degraded by the gastrointestinal microflora and that most is absorbed from the small intestine (Yuyama 1999).

Regarding potential biological activities, trigonelline has been shown to possess, hypoglycaemic, neuroprotective, anti-invasive, estrogenic, and antibacterial activities. After a four week treatment of diabetic rats with trigonelline (50 mg day^{-1} kg^{-1} b.w.), blood glucose, total cholesterol, and triglycerides were significantly reduced.

Table 24.2 Plasma pharmacokinetic parameters of trigonelline and 1-methylpyridinium circulating in plasma of male and female volunteers after the ingestion of 350 ml of filter coffee containing 808 μmol (110 mg) of trigonelline, and 172 μmol (16 mg) of 1-methylpyridinium contained in a 200 ml serving of instant coffee.

Pyrimidine derivatives	C_{max} (μM)		T_{max} (h)		$T_{1/2}$ (h)		0–8 h urinary excretion (% of intake)	
	Male	Female	Male	Female	Male	Female	Male	Female
Trigonelline	5.5 ± 0.7	6.5 ± 0.1	2.3 ± 1.0	3.2 ± 1.0	4.6 ± 1.6	5.5 ± 2.1	57.4 ± 4.0	46.2 ± 3.4
1-Methylpyridinium	0.8 ± 0.2	0.8 ± 0.1	1.0 ± 0.4	1.7 ± 0.5	2.3 ± 0.5	2.1 ± 0.3	69.1 ± 3.8	61.9 ± 7.4

Data expressed as mean values ± SD (males $n = 7$; females n = 6). C_{max} – peak plasma concentration; T_{max} – time after ingestion to reach C_{max}; $T_{1/2}$ – elimination half-life.
Source: Based on the data of Lang et al. (2010).

Furthermore, there were altered insulin levels, insulin sensitivity index, and insulin content in the pancreas, all showing values similar to a non-diabetic control group (Zhou et al. 2011). Trigonelline regulates key enzymes of glucose and lipid metabolism, such as glucokinase, glucose-6-phosphatase, fatty acid synthase, and carnitine palmitoyl transferase in diabetic rats (Yoshinari et al. 2009). In a randomized crossover study with healthy, male humans a 500 mg dose of trigonelline led to significantly lower glucose and insulin levels 15 minutes after an oral glucose load compared to a placebo (van Dijk et al. 2009).

Trigonelline at 30 and 100 μM prevented, in a dose dependent manner, dendritic and axonal atrophy in rat cortical neurons induced by the neurotoxin amyloid-β-peptide [Aβ (25–35)]. Moreover, orally administered trigonelline (500 mg kg^{-1} b.w.) improved memory in mice treated with Aβ (25–35) (Tohoda et al. 2005). Gaur et al. (2013) studied the neuroprotective effects of a fenugreek extract, containing 82% trigonelline, in rodent models of Parkinson's disease. There was a significant increase in the number of ipsilateral rotations in 6-hydroxydopamine-induced unilateral lesioned rats that orally ingested the extract (30 mg kg^{-1} b.w.) compared to the control group. The extract also induced a significant reversal of motor dysfunction caused by 1-methyl-4-phenyl-1,2,3,6-tetrahydropyridine-induced lesions in C57BL/6 mice in pre-treatment (1 hour) but not in post-treatment (1 hour) experiments.

Cell invasion is one of the key steps in the development of cancer metastasis (Al-Mehdi et al. 2000). The effects of trigonelline on the proliferation and invasion of cancer cells were studied using rat ascites hepatoma cell line AH109A. At concentrations of 2.5–4.0 μM trigonelline inhibited the invasion of hepatoma cells in a dose responsive manner but did not impact on proliferation. Anti-invasive activity was also determined by incubating AH109A cells with serum from rats fed 100 μmol trigonelline per 100 g b.w. by gavage. Serum obtained 2 hours after intake most strongly inhibited invasion of AH109A cells and significant inhibition was also measured in serum collected 3 hours and 6 hours after intake. These results suggest that trigonelline, and/or its metabolites, absorbed from the gastrointestinal tract retain their effectiveness in the bloodstream. Both trigonelline and trigonelline-loaded rat serum suppressed reactive-oxygen-species (ROS)-potentiated invasive activity of pre-treated cells but did not scavenge intracellular peroxides. This indicates that a mechanism other than a direct antioxidant effect is involved in trigonelline-mediated suppressive effects on ROS-potentiated invasion (Hirakawa et al. 2005).

Trigonelline stimulates proliferation of MCF-7 breast cancer cells and significantly enhances cell growth *in vitro* at concentrations as low as 100 pM. This effect was attributed to the activation of the oestrogen receptor which classifies trigonelline as a novel phytoestrogen. Soy phytoestrogens play a dual role in the development and progression of breast cancer and may exhibit estrogenic and antiestrogenic effects (Bouker and Hilakivi-Clarke 2000). It will be interesting to determine if trigonelline exerts similar effects.

The transcription factor Nrf2 plays a key role in cancer development and chemoresistance as its overexpression confers stability to cells and it is thought to be an adaptive cell reaction to chemical and oxidative insults. This makes NrF2 activation a potential target in anticancer therapy (Singh et al. 2010) as it leads to protection against DNA damage and a reduction in tumorigenesis (Osbum and Kensler 2008) but NrF2 activation is also a means by which cancer cells resist chemotherapy. Trigonelline is an effective inhibitor of Nrf2 but in the process it has been reported to increase the sensitivity of chemoresistant pancreatic and colon cell lines to anticancer drugs (Arlt et al. 2013).

Trigonelline impacts on the adhesive properties of *Streptococcus mutans*, the major causative agent of dental caries in humans, by reducing the ability of the bacterium to adsorb onto saliva-coated hydroxyapatite beads. Trigonelline is also a potential antimicrobial agent against the highly invasive pathogen *Salmonella enterica* (Almeida et al. 2006).

24.4 Summary

The balance between beneficial and negative impacts of caffeine on well-being, depend largely on the individual susceptibility to the purine alkaloid. Like caffeine, there are reports of both positive and negative effects of theobromine on human health. Theobromine can however be toxic to animals, including dogs. Trigonelline has therapeutic potential as a hypoglycaemic and neuroprotective agent. It also has anticariogenic effects.

References

Al-Mehdi, B., Tozawa, K., Fisher, K. et al. (2000). Intravascular origin of metastasis from the proliferation of endothelium-attached tumor cells: a new model for metastasis. *Nat. Med.* 6: 100–102.

Almeida, A.A.P., Farah, A., Silva, D.A. et al. (2006). Antibacterial activity of coffee extracts and selected coffee chemical compounds against enterobacteria. *J. Agric. Food Chem.* 54: 8738–8743.

Arlt, A., Sebens, S., Krebs, S. et al. (2013). Inhibition of the Nrf2 transcription factor by the alkaloid trigonelline renders pancreatic cancer cells more susceptible to apoptosis through decreased proteasomal gene expression and proteasome activity. *Oncogene* 32: 4825–4835.

Arnaud, M.J. (2011). Pharmacokinetics and metabolism of natural methylxanthines in animal and man. In: *Methylxanthines (Handbook of Experimental Pharmacology)*, vol. 200 (ed. M.J. Arnaud), 33–91. Heidelberg: Springer.

B.B.C., 2001. Red Bull in suspected link to deaths. (12 July 2001). http://news.bbc.co.uk/1/hi/1435409.stm.

B.B.C., 2007. Girl overdoses on espresso coffee. (13 August 2007). http://news.bbc.co.uk/1/hi/england/wear/6944026.stm.

B.B.C., 2013. Man died after eating Hero Instant Energy Mints, says coroner. (11 October 2013). www.bbc.co.uk/news/uk-england-birmingham-24492833.

Bech, B.H., Obel, C., Henriksen, T.B., and Olsen, J. (2007). Effect of reducing caffeine intake on birth weight and length of gestation: randomised controlled trial. *BMJ* 334: 409–412.

Borges, G., Ottaviani, J.I., van der Hooft, J.J.J. et al. (2018). Absorption, metabolism, distribution and excretion of (–)-epicatechin: a review of recent findings. *Mol. Aspects Med.* 61: 18–30.

Bouker, K.B. and Hilakivi-Clarke, L. (2000). Genistein: does it prevent or promote breast cancer? *Environ. Health Perspect.* 108: 701–708.

Boylan, S., Cade, J.E., Dolby, V.A. et al. (CARE Study Group) (2008). Maternal caffeine intake during pregnancy and risk of fetal growth restriction: a large prospective observational study. *BMJ* 337: a2332. https://doi.org/10.1136/bmj.a2332.

Casal, S., Oliveira, B., and Margarida, A.F. (2000). HPLC/diode array applied to thermal degradation of trigonelline, nicotinic acid and caffeine in coffee. *Food Chem.* 68: 481–485.

Clifford, M.N. (1985). Chemical and physical aspects of green coffee and coffee products. In: *Coffee: Botany, Biochemistry and Production of Beans and Beverage* (eds. M.N. Clifford and K.C. Willson), 305–374. London: Croom Helm.

Clifford, M.N., Ludwig, I.A., and Crozier, A. (2018). Chemical composition of coffee beans. In: *Achieving Sustainable Cultivation of Coffee. Breeding and Quality Traits*, vol. 1 (ed. P. Lashemes), 661–675. Cambridge: Burleigh Dodds Science Publishing.

Corol, D.T., Ravel, C., Raksegi, M. et al. (2012). Effects of genotype and environment on the contents of betaine, choline, and trigonelline in cereal grains. *J. Agric. Food Chem.* 60: 471–481.

Crozier, A., Yokota, T., Jaganath, I.B. et al. (2006). Secondary metabolites as dietary components in plant-based foods and beverages. In: *Plant Secondary Metabolites: Occurrence, Structure and Role in the Human Diet* (eds. A. Crozier, M.N. Clifford and H. Ashihara), 208–302. Oxford: Blackwell Publishing.

Crozier, T.W.M., Stalmach, A., Lean, M.E.J., and Crozier, A. (2012). Espresso coffees, caffeine and chlorogenic acid intake: potential health implications. *Food Funct.* 3: 30–33.

Devasagayam, T., Kamat, J., Mohan, H., and Kesavan, P.C. (1996). Caffeine as an antioxidant: inhibition of lipid peroxidation induced by reactive oxygen species. *Biochim. Biophys. Acta* 1282: 63–70.

van Dijk, A.E., Olthof, M.R., Meeuse, J.C. et al. (2009). Acute effects of decaffeinated coffee and the major coffee components chlorogenic acid and trigonelline on glucose tolerance. *Diabetes Care* 32: 1023–1025.

European Food Safety Authority (2015). Scientific opinion on the safety of caffeine: EFSA panel on dietetic products, nutrition and allergies. *EFSA J.* 13: 4102.

Farah, A. (2012). Coffee constituents. In: *Coffee: Emerging Health Effects and Disease Prevention* (ed. Y.F. Chu), 21–58. Oxford, UK: Wiley Blackwell.

Fenster, L., Hubbard, A., Swan, S.H. et al. (1997). Caffeinated beverages, decaffeinated coffee, and spontaneous abortion. *Epidemiology* 8: 515–523.

Fortier, I., Marcoux, S., and Beaulac-Baillargeon, L. (1993). Relation of caffeine intake during pregnancy to intrauterine growth retardation and preterm birth. *Am. J. Epidemiol.* 137: 931–940.

Fredholm, B.B., Bäettig, K., Holmén, J. et al. (1999). Actions of caffeine in the brain with special reference to factors that contribute to its widespread use. *Pharmacol. Rev.* 51: 83–133.

Gartrell, B.D. and Reid, C. (2007). Death by chocolate: a fatal problem for an inquisitive wild parrot. *N. Z. Vet. J.* 55: 149–151.

Gaur, V., Bodhankar, L., Mohan, V., and Thakurdesai, P.A. (2013). Neurobehavioral assessment of hydroalcoholic extract of *Trigonella foenum-graecum* seeds in rodent models of Parkinson's disease. *Pharm. Biol.* 51: 550–557.

Giannelli, M., Doyle, P., Roman, E. et al. (2003). The effect of caffeine consumption and nausea on the risk of miscarriage. *Paediatr. Perinat. Epidemiol.* 17: 316–323.

Guieu, R., Devaux, C., and Henry, H. (1998). Adenosine and migraine. *Can. J. Neurol. Sci.* 25: 55–58.

Heckman, M.A., Weil, J., and Gonzalez de Mejia, E. (2010). Caffeine (1,3,7-trimethylxanthine) in foods: a comprehensive review on consumption, functionality, safety, and regulatory matters. *J. Food Sci.* 75: R77–R87.

Heiss, C., Dejam, A., Kleinbongard, P. et al. (2003). Vascular effects of cocoa rich in flavan-3-ols. *JAMA* 290: 1030–1031.

Heiss, C., Jahn, S., Taylor, M. et al. (2010). Improvement of endothelial function with dietary flavanols is associated with mobilization of circulating angiogenic cells in patients with coronary artery disease. *J. Am. Coll. Cardiol.* 56: 218–224.

Heiss, C., Sansone, R., Karimi, H. et al. (Flaviola Consortium, Eurpoean Union 7th Framework program) (2015). Impact of cocoa flavanol intake on age-dependent vascular stiffness in healthy men: a randomized, controlled, double-masked trial. *Age* 37: 9794.

Higdon, J.V. and Frei, B. (2006). Coffee and health: a review of recent human research. *Crit. Rev. Food Sci. Nutr.* 46: 101–123.

Hirakawa, N., Okauchi, R., Miura, Y., and Yagasaki, K. (2005). Anti-invasive activity of niacin and trigonelline against cancer cells. *Biosci. Biotechnol. Biochem.* 69: 653–658.

Jannsson, D.S., Galgan, V., Schubert, B., and Segerstad, C.H. (2001). Theobromine intoxication in a red fox and a European badger in Sweden. *J. Wildl. Dis.* 37: 362–365.

Juliano, L.M. and Griffith, R.R. (2004). A critical review of caffeine withdrawal: empirical validation of symptoms and signs, incidence, severity, and associated features. *Psychopharmacology* 176: 1–29.

Juliano, L.M., Ferre, S., and Griffiths, R.R. (2009). The pharmacology of caffeine. In: *Principles of Addiction Medicine* (eds. R.K. Ries, D.A. Fiellin, S.C. Miller and R. Saitz), 159–178. Philadelphia: Lippincott Williams & Wilkins.

Karatzis, E., Papaioannou, T.G., Aznaouridis, K. et al. (2005). Acute effects of caffeine on blood pressure and wave reflections in healthy subjects: should we consider monitoring central blood pressure? *Int. J. Cardiol.* 98: 425–430.

Kitagawa, M., Houzen, H., and Tashiro, K. (2007). Effects of caffeine on the freezing of gait in Parkinson's disease. *Mov. Disord.* 22: 710–712.

Koshiro, Y., Zheng, X.Q., Wang, M. et al. (2006). Changes in content and biosynthetic activity of caffeine and trigonelline during growth and ripening of *Coffea arabica* and *Coffea canephora* fruits. *Plant Sci.* 171: 242–250.

Ky, C.L., Louarn, J., Dussert, S. et al. (2001). Caffeine, trigonelline, chlorogenic acids and sucrose diversity in wild *Coffea arabica* L. and *C. canephora* P. accessions. *Food Chem.* 75: 223–230.

Lang, R., Wahl, A., Skurk, T. et al. (2010). Development of a hydrophilic liquid interaction chromatography–high-performance liquid chromatography–tandem mass spectrometry based stable isotope dilution analysis and pharmacokinetic studies on bioactive pyridines in human plasma and urine after coffee consumption. *Anal. Chem.* 82: 1486–1497.

Lang, R., Dieminger, N., Beusch, A. et al. (2013a). Bioappearance and pharmacokinetics of bioactives upon coffee consumption. *Anal. Bioanal. Chem.* 405: 8487–8503.

Lang, R., Yagar, E.F., Wahl, A. et al. (2013b). Quantitative studies on roast kinetics for bioactives in coffee. *J. Agric. Food Chem.* 61: 12123–12128.

Laville, S. 2000. Student died after sharing three cans of Red Bull. *The Telegraph* (15 November 2000). http://www.telegraph.co.uk/news/health/1374291/Student-died-after-sharing-three-cans-of-Red-Bull.html.

Lean, M.E.J., Ashihara, H., Clifford, M.N., and Crozier, A. (2012). Purine alkaloids: a focus on caffeine and related compounds in beverages. In: *Teas, Cocoa and Coffee: Plant Secondary Metabolites and Health* (eds. A. Crozier, Ashihara and F. Tomás-Barberán), 25–44. Oxford: Wiley Blackwell.

Lee, C. (2000). Antioxidant ability of caffeine and its metabolites based on the study of oxygen radical absorbing capacity and inhibition of LDL peroxidation. *Clin. Chim. Acta.* 295: 141–154.

Ludwig, I.A., Mena, P., Calani, L. et al. (2014). Variations in caffeine and chlorogenic acid contents of coffees: what are we drinking. *Food Funct.* 5: 1718–1726.

MacDonald, T.M., Sharpe, K., Fowler, G. et al. (1991). Caffeine restriction: effect on mild hypertension. *BMJ* 303: 1235–1238.

Maeztu, L., Andueza, S., Ibañez, C. et al. (2001). Multivariate methods for characterization and classification of espresso coffees from different botanical varieties and types of roast by foam, taste, and mouthfeel. *J. Agric. Food Chem.* 49: 4743–4747.

Mahoney, C.R., Giles, G.E., Marriott, B.P. et al. (2018). Intake of caffeine from all sources and reasons for use by college students. *Clin. Nutr.* https://doi.org/10.1016/j.clnu.2018.04.004.

Mazzafera, P. (1991). Trigonelline in coffee. *Phytochemistry* 30: 2309–2310.

McCusker, R.R., Goldberg, B.A., and Cone, E. (2002). Caffeine content of specialty coffees. *J. Anal. Toxicol.* 27: 520–522.

Metro, 2008. Man killed 'by 4 cans of Red Bull a day'. (24 April 2008). https://metro.co.uk/2008/04/24/man-killed-by-4-cans-of-red-bull-a-day-104714.

Mills, J.L., Holmes, L.B., Aarons, J.H. et al. (1993). Moderate caffeine use and the risk of spontaneous abortion and intrauterine growth retardation. *JAMA* 569: 593–597.

Mitchell, E.S., Slettenaar, M., van der Meer, N. et al. (2011). Differential contributions of theobromine and caffeine on mood, psychomotor performance and blood pressure. *Physiol. Behav.* 104: 816–822.

Osbum, W.O. and Kensler, T.W. (2008). Nrf2 signalling: an adaptive response pathway for protection against environmental toxic insults. *Mutat. Res.* 659: 31–39.

Perrone, D., Farah, A., and Donangelo, C.M. (2008). Fast simultaneous analysis of caffeine, trigonelline, nicotinic acid and sucrose in coffee by liquid chromatography-mass spectrometry. *Food Chem.* 110: 1030–1035.

Radwan, S.S. and Kokate, C.K. (1980). Production of higher levels of trigonelline by cell cultures of *Trigonella foenum-graecum* than by the differentiated plant. *Planta* 147: 340–344.

Rasch, V. (2003). Cigarette, alcohol, and caffeine consumption: risk factors for spontaneous abortion. *Acta Obstet. Gynecol. Scand.* 82: 182–188.

Riksen, N.P., Smits, P., and Rongen, G.A. (2011). The cardiovascular effects of methylxanthines. *Handb. Exp. Pharmacol.* 200: 413–437.

Rodrigues-Mateos, A., Weber, T., Skene, S..S. et al. (2018). Assessing the respective contributions of dietary flavonols and procyanidins in mediating cardiovascular effects in humans: a role for the gut microbiome? *Am. J. Clin. Nutr.* 108: 1229–1237.

Rogers, P., Hohoff, C., Heatherley, S.V. et al. (2010). Association of the anxiogenic and alerting effects of caffeine with ADORA2A and ADORA1 polymorphisms and habitual level of caffeine consumption. *Neuropsychopharmacology* 35: 1973–1982.

Rudelle, S., Ferruzzi, M.G., Cristiani, I. et al. (2007). Effect of a thermogenic beverage on 24-hour energy metabolism in humans. *Obesity* 15: 349–355.

Sansone, R., Ottaviani, J.I., Rodrigues-Mateos, A. et al. (2017). Methylxanthines enhance the effects of cocoa flavanols on cardiovascular function. *Am. J. Clin. Nutr.* 105: 352–360.

Savitz, D.A., Chan, R.L., Herring, A.H. et al. (2008). Caffeine and miscarriage risk. *Epidemiology* 19: 55–62.

Schroeter, H., Heiss, C., Balzer, J. et al. (2006). (–)-Epicatechin mediates beneficial effects of flavanol-rich cocoa on vascular function in humans. *Proc. Natl. Acad. Sci. USA* 103: 1024–1029.

Sengpiel, V., Elind, E., Bacelis, J. et al. (2013). Maternal caffeine intake during pregnancy is associated with birth weight but not with gestational length: results from a large prospective observational cohort study. *BMC Med.* 11 (42) https://doi.org/10.1186/1741-7015-11-42.

Singh, S., Vrishni, S., Singh, B.K. et al. (2010). Nrf2-ARE stress response mechanism: a control point in oxidative stress-mediated dysfunctions and chronic inflammatory diseases. *Free Radical Res.* 44: 1267–1288.

Stadler, R.H., Varga, N., Hau, J. et al. (2002a). Alkylpyridiniums. 1. Formation in model systems via thermal degradation of trigonelline. *J. Agric. Food Chem.* 50: 1192–1199.

Stadler, R.H., Varga, N., Milo, C. et al. (2002b). Alkylpyridiniums. 2. Isolation and quantification in roasted and ground coffee. *J. Agric. Food Chem.* 50: 1200–1206.

Taguchi, H., Sakaguchi, M., and Shimabayashi, Y. (1985). Trigonelline content in coffee beans and the thermal conversion of trigonelline into nicotinic acid during the roasting of coffee beans. *Agric. Biol. Chem.* 49: 3467–3471.

Tohoda, C., Kuboyama, T., and Komatsu, K. (2005). Search for natural products related to regeneration of neuronal network. *Neurosignals* 14: 34–45.

Trevitt, J., Kawa, K., Jalali, A., and Larsen, C. (2009). Differential effects of adenosine antagonists in two models of Parkinsonian tremor. *Pharmacol. Biochem. Behav.* 94: 24–29.

Usmani, O.S., Belvisi, M.G., Patel, H.J. et al. (2005). Theobromine inhibits sensory nerve activation and cough. *FASEB J.* 19: 231–232.

Viani, R. and Horman, I. (1974). Thermal behaviour of trigonelline. *J. Food Sci.* 39: 1216–1217.

Vickroy, T.W., Chang, S.K., and Chou, C.C. (2008). Caffeine-induced hyperactivity in the horse: comparisons of drug and metabolite concentrations in blood and cerebrospinal fluid. *J. Vet. Pharmacol. Ther.* 31: 156–166.

Vlachopoulos, C.V., Hirata, K., and O'Rourke, M.F. (2003). Effect of caffeine on aortic elastic properties and wave reflection. *J. Hypertens.* 21: 563–570.

Vlachopoulos, C.V., Aznaouridis, K., Alexopoulos, N.A. et al. (2005). Effect of dark chocolate on arterial function in healthy individuals. *Am. J. Hypertens.* 18: 785–791.

Vlachopoulos, C.V., Alexopoulos, N.A., Aznaouridis, K.A. et al. (2007). Relation of habitual cocoa consumption to aortic stiffness and wave reflections, and to central hemodynamics in healthy individuals. *Am. J. Cardiol.* 99: 1473–1475.

Vlajinac, H.D., Petrovic, R.R., Marinkovic, J.M. et al. (1997). Effect of caffeine intake during pregnancy on birth weight. *Am. J. Epidemiol.* 145: 335–338.

Wadge, A. (2009). Food standard Agency's advice on caffeine. *BMJ* 338: b299.

Wen, W., Shu, X.O., Jacobs, R.D., and Brown, J.E. (2001). The association of maternal caffeine consumption and nausea with spontaneous abortion. *Epidemiology* 12: 38–42.

Weng, X., Odouli, R., and Li, D.-K. (2008). Maternal caffeine consumption during pregnancy and the risk of miscarriage: a prospective cohort study. *Am. J. Obstet. Gynecol.* 198: 279.e1–279.e8.

Yamamoto, T., Takahashi, H., Suzuki, K. et al. (2014). Theobromine enhances absorption of cacao polyphenol in rats. *Biosci. Biotechnol. Biochem.* 78: 2059–2063.

Yoshinari, O., Sato, H., and Igarashi, K. (2009). Anti-diabetic effects of pumpkin and its components, trigonelline and nicotinic acid on Goto-Kakizaki rats. *Biosci. Biotechnol. Biochem.* 73: 1033–1041.

Yuyama, S. (1999). Absorption of trigonelline from the small intestine of the specific pathogen-free (SPF) and germ-free (GF) rats *in vivo*. *Adv. Exp. Med. Biol.* 467: 723–727.

Zhou, J., Zhou, S., and Zeng, S. (2011). Experimental diabetes treated with trigonelline: effect on ß-cell. Fund. *Clin. Pharmacol.* 27: 279–287.

Index

Note: Page references in *italics* refer to Figures; those in **bold** refer to Tables

Plant Nucleotide Metabolism – Biosynthesis, Degradation, and Alkaloid Formation, First Edition.
Hiroshi Ashihara, Iziar A. Ludwig and Alan Crozier.

d

g

www.ingramcontent.com/pod-product-compliance
Lightning Source LLC
LaVergne TN
LVHW061736150926
848065LV00005B/32